INTEGRATED STRATEGIES FOR DRUG DISCOVERY USING MASS SPECTROMETRY

INTEGRATED STRATEGIES FOR DRUG DISCOVERY USING MASS SPECTROMETRY

MIKE S. LEE
Milestone Development Services

A JOHN WILEY & SONS, INC., PUBLICATION

Published by John Wiley & Sons, Inc., Hoboken, New Jersey.
Published simultaneously in Canada.

For general information on our other products and services please contact our Customer Care Department within the U.S. at 877-762-2974, outside the U.S. at 317-572-3993 or fax 317-572-4002.

Wiley also publishes its books in a variety of electronic formats. Some content that appears in print, however, may not be available in electronic format.

Library of Congress Cataloging-in-Publication Data:

Integrated strategies for drug discovery using mass spectrometry / [edited by] Mike S. Lee.
p. cm.
ISBN-13 978-0-471-46127-2
ISBN-10 0-471-46127-X (cloth)
1. Mass spectrometry. 2. Drugs–Analysis. 3. High throughput screening (Drug development)
I. Lee, Mike S., 1960–
RS189.5.S65155 2005
615′.19–dc22 2005003235

Printed in the United States of America

10 9 8 7 6 5 4 3 2 1

CONTENTS

PREFACE

It could be argued that drug discovery strategies of the past were analogous to fishing with a line and hook. A systematic series of trial and error experiments such as quantitative structure-activity relationship (QSAR) methodologies, for example, were commonplace. The process was iterative and labor intensive. A chemist or biologist had relatively few restrictions, particularly with time, to explore ideas and test hypotheses. The overall endeavor was more craft than process.

Today's preference for drug discovery is on high throughput approaches to accelerate the generation, identification, and optimization of molecules with desirable drug properties. A large number of samples and corresponding analytical measurements are required to make quick and confident decisions. Certainly, the need for a faster pace of research is warranted. However, the current strategies that emphasize fast cycle times make it appear that new drug candidates are discovered not with a line and hook approach but harvested with nets. The craft still remains, but it is often difficult to see in the midst of production-scale events.

Fueled by the prospects of a highly efficient drug discovery process and the need for an expanded drug development pipeline, the pharmaceutical industry has aggressively embraced technology over the past decade. Medicinal chemistry has played a central role with the integration of automated synthesis technologies. New drug discovery paradigms were subsequently born and these high throughput medicinal chemistry approaches helped to serve as the gateway for novel biomolecular screening and proteomics techniques. During this time, high throughput screening approaches for drug metabolism and pharmacokinetic characteristics became an essential part of the drug discovery landscape.

The swift integration of highly productive technologies that focused on sample generation created bottlenecks in drug discovery. Simply put, medicinal chemists and their collaborators were generating samples at a faster rate than they were capable of analyzing them. New formats for automated synthesis combined with a faster pace of drug discovery led to a shift in sample analysis requirements from a relatively pure sample type to a trace-mixture. Traditional methods of analysis became antiquated. New analytical strategies and techniques were necessary to meet sample throughput requirements and manpower constraints. Technologies with appropriate sensitivity, selectivity, and speed were quickly integrated into mainstream drug discovery paradigms to effectively handle the predominant sample type, a trace-mixture. Drug discovery was forced to become more dependent on technology and become more agile. Fortunately, scientists became more open about the notion of using mass spectrometry as a front-line tool throughout drug discovery.

This book is based on a special issue of *Curr. Top. Med. Chem.* published in January 2002. The focus is primarily on mass spectrometry-based applications in drug discovery that require trace-mixture analysis. The selection of topics is not intended to be comprehensive, but rather highlight the creativity of analytical scientists and their collaborators that have led to current standards for analysis in drug discovery. Innovators in the field describe their unique perspectives on integrated strategies for analysis and share future prospects. The topics represent current industry benchmarks in specific drug discovery activities that deal with proteomics, biomarker discovery, metabonomic approaches for toxicity screening, lead identification, compound libraries, quantitative bioanalytical support, biotransformation, reactive metabolite characterization, lead optimization, pharmaceutical property profiling, sample preparation strategies, and automation.

Interestingly, the traditional definition and core focus of drug discovery is still applicable today. A significant change has been the preference for high throughput approaches and the explosive development of analytical instrumentation that allows for the facile interrogation (qualitative and quantitative) of a trace-mixture. Analytical technologies now provide an integral component for modern drug discovery practices. Similar to the location mantra prescribed in real estate, the fundamental merits of sensitivity are a requisite for high throughput analytical measurements in drug discovery. Due to limited sample quantities and the need to interrogate compounds of interest at low concentration, "sensitivity, sensitivity, sensitivity" will most likely remain as the analytical mantra in drug discovery. However, highly sensitive analytical platforms such as mass spectrometers will continue to benefit from orthogonal techniques such as chromatography, sample preparation, and informatics.

These techniques will enable the analytical scientist to demonstrate control of the analytical method and provide the necessary selectivity, speed, reproducibility, and robustness.

It is hoped that this book will provide historical perspective on analytical strategies in drug discovery and illustrate the widened scope of mass spectrometry applications.

Mike S. Lee

CONTRIBUTORS

Bradley L. Ackermann, Department of Drug Disposition, Lilly Research Laboratories, Eli Lilly and Company, Indianapolis, IN 46285

David Bar-Or, Trauma Research, Swedish Medical Centre, Englewood, CO 80110

Michael J. Berna, Department of Drug Disposition, Lilly Research Laboratories, Eli Lilly and Company, Indianapolis, IN 46285

Josip S. Blonder, SAIC-Frederick, National Cancer Institute at Frederick, Frederick, MD 21702

Andrew W. Carr, Dow AgroSciences, 9330 Zionsville Road, Indianapolis, IN 46268

Shuo Chen, The Scripps Research Institute, La Jolla, CA 92037

Xueheng Cheng, R4PN, AP9A, Abbott Laboratories, 100 Abbott Park Road, Abbott Park, IL 60064

Ben Chien, Quest Pharmaceuticals Services, L.L.C., 3 Innovation Way, Suite 240, Delaware Technology Park, Newark, DE 19711

Thomas P. Conrads, SAIC-Frederick, National Cancer Institute at Frederick, Frederick, MD 21702

Lendell L. Cummins, Ibis Therapeutics, A Division of Isis Pharmaceuticals, 1891 Rutherford Road, Carlsbad, CA 92008

C. Gerald Curtis, Bowman Research (UK) Ltd., Imperial House, Imperial Park, Newport, Gwent, NP 10-8UH, UK

Margaret Davis, Biotransformation Division, Department of Drug Safety and Metabolism, Wyeth Research, 500 Arcola Road, Collegeville, PA 19426

William DeMaio, Biotransformation Division, Department of Drug Safety and Metabolism, Wyeth Research, 500 Arcola Road, Collegeville, PA 19426

Li Di, Wyeth Research, Chemical and Screening Sciences, CN8000, Princeton, NJ 08543

Jared J. Drader, Ibis Therapeutics, A Division of Isis Pharmaceuticals, 1891 Rutherford Road, Carlsbad, CA 92008

Dennis O. Duebelbeis, Dow AgroSciences, 9330 Zionsville Road, Indianapolis, IN 46268

Carmen L. Fernández-Metzler, Merck Research Laboratories, Department of Drug Metabolism, Sumneytown Pike, West Pike, PA 19486

H. Mario Geysen, University of Virginia, Department of Chemistry, McCormick Road, P.O. Box 400319, Charlottesville, VA 22904

Jeffrey R. Gilbert, Dow AgroSciences, 9330 Zionville Road, Indianapolis, IN 46268

Thomas Hartmann, Amgen, Inc., One Amgen Center Drive, Thousand Oaks, CA 91320

Jill Hochlowski, Global Pharmaceutical Research and Development, 100 Abbott Park Road, Abbott Laboratories, Abbott Park, IL 60064

Steven A. Hofstadler, Ibis Therapeutics, A Division of Isis Pharmaceuticals, 2292 Faraday Avenue, Carlsbad, CA 92008

Elaine Holmes, Biological Chemistry, Biomedical Sciences Division, Imperial College, SAF Building, Exhibition Road, South Kensington, London, SW7 2AZ, UK

Haleem J. Issaq, SAIC-Frederick, National Cancer Institute at Frederick, Frederick, MD 21702

Jonathan L. Josephs, Bristol-Myers Squibb, Pharmaceutical Research Institute, Hopewell, NJ 08534

Neil L. Kelleher, Department of Chemistry, University of Illinois, 53 Roger Adams Laboratory, 600 South Mathews Avenue, Urbana, IL 61801

Edward H. Kerns, Wyeth Research, Chemical and Screening Sciences, CN8000, Princeton, NJ 08543

Richard C. King, Merck Research Laboratories, Department of Drug Metabolism, Sumneytown Pike, West Pike, PA 19486

Walter A. Korfmacher, Director, Exploratory Drug Metabolism, Department of Drug Metabolism and Pharmacokinetics, Schering-Plough Research Institute, 2015 Galloping Hill Road, Kenilworth, NJ 07033

John D. Laycock, Amgen, Inc., One Amgen Center Drive, Thousand Oaks, CA 91320

Paul Lewer, Dow AgroSciences, 9330 Zionsville Road, Indianapolis, IN 46268

Anthony T. Murphy, Department of Drug Disposition, Lilly Research Laboratories, Eli Lilly and Company, Indianapolis, IN 46285

Kumar Ramu, Quest Pharmaceuticals Services, L.L.C., 3 Innovation Way, Suite 240, Delaware Technology Park, Newark, DE 19711

Mark Sanders, Bristol-Myers Squibb, Pharmaceutical Research Institute, Princeton, NJ 08543

Kristin A. Sannes-Lowery, Ibis Therapeutics, A Division of Isis Pharmaceuticals, 2292 Faraday Avenue, Carlsbad, CA 92008

JoAnn Scatina, Biotransformation Division, Department of Drug Safety and Metabolism, Wyeth Research, 500 Arcola Road, Collegeville, PA 19426

Frank Schoenen, Glaxo Smith Kline, 5 Moore Drive, RTP 27709

David Semin, Discovery Analytical Sciences, Amgen Inc., One Amgen Center Drive, Mail Stop 29-M-B, Thousand Oaks, CA 91320

John P. Shockcor, Waters Corporation, 34 Maple St., Milford, MA 01757

Marshall M. Siegel, Wyeth Research, Discovery Analytical Chemistry, 401 N. Middletown Road, Building 222, Room 1043, Pearl River, NY 10965

Raju Subramanian, Merck Research Laboratories, Department of Drug Metabolism, Sumneytown Pike, West Pike, PA 19486

Rasmy Talaat, Biotransformation Division, Department of Drug Safety and Metabolism, Wyeth Research, 500 Arcola Road, Collegeville, PA 19426

Gary A. Valaskovic, New Objective, Inc., 2 Constitution Way, Woburn, MA 01801

Timothy D. Veenstra, Biomedical Proteomics Program, SAIC-Frederick, National Cancer Institute at Frederick, Frederick, MD 21702

David S. Wagner, Glaxo Smith Kline 5 Moore Drive, RTP, NC 27709

Richard W. Wagner, Glaxo Smith Kline 5 Moore Drive, RTP, NC 27709

Jack Wang, Biotransformation Division, Department of Drug Safety and Metabolism, Wyeth Research, 500 Arcola Road, Collegeville, PA 19426

Jim Wang, Biotransformation Division, Department of Drug Safety and Metabolism, Wyeth Research, S3415, 500 Arcola Road, Collegeville, PA 19426

David A. Wells, Sample Prep Solutions Company, Saint Paul, MN 55119

Yining Zhao, Discovery Analytical Sciences, Amgen Inc., One Amgen Center Drive, Mail Stop 29-M-B, Thousand Oaks, CA 91320

ACKNOWLEDGEMENTS

The genesis of this book started when Allen Reitz, Editor-in-Chief of *Current Topics in Medicinal Chemistry*, contacted me to follow up my book *LC/MS Applications in Drug Development* with a special issue devoted to drug discovery. I am grateful to Allen for his interest in the topic and allowing me to highlight specific applications of mass spectrometry in drug discovery.

The staff at John Wiley & Sons has provided tremendous support. In particular, I would like to thank my editor, Bob Esposito, who has been an invaluable resource during this project as well other projects.

I am indebted to all of the authors for their willingness share their experiences, knowledge, and perspectives on drug discovery. The contributions as well as the many conversations and interactions were fantastic! Special thanks go to Brad Ackermann and Marshall Siegel for all their advice and recommendations.

The growth of mass spectrometry in the pharmaceutical industry has been truly remarkable. I could not have predicted such an outcome and response. I am indebted to the many scientists and "blue helmets" who have paved the way. I am thankful to have had the opportunity to interact with so many interesting people who shared a common passion for the analytical sciences and drug discovery.

Finally, I thank my wife and family for their encouragement and support for everything I do.

1

MINIATURIZED FORMATS FOR EFFICIENT LIQUID CHROMATOGRAPHY MASS SPECTROMETRY–BASED PROTEOMICS AND THERAPEUTIC DEVELOPMENT

GARY A. VALASKOVIC AND NEIL L. KELLEHER

INTRODUCTION

In recent years, many scientists and engineers, in both academia and industry, have turned their attention to the miniaturization of laboratory methods. Many labor to develop lab-on-a-chip devices that promise to do for chemistry what integrated circuits have done for electronics [1]. Properly engineered, miniaturization offers significant improvements in sensitivity and speed at reduced cost when compared to conventional laboratory-scale methods. While much of this general promise is yet unrealized, there has been significant progress in specific niche areas of analytical chemistry. Given the central role that liquid chromatography coupled to mass spectrometry (LC-MS) plays in drug discovery and development [2], much attention has been applied to the miniaturization of sample delivery with electrospray ionization mass spectrometry (ESI-MS). Historically, the need to analyze and identify picomole (10^{-12} mol) to femtomole (10^{-15} mol) quantities of protein samples coincided with breakthroughs in both off- and on-line miniaturized sample handling for

Integrated Strategies for Drug Discovery Using Mass Spectrometry, Edited by Mike S. Lee

ESI-MS. As a result, perhaps no other area in measurement science has benefited from the miniaturization of analytical methods more than the nascent field of proteomics. Indeed, such miniaturization is a requirement for success [3].

Contemporary MS has evolved rapidly since the first reports of matrix-assisted laser-desorption/ionization (MALDI) [4] and electrospray ionization (ESI) [5]. The pharmaceutical industry is in a phase where an engineering ethos is center stage for characterization of small and large molecules alike. As MS tools proliferate, increased efficiency, such as those applications reviewed in this book, will enable an increased number of analyses per day, decreased sample volumes, and higher molecular weight analytes [6,7].

This review endeavors to illuminate the major trends in miniaturized sample preparation and delivery for ESI-MS coupled to nanobore LC-MS. Particular emphasis is placed on applications in proteomics. The first successful miniaturized ESI-MS format is now commonly referred to as *static nanospray*, originally conceptualized by Wilm and Mann in 1989 [8]. The method was subsequently refined for use as a protein identification tool by Wilm and Mann at the European Molecular Biology Laboratory in the mid 1990s [9]. At the same time, a number of academic and industrial groups developed methods to interface separation science tools, such as nanobore LC and capillary electrophoresis (CE), with ESI-MS [10–13]. Although the on-line methods were slower to develop, the need for higher sensitivity and throughput (again being driven largely by proteomics) has accelerated their maturation. Proteomics researchers now commonly use a range of techniques that combine aspects of both off- and on-line approaches to nanospray and nanobore LC. This review does not cover the miniaturization of MS instrumentation, which is currently an active area of research [14,15]. A discussion of other separation methods, such as CE coupled to nanospray, appears in a review by Wood et al. [16]. Foret and Preisler have reviewed interfacing nanobore LC to MALDI [17].

OFF-LINE METHODOLOGY: STATIC NANOSPRAY

In comparison to the sophisticated engineering associated with ESI-MS at conventional (mL/min) flow rates, static nanospray is a method of elegant simplicity. Wilm and Mann [8] first observed that the reduced diameter of the ESI emitter to the micrometer scale reduces sample consumption to a flow rate typically within 20–40 nL/min and generates signal-to-noise (S/N) levels comparable to conventional ESI. Furthermore, the electric field that is generated by the high voltage within the source creates sufficient electrostatic pressure to pull the mobile phase through the emitter orifice and obviates the need for any additional pumping. In this pure electrospray mode, sample consumption is self-regulating; the volume consumed is near optimal for ESI source operation.

Interestingly, the implementation of nanospray was strikingly similar to the first published observations of electrostatic spraying of liquids by Zeleny in 1914 [18]. This self-limiting characteristic gives static nanospray its principal analytical advantage: reduced sample consumption (by 100:1 or more) without any significant loss or compromise in MS ion intensity. For example, at 20 nL/min a 1-μL sample will yield a signal for roughly one hour before expiration. The extended run time yields sufficient signal for even the most ambitious tandem mass spectrometry (MS/MS) or MS^n experiments. Extended run time gives the static format nearly universal application across a broad range of MS analyzers, which include triple-quadrupole, time-of-flight (TOF), ion-trap, Fourier-transform (FT), and hybrid instruments. The key feature of static nanospray, an ultralow flow rate, illustrates the behavior of ESI-MS as a (primarily) concentration sensitive detector [19]. Concentration-dependant sensitivity is the foundation for the success of ESI miniaturization, since sample volumes can be reduced by orders of magnitude (to the nanoliter/picoliter scale) without concurrent loss of S/N.

A schematic of a typical nanospray source and photomicrograph of a nanospray emitter are shown in Figure 1.1. The commonly accepted scheme for static nanospray employs a glass needle or nanovial. The typical nanospray emitter is fabricated from borosilicate, aluminosilicate, or quartz tubing with a large internal bore (0.5–1 mm) that is finely tapered at one end to a 1–4-μm inside-diameter (ID) opening. Electrical contact with the mobile phase, a necessary requirement for operation, is usually made through a conductive coating that is applied to the outside of the emitter or via a platinum or gold wire inserted within. MS data from a typical glass nanospray tip of a peptide standard solution is shown in Figure 1.2. The sample loading requirement is typically 0.1 to 2 μL, and the needles are normally used once per analysis, since adequate rinsing of the small orifice is not practical. Flow

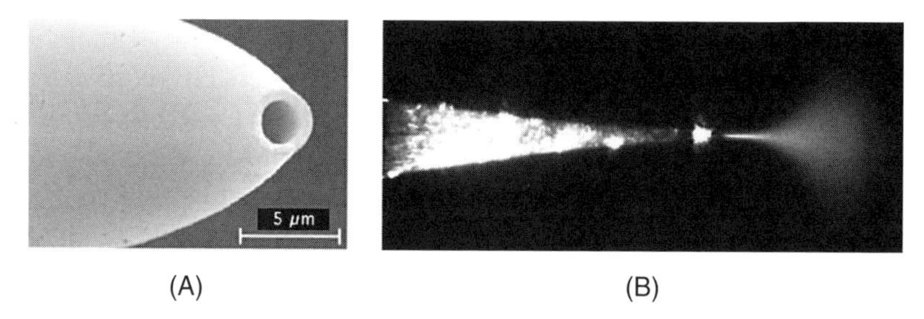

FIGURE 1.1 (A) Schematic of a typical static nanospray apparatus that uses a glass needle tapered to a 1–4-μm ID tip. The high voltage for ESI is either applied to a conductive coating on the tip, or placed inside the tip with a metal wire. (B) Reflected light photomicrograph of a spray plume from a 4-μm ID nanospray tip at approximately 60 nL/min. The ESI spray plume is visible on the right-hand side of the image. (Photograph courtesy of New Objective, Inc.)

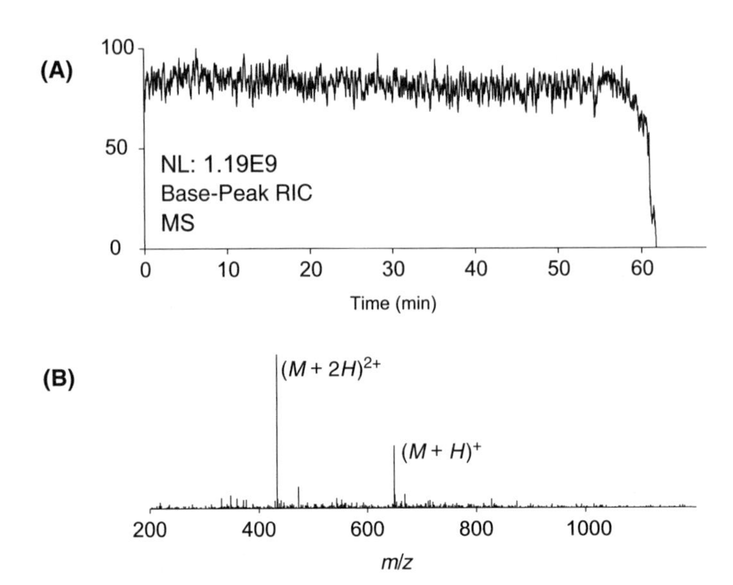

FIGURE 1.2 (A) Reconstructed base-peak ion current from a platinum-coated, 4-μm-ID glass nanospray emitter. The emitter was loaded with 5 μL of peptide standard (Angiotensin I, 10 μM in 50% methanol, 2% acetic acid) with a spray voltage of 1100 V. The 60 min of ion current translates into a flow rate of approximately 80 nL/min. (B) Randomly selected single-scan mass spectrum from the data set used in (A).

rates are typically 20–30 nL/min [9] for 1 to 2 μm glass needles. Flow rates as low as 1.6 nL/min [20] and 0.1 nL/min [21] have been reported for fine fused-silica needles. The mass limit of detection (MLOD) is typically in the low picomole to tens of femtomole range for proteins and peptides [9,20,22], although a specialized variant using nanoliter volume silica needles has been reported in the attomole range [21]. Nanospray offers a wider range of suitable mobile phases for electrospray [9], and offers a higher tolerance to salt contaminants than conventional ESI does [23,24]. The concentration limit of detection (CLOD) for peptides is typically on the order of low picomole/μL to tens of femtomole/μL [22]. Geronmanos and co-workers [20] used extended acquisition and summing, techniques that lowered the CLOD to better than 1 femtomole/μL on a triple quadrupole instrument in MS/MS mode. Under real-world operating conditions, one of the principal factors limiting CLOD is the need for clean, desalted samples, which significantly reduces chemical background noise [22]. A number of microdesalting solid-phase extraction (SPE) techniques have been developed that use either submicroliter volume spin traps [9,25] or microsampling pipette tips [26,27].

Traditional static nanospray can be a time-intensive effort and typically requires the expertise of a skilled operator. Automated methods can increase throughput and involve either traditional capillaries or microfluidic chips.

Geronmanos and co-workers [20] combined a fused-silica capillary emitter with an automated *X*, *Y*, *Z* positioning system. An autosampler-like mechanism was directly incorporated into the source, and the distal end of the emitter capillary was directly inserted into the sample vial. Karger and co-workers [28] constructed an array of capillary emitters, with an individual emitter that corresponds to a given sample in a 96-well plate. Sample flow was driven by the pressure differential between the distal and proximal ends of the emitter. Each array element (sample) was indexed in front of the MS inlet, and a custom software suite was used for array positioning and sample tracking. Schultz and co-workers [29,30] have developed an automated emitter array based on silicon wafers processed by deep reactive ion etching (DRIE). Each identical nozzle of the array has an inside diameter in the 8- to 15-μm range; the typical array size is 10×10. The sample is delivered to the back end of a given nozzle by an electrically conductive micropipette tip. The relatively larger diameter of the silica emitters translates into a somewhat higher operating flow rate, typically within the 100- 300-nL/min range [29,30]. To further reduce orifice size and wall thickness, an improved version of the DRIE process was developed by Roeraade and colleagues [31]. The refined shape of the emitter appears to improve S/N reproducibility. When used in a disposable manner, the array-based systems share the advantage of eliminating detectable sample carryover from previous analysis.

ON-LINE METHODOLOGY: NANOSCALE LC-MS

There is no doubt that the combination of high-performance liquid-chromatographic (HPLC) separations with MS affords one of the most powerful, centrally applied tools in drug discovery and development [2]. While there is a clear trend toward incrementally smaller ID columns from the 4.6-mm format, proteomics illustrates the many benefits of LC miniaturization. Given both the complexity (thousands to tens of thousands) and dynamic range (from 10^1 to 10^6 copies per cell) of proteins expressed in a cellular system [32], exceptional demands are placed on the analytical system in terms of speed, sensitivity, dynamic range, and specificity [33]. With the relatively high CLOD of LC-MS, relative to fluorescence or radio assays, conventional (mm diameter) formats for LC-MS have inadequate sensitivity for protein analysis. Consider 100 femtomole (10^{-15} mol) of protein isolated from a two-dimensional (2D) electrophoretic gel slice where the typical minimum molar concentration must typically range between 0.1 and 1 μM for analysis by ESI-MS/MS. The total sample must therefore be contained within a volume no larger than 1 μL (and preferably 0.1 μL) as the sample flows through the end of the ESI emitter into the mass spectrometer. A 1-mm-ID microbore column with a typical column volume of 200 μL at a flow rate of 50 μL/min yields an elution volume of roughly 12 μL for a 15-s-wide peak. These conditions are significantly different than a 75-μm-ID nanobore column. A 75-μm-ID nanobore column would have a typical column volume of 0.5 μL at a flow rate of 200 nL/min and yield an elution volume of roughly 50 nL for a 15-s-wide peak. When calculated on a volume basis, a 75-μm-ID nanobore column yields a relative concentration improvement of over 3700-fold compared to a 4.6-mm-ID column [12].

Sample Injection

Since optimal injection volumes are less than 100 nL for isocratic elution [12], nanobore columns may appear to be impractical. Because this volume is difficult to handle by conventional means, gradient elution is commonly used in conjunction with large-volume sample injection. A well-accepted strategy is to load a microliter volume (1–5 μL) directly on-column with a pressure-bomb apparatus [12] or from an injection loop with a suitable low-dispersion, six- or ten-port valve. An alternative approach is to place a low-volume (μL), reverse-phase (RP) trapping cartridge within the sample loop of a stand-alone injection valve or autosampler [34,35]. A large volume (typically 1 to 25 μL) is loaded and washed under aqueous conditions onto the sample trap. Because the trap is typically much shorter than an analytical column, the low back pressure enables sample loading at a higher flow rate than on-column injection. A significant increase in throughput is gained with the use of higher loading rates in combination with automated sampling tools. Throughput is further enhanced with the use of an auxiliary loading pump and 10-port valve. This format provides for the loading and conditioning of multiple traps in a column switching format. The trap cartridge also serves as an effective guard column that greatly extends the lifetime of the analytical column, typically to many hundreds of injections. A disadvantage of this approach is the potential loss of analytes that are not efficiently retained by the cartridge.

When trapped injection techniques are used, care must be taken to avoid sloppy, unnecessary connections and to minimize the volume between the trap and analytical column. Meiring demonstrated the benefits of minimizing trap-to-column volume by placing the trap close to the column through a modified tee connection that significantly improved peak shape and sensitivity [36,37]. Licklider has developed this concept further with the vented column approach [38]. By adding a fluidic valved venting port (again built from a tee connection) within the first few centimeters of the packed column bed, the back pressure through that portion of the bed that precedes the vent can be decreased. The column is fabricated by packing the bed through and into the venting tee. Therefore the interior tee volume, which would normally contribute a significant swept volume between the trap and column, contributes to the separation. As a result, injection and washing flow rates can be an order of magnitude higher than the elution flow rate. The arrangement significantly reduces potential loss of sample, since interaction with nonchromatographic surfaces is kept to a minimum.

Mobile-Phase Delivery

Flow rates in the range of 100 to 500 nL/min are required for optimal separations for nanoscale columns (ID < 100 μm). Given the limitations of conventional HPLC instrumentation, the common solution has been to use precolumn flow splitting [12]; pump technology is rapidly improving, however, and commercial systems that directly generate nanoliter-per-minute flow rates are now available. When using a traditional HPLC pump, it is common to use a split ratio of 1000:1, with the gradient HPLC system operating at a flow rate of 200 μL/min. Suitable flow splitters range from various tee arrangements [12,39] to commercially available units based on balanced splitting or mass-flow control. While a simple tee arrangement is considered by many to be quite suitable for qualitative analysis in identification and discovery, poor reproducibility limits productivity in a quantitative environment. Newer, commercially available, feedback-controlled pumping technologies that involve either flow splitting or direct generation overcome these limitations. Binary gradient delivery is a requirement for basic RP operation, with dual-gradient operation preferred for multidimensional methods.

Column-to-MS Interface

Many schemes have been used for the integration of nanoscale LC columns with ESI-MS. Before it was widely known that ESI is stable at low flow rates, early coupling efforts used coaxial sheath flow. In this arrangement, column effluent is combined with a second makeup mobile phase, with the total flow rate into a range (0.1 to 1 mL/min) accommodated by commercial hardware [40,41]. Once it was established that flow rate could be reduced without a concurrent loss in S/N [42], researchers began to directly couple nanoscale columns to a variety of sharpened (pulled, ground, or etched) narrow-bore ESI emitters [43–45]. High-voltage contact for ESI is typically established between the column and tip using a junction contact with a metal fitting or sleeve [43], or directly at the end of the flow stream using an emitter coated with an electrical conductor [45]. Much attention has been paid to different thin-film formulations for use in either LC or CE-MS, and include noble-metal films [46], overcoated films [47], conductive polymers [48], and composite conductive materials [49,50].

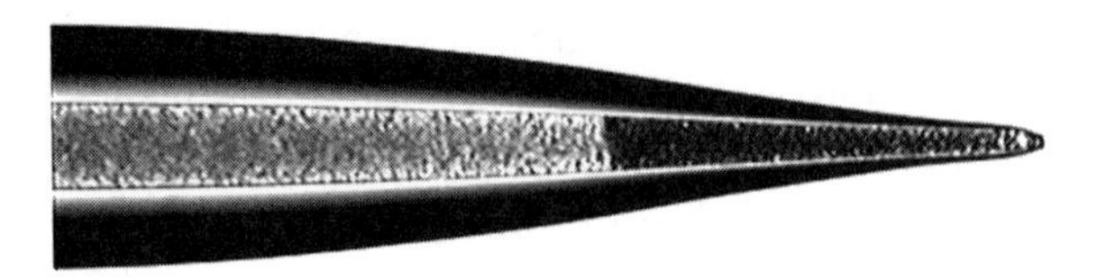

FIGURE 1.3 Photomicrograph of a tapered fused-silica tip with an integral fit and packed with 5-μm-diameter C18 reverse-phase media. The ID of the tip is 15 μm and the ID of the packed chromatography bed is 75 μm. (Photograph courtesy of New Objective, Inc.)

Emmett and co-workers eliminated the column-to-emitter connection and fabricated the column and emitter as an integral unit [10]. The electrospray is formed directly on the column outlet. Postcolumn band broadening is eliminated by eliminating the connection volume between column and emitter. The mobile phase carries the high voltage to the end of the emitter for spray formation; a moderately conductive mobile phase is a key requirement for ESI operation [19]. The integral approach has proved to be very robust, since tip clogging (the column acts as a filter) is virtually eliminated. A junction-style contact also obviates the need for the application of a conductive coating to the emitter and keeps the emitter design simple and offers significant resistance to arcing. Importantly, this design moves the high-voltage contact to the high-pressure end (the inlet side) of the column [10,51]. Electrolysis products forming at the electrode remain in solution, since the contact is made under high pressure (typically 1000–3000 psi). This design has also proved to be compatible with polymer-based monolithic beds that enable both smaller column formats [52] and more versatile chemistry [53]. A photo of an integral ESI column/emitter is shown in Figure 1.3 and the typical implementation for nanoscale LC-MS is shown in Figure 1.4.

A modest example of the sensitivity possible for the integral column/tip LC-MS configuration is shown in Figure 1.5. A 75-μm-ID column that terminates in a 15-μm-ID tip was packed with RP material to a length of 10 cm. A split-flow gradient mobile phase of approximately 1000:1 was delivered

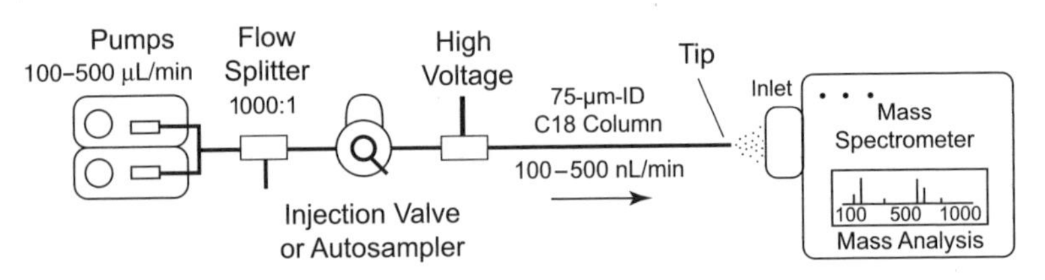

FIGURE 1.4 Schematic of a typical nanoscale LC-MS interface system. A precolumn flow splitter is used to reduce to flow rate of a conventional pumping system to the 100–500 nL/min. The voltage for ESI is applied on the high-pressure distal side of the column using a junction electrical contact.

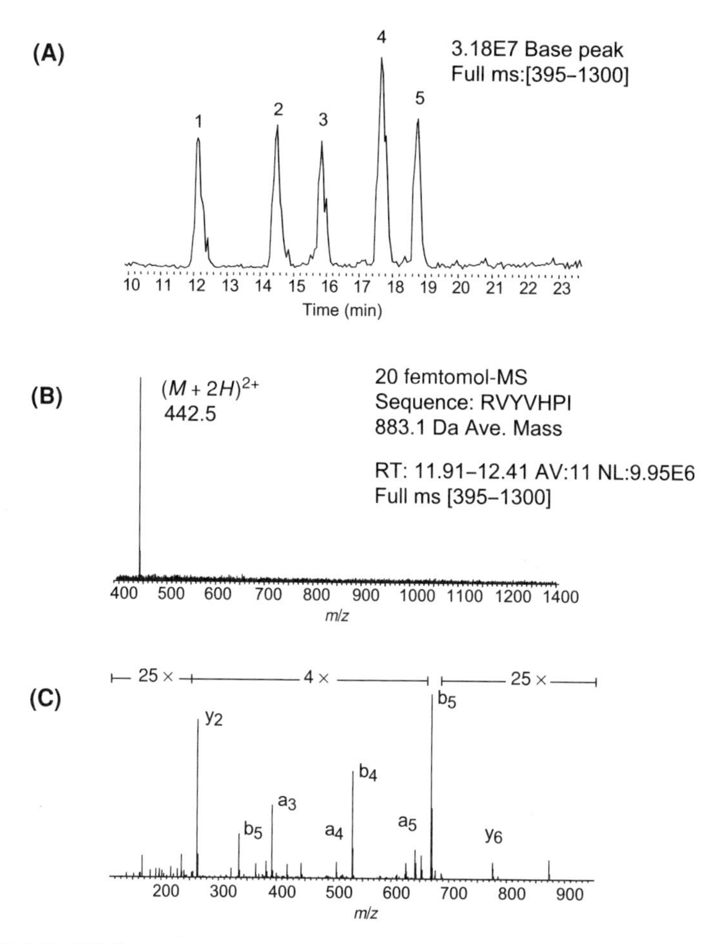

FIGURE 1.5 (A) Base-peak reconstructed ion chromatogram from a mixture of five standard peptides from a 20 femtomole/peptide on-column injection using gradient elution from a 75-μm-ID tip column as shown Figure 1.4. (B) MS spectrum of peak number 1 showing a doubly charged molecular ion of high S/N. (C) MS/MS spectrum of peak number 3 showing labeled N-terminal (a, b) and C-terminal (y) fragment ions. Scale expansion factors are denoted at the top of the figure. (Data courtesy of William S. Lane, Harvard University.)

for a through-column flow rate of approximately 200 nL/min. A 200-nL on-column injection of a standard mixture of five peptides (20 femtomole/peptide) was delivered prior to water/acetonitrile (5% to 95% acetonitrile in 0.05 M acetic acid) gradient elution. The column was positioned within 1–2 mm of an ion trap mass-spectrometer inlet. High voltage (1–2 kV) was applied on the inlet side of the column directly to the mobile phase through a wire electrode assembly. Excellent peak shape (no tailing) and exceptional MS and MS/MS S/N of greater than 50:1 at the 20 femtomole level was obtained. In contrast to the single-use nanospray tips, such columns can perform well for hundreds of injections. The approach has proved itself in extended time scale experiments, such as large-scale proteome complex analysis [54,55] and quantitative proteomics [56,57].

PEPTIDE-DRIVEN PROTEOMICS

With the realization of these new on-line and off-line tools for MS, partial characterization of mammalian proteomes is now feasible. While an initial analysis of ~150 yeast proteins was achieved in 1996 [25], identification of ~1500 from a single experiment was reported only five years later [54]. The classic proteomic analysis, represented schematically in Figure 1.6, involves 2D gel separation and protein identification of stained spots accomplished by in-gel digestion, with subsequently matching peptide molecular weight (MW) values with those of deoxyribonucleic acid (DNA)-predicted peptides from a database of expressed sequence tags (ESTs) or predicted open reading frames (ORFs). When an automated MALDI-based approach is used for rapid peptide fingerprinting [58], followed by either nanospray MS/MS or nanoscale LC/MS/MS on samples that yield poor "hits" by MALDI, it is conceivable to analyze thousands of proteins per day on a single instrument. Such schemes are now being implemented in a number of industrialized proteomic facilities [55,59]. There are several academic and commercial databases and algorithms to support such efforts that include Protein Prospector [60], Profound [58], Prowl [61], and Mowse [62]. Prowl and Mowse have probability-based score reporting like BLAST searches for sequence homology.

Areas of intense research are to increase proteome coverage (especially at low copy numbers per cell) while increasing throughput, and to do so in a quantitative manner. Yates and co-workers have developed a large-scale analytical scheme that combines multidimensional nanoscale LC, MS/MS, and automated sequence analysis referred to as multidimensional protein identification

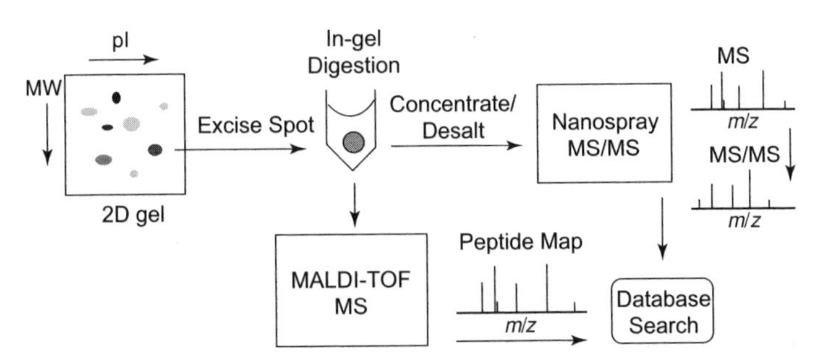

FIGURE 1.6 Schematic of the conventional method of proteomics analysis that incorporates 2D gel electrophoresis, in gel digestion with subsequent MS analysis and identification by either MALDI MS or nanospray MS/MS.

technology (MudPIT) [54]. Application of MudPIT to *Saccharomyces cerevisiae* was successful in the identification of 1484 proteins, of which 131 contained predicted transmembrane domains, and make the MudPIT approach superior to 2D gels for throughput. The authors used integral column/emitters packed with a biphasic bed (i.e., the columns were first packed with a RP, and then with a strong cation exchange (SCX), material). After initial sample loading of a shotgun (whole organism) proteome digest, peptides were step-eluted from the SCX portion of the column onto the RP bed. RP elution then followed, using a traditional water/acetonitrile gradient. This approach minimizes sample handling, a necessary requirement for the detection of low-abundance proteins. A number of variations of the multidimensional LC have been implemented [63]. Lubman and co-workers employed isoelectric focusing as the first dimension with RP-HPLC as the second [64]. Gygi and co-workers [65] used SCX as the first separation dimension with fraction collection. The second dimension was RP-HPLC with the vented column approach as described earlier. When the SCX is moved off-line from the RP-HPLC, the use of large-diameter SCX columns for milligram-scale sample injections is enabled and the dynamic range is increased.

Smith and co-workers have combined FTMS with ultrahigh-pressure LC for the use of 80-cm-long nanoscale columns for high throughput analysis ($\sim 10^5$ peptide masses measured in one run) [66] with reliance on an accurate mass tag formalism. The relative protein expression ratios can be measured with the isotope-coded affinity-tag (ICAT) reagent and method [56]. The ICAT strategy allows for relative comparison of proteomes from different cellular states and has been reviewed extensively elsewhere [67–69]. These contemporary approaches to MS-based proteomics will be especially powerful (and accessible to more nonexperts) for drug discovery efforts and in initiating studies that focus on a smaller scale through interplay with genetic data [70] and more sophisticated analysis of polymorphism [71]. Further, the new nongel methods will enable more powerful biomarker diagnosis of (disease) phenotypes and facilitate a fundamental understanding of biological systems [63,69]. Much of this future work will be driven by the measurement of two-to-five peptides per protein and provide direct mass information on 5–50% of the primary sequence.

Although mass spectrometry has made significant strides in sensitivity and proteome coverage, a recent report by Weissman and co-workers indicates that significant challenges remain [72]. With the use of a complete fusion, tandem-affinity-purification (TAP)-tag library for the genome of *Saccharomyces cerevisiae*, protein expression levels were quantitatively determined with Western blots by chemiluminescent detection. These results were compared to MS-based (MudPIT) results [54,73], and the MS data were strongly biased toward the detection of abundant proteins. For the 75% of the proteome that is represented by proteins present at fewer than 5000 copies per cell, only 8% were observed by MS. This landmark study will serve as an important benchmark for future MS method development.

THE NEXT WAVE: POSTTRANSLATIONAL MODIFICATIONS AND INTACT PROTEINS

Over 30 families of covalent modifications can occur for proteins; these include, among others, N-terminal lipoylation, acetylation, and proteolytic processing. The most comprehensive posttranslational modification (PTM) database now has over 200 entries [74]. Most posttranslational events generate proteins of different MW than that predicted from their respective ORFs. In principle, MS is ideally suited to detect events not easily predicted from sequence or that are yet unknown. In the classic 2D gel approach, the proteolytic digestion and extraction procedures outlined earlier constitute a bottom-up strategy (Figure 1.7A) and often leaves much of the protein uncharacterized. This approach to protein analysis of 381 *Escherichia coli* spots [75] and 303 *Haemophilus influenza* spots [76] in bacteria suggest that at least 18% and 22%, respectively, are modified. (After identification, these proteins did not appear near their predicted isoelectric point (pI) or MW values on the 2D gel.) Little or no direct evidence of the source of these discrepancies is usually obtained.

Some believe that in mammalian systems, nearly every protein has an interesting feature of primary structure associated with it. Thus, considerable research attention is focused on PTMs, with recent software developed to provide some characterization [77]. For analysis of the "phosphoproteome," the Chait [78], Smith [79], and Aebersold [80] laboratories have published targeted approaches that use the specific chemistry of phosphorylated

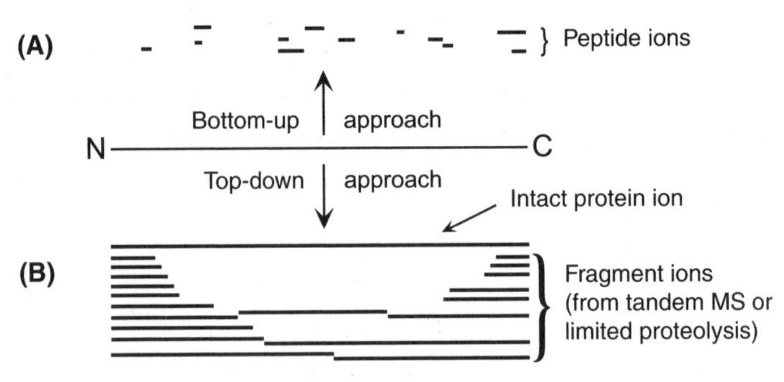

FIGURE 1.7 (A) Bottom-up vs. (B) top-down approaches to proteomic analysis. Top-down approaches use mass analysis of intact protein molecular ions with subsequent fragment analysis by tandem MS or limited proteolytic digestion. The bottom-up strategy uses analysis of peptide fragments from exhaustive proteolysis.

hydroxyls. Some PTMs, such as histidine phosphorylation occur, but are present on <0.1% of a particular protein in vivo. This occupancy can be addressed somewhat by the use of targeted strategies for selective detection of modified peptides, such as precursor ion scanning [81] and affinity chromatography [27,82]. It is essential that both the ionization method for MS and subsequent activation for MS/MS leave the PTM intact on the protein or peptide fragment. If the PTM survives intact into the mass spectrometer, then promising new MS/MS fragmentation methods such as electron-capture dissociation [68] appear to leave labile PTMs attached due to a nonergodic ion dissociation mechanism [83].

Despite much recent effort, a perspective by Mann [84] identified a largely uncharted area of high-throughput proteomic analyses: the collection of data reflects the entire covalent state of gene products. Such analyses with 100% sequence coverage for the efficient survey of the complete DNA-predicted sequence of a protein are possible via a top-down strategy (Figure 1.7B) [85]. The top-down approach can be realized by in-source decay (ISD) on a MALDI instrument [86] and charge reduction on an ESI ion trap MS [87], but this approach is most robust with ESI and the corresponding fragmentation inside a >6-tesla fourier transform mass spectrometry (FTMS) instrument. The top-down approach that uses ESI/FTMS generates larger protein fragments (5–40 kDa) from direct MS/MS of intact protein ions or limited proteolysis. Direct MS/MS of protein ions usually enables 100% sequence coverage below 50 kDa using a 6-tesla instrument. Such MS/MS data allow (1) protein identification by database retrieval [88], (2) verification of the N- and C-terminal positions, (3) confirmation of large sections of DNA-predicted sequence, and (4) localization of regions of error or modification [85].

Proper sample preparation is an absolute requirement for success for intact protein analysis. To obtain accurate molecular weights, the protein must be presented to the mass spectrometer virtually free of salts and detergents, which would otherwise cause adduct formation [89]. Recently, work has begun to establish efficient MS sampling methods devoted to intact proteins by coupling chromatographic [66] and electrophoretic [11,90,91] separations to FTMS. Valaskovic and co-workers [11] were able to detect intact proteins with narrow bore CE in MS and MS/MS mode with sub-attomole and attomole limits of detection, respectively. Smith and coworkers combined capillary isoelectric focusing with FTMS to allow for the analysis of hundreds of proteins [91], although MS/MS was not achieved. The Marshall group used an approach to analyze ~10 proteins by on-line RP-HPLC, and MS/MS was achieved on proteins up to 30 kDa [92]. On-line fragmentation of ~40 proteins was achieved with an infrared laser and capillary-LC on a 9.4-tesla instrument [92] with the external accumulation of ions that allow a near 100% LC duty cycle [93]. The recent addition of hybrid instruments, such as the quadrupole time-of-flight

[94] (Q-TOF) and quadrupole FTMS (Q-FTMS), significantly increases the specificity and sensitivity of LC-MS and provides for the analysis of highly complex mixtures. This idea has existed in the FTMS community for some time and was recently achieved by the Smith [95] and Marshall Laboratories [96]. The Q-FTMS hybrid allows for greatly enhanced dynamic range via gas-phase selection and concentration of analyte in a quadrupole trap, with subsequent high-resolution MS and MS/MS analysis in FT electromagnetic trap.

PROSPECTS FOR NANOBORE LC-MS IN "SMALL MOLECULE" DRUG DEVELOPMENT

The success of nanoscale formats for MS and LC-MS/MS has been driven primarily by the minute sample quantities demanded of proteomics analysis. While sample size may not be strictly limited in traditional small-molecule applications, an exhaustive LC-MS review by Lee and Kerns points to miniaturization as a key area of future development [2]. As 384 well plates, or even higher-density plates, gain acceptance along with the industry standard 96-well format, most aspects of sample manipulation, delivery, and analysis will require scale-down for success. In addition to having a significant positive impact on sensitivity, miniaturization may also significantly speed high throughput analysis [97].

The trend toward miniaturization for small-molecule analysis is driven in part by the desire to fully characterize a promising drug candidate as early in the discovery process as possible to reduce the cost of development failure [2]. To reduce the number of new chemical entities (NCEs) that fail in the late stages of drug development, a wide array of in vitro and in vivo assays are implemented early in the discovery phase to eliminate compounds with unfavorable characteristics [2,98,99]. Methods that provide information relevant to metabolism and metabolite identification [98,99] show particular promise. As compounds go through a wider variety of assays, the amount of compound available for testing and analysis decreases. At later stages of drug development, multiple assays pose no problem, since large-scale (gram-to-kilogram) synthesis creates sufficient compound for full characterization. The implementation of assays earlier in the discovery phase, however, especially within today's high-throughput, automated-synthesis environment, places limits on the amount of compound available for characterization.

Rourick and co-workers endeavored to test the feasibility of nanoscale LC-MS on metabolite analysis with an in vitro assay based on human liver microsomal incubation [100]. Buspirone, an anxiolytic drug with a well-characterized metabolic profile [101], was chosen as a test compound. At an initial concentration of 4 μM, buspirone was incubated with microsomal protein (1 mg/mL) and 4 mM NADPH in 50 mM sodium phosphate buffer at 37°C for 30 minutes. After quenching with the addition of 60 μL of 0.3 M TCA, samples were centrifuged for 15 minutes with precipitation of the protein pellet. Aliquots of supernate were injected onto a 2-mm-ID C18 column for conventional LC-MS or through a nanospray LC-MS system (Figure 1.4) with a 75-μm nanobore ID column-emitter with a 15-μm-ID tip. For the 2-mm-ID column, a 30-μL injection was followed by a 10-minute, 5% to 95% acetonitrile gradient in 0.05% formic acid. Figure 1.8A shows the reconstructed ion chromatograms (RICs) for the total ion current, the parent mass, the parent mass + 16 Da, and the parent mass + 32 Da. Little of the

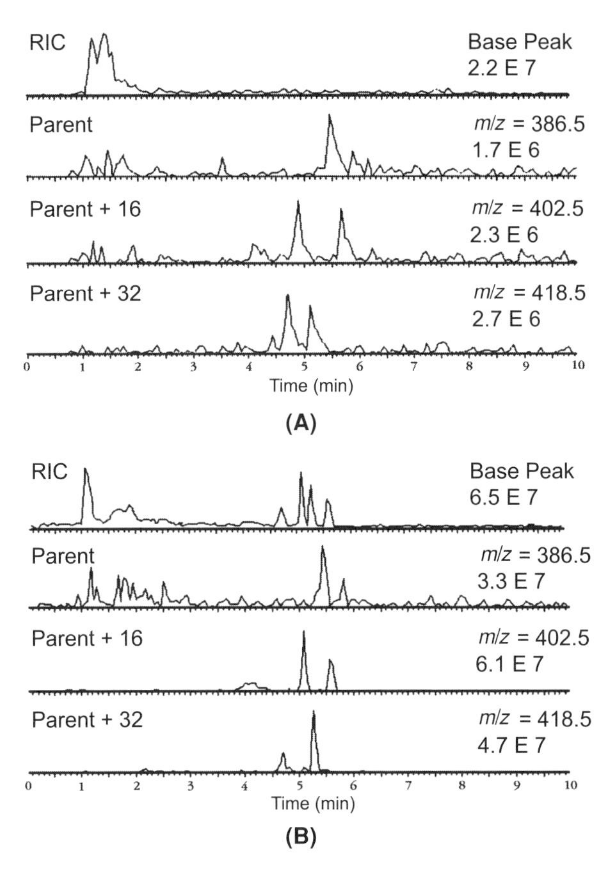

FIGURE 1.8 Reconstructed ion chromatograms (RICs) from LC-MS analysis of human liver microsomal incubations ($T = 30$ minutes) of Buspirone (4-μM initial concentration) for (A) a 30-μL injection on a 2-mm-ID column, and (B) a 3-μL injection on a 75-μm-ID column. The order of magnitude increase in ion current and superior S/N for the 75-μm column results even though an order of magnitude less sample is consumed in the analysis. The top trace for each figure is a base-peak RIC, while the lower traces are selected ion chromatograms (SICs) for the masses (mass-to-charge ratio (m/z)) indicated on the right-hand side. The number below each m/z value is the total peak ion current. The experimental arrangement of the column is shown in Figure 1.4. (Data courtesy of Robyn Rourick, Kalypsys, Inc., and Daniel Kassel, Syrrx, Inc.)

parent drug remains after 30 min, with much of the drug being hydroxylated at one or two sites [101]. The injection volume was reduced to 3 μL for the 75-μm-ID column. Elution was carried out with the identical mobile-phase gradient, but the through-column flow rate was reduced to 500 nL/min with the flow splitter. Figure 1.8B shows the same RICs as in Figure 1.8A, but with a superior S/N and greater ion current for the analyte. The high-quality MS and MS/MS data obtained from the chromatographic peak at 5.25 minutes of Figure 1.8B is shown in Figure 1.9. The 75-μm integral column/emitter offers a two orders of magnitude improvement in sensitivity. Sample volumes were successfully reduced by one order of magnitude, while the mass spectrum S/N was improved.

Vouros and co-workers recently demonstrated the effectiveness of nanospray to improve both the S/N quality concurrent with a reduction in matrix effects such as ion suppression [102]. Rather than use a nanobore LC column, the authors constructed a unique postcolumn flow that allowed for nanospray flow rates with conventional column formats. A cylindrically symmetric flow splitter was shown to preserve chromatographic peak shape at split ratios as high as 2000:1; thus, effluent from a conventional 2.1-mm C18 column at 200 μL/min was reduced to 100 nL/min. The heart of the flow splitter was a fused-silica needle emitter with an orifice diameter of 5 μm. The nanoflow splitter was applied to the in vitro metabolite analysis of a test

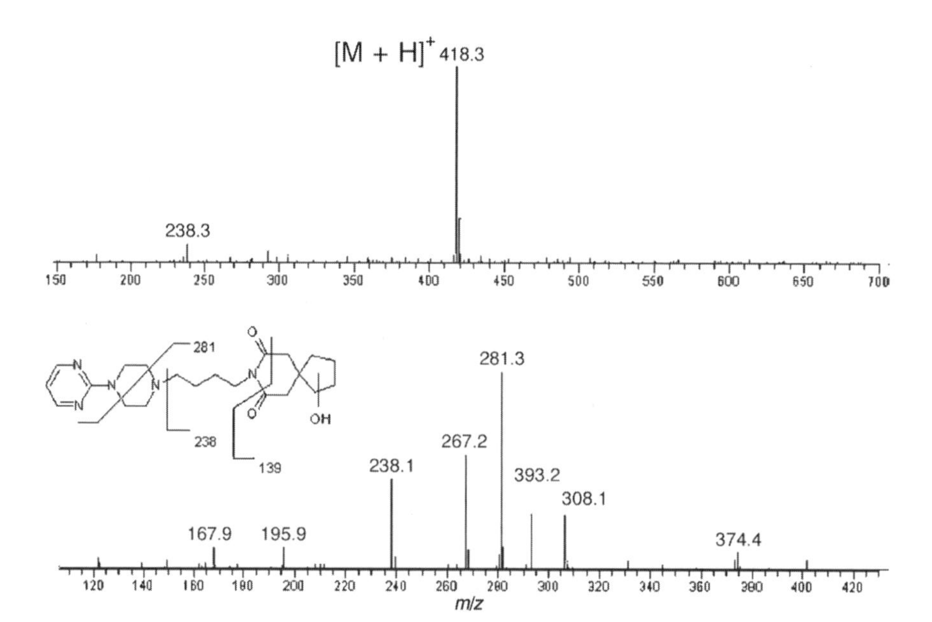

FIGURE 1.9 MS and MS/MS data from the run shown in Figure 1.8B for $T = 5.3$ minutes The inset shows probable assignments of the MS/MS fragments based on hydrolysis of the parent molecule. (Data courtesy of Robyn Rourick, Kalypsys, Inc., and Daniel Kassel, Syrrx, Inc.)

compound, Indinavir (MW = 613) on a triple quadrupole mass spectrometer. For the metabolites positively identified by MS/MS, operation at 100 nL/min improved the signal by an average value of 145%.

Furthermore, as shown in Figure 1.10, the 100-nL/min data identified a hydroxylated metabolite with an elution profile close to the parent drug molecule. This hydroxylated metabolite had not been previously observed at 200 μL/min. The authors propose that the increase in S/N, and decrease in ion suppression, is due to the smaller droplets and higher surface area generated with nanospray. Further support of reduced ion suppression with nanospray has been observed by Schmidt and co-workers [24], provided, however, that

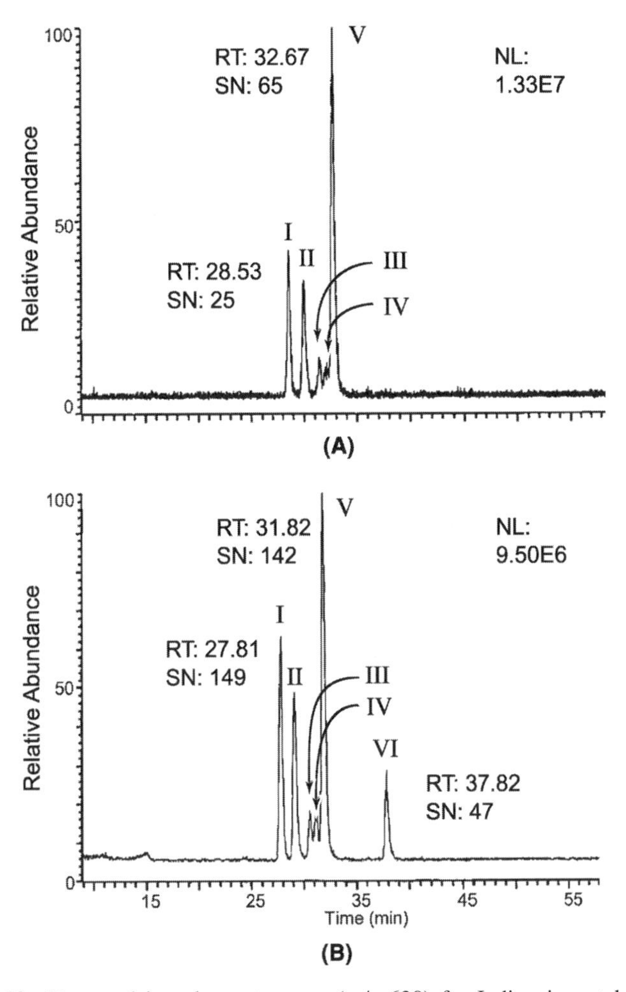

FIGURE 1.10 Extracted ion chromatograms (*m/z* 630) for Indinavir metabolites generated from a rat liver microsomal incubation at (A) 200 μL/min through a conventional ESI source, and (B) 0.1 μL/min from a 1000:1 postcolumn nanosplitter with a 5-μm-ID fused-silica nanospray emitter. (Reprinted from Gangl et al. [102], used with permission.)

the emitter design generates a suitable spray at flow rates below 100 nL/min. The authors reported a virtual elimination of ion suppression at flow rates below 20 nL/min and suggested that operation at the lowest possible flow rate is beneficial.

In these examples, nanospray formats offer equal, if not superior, separation performance, with significant improvements in sensitivity. To achieve acceptance within most discovery and development environments, such methods will need to demonstrate a high level of robustness, and a tighter relative standard deviation than that required of typical qualitative proteomics. Quantitative applications requiring the unique sensitivity of nanospray will need to be tightly controlled. Continual improvements in column technology, source technology, and sample preparation will be required. Toward this end, research in one of the author's laboratories (GAV) is focused-on the development of an orthogonal feedback-controlled nanospray source [103]. This source uses spray-mode [104,105] control to define, specify, and qualify nanospray operation.

CONCLUSION

Advances in LC-MS technology during the last decade have provided bioanalytical science with unmatched chemical selectivity, specificity, and sensitivity. Protein analysis and identification has been revolutionized, with experiments once requiring months to isolate protein now completed in less than one day. The mere problem of (pure) protein identification has been solved through the coupling of nanoscale-sample preparation and separation with MS/MS instrumentation. Despite these advances, the general analytical challenge of identification and characterization of proteins on a proteomic scale has not been solved. Undoubtedly, as research progresses beyond identification toward understanding biological function, proteomics by MS will move from a bottom-up (peptide driven) to top-down (protein driven) ideology. Proteomics has driven analysts to continually improve LC-MS sensitivity without compromise. As the need for higher throughput on smaller samples develops, these tools will find broader applicability to the varied applications of LC-MS within drug discovery and development.

ACKNOWLEDGMENTS

GAV thanks James P. Murphy for the static nanospray data and the staff of New Objective in the preparation of this manuscript. One of the authors (N.L.K.) acknowledges the support of the National Institutes of Health, grant number GM 067193.

REFERENCES

1. Figeys, D.; Pinto, D. "Lab-on-a-Chip: A Revolution in Biological and Medical Sciences," *Anal. Chem.* **72,** 330A–335A (2000).
2. Lee, M.S.; Kerns, E.H. "LC/MS Applications in Drug Development," *Mass Spectrom. Rev.* **18,** 187–279 (1999).
3. Laurell, T.; Marko-Varga, G. "Miniaturisation is Mandatory Unravelling the Human Proteome," *Proteomics* **2,** 345–351 (2002).
4. Hillenkamp, F.; Karas, M.; Beavis, R.C.; Chait, B.T. "Matrix-Assisted Laser Desorption/Ionization Mass Spectrometry of Biopolymers," *Anal. Chem.* **63,** 1193A–1203A (1991).
5. Fenn, J.B.; Mann, M.; Meng, C.K.; Wong, S.F.; Whitehouse, C.M. "Electrospray Ionization for Mass Spectrometry of Large Molecules," *Science* **246,** 64–71 (1989).
6. McLafferty, F.W. "Analytical Information from Mass Spectrometry, Past and Future," *J. Am. Soc. Mass Spectrom.* **1,** 1–5 (1990).
7. McLafferty, F.W.; Fridriksson, E.K.; Horn, D.M.; Lewis, M.A.; Zubarev, R.A. "Biomolecule Mass Spectrometry," *Science* **284,** 1289–1290 (1999).
8. Wilm, M.S.; Mann, M. "Electrospray and Taylor-cone Theory, Dole's Beam of Macromolecules at Last?" *Int. J. Mass Spectrom. Ion Processes* **136,** 167–180 (1994).
9. Wilm, M.S.; Mann, M. "Analytical Properties of the Nanoelectrospray Ion Source," *Anal. Chem.* **68,** 1–8 (1995).
10. Emmett, M.R.; Caprioli, R.M. "Micro-Electrospray Mass Spectrometry: Ultra-High-Sensitivity Analysis of Peptides and Proteins," *J. Am. Soc. Mass Spectrom.* **5,** 605–613 (1994).
11. Valaskovic, G.A.; Kelleher, N.L.; McLafferty, F.W. "Attomole Protein Characterization by Capillary Electrophoresis-Mass Spectrometry," *Science* **273,** 1199–1202 (1996).
12. Tomer, K.B.; Moseley, M.A.; Deterding, L.J.; Parker, C.E. "Capillary Liquid Chromatography/Mass Spectrometry," *Mass Spectrom. Rev.* **13,** 431–457 (1994).
13. Wahl, J.H.; Goodlet, D.R.; Udseth, H.R.; Smith, R.D. "Attomole Level Capillary Electrophoresis-Mass Spectrometric Protein Analysis Using 5 μm i.d. Capillaries," *Anal. Chem.* **64,** 3194–3196 (1992).
14. Taylor, S.; Srigengan, B.; Gibson, J.R.; Tindall, D.; Syms, R.; Tate, T.J.; Ahmad, M.M. "A Miniature Mass Spectrometer for Chemical and Biological Sensing," *SPIE Proc.* **4036,** 187–193 (2000).
15. Henry, C.M. "The Incredible Shrinking Mass Spectrometers," *Anal. Chem.* **71,** 264A–268A (1999).
16. Wood, T.D.; Moy, M.A.; Dolan, A.R.; Bigwarfe, P.M.; White, T.P.; Smith, D.R.; Higbee, D.J. "Miniaturization of Electrospray Ionization Mass Spectrometry," *Appl. Spec. Rev.* **38,** 187–244 (2003).

17. Foret, F.; Preisler, J. "Liquid Phase Interfacing and Miniaturization in Matrix-assisted Laser Desorption/Ionization Mass Spectrometry," *Proteomics* **2,** 360–372 (2002).

18. Zeleny, J. "The Electrical Discharge from Liquid Points, and A Hydrostatic Method of Measuring the Electric Intensity at Their Surfaces," *Phys. Rev.* **3,** 69–91 (1914).

19. Kebarle, P.; Tang, L. "From Ions in Solution to Ions in the Gas Phase: The Mechanism of Electrospray Ionization Mass Spectrometry," *Anal. Chem.* **65,** 972A–986A (1993).

20. Geronmanos, S.; Freckleton, G.; Tempst, P. "Tuning of an Electrospray Ionization Source for Maximum Peptide-Ion Transmission into a Mass Spectrometer," *Anal. Chem.* **72,** 777–790 (2000).

21. Valaskovic, G.A.; Kelleher, N.L.; Little, D.P.; Aaserud, D.J.; McLafferty, F.W. "Attomole-Sensitivity Electrospray Source for Large-Molecule Mass Spectrometry," *Anal. Chem.* **67,** 3802–3805 (1995).

22. Neugebauer, J.; Moseley, M.A.; Vissers, J.P.C.; Moyer, M. "Comparison of Nanospray and Fully Automated Nanoscale Capillary LC/MS/MS for Protein Identification," in *Proceedings of the 47th ASMS Conference on Mass Spectrometry and Allied Topics*, Dallas, TX, June 13–17, 1999, 2877–2878.

23. Juraschek, R.; Dulcks, T.; Karas, M. "Nanoelectrospray—More Than Just a Minimized-Flow Electrospray Ionization Source," *J. Am. Soc. Mass Spectrom.* **10,** 300–308 (1999).

24. Schmidt, A.; Karas, M.; Dulcks, R. "Effect of Different Flow Rates on Analyte Ion Signals in Nano-ESI MS, or: When Does ESI Turn into Nano-ESI?," *J. Am. Soc. Mass Spectrom.* **14,** 492–500 (2003).

25. Shevchenko, A.; Jensen, O.N.; Podtelejnikov, A.V.; Sagliocco, F.; Wilm, M.; Vorm, O.; Mortensen, P.; Shevchenko, A.; Boucherie, H.; Mann, M. "Linking Genome and Proteome by Mass Spectrometry: Large-Scale Identification of Yeast Proteins from Two Dimensional Gels," *Proc. Natl. Acad. Sci. USA* **93,** 14440–14445 (1996).

26. Erdjument-Bromage, H.; Lui, M.; Lacomis, L.; Grewal, A.; Annan, R.S.; McNulty, D.E.; Carr, S.A.; Tempst, P. "Examination of Micro-tip Reversed-phase Liquid Chromatographic Extraction of Peptide Pools for Mass Spectrometric Analysis," *J. Chromatog. A* **826,** 167–181 (1998).

27. Posewitz, M.C.; Tempst, P. "Immobilized Gallium(III) Affinity Chromatography of Phosphopeptides," *Anal. Chem.* **71,** 2883–2892 (1999).

28. Liu, H.; Felten, C.; Xue, Q.; Zhang, B.; Jedrzejewski, P.; Karger, B.L.; Foret, F. "Development of Multichannel Devices with an Array of Electrospray Tips for High-Throughput Mass Spectrometry," *Anal. Chem.* **72,** 3303–3310 (2000).

29. Schultz, G.A.; Corso, T.N.; Prosser, S.J.; Zhang, S. "A Fully Integrated Monolithic Microchip Electrospray Device for Mass Spectrometry," *Anal. Chem.* **72,** 4058–4063 (2000).

30. Dethy, J.M.; Ackermann, B.L.; Delatour, C.; Henion, J.D.; Schultz, G.A. "Demonstration of Direct Bioanalysis of Drugs in Plasma Using Nanoelectrospray Infusion from a Silicon Chip Coupled with Tandem Mass Spectrometry," *Anal. Chem.* **75,** 805–811 (2003).
31. Sjodahl, J.; Melin, J.; Griss, P.; Emmer, A.; Stemme, G.; Roeraade, J. "Characterization of Micromachined Hollow Tips for Two-dimensional Nanoelectrospray Mass Spectrometry," *Rapid Comm. Mass Spec.* **17** (2003).
32. Yeung, E.S. "Chemical Analysis of Single Human Erythrocytes," *Acc. Chem. Res.* **27,** 409–414 (1994).
33. Haynes, P.A.; Yates, J.R. "Proteome Profiling—Pitfalls and Progress," *Yeast* **17,** 81–87 (2000).
34. Salzmann, J.P.; vanSoest, R.E.J.; Vissers, H.; Chervet, J.P. "Automated On-line Sample Handling and Analysis by Micro or Capillary LC prior to MS," in *Proceedings of the 44th Annual Conference on Mass Spectrometry and Allied Topics*, Portland, Oregon, May 12–16, 1996, 200.
35. Heeft, E.v.d.; Hove, G.J.t.; Herberts, C.A.; Meiring, H.D.; Els, C.A.C.M.v.; Jong, A.P.J.M.d. "A Microcapillary Column Switching HPLC-Electrospray Ionization MS System for the Direct Identification of Peptides Presented by Major Histocompatibility Complex Class I Molecules," *Anal. Chem.* **70,** 3742–3751 (1998).
36. Meiring, H.D.; Barroso, B.M.; Heeft, E.v.d.; Hove, G.J.t.; Jong, A.P.J.M.d. "Sheathless Nanoflow HPLC-ESI/MS in Proteome Research and MHC Bound Peptide Identification," in *Proceedings of the 47th ASMS Conference on Mass Spectrometry and Allied Topics*, Dallas, Texas, June 13–17, 1999.
37. Meiring, H.D.; Heeft, E.v.d.; Hove, G.J.t.; Jong, A.P.J.M.d. "Nanoscale LC-MS: Technical Design and Applications to Peptide and Protein Analysis," *J. Sep. Sci.* **25,** 557–568 (2002).
38. Licklider, L.J.; Thoreen, C.C.; Peng, J.; Gygi, S.P. "Automation of Nanoscale Microcapillary Liquid Chromatography–Tandem Mass Spectrometry with a Vented Column," *Anal. Chem.* **74,** 3076–3083 (2002).
39. Davis, M.T.; Stahl, D.C.; Swiderek, K.M.; Lee, T.D. "Capillary Liquid Chromatography/Mass Spectrometry for Peptide and Protein Characterization," *Methods: A Companion to Methods in Enzymology* **6,** 304–314 (1994).
40. Moseley, M.A.; Jorgenson, J.W.; Shabanowitz, J.; Hunt, D.F.; Tomer, K.B. "Optimization of Capillary Zone Electrophoresis/Electrospray Ionization Parameters for the Mass Spectrometry and Tandem Mass Spectrometry Analysis of Peptides," *J. Am. Soc. Mass Spectrom.* **3,** 289–300 (1992).
41. Parker, C.E.; Perkins, J.R.; Tomer, K.B.; Shida, Y.; O'Hara, K.; Kono, M. "Application of Nanoscale Packed Capillary Liquid Chromatography (75 μm id) and Capillary Zone Electrophoresis/Electrospray Ionization Mass Spectrometry to the Analysis of Macrolide Antibiotics," *J. Am. Soc. Mass Spectrom.* **3,** 563–574 (1992).

42. Gale, D.C.; Smith, R.D. "Small Volume and Low Flow-Rate Electrospray Ionization Mass Spectrometry of Aqueous Samples," *Rapid Commun. Mass Spectrom.* **7,** 1017–1021 (1993).
43. Davis, M.T.; Stahl, D.C.; Hefta, S.A.; Lee, T.D. "A Microscale Electrospray Interface for On-line, Capillary Liquid Chromatography/Tandem Mass Spectrometry of Complex Peptide Mixtures," *Anal. Chem.* **67,** 4549–4556 (1995).
44. Warriner, R.N.; Craze, A.S.; Games, D.E.; Lane, S.J. "Capillary Electrochromatography/Mass Spectrometry—A Comparison of the Sensitivity of Nanospray and Microspray Ionization Techniques," *Rapid Commun. Mass Spectrom.* **12,** 1143–1149 (1998).
45. Kelly, J.F.; Ramaley, L.; Thibault, P. "Capillary Zone Electrophoresis-Electrospray Mass Spectrometry at Submicroliter Flow Rates: Pratical Considerations and Analytical Performance," *Anal. Chem.* **69,** 51–60 (1996).
46. Kriger, M.S.; Cook, K.D.; Ramsey, R.S. "Durable Gold-Coated Fused Silica Capillaries for Use in Electrospray Mass Spectrometry," *Anal. Chem.* **67,** 385–389 (1995).
47. Valaskovic, G.A.; McLafferty, F.W. "Long-Lived Metalized Tips for Nanoliter Electrospray Mass Spectrometry," *J. Am. Soc. Mass Spectrom.* **7,** 1270–1272 (1996).
48. Maziarz, E.P.; Lorenz, S.A.; White, T.P.; Wood, T.D. "Polyaniline: A Conductive Polymer Coating for Durable Nanospray Emitters," *J. Am. Soc. Mass Spectrom.* **11,** 659–663 (2000).
49. Barnidge, D.R.; Nilsson, S.; Markides, K.E. "A Design for Low-Flow Sheathless Electrospray Emitters," *Anal. Chem.* **71,** 4115–4118 (1999).
50. Wetterhall, M.; Nillson, S.; Markides, K.E.; Bergquist, J. "A Conductive Polymeric Material Used for Nanospray Needle and Low-Flow Sheathless Electrospray Ionization Applications," *Anal. Chem.* **74,** 239–245 (2002).
51. Gatlin, C.L.; Kleeman, G.R.; Hays, L.G.; Link, A.J.; Yates, J.R. "Protein Identification at the Low Femtomole Level from Silver-Stained Gels Using a New Fritless Electrospray Interface for Liquid Chromatography-Microspray and Nanospray Mass Spectrometry," *Anal. Biochem.* **263,** 93–101 (1998).
52. Ivanov, A.R.; Zang, L.; Karger, B.L. "Low-attomole Electrospray Ionization MS and MS/MS Analysis of Protein Tryptic Digests Using 20-μm i.d. Polystyrene-Divinylbenzene Monolithic Capillary Columns," *Anal. Chem.* **75,** 5306–5316 (2003).
53. Peterson, D.S.; Rohr, T.; Svec, F.; Frechet, J.M.J. "Dual-function Microanalytical Device by In Situ Photolithographic Grafting of Porous Polymer Monolith: Integrating Solid-Phase Extraction and Enzymatic Digestion for Peptide Mass Mapping," *Anal. Chem.* **75,** 5328–5335 (2003).
54. Washburn, M.P.; Wolters, D.; Yates, J.R. "Large-Scale Analysis of the Yeast Proteome by Multidimensional Protein Identification Technology," *Nat. Biotechnol.* **19,** 242–247 (2001).

55. Ho, Y.; Gruhler, A.; Heilbut, A.; Bader, G.D.; Moore, L.; Adams, S.L.; Miller, A.; Taylor, P.; Bennett, K.; Boutlier, K.; Yang, L.; Wolting, C.; Donaldson, A.; Schandorff, S.; Shewnarane, J.; Vo, M.; Taggart, J.; Goudreault, M.; Muskat, B.; Alfarano, C.; Dewar, D.; Lin, Z.; Michalickova, K.; Willems, A.R.; Sassi, H.; Nielsen, P.A.; Moran, M.F.; Durocher, D.; Mann, M.; Hoque, C.W.V.; Figeys, D.; Tyers, M. "Systematic Identification of Protein Complexes in *Saccharomyces cerevisiae* by Mass Spectrometry," *Nat.* **415,** 181–183 (2002).
56. Gygi, S.P.; Rist, B.; Gerber, S.A.; Turecek, F.; Gelb, M.H.; Aebersold, R. "Quantitative Analysis of Complex Protein Mixtures Using Isotope-Coded Affinity Tags," *Nat. Biotechnol.* **17,** 994–999 (1999).
57. Wang, W.; Zhou, H.; Lin, H.; Roy, S.; Shaler, T.A.; Hill, L.R.; Norton, S.; Kumar, P.; Anderle, M.; Becker, C.H. "Quantification of Proteins and Metabolites by Mass Spectrometry without Isotopic Labeling or Spiked Standards," *Anal. Chem.* **75,** 4818–4826 (2003).
58. Binz, P.A.; Muller, M.; Walther, D.; Bienvenut, W.V.; Gras, R.; Hoogland, C.; Bouchet, G.; Gasteiger, E.; Fabbretti, R.; Gay, S.; Palagi, P.; Wilkins, M.R.; Rouge, V.; Tonella, L.; Paesano, S.; Rossellat, G.; Karmime, A.; Bairoch, A.; Sanchez, J.C.; Appel, R.D.; Hochstrasser, D.F. "A Molecular Scanner to Automate Proteomic Research and to Display Proteome Images," *Anal. Chem.* **71,** 4981–4988 (1999).
59. Gavin, A.C.; Bosche, M.; Krause, R.; Grandi, P.; Marzioch, M.; Bauer, A.; Schultz, J.; Rick, J.M.; Michon, A.M.; Cruciat, C.M.; Remor, M.; Hofert, C.; Schelder, M.; Raida, M.; Bouwmeester, T.; Bork, P.; Seraphin, B.; Kuster, B.; Neubauer, G.; Superti-Furga, G. "Functional Organization of the Yeast Proteome by Systematic Analysis of Protein Complexes," *Nat.* **415,** 143–147 (2002).
60. Clauser, K.R.; Baker, P.; Burlingame, A.L. "Role of Accurate Mass Measurement (+/− 10 ppm) in Protein Identification Strategies Employing MS or MS/MS and Database Searching," *Anal. Chem.* **71,** 2871–2882 (1999).
61. Fenyö, D.; Qin, J.; Chait, B.T. "Protein Identification Using Mass Spectrometric Information," *Electrophoresis* **19,** 998–1005 (1998).
62. Pappin, D.J.C.; Hojrup, P.; Bleasby, A.J. "Rapid Identification of Proteins by Peptide-Mass Fingerprinting," *Curr. Biol.* **3,** 327–332 (1993).
63. Figeys, D. "Proteomics in 2002: A Year of Technical Development and Wide-Ranging Applications," *Anal. Chem.* **75,** 2891–2905 (2003).
64. Wall, D.B.; Kachman, M.T.; Gong, S.; Hinderer, R.; Parus, S.; Misek, D.E.; Hanash, S.M.; Lubman, D. "Isoelectric Focusing Nonporous RP HPLC: A Two-Dimensional Liquid-Phase Separation Method for Mapping of Cellular Proteins with Identification Using MALDI-TOF Mass Spectrometry," *Anal. Chem.* **72,** 1099–1111 (2000).
65. Peng, J.; Elias, J.E.; Thoreen, C.C.; Licklider, L.J.; Gygi, S.P. "Evaluation of Multidimensional Chromatography Coupled with Tandem Mass Spectrometry (LC/LC-MS/MS) for Large-Scale Protein Analysis: The Yeast Proteome," *J. Proteome Res.* **2,** 43–50 (2003).

66. Shen, Y.; Zhao, R.; Belov, M.E.; Conrads, T.P.; Anderson, G.A.; Tang, K.; Pasa-Tolic, L.; Veenstra, T.D.; Lipton, M.S.; Udseth, H.R.; Smith, R.D. "Packed Capillary Reverse-Phase Liquid Chromatography with High-Performance Electrospray Ionization Fourier Transform Ion Cyclotron Resonance Mass Spectrometry for Proteomics," *Anal. Chem.* **73,** 1766–1775 (2001).
67. Mann, M. "Quantitative Proteomics?" *Nature Biotechnol.* **17,** 954–955 (1999).
68. Kelleher, N.L. "From Primary Structure to Function: Biological Insights from Large Molecule Mass Spectra," *Chem. Biol.* **7,** R37–R45 (2000).
69. Aebersold, R.; Mann, M. "Mass Spectrometry-Based Proteomics," *Nature* **422,** 198–207 (2003).
70. Roses, A.D. "Pharmacogenetics and the Practice of Medicine," *Nature Biotechnol.* **405,** 857–865 (2000).
71. Gura, T. "Can SNPs Deliver on Susceptibility Genes?" *Science* **293,** 593–595 (2001).
72. Ghaemmaghami, S.; Huh, W.K.; Bower, K.; Howson, R.W.; Belle, A.; Dephoure, N.; O'Shea, E.K.; Weissman, J.S. "Global Analysis of Protein Expression in Yeast," *Nature* **425,** 737–740 (2003).
73. Washurn, M.P.; Koller, A.; Oshiro, G.; Ulaszek, R.R.; Plouffe, D.; Deciu, C.; Winzeler, E.; Yates, Y.R. "Protein Pathway and Complex Clustering of Correlated mRNA and Protein Expression Analysis in *Saccharomyces cerevisiae*," *Proc. Natl. Acad. Sci. USA* **100,** 3107–3112 (2003).
74. Garavelli, J.S. "The RESID Database of Protein Structure Modifications: 2000 Update," *Nucleic Acids Res.* **28,** 209–211 (2000).
75. Link, A.J.; Robison, K.; Church, G.M. "Comparing the Predicted and Observed Properties of Proteins Encoded in the Genome of *Escherichia coli* K-12," *Electrophoresis* **18,** 1259–1313 (1997).
76. Link, A.J.; Hays, L.G.; Carmack, E.B.; Yates, J.R. "Identifying the Major Proteome Components of *Haemophilus influenza* Type-strain NCTC 8143," *Electrophoresis* **18,** 1314–1334 (1997).
77. Wilkins, M.R.; Gasteiger, E.; Gooley, A.A.; Herbert, B.R.; Molloy, M.P.; Binz, P.A.; Ou, K.; Sanchez, J.C.; Bairoch, A.; Williams, K.L.; Hochstrasser, D.F. "High-Throughput Mass Spectrometric Discovery of Protein Post-Translational Modifications," *J. Mol. Biol.* **289,** 645–657 (1999).
78. Oda, Y.; Nagasu, T.; Chait, B.T. "Enrichment Analysis of Phosphorylated Proteins as a Tool for Probing the Phosphoproteome," *Nature Biotechnol.* **19,** 379–382 (2001).
79. Goshe, M.B.; Conrads, T.P.; Panisko, E.A.; Angell, N.H.; Veenstra, T.D.; Smith, R.D. "Phosphoprotein Isotope-Coded Affinity Tag Approach for Isolating and Quantifying Phosphopeptides in Proteome-wide Analyses," *Anal. Chem.* **73,** 2578–2586 (2001).
80. Zhou, H.; Watts, J.D.; Aebersold, R. "A Systematic Approach to the Analysis of Protein Phosphorylation," *Nature Biotechnol.* **19,** 375–378 (2001).

81. Annan, R.S.; Huddleston, M.J.; Verma, R.; Deshaies, R.J.; Carr, S.A. "A Multidimensional Electrospray MS-Based Approach to Phosphopeptide Mapping," *Anal. Chem.* **73,** 393–404 (2001).

82. Nuwaysir, L.M.; Stults, J.T. "Electrospray Ionization Mass Spectrometry of Phosphopeptides Isolated by On-Line Immobilized Metal-Ion Affinity Chromatography," *J. Am. Soc. Mass Spectrom.* **4,** 662–669 (1993).

83. Kelleher, N.L.; Zubarev, R.A.; Bush, K.; Furie, B.; Furie, B.C.; McLafferty, F.W.; Walsh, C.T. "Localization of Labile Posttranslational Modifications by Electron Capture Dissociation: The Case of g-Carboxyglutamic Acid," *Anal. Chem.* **71,** 4250–4253 (1999).

84. Pandey, A.; Mann, M. "Proteomics to Study Genes and Genomes," *Nature* **405,** 837–846 (2000).

85. Kelleher, N.L.; Lin, H.Y.; Valaskovic, G.A.; Aaseruud, D.J.; Fridriksson, E.K.; McLafferty, F.W. "Top Down Versus Bottom Up Protein Characterization by Tandem High-Resolution Mass Spectrometry," *J. Am. Chem. Soc.* **121,** 806–812 (1999).

86. Reiber, D.C.; Grover, T.A.; Brown, R.S. "Identifying Proteins Using Matrix-Assisted Laser Desorption/Ionization In-Source Fragmentation Data Combined with Database Searching," *Anal. Chem.* **70,** 673–683 (1998).

87. Schaaff, T.G.; Cargile, B.J.; Stephenson, J.L.; J.; McLuckey, S.A. "Ion Trap Collisional Activation of the (M+2H)2+–(M + 17H)17+ Ions of Human Hemoglobin-Chain," *Anal. Chem.* **72,** 899–907 (2000).

88. Mørtz, E.; O'Connor, P.B.; Roepstorff, P.; Kelleher, N.L.; Wood, T.D.; McLafferty, F.W.; Mann, M. "Sequence Tag Identification of Intact Proteins by Matching Tandem Mass Spectral Data Against Sequence Data Bases," *Proc. Natl. Acad. Sci. USA* **93,** 8264–8267 (1996).

89. Little, D.P.; Speir, J.P.; Senko, M.W.; O'Connor, P.B.; McLafferty, F.W. "Infrared Multiphoton Dissociation of Large Multiply Charged Ions for Biomolecule Sequencing," *Anal. Chem.* **66,** 2809–2815 (1994).

90. Yang, L.; Lee, C.S.; Hofstadler, S.A.; Pasa-Tolic, L.; Smith, R.D. "Capillary Isoelectric Focusing-Electrospray Ionization Fourier Transform Ion Cyclotron Resonance Mass Spectrometry for Protein Characterization," *Anal. Chem.* **70,** 3235–3241 (1998).

91. Jensen, P.K.; Pasa-Tolic, L.; Anderson, G.A.; Horner, J.A.; Lipton, M.S.; Bruce, J.E.; Smith, R.D. "Probing Proteomes Using Capillary Isoelectric Focusing-Electrospray Ionization Fourier Transform Ion Cyclotron Resonance Mass Spectrometry," *Anal. Chem.* **71,** 2076–2084 (1999).

92. Li, W.; Hendrickson, C.L.; Emmett, M.R.; Marshall, A.G. "Identification of Intact Proteins in Mixtures by Alternated Capillary Liquid Chromatography Electrospray Ionization and LC ESI Infrared Multiphoton Dissociation Fourier Transform Ion Cyclotron Resonance Mass Spectrometry," *Anal. Chem.* **71,** 4397–4402 (1999).

93. Senko, M.W.; Hendrickson, C.L.; Emmett, M.R.; Shi, S.D.-H.; Marshall, A.G. "External Accumulation of Ions for Enhanced Electrospray Ionization Fourier Transform Ion Cyclotron Resonance Mass Spectrometry," *J. Am. Soc. Mass Spectrom.* **8,** 970–976 (1997).
94. Blackburn, R.K.; Moseley, M.A. "Quadrupole Time-of-Flight Mass Spectrometry: A Powerful New Tool for Protein Indentification and Characterization," *Am. Pharm. Rev.* **2,** 49–59 (1999).
95. Belov, M.E.; Nikolaev, E.N.; Anderson, G.A.; Auberry, K.J.; Harkewicz, R.; Smith, R.D. "Electrospray Ionization-Fourier Transform Ion Cyclotron Mass Spectrometry Using Ion Preselection and External Accumulation for Ultrahigh Sensitivity," *J. Am. Soc. Mass Spectrom.* **12,** 38–48 (2001).
96. Hendrickson, C.H.; Emmett, M.R.; Quinn, J.P.; Marshall, A.G. "Quadrupole Mass Filtered External Accumulation for Fourier-Transform Ion Cyclotron Resonance Mass Spectrometry." In Proceedings of the *48th ASMS Conference on Mass Spectrometry and Allied Topics*, Long Beach, CA, June 11–15, 2000.
97. Henion, J.; Prosser, S.J.; Corso, T.N.; Schultz, G.A. "Sample Preparation and Analysis Strategies for High Throughput LC/MS/MS Analysis of Biological Samples," *Am. Pharm. Rev.* **3,** 19–29 (2000).
98. Cox, K.A.; Clarke, N.J.; Rindgen, D.; Korfmacher, W.A. "Higher Throughput Metabolite Identification in Drug Discovery: Current Capabilities and Future Trends," *Am. Pharm. Rev.* **4,** 45–52 (2001).
99. Rourick, R.; Xu, R.; Kassel, D. "High Throughput Metabolic Structure Elucidation Strategies in Drug Discovery," in *Proceedings of the 48th Conference on Mass Spectrometry and Allied Topics*, Long Beach, CA, June 11–15, 2000.
100. Rourick, R.; Kassel, D.; Lee, M.; Valaskovic, G.A. "Capillary LC/MS and LC/MS/MS Strategies Applied to High Throughput Structure Elucidation in Drug Discovery," in *Proceedings of the 49th Conference on Mass Spectrometry and Allied Topics*, Chicago, IL, May 27–31, 2001.
101. Kerns, E.H.; Rourick, R.A.; Volk, K.J.; Lee, M.S. "Buspirone Metabolite Structure Profile using a Standard Liquid Chromatographic-Mass Spectrometric Protocol," *J. Chromatog. B* **698,** 133–145 (1997).
102. Gangl, E.T.; Annan, A.; Spooner, N.; Vouros, P. "Reduction of Signal Supression Effects in ESI-MS Using a Nanosplitting Device," *Anal. Chem.* **73,** 5635–5644 (2001).
103. Valaskovic, G.A.; Lee, M.S. "Orthogonal Control Sytems for Electrospray Ionization Mass Spectrometry," in *Proceedings of the 50th ASMS Conference on Mass Spectrometry and Allied Topics*, Orlando, FL, June 2–6, 2002.
104. Jaworek, A.; Krupa, A. "Classification of the Modes of EHD Spraying," *J. Aerosol Sci.* **30,** 873–893 (1999).
105. Juraschek, R.; Schmidt, A.; Karas, M.; Rollgen, F.W. "Dependence of Electrospray Ionization Efficiency on Axial Spray Modes," *Adv. Mass Spectrom.* **14,** 1–15 (1998).

2

MASS-SPECTROMETRY-BASED DRUG SCREENING ASSAYS FOR EARLY PHASES IN DRUG DISCOVERY

MARSHALL M. SIEGEL

INTRODUCTION

In the exploratory stage of traditional drug discovery, two steps are taken to evaluate promising drug candidates. The first step is to determine whether the new chemical entity (NCE) interacts with an active site of a biopolymer of therapeutic interest. The second step is to determine whether the new chemical entity interferes with a cell-based biological process. The first approach is generally evaluated with an in vitro assay involving an isolated biopolymer, generally a protein, deoxyribonucleic acid (DNA) model, or ribonucleic acid (RNA) model, while the second approach is generally evaluated with an in vivo cell-based assay. In vitro screening of new drug candidates in low and high throughput formats with mass spectrometry (MS) as the preferred detector is the subject of this review. The mass spectrometric techniques currently in vogue for drug screening are principally electrospray ionization (ESI) MS, and to a much lesser extent, matrix-assisted laser-desorption/ionization (MALDI) MS.

A general goal of in vitro screening is to study the noncovalent interactions between a biopolymer and drug candidate, and to identify those components

Integrated Strategies for Drug Discovery Using Mass Spectrometry, Edited by Mike S. Lee

that selectively bind to the active site of the biopolymer. In many applications, the drug candidates consist of a library of known chemical entities, which can be analyzed individually or as mixtures. Likewise, the biopolymers used in these studies are generally fully characterized and can be used either as single components or as mixtures together with the drug candidates for studying the biopolymer–drug noncovalent interactions. The components that react with the active site may inhibit or activate the biological activity of the biopolymer. Nevertheless, the identification of compounds that bind specifically to a biopolymer does not certify activity in a biological system. Activity can only be demonstrated with an in vivo biological assay.

Three mass spectrometric approaches have been taken to study biopolymer–drug noncovalent complexes produced under native conditions. One approach is to study the biopolymer–drug complex directly in the gas phase by ESI-MS [1–7] and to assay the complex with tandem MS (MS/MS) techniques. The second approach uses condensed-phase separation techniques to resolve the biopolymer–drug complex from unreacted drug or biopolymer, followed by the analysis of the drug by MS techniques [8]. In the first approach, the ESI-MS analysis is conducted under less sensitive native conditions, while in the second approach the complex can be analyzed by ESI-MS under more sensitive denaturing conditions, and in special cases, even by MALDI-MS. The third approach is the study of screened compounds that were noncovalently bound to an active surface or even covalently bound to a bead surface by releasing and analyzing the small molecules with MALDI-MS techniques.

In this review, the in vitro drug-screening method is illustrated with the direct gas-phase ESI-MS analysis method for peptide–drug, protein–drug, RNA–drug, and DNA–drug complexes. The indirect drug-screening method, with separation techniques coupled with ESI-MS, is illustrated principally for protein–drug complexes using the ancillary techniques of affinity chromatography, ultrafiltration, ultracentrifugation, gel permeation chromatography (GPC) spin-columns, GPC reverse-phase high-performance liquid chromatographic (HPLC) and capillary electrophoretic methods. MALDI-MS methods for solid-phase samples are illustrated for samples bound to affinity probe tips and for the analysis of sorted combinatorial chemistry beads. Figure 2.1 summarizes schematically all the strategies used today for screening noncovalent biopolymer–ligand complexes with mass-spectrometric detection for exploratory and early drug discovery.

ESI-MS: GAS-PHASE DRUG-SCREENING STUDIES

Peptide–Drug Complexes

One of the earliest contributions with ESI-MS for the study of noncovalent interactions in the gas phase was the study by Henion and co-workers [9] of

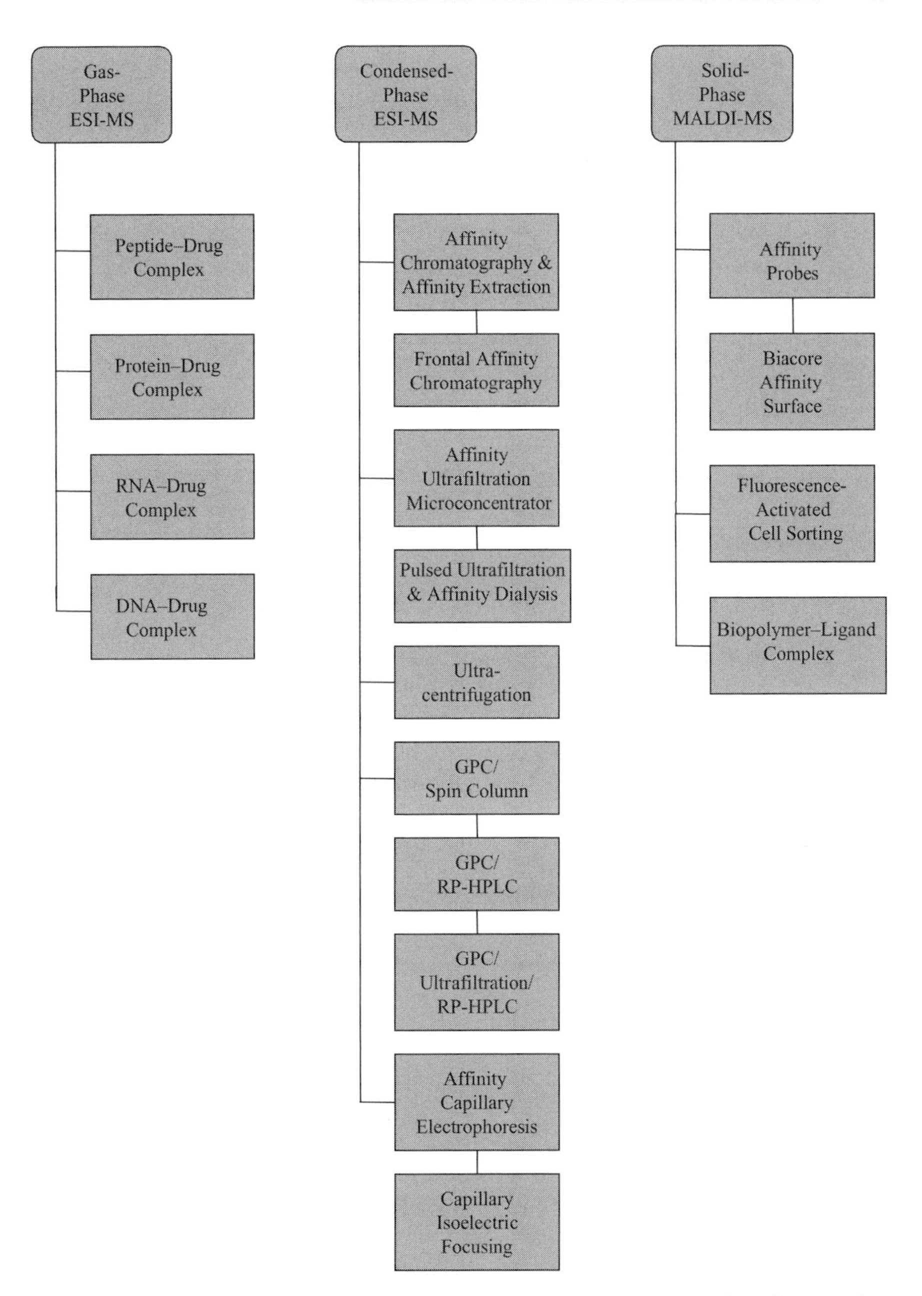

FIGURE 2.1 Schematic summary of all the strategies used today for screening of noncovalent biopolymer–ligand complexes with mass spectrometric detection for exploratory and early drug discovery.

model cell-wall peptides with members of the glycopeptide family of antibiotics, e.g., vancomycin and ristocetin. The antibiotics are potent inhibitors of cell-wall biosynthesis since they form strong noncovalently bound complexes with the peptidoglycan precursor C-terminus residues *X*-D-Ala-D-Ala, where *X* is L-lysine, L-diaminopimelic acid, L-alanine, or L-homoserine. This process disrupts bacterial cell-wall synthesis causing lysis of the bacteria. Typical mass spectra of glycopeptides with cell-wall models exhibit abundant noncovalent adducts of the antibiotic with the cell-wall models under native conditions (pH 7 with and without NH_4Ac buffer, RT), and even under denaturing conditions (pH 3–9, up to 8 M urea, 4 M KCl, and 1% SDS, 60°C), due to the strong binding between the two reagents. This is illustrated for the equimolar (25-mM) mixture of vancomycin and Ac_2KAA (Figure 2.2A). Verification of the assignments was confirmed by the formation of principally free vancomycin and free Ac_2KAA observed in the MS/MS spectrum of the isolated vancomycin-Ac_2KAA complex (Figure 2.2B). Binding constants were computed from the ESI-MS data as a function of concentration with Scatchard plots. Heck and co-workers [7,10–15] extended these studies to the additional glycopeptide antibiotics avoparcin, eremomycin, and other naturally occurring and chemically modified forms of members of the vancomycin class of antibiotics. These studies demonstrate that screening for new cell-wall binding compounds can be achieved by studying the noncovalent complexes formed between model cell-wall precursor peptides and compound libraries with ESI-MS detection for the model cell-wall–drug complexes. It may be possible as well to use the same strategy to screen for drugs, that bind to drug-resistant organisms that generate peptidoglycan precursor C-terminus residues *X*-D-Ala-D-Lac and *X*-D-Ala-D-Ser, where the terminating peptides are D-lactate and D-serine, respectively, instead of the normal D-alanine [7,11,16].

Protein–Drug Complexes

Smith and co-workers [17–20] demonstrated an elegant gas-phase electrospray approach for studying noncovalent interactions between a library of compounds and a protein to identify tight binding complexes. The method exploits the unique capabilities of Fourier transform ion cyclotron resonance (FTICR) MS. The protein used was carbonic anhydrase II (CAII, MW 28,996), and the model drugs were two combinatorial libraries (containing 289 and 256 compounds, respectively), each with a general structure possessing two amino acid residues flanked with a benzenesulfonamide group at the N-terminus and β-alanine at the C-terminus. Figure 2.3 illustrates the general ESI-FTICR-MS methodology useful for screening. The CAII–combinatorial library complex was sprayed in the negative-ion mode under native conditions, accumulated in the FTICR cell, and detected (Figure 2.3A). The major broad unresolved

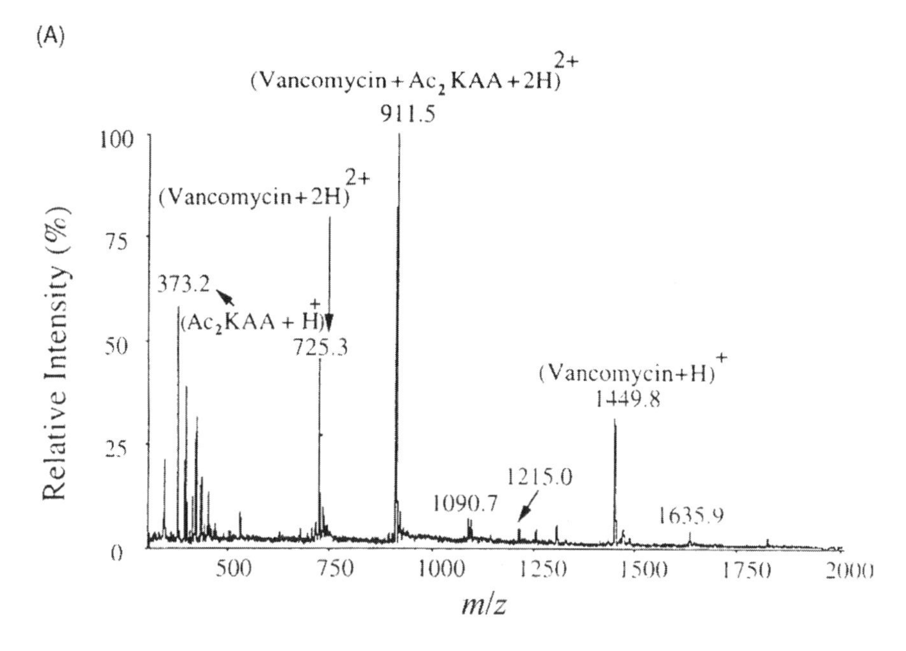

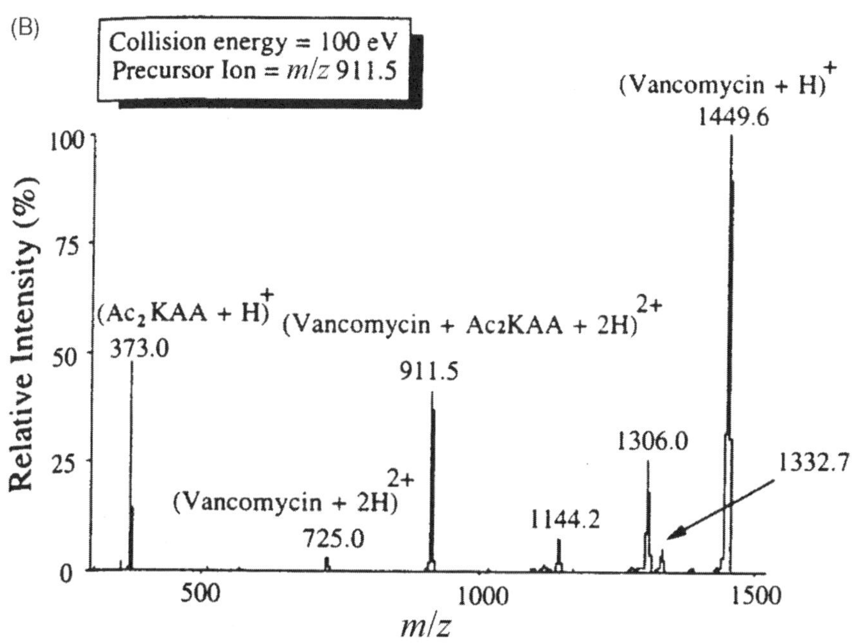

FIGURE 2.2 (A) ESI mass spectrum of an equimolar (25 mM) mixture of vancomycin and Ac_2KAA. Note the intense doubly charged ion at mass-to-charge ratio (m/z) 911.5 corresponding to the non-covalent vancomycin-Ac_2KAA complex. (B) ESI-MS-MS mass spectrum (collision energy 100 eV) of the isolated doubly charged precursor ion at m/z 911.5 confirming the structure of the complex consisting of the singly charged vancomycin (m/z 1449.6) and Ac_2KAA (m/z 373.0) ions. (Reprinted from Lim et al. [9], used with permission. Copyright 1995 by John Wiley & Sons, Ltd., UK.)

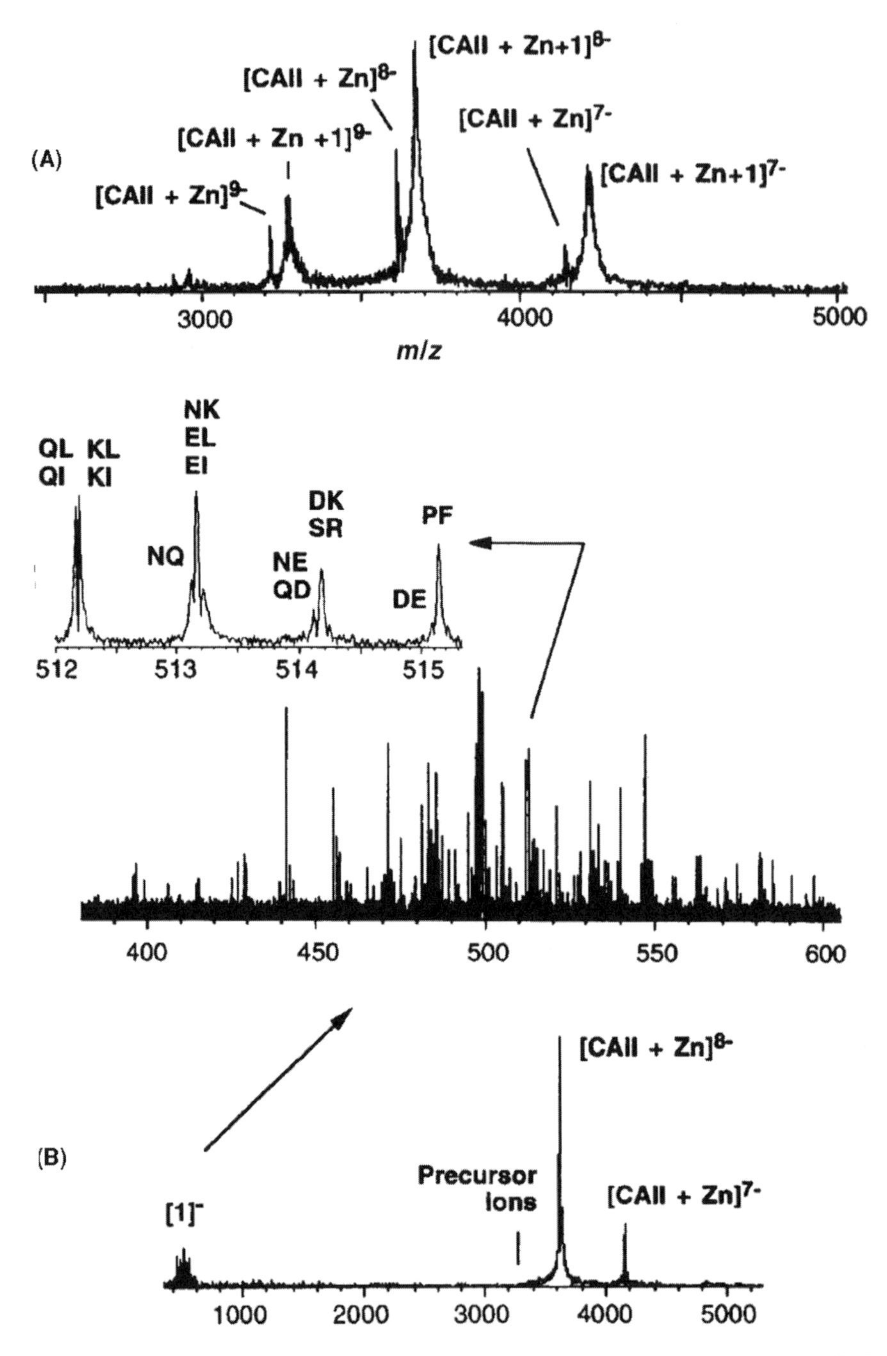

FIGURE 2.3 (A) ESI-FTICR mass spectrum of a mixture of carbonic anhydrase II (CAII) and combinatorial library (289 components, **1**). (B) MS-MS spectrum of the isolated complex of ions $[\text{CAII} + \text{Zn} + \mathbf{1}]^{9-}$. The expanded view shows the region of the singly charged inhibitors $\mathbf{1}^{1-}$, and the doubly expanded region shows the high resolution achieved in the experiment and the amino acid residue composition of the inhibitor ions. (Reprinted from Gao et al. [18], used with permission. Copyright 1996 by the American Chemical Society.)

peaks correspond to the CAII–drug complexes (labeled CAII + Zn + **1**), with charge states –7 to –9. The weaker sharp peaks with similar charge states correspond to uncomplexed CAII (labeled CAII + Zn). The –9 charge state was then isolated and collisionally dissociated in a MS/MS experiment. The resulting spectrum (Figure 2.3B) exhibits the –7 and –8 charge states of the uncomplexed free CAII protein and the singly negatively charged uncomplexed drugs $\mathbf{1}^{1-}$ in the region 400–600 mass-to-charge ratio (m/z) (see expanded view for Figure 2.3B). The individual components in the library were identified with the high-resolution exact-mass capability of the FTICR-MS (see doubly expanded view of Figure 2.3B). The relative binding affinities of drug to protein should be linearly related to the relative ion abundances observed for the different components because the initial concentrations of each of the components were assumed to be identical. The ionization efficiencies of each of the components are assumed to be identical. For increased low mass MS/MS sensitivity, a number of multiply charged states of the protein–drug complex could be isolated and dissociated simultaneously. However, there is no way to guarantee that upon dissociation the freed drug would be charged, since there are multiple competing dissociation pathways. The drug would not be detectable by MS if it dissociated as a neutral species. In addition, the drug could not be easily identified if it fragmented or the protein fragmented complexed with the drug.

Similar experiments were recently performed with FTICR-MS to screen a 324-member peptide library for gas-phase noncovalent complexes of Hck Src homology 2-domain receptor [21]. A multiply charged ligand receptor complex was isolated and then irradiated with a carbon dioxide infrared (IR) laser to dissociate the bound ligands. The most abundant ligands identified in the gas phase were found to correlate with ligands isolated with condensed-phase methods (affinity chromatography and centrifugal ultrafiltration); however, some discrimination against hydrophobic ligands was reported in the gas-phase studies.

RNA–Drug Complexes

Hofstadler and coworkers [22–27] developed a high throughput screening method for the study of noncovalently bound drug candidates to model RNA sequences with FTICR-MS. Mixtures of three RNA targets with 25 compounds, equivalent to the analysis of 75 reactions in one well of a 96 well-plate assay, were analyzed under native conditions by ESI-FTICR-MS with flow injection analysis (7.5 μL/well) in under one minute cycle time. In the absence of active drug, only responses to the RNA targets were observed (Figure 2.4A). However, with the addition of lividomycin, a known aminoglycoside binder to the 16S RNA binding site, an abundant new peak emerged

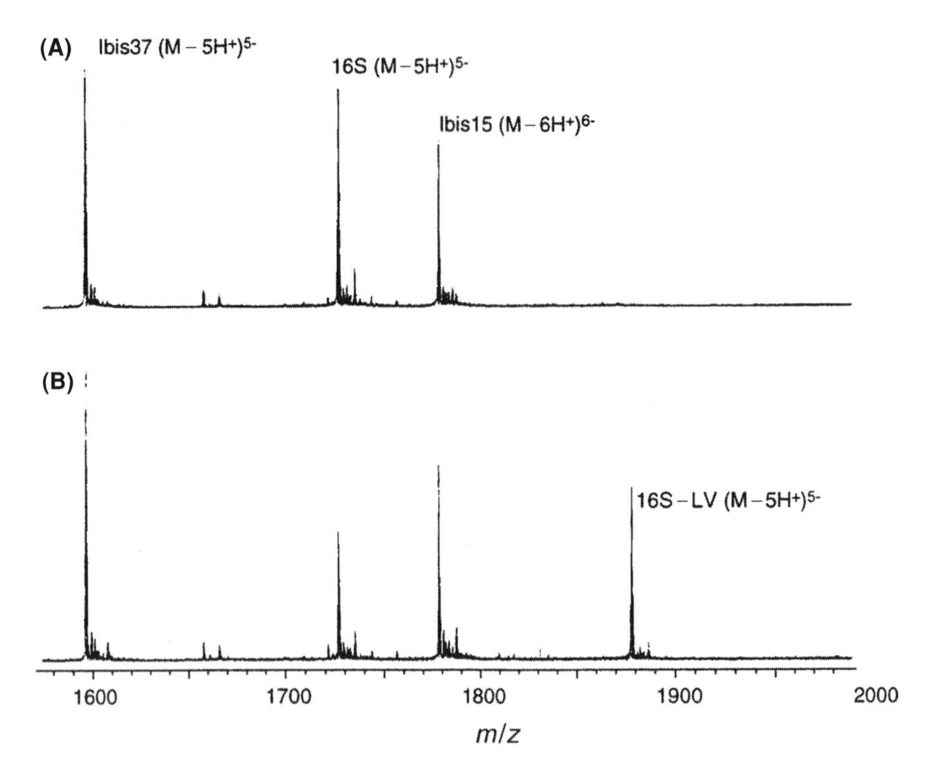

FIGURE 2.4 (A) ESI-FTICR mass spectrum of a mixture of three RNA targets (Ibis37, 16S, and Ibis15) with a mixture of 25 compounds. No complexes with significant ion abundances were observed. (B) ESI-FTICR mass spectrum of the original mixture, spiked with lividomycin (LV) as in (A). Note the presence of the abundant 16S–LV complex in the presence of the 25 nonbinding compounds and the specificity of the LV for the 16S RNA target vs. the Ibis37 and Ibis15 RNA targets. (Reprinted from Hofstadler and Griffey [22], used with permission. Copyright 2000 by PharmaPress Ltd., UK.)

corresponding to the noncovalent complex of the 16S RNA with lividomycin (Figure 2.4B). Under the low-concentration conditions used (2.5 μM/target and 50 μM/ligand) and the low likelihood of a hit (0.01% hit rate), the ligands in the mixture do not interfere with each other in forming the RNA–ligand complexes. This approach for high throughput RNA–drug screening is fully automated, including sample injection, data acquisition, data processing, and data interpretation, and is referred to as multitarget affinity/specificity screening (MASS). In this FTICR-MS experiment, the mass axis can be accurately calibrated (~1-ppm relative error) from the known multiply charged states, and elemental composition of the RNA species and the elemental compositions of the drugs present in the RNA–drug complexes can be confirmed from their exact masses. In addition, the relative binding affinities of drug candidates to RNA can be estimated from the relative abundances of the RNA–drug complex signals, single-point measurements, and can range from very strong (nanomolar) to very weak (millimolar) interactions. Key features for success in the high throughput analysis of noncovalent RNA–drug complexes is the use of volatile buffers (NH_4Ac, NH_4HCO_3), the absence of nonvolatile cations (Na^{1+}, K^{1+}), and optimization of the ESI interface to operate under the most gentle condition to prevent dissociation of the noncovalent complexes. The MASS method was used to screen 100,000 compounds against three RNA targets in less than two weeks by pooling 11 compounds and three targets per well. Recently, a similar high throughput methodology was developed with ion-trap MS instrumentation [28].

An extension of the RNA–drug screening program just described is the use of the hits to identify common structural motifs [29–31]. Derivatives of these compounds are prepared and screened by MASS. Using an iterative process, a structure–activity relationship (SAR) pattern emerges that serves as a guide to elaborate higher-affinity ligands (Figure 2.5). This approach for finding higher-affinity RNA ligands with SAR and MS is referred to as *SAR by MS*. Despite the fact that structures and activities are correlated, MS alone cannot easily distinguish between specific and nonspecific noncovalent binding. For

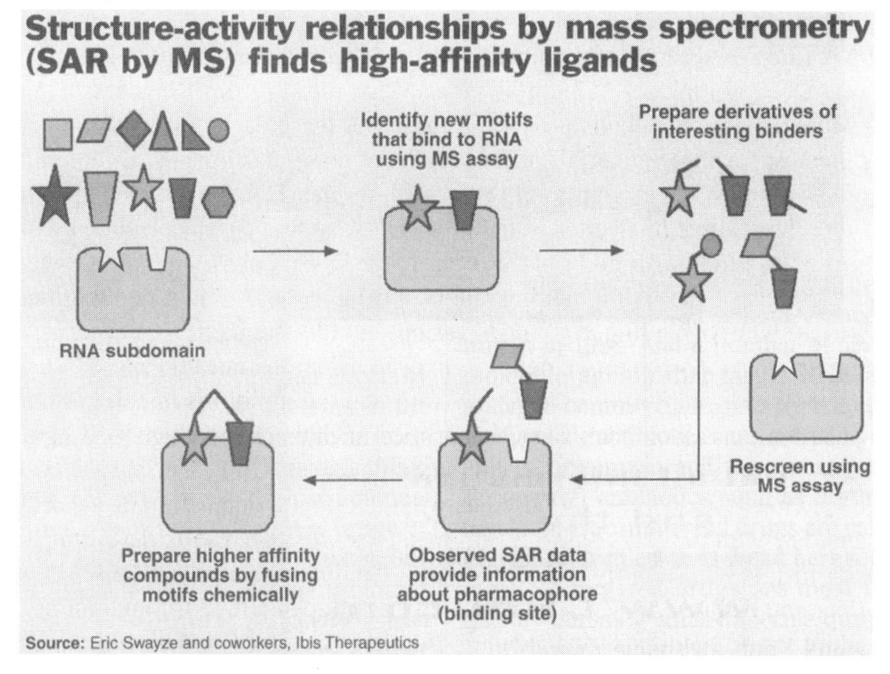

FIGURE 2.5 Schematic diagram for iteratively finding higher-affinity RNA ligands with SARs, with FTICR-MS as the detector, a technique referred to as SAR by MS. (Reprinted from Borman [29], used with permission. Copyright 2000 by the American Chemical Society.)

confirmation of these effects, NMR and X-ray crystallography methods are used.

DNA–Drug Complexes

Recently, a thorough review of noncovalent interaction studies of DNA and RNA with MS was prepared by Hofstadler and Griffey [5]. Compounds that bind to DNA are useful as antitumor, antiviral, and antibiotic drugs. DNA binding drugs interact with duplex DNA principally as minor groove binders (at the rich A and T regions of DNA), as intercalators (at the rich G and C regions of DNA), or in a mixed mode of the two processes. In recent work by Gross [32–34], Gabelica [35,36], and their co-workers, ESI-MS has been shown to be a simple, sensitive, and reliable tool for characterizing noncovalent DNA–drug interactions in the gas phase. (These MS studies were shown to be consistent with similar studies in the condensed phase with other spectroscopic tools.) The authors claim that the native state of the annealed duplex DNA was maintained in solutions prepared for ESI-MS containing 50–100 mM NH_4Ac and up to 50% methanol to aid in spraying. The optimum concentration for a model dodecamer DNA was found to be 10 μM, where the signal-to-noise (S/N) of the duplex/single-strand DNA intensity ratio was maximum and the association of noncomplementary strands was absent. Such DNA samples when mixed with equimolar concentrations of drugs form 1:1 drug–DNA complexes of high abundance for minor groove binders, while for intercalators the abundance is a function of the polarity of the drug, and is greatest for ionic materials and weakest for less polar ligands. Competion studies, employing a mixture of drugs with duplex-DNA, generated spectra where the relative intensities of the drug–DNA adduct peaks correlated with the relative binding affinities of the drugs to the DNA. Gross et al. also showed that models of the likely binding site of a drug to DNA could be identified from the relative binding of a drug to various double-stranded DNA hexamers, ranging from GC-rich to AT-rich. These results demonstrate that ESI-MS can be used as a rapid drug-screening tool for potential inhibitors of cell growth.

Certain classes of drugs may be detrimental as therapeutic agents if binding to DNA is an undesirable secondary side effect. Often the inhibition of an enzyme in an in vivo cell-based assay is due to the interaction of the drug candidate with the cell's DNA, resulting in false-positive results. To eliminate such results and identify only those drug candidates that react with DNA, Greig and Robinson [37,38] have recently developed a high throughput procedure for screening pharmaceutical candidates by studying their interaction with model duplex and single-stranded DNAs using ESI-MS. The same procedure described in the preceding paragraph can be used to profile drugs for undesirable side reactions with DNA.

ESI-MS: CONDENSED-PHASE DRUG-SCREENING STUDIES

Affinity Chromatography/ESI-MS and Affinity Extraction/ESI-MS

The traditional use of affinity chromatography for the screening of drugs is illustrated by the work of Henion and co-workers [39]. An immunoaffinity protein-G column was prepared with benzodiazapine antibodies. Drug mixtures were passed through the column and unbound drugs were eluted. The retained noncovalent antibody–drug complexes were then dissociated and eluted unto a C-18 restricted-access media (RAM) column and the antibody was eluted from the retained drugs. The drugs were eluted onto a C-8 reversed-phase column where they were separated and then finally detected with ESI-MS. This procedure was automated and could be applied to the rapid screening of chemical libraries. The use of affinity chromatography has great potential, especially off-line as an extraction technique, for prescreening very complex mixtures for selecting the tightest binders in the mixture while reusing the precious quantities of receptor bound to the affinity column. This was demonstrated by Kelly et al. [40], where an affinity gel was prepared by attaching the Src homology 2 (SH2) domain of a kinase to a glutathione-agarose gel. A library of synthetic peptides was incubated with the affinity gel. Bound peptides were extracted at decreasing pHs and the strongest binders were identified using ESI-MS and ESI-MS/MS. The challenges in preparing affinity gels and affinity columns hamper the acceptance of the procedure as a simple and rapid screening methodology. It should be noted that the use of antibodies in an affinity screening program is generally limited to the binding of compounds from the chemical class to which the antibody was raised.

Frontal Affinity Chromatography/ESI-MS

Schriemer and co-workers [8,41–47] developed a method for high throughput screening of drug candidates noncovalently bound to receptors that is referred to as frontal affinity chromatography (FAC) with ESI-MS detection. The method uses a miniature affinity chromatographic column, where a receptor of interest is immobilized on the chromatographic support, and ligands are continuously infused into the column. The ligands pass through the column under equilibrium conditions such that nonbinding compounds pass through in the void volume of the column, weak noncovalently binding materials pass through next, and strongly noncovalently binding compounds pass through last. Since the mass of each compound is known, ESI-MS can be used as a multidimensional detector for obtaining the responses of each of the components in the mixture as a function of time (or flow volume) to produce mass

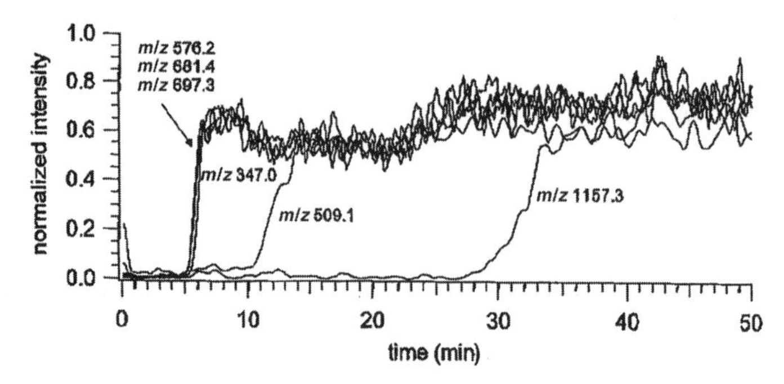

FIGURE 2.6 FAC/ESI mass chromatograms of a six-component oligosaccharide library, which passed through an affinity column, directed against a Salmonella polysaccharide. Selected ion chromatograms of the indicated masses correspond to unretained components (*m*/*z* 576.6, 681.4, 697.3) and to components with increasing degrees of binding (*m*/*z* 347.0, 509.1 and 1157.3, respectively). (Reprinted from Schriemer et al. [41], used with permission. Copyright 1998 by Wiley-VCH Verlag GmbH, Weinheim, Germany.)

chromatograms for each of the components. An example of FAC/ESI-MS is illustrated in Figure 2.6 for the resolution of polysaccharides with a microaffinity column containing an immobilized monoclonal antibody directed against a Salmonella polysaccharide. The three nonactive oligosaccharides (*m*/*z* 576.2, 681.4, 697.3) break through in the void volume, while the successive ion chromatograms corresponding to oligosaccharides with *m*/*z* 347.0, 509.1 and 1157.3 indicate weak, moderate, and fairly strong binding compounds, assuming competitive binding to the same binding site. In general, however, these results can also be interpreted as due to multiple binding sites with different binding affinities for the different components. Advantages of the FAC/ESI-MS approach for screening include the following: large numbers of compounds can be easily screened; the amounts of drugs and substrates used are relatively low; the affinity column can be reused repeatedly, further reducing the consumption of substrates; relative binding strengths can be ranked, since they are proportional to the breakthrough times; throughput can be increased with parallel affinity columns and multisprayers; isobars can be distinguished with MS/MS detection methods; and dissociation constants (K_d) can be obtained from concentration studies. Mixtures of up to 100 components have been analyzed by FAC/ESI-MS, however, as the size of the mixture increases, charge suppression can occur, which reduces or prevents the ionization of components in the mixture.

An extension of the FAC/ESI-MS method, described in the preceding paragraph, is the prescreening of a compound library for unknown ligands. This can be accomplished by use of an "indicator" compound, viz., a known competitive ligand preequilibrated with the FAC affinity column and added to a mixture prior to FAC analysis [43]. Under these conditions, the breakthrough curve for the known indicator is shifted toward the breakthrough curve for the void volume, if a competitive binder is present in the mixture. The shifting of the indicator breakthrough curve is evidence that the mixture contains components that are binding specifically to the substrate, since they compete with the indicator. This approach is used to rapidly prescreen a library for new ligands by use of a weaker indicator ligand that has a short breakthrough time as a means for indicating the presence of a stronger binder, which may have a much longer breakthrough time under normal FAC conditions. Such a mixture would have to be deconvoluted with normal FAC/ESI-MS conditions to identify the unknown binder. However, the savings in time by use of the prescreening methodology is justified, since most mixtures are not likely to contain strong binders.

Affinity Centrifugal Ultrafiltration (Microconcentrator)/ESI-MS

An ultrafiltration microconcentrator is a useful tool for separating high-molecular-weight biopolymers from small molecules. These devices consist of a centrifuge tube, which has mounted on its base an ultrafiltration membrane, which is designed to retain molecules above a selected molecular-weight cut-off. On centrifugation, the higher-molecular-weight components are retained and concentrated in a small volume of retained solvent. These microconcentrator devices have been evaluated and used to rapidly and efficiently screen drug candidate libraries by centrifuging incubated drug candidates and protein targets whereby the proteins and noncovalently bound protein–drug complexes are retained and the free drug candidates pass through the ultrafiltration membrane to waste. The protein–drug complexes are then denatured. The formerly bound low-molecular-weight drug candidates are finally separated from the protein and identified by ESI-MS. Henion and co-workers [48] developed an affinity ultrafiltration/ESI-MS technique that used polyclonal antibodies raised against benzodiazepine drugs to capture benzodiazepine-related drug candidates from a chemical library by forming noncovalent immunoaffinity complexes. These complexes were retained by the microconcentrator upon centrifugation, and the other components passed through the membrane to waste. The drugs were liberated from the antibody complexes by acidification and the resulting solution analyzed for the freed drugs by HPLC/ESI-MS. Siegel and co-workers [49] took a similar approach with microconcentrators to identify noncovalently bound inhibitors of human cytomegalovirus proteases with flow injection ESI-MS. To reduce false-positive results from compounds that bind nonspecifically to the protein and to the wall and membrane of the microconcentrator, the retained protein, protein-complex, and the microconcentrator have to be washed to remove the extraneous nonspecifically bound drugs. Care must be taken to minimize drug losses from the protein–drug complex, since upon washing equilibrium is reestablished between the complex and its environment, causing dissociation of the complex. These effects are less serious for strongly bound complexes and more significant for weakly bound complexes. As a control for false positives, the screening of a mixture in the absence of a protein could be performed where the components present in the last wash of the microconcentrator are identified by ESI-MS. High throughput could be achieved by use of a 96-well-plate ultrafiltration array with parallel multisprayer ESI-MS analysis [50]. Yang and co-workers [51] reported a related affinity selection method employing membranes and ESI-MS for drug screening. Recently, Comess and co-workers [52,53] used centrifugal ultrafiltration with ESI-MS detection to identify library components that promiscuously bind noncovalently to a variety of proteins, producing false-positive drug screening hits. Mixtures were analyzed containing up to 3000 components with fetal calf serum albumin, a model protein useful for identifying these promiscuous library members. After centrifugal ultrafiltration, the retained protein–ligand complex was washed three times with buffer to remove excess free ligands. The retentate was then collected and treated with methanol to denature and precipitate the protein and with CH_2Cl_2 to free the bound ligands from the denatured protein. After an additional centrifugation step, the supernate containing the protein free ligand mixture was analyzed by direct infusion ESI-MS. In the event of a large number of hits, compounds were identified by dereplication of the screen with subsets of the original library.

Pulsed Ultrafiltration/ESI-MS and Affinity Dialysis/ESI-MS

An extension in the use of ultrafiltration membranes for separation of small molecules from noncovalently complexed biomolecules is their use as in-line filters for mass separation. Van Breemen and co-workers [54–57] developed a method, referred to as pulsed ultrafiltration, in which a flowthrough cell was designed with an ultrafiltration membrane in place of a filter membrane. A library of ligand candidates was introduced as a "pulse" via flow injection into an ultrafiltration cell with trapped receptor already present. After the unbound and weakly bound compounds were washed away, the ligand–receptor complexes were disrupted with a denaturing solution introduced into the flow system. The released ligand mixture was then analyzed directly by ESI-MS or resolved with an HPLC column and analyzed by ESI-MS. Samples of ligand mixtures incubated with receptors can also be infused into the ultrafiltration cell and the ligands analyzed as just described. This latter scheme was used for the identification of cyclooxygenase-2 (COX-2) inhibitors [58] present in a fortified plant extract with an ultrafiltration cell fitted with a mechanical stirrer

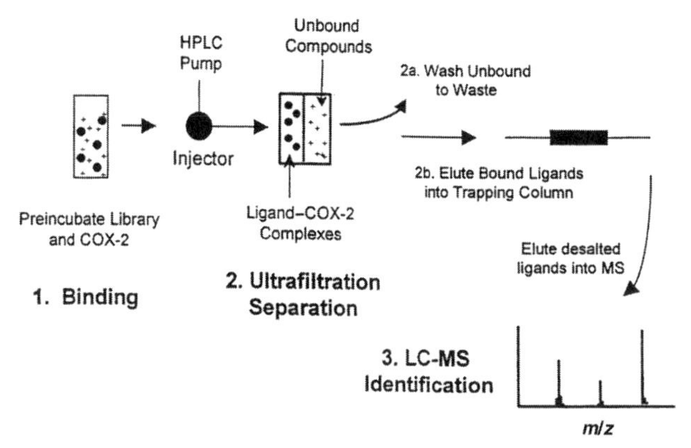

FIGURE 2.7 Scheme illustrating the pulsed ultrafiltration ESI-MS assay for screening COX-2 inhibitors. (Reprinted from Nikolic et al. [58], used with permission. Copyright 2000 by the American Chemical Society.)

(see Figure 2.7). The filtrate from the ultrafiltration membrane was trapped with a short HPLC column and then chromatographically eluted and analyzed by ESI-MS. Ion chromatograms were generated from the plant extract samples with and without COX-2 present. As illustrated in Figure 2.8, signal enhancements observed in the fortified sample relative to the control sample indicate strong responses for inhibitors producing $(M - H)^{1-}$ negative ions at

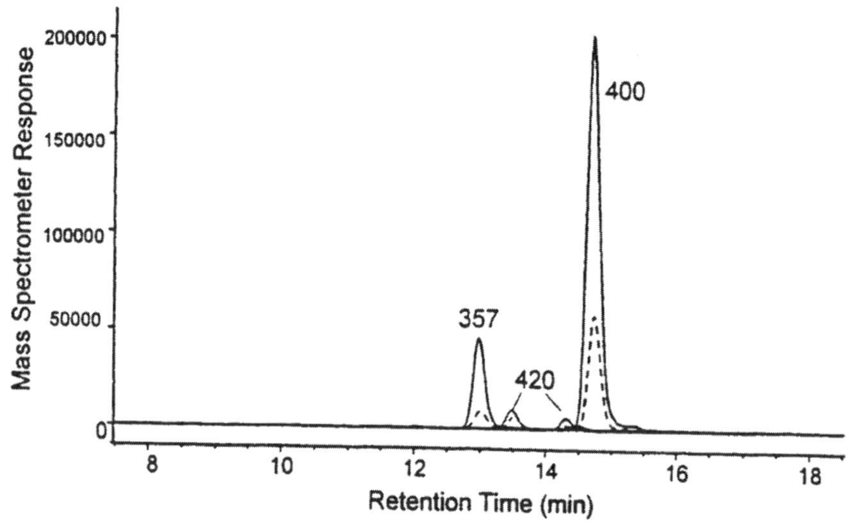

FIGURE 2.8 Computer-reconstructed HPLC/ESI ion chromatograms from eluted components in the COX-2 pulsed ultrafiltration screening assay with indicated $[M - H]^{1-}$ m/z values. The solid line corresponds to the experiments with COX-2 present, and the dotted lines correspond to the control experiment with COX-2 absent. (Reprinted Nikolic et al. [58], used with permission. Copyright 2000 by the American Chemical Society.)

m/z 357 and 400, and weaker responses for two isomers each at *m/z* 420. Additional examples of pulsed ultrafiltration/ESI-MS used for screening have been to identify ligands that noncovalently bind to adenosine deaminase [55,56], dihydrofolate reductase [59], and human serum albumin [60]. Other applications demonstrated the use of pulsed ultrafiltration/ESI-MS for metabolite screening [61,62] and for estimating dissociation constants [54]. Advantages of pulsed ultrafiltration include the fact that receptors are unmodified, studied in the condensed phase, and are reusable. High throughput screening could be achieved with miniature ultrafiltration cells, especially when mounted in parallel for ESI-MS analysis. The screening of large chemical libraries containing a variety of chemical entities for receptor–ligand complexes has not yet been demonstrated. Care has to be taken in such studies to prevent false-positive hits, since the ESI-MS responses with and without a receptor in the ultrafiltration cell have to be compared quantitatively.

Jiang and Lee [63] proposed a screening method, referred to as *affinity dialysis*, whereby a hollow-fiber dialysis membrane (e.g., with a 13,000 molecular-weight cutoff) was used in a flowthrough format to identify ligands strongly bound noncovalently to receptors. Ligands that bind to a protein receptor selectively pass through the hollow fiber for ESI-MS detection, while other low-molecular-weight components pass through the walls of the membrane and are removed by the dialysis buffer flowing around the outside of the fiber in the countercurrent direction. Subsequently, the ligands from the concentrated complex were dissociated from the receptor and identified by ESI-MS. As in the pulsed ultrafiltration approach, ESI-MS responses with this affinity dialysis method have to be compared with and without protein receptor in order to characterize the degree of binding of each ligand to the receptor. A microfluidic device was fabricated to efficiently perform affinity dialysis and concentration of the sample to thereby reduce sample consumption and improve sensitivity for ESI-MS detection [64]. For competitive binding studies, Benkestock and co-workers [65] injected a ligand–protein complex into the hollow-fiber dialysis membrane while a competitive ligand was administered in the dialysis buffer outside of the fiber. The displacement of the first ligand for the second was monitored by ESI-MS (in a gas-phase drug-screening study).

Ultracentrifugation/ESI-MS

A unique screening procedure developed for isolating drugs noncovalently bound to proteins uses ultracentrifugation coupled with ESI-MS [66]. The centrifugal force exerted on a protein–drug mixture under native conditions with equilibrium analytical ultracentrifugation creates a concentration gradient for the free protein and protein–drug complex, but not for the free drug [67]. The ligand concentration at various depths of the centrifuge tube is measured by ESI-MS and the drugs with the highest gradient are the strongest binders, since they form the strongest protein–drug complex. Drug mixtures can be analyzed with this method; however, multiple measurements have to be made for each sample at various depths to determine the gradients for each component, and so the technique is limited to relatively low throughput.

GPC-Spin Column/ESI-MS

A procedure recently developed for screening mixtures of potential drug candidates by their noncovalent interaction with proteins uses size-exclusion gel permeation chromatography in a spin-column format whose eluent is then analyzed by ESI-MS [49,68]. A spin column is a short column packed with GPC material that is centrifuged. This methodology rapidly separates mixtures of free ligands from protein and protein–ligand complexes under native conditions. The eluent, consisting of protein and protein–ligand complexes, is denatured and the freed ligand analyzed under denaturing conditions by flow-injection analysis with ESI-MS. This procedure decouples the separation step from the analysis step for optimum flexibility with available instrumentation. An example of the GCP–spin-column/ESI-MS methodology for drug screening of an impure drug, peptidic difluoromethylene ketone (DFMK), before and after spin-column/ESI-MS analysis is illustrated in Figure 2.9A and 2.9B,

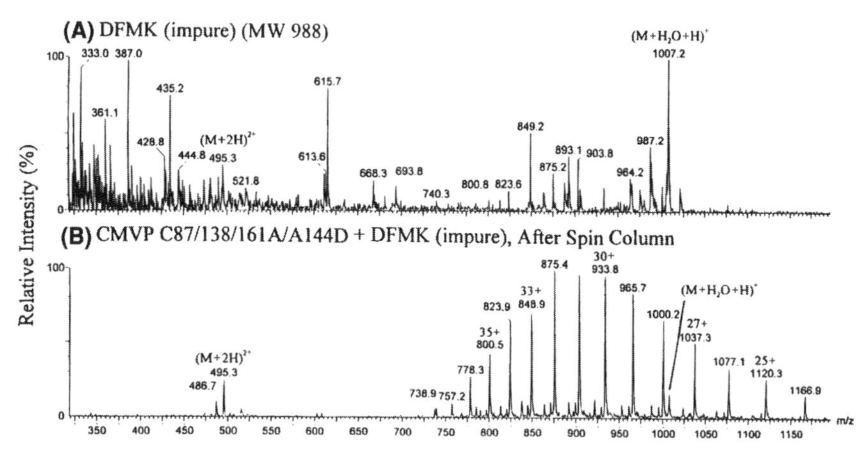

FIGURE 2.9 Illustration of the GPC–spin-column/ESI-MS methodology for drug screening of an impure drug peptidic difluoromethylene ketone (DFMK, MW 988.5) with cytomegalovirus (CMV) protease (A144D/C87A/C138A/C161A, MW 27,957). (A) ESI mass spectrum of the impure DFMK drug. (B) ESI mass spectrum of a mixture of the impure DFMK and CMV protease after spin-column/ESI-MS analysis. The DFMK-CMV protease complex passed through the spin column and the DFMK impurities were retained by the GPC column, generating a spectrum of the purified DFMK ($[M + 2H]^{1+}$, m/z 495.3; $[M + H_2O + H]^{1+}$, m/z 1007.2); and of the CMV protease. (Reprinted Siegel et al. [49], used with permission. Copyright 1998 by John Wiley & Sons, Ltd., UK.)

respectively. The impure drug, upon passing through the spin column, emerges as a major component together with the protease, with which it formed a noncovalent complex. The gel retained all other impurity components. In this way, large numbers of components can be routinely screened. Recently, the GPC–spin-column/ESI-MS method has been applied to the screening of drug candidates that noncovalently bind to RNA [69].

The use of size-exclusion chromatography/ESI-MS for the identification of ligands from protein–ligand complexes in the excluded volume of the column was one of the earliest methods developed for screening drug candidates [70] and is essentially equivalent to the GPC–spin-column/ESI-MS methodology. An application of GPC–spin-column/ESI-MS methodology is to reequilibrate the eluent of a spin column and pass it again through a second spin-column to detect by ESI-MS the enriched tighter binding ligands at the expense of the weaker binding ligands [71]. A proposed method for screening potential drug candidates with the GPC–spin columns is to analyze by ESI-MS the compounds retained after elution from the spin column [72]. In such studies, the absence of a ligand indicated strong affinity toward a target, while the presence of a ligand indicated a much weaker or no affinity.

An ESI-MS/nuclear-magnetic-resonance (NMR) [73] screening assay has been described that combines the inherent strengths of size-exclusion gel chromatography, mass spectrometry, and NMR to identify bound complexes in a relatively universal high-throughput screening approach. The combination of ESI-MS with NMR also provides a mechanism to overcome limitations associated with each individual technique. The ESI-MS/NMR assay is a structure-based approach for discovering protein ligands and for drug design. The spin-column/ESI-MS analysis provides evidence for ligand binding to the target protein at low sample concentrations with relatively short acquisition times. NMR analysis provides confirmation of direct ligand binding with the protein target while distinguishing between stoichiometric and nonspecific binding. Additionally, NMR can identify the ligand binding site on the protein surface that can be correlated with the known active site for inferring biological activity. Also, NMR can be used to, or, in combination with molecular modeling and X-ray crystallography, generate a co-structure of the protein with the identified ligand.

An important prerequisite for the ESI-MS/NMR assay is the appropriate design of the compound library to take full advantage of the assay's strengths. Fundamentally, the library mixtures are prepared to contain compounds with different molecular weights, since the molecular weight is used as an identity tag to eliminate the necessity of a deconvolution step. Additional considerations include structural diversity, druglike qualities, and solubility. Typically, each well of a 96-well-plate array is prepared as a spin column, loaded with incubated protein and a mixture of 10 components, and centrifuged.

The 96 eluents are then analyzed by ESI-MS either individually or in parallel (eight at a time) with a multisprayer. Currently, ESI-MS acquisition speeds of up to about 30 minutes per plate with a multisprayer system can be achieved. Customized software has been designed to also automatically evaluate the data and search for the compounds that noncovalently bind to the protein [74]. The hits from the mass spectrometry analysis of the gel–filtration filtrate are then evaluated by 2D ^{1}H-^{15}N HSQC NMR experiments, where typical acquisition times are on the order of 15 minutes requiring 0.1–0.2-mM protein samples. The observation of chemical shift perturbations in the 2D ^{1}H-^{15}N HSQC spectra confirms binding of the compound to the protein. By mapping the protein residues that incurred a chemical shift perturbation in the presence of the compound onto the protein surface, the binding site of the ligand can be identified. The proximal location of the compound binding site with known active or functional sites may also infer biological activity of the compound. Conversely, the compound is a nonspecific binder in the absence of any chemical shift perturbations or if the protein residues that incur chemical shift perturbations do not cluster in a unique location, but are randomly scattered on the protein surface. Compounds that demonstrate the preferred activity in the NMR experiments are then further evaluated by determining a co-structure of the compound with the target protein by any combination of NMR, X-ray crystallography, and modeling techniques. The ESI-MS/NMR assay is an iterative process where a new directed library, prepared based on the results of previous hits, can be screened employing the ESI-MS/NMR assay. Using the GPC–spin-column/ESI-MS/NMR drug screening protocol, $\sim$200 compounds were analyzed with the MMP-1 protein, and $\sim$32,000 compounds were analyzed with the RGS4 protein with IC_{50} ranging from 9 nM to 100 μM. Hits with biological activity were identified for both proteins.

Recently, Schnier and co-workers [75] extended the GPC-spin-column/ESI-MS assay for the analysis of a target protein with mixtures of 80 components (5-μM protein, 1 μM per compound). The compounds were pooled using two different procedures so that a specific compound is found in two wells with completely different well-mates. For high throughput screening, eight spin-column eluants were injected into eight ballistic gradient HPLC systems, which fed into an ESI-MS equipped with an eight-channel multisprayer system. The cycle time for each run of eight wells containing 640 samples was 2 minutes with injections in the overlay mode. Using this procedure, a GPC spin column kinase receptor ESI-MS assay was performed on 25,000 compounds, pooled twice, in 2.6 hours. A total of 320 overlapping hits were observed between the two sample pools. In a very recent report [76], mixtures of 400 compounds were incubated with a protein and successfully analyzed for noncovalent binders with an array of GPC-spin columns with ESI-MS detection in a very high throughput manner.

Tandem Chromatography GPC/RP-HPLC/MS

The tandem chromatographic method of GPC coupled with reversed-phase (RP) HPLC as a sample introduction method for ESI-MS has been proposed as a high throughput approach for studying noncovalent protein–drug complexes for drug screening [77]. This method of screening can be viewed as a tandem chromatographic extension of the GPC–spin-column/ESI-MS method. The GPC chromatographic analysis of an incubated sample of protein and a mixture of drug candidates produces two broad peaks corresponding to the unreacted drug candidates and the protein complexed with the drug candidates. A valving system is activated to divert the later peak to an RP-HPLC column, which under denaturing conditions resolves the individual drug candidates and the remaining protein. Using this approach, Annis [78,79] claimed that protein mixtures containing up to 2500–5000 drug candidates can be screened in 6 minutes per sample, which corresponds to screening of about 500,000 samples per day. This ultrahigh-capacity screening method typically used protein concentrations of 10 μM and each drug candidate at a concentration of 1 μM. Under these screening conditions, the most likely observed hits will be those compounds that most strongly bind to the protein since the weaker binders would be competitively displaced. A related multidimensional chromatography method coupled to mass spectrometry for target-based screening was proposed by Hsieh and his co-workers [80,81] and van der Greef and Irth [82].

Instead of using GPC and RP columns in tandem for drug screening, van Elswijk and co-workers [83] used two short RAM columns, each preceded by a reaction coil. Ligands and protein were fed into the first reactor coil to generate noncovalent complexes, which flowed into the first RAM column where free ligands were trapped, while the noncovalent complexes passed into the second reactor coil. There the complexes were subjected to a low pH buffer that disrupted the complex. The freed ligands were then trapped in the second RAM column, while the protein flowed to waste. The carrier solvent was then switched to remove the trapped ligands, which were detected by ESI-MS. This in-line flowthrough methodology was applied to a variety of drug-screening assays and has the advantage of using small amounts of ligand and protein with moderate ESI sensitivity due to the absence of protein ion suppression effects.

GPC/Ultrafiltration/RP-HPLC/ESI-MS

Dunayevskiy and Hughes [84] proposed the ultimate in multidimensional chromatographic methods for the screening of drug candidates noncovalently bound to receptors by combining serially GPC, ultrafiltration, and RP-HPLC in on-line and off-line modes with ESI-MS detection in a presumably high throughput format. The on-line mode incorporates column switching and sample trapping, while the off-line mode incorporates a lyophilization–concentration–resuspension step. An alternative unreported approach would be to replace the ultrafiltration step with turbulent flow chromatography to remove the high mass receptor (protein) from the low mass ligands.

Capillary Electrophoretic/ESI-MS Methods: Affinity Capillary Electrophoresis/ESI-MS and Capillary Isoelectric Focusing/ESI-MS

Capillary electrophoresis (CE) separations are produced by differential migration of solutes confined to narrow-bore capillaries subjected to high electric fields. The mobility of a species is related directly to its net charge and is inversely related to its hydrodynamic drag (mass and shape). Affinity capillary electrophoresis (ACE) [85] is achieved when a migration change for a ligand (or receptor) occurs due to the interaction with a receptor (or ligand) present in the electrophoretic buffer. A procedure often used in ACE studies is to maintain in the electrophoretic buffer the receptor as a neutral species (by adjusting the pH to the isoelectric point (pI) of the receptor) and the ligands as charged species. The mobility of the ligand–receptor complex is lower than the free ligand due to the additional hydrodynamic drag of the complex vs. the free ligand, while the net charges of both species are identical. This approach is very useful for screening components of chemical libraries, which selectively bind to receptors due to the high resolution and high sensitivity achievable with ACE. The most common in-line detectors used have been UV-Vis and laser-induced fluorescence; however, when the column output is interfaced to an ESI-MS, the resolved components could be identified by their molecular weights and their structures elucidated by MS/MS.

Karger and coworkers [86–88] demonstrated the use of ACE/ESI-MS for the identification of new peptide constructs that bind more strongly to vancomycin than the known ligand D-Ala-D-Ala (AA). The procedure used was to adjust the electrophoretic buffer pH to that of the pI of the vancomycin receptor (pI 8.1). The ligands used were members of a peptide library with the structure Fmoc-DDX_1X_2, where X was one of 10 selected D-amino acids. Figure 2.10A and 2.10B, respectively, illustrate electropherograms obtained without and with vancomycin in the electrophoretic buffer. Based on the ESI-MS analysis, Figure 2.10A exhibits three distinct peaks (I, II, III), which from ESI-MS analysis contained peptides with two, one, and no glutamic acid residues corresponding to net charges of −5, −4, and −3, respectively. Figure 2.10B exhibits a similar group of three peaks, but shows in addition a series of peaks with longer migration times corresponding to noncovalent

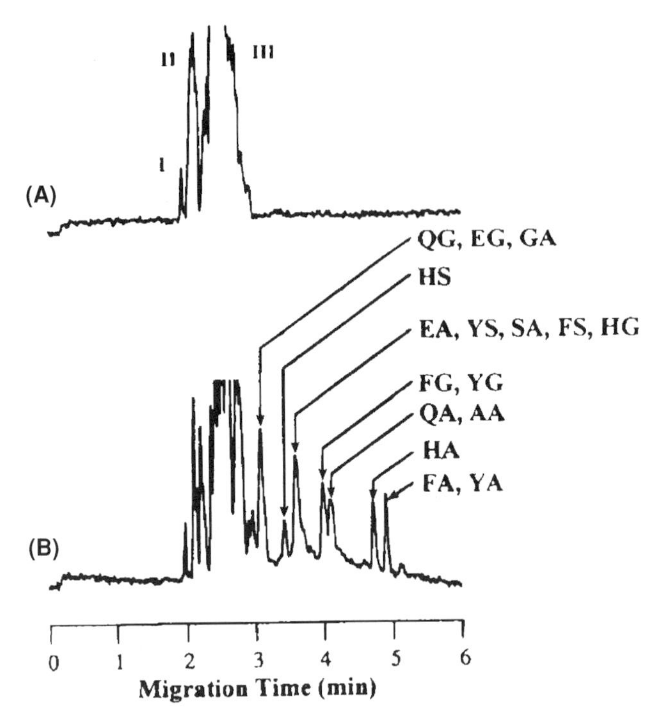

FIGURE 2.10 ACE-UV electropherograms of the Fmoc-DDX_1X_2 library (100 peptides) (A) without, and (B) with vancomycin. See text for discussion. (Reprinted from Chu et al. [87], used with permission. Copyright 1996 by the American Chemical Society.)

vancomycin-Fmoc-DDX_1X_2 complexes, where the residues X_1 and X_2, listed in Figure 2.10B, were identified by ESI-MS and ESI-MS/MS. The interesting feature is that a number of peptides have migration times greater than that of AA corresponding to binding affinities greater than that of AA. In addition, these stronger binding peptides, all C-terminating with A, have penultimate C-terminal residues F, Y, and H, each containing aromatic moieties. ACE/ESI-MS has also been used for evaluating the binding to vancomycin of larger peptide libraries, and it appears that the upper limit in a single assay of this type is about 1000 peptides. To extend such ACE/ESI-MS assays to even larger numbers of components, prescreening of the mixtures prior to ACE with affinity columns [87] or affinity coated magnetic beads [89] has been demonstrated. This approach for in vitro exploratory and early-discovery drug screening appears to be very promising due to the small quantities of precious receptors needed, the high resolution achieved, and the short analysis times required. However, the sensitivities achieved with non-mass-spectrometric detectors and their reliability and robustness is greater than that often achievable with ESI-MS. In general, special ACE methods have to be developed for the analysis of receptor complexes formed with uncharged ligands. Other ACE/ESI-MS applications, potentially useful for drug screening, are epitope mapping [90], analysis of DNA complexes [91], FK506-Binding Protein [FKBP]–rapamycin complexes [92], and binding constants of vancomycin–peptide complexes [93].

Recently, capillary isoelectric focusing/ESI-MS has been demonstrated as a potential method for in vitro screening of compounds that noncovalently bind to active proteins [94,95]. The procedure used for this assay is to load into a capillary a mixture of the receptor of interest (generally a protein) ligands, and an ampholyte. Upon applying a high voltage across the capillary, a pH gradient is formed and a current is generated which decreases as the components separate and focus at their isoelectric points. This focusing step can concentrate the sample up to two orders of magnitude. At this point, the resolved components, which could include the receptors, ligands, and receptor–ligand complexes, can be mobilized by applying a low external static pressure to transport the resolved components from the capillary to an ESI-MS for detection.

Tethering with HPLC/ESI-MS Detection

An ingenious screening method was developed using the covalent disulfide exchange reaction between members of a disulfide-containing chemical library with a native or engineered cysteine in the region of the active site of an enzyme protein [96,97] or the surface region of a protein–protein interaction [98,99]. The formation of the disulfide bond with the protein, referred to as *tethering*, is stabilized by the noncovalent interaction of the moiety attached to the disulfide with the active site in the protein. To evaluate the stability of the protein–disulfide product, the reaction was conducted under reducing conditions (1 mM 2-mercaptoethanol) in buffer under native conditions. The protein-captured disulfide ligands were identified with a ballistic gradient HPLC/ESI-MS method [100,101]. Libraries of disulfide compounds were screened in this manner as mixtures of 10 compounds. Approximately 2.5 million compounds were analyzed annually with this methodology to screen for small-molecule antagonists of enzyme and protein–protein interactions. The tethering methodology has also been extended to search for more than one adjacent active site in a protein with HPLC/ESI-MS for the detection of the covalently bound disulfide protein–ligand product [102].

Condensed-Phase Competitive Binding Assays

Two approaches using mass spectrometry have been reported to determine the binding affinities of ligands with proteins in the condensed phase. One method determines the relative binding affinities by analyzing the GPC–spin-column eluent of a mixture of ligands incubated with the protein by determining the abundances of the individual components [49]. The ESI-MS response of each of the ligands present in the eluent, normalized to the response of a fixed concentration of each of the ligands, should be proportional to the relative binding affinities for each of the ligands. A more sophisticated approach was taken by Wanner and co-workers [103–105]. A compound known to have a high binding affinity with a protein was incubated with a ligand of unknown binding affinity. Various concentrations of the ligand of unknown binding affinity were incubated with the strongly bound ligand–protein complex. For each sample, the protein was removed by ultracentrifugation of filtration and from the measured mass-spectral response for the strong binder, the concentration of the competing drug that inhibits 50% of the specific binding (IC_{50}) is computed. This approach mimics radiological assays [106] commonly used for determining IC_{50}, but is simple and straightforward without requiring radioligands.

MALDI-MS: SOLID-PHASE DRUG-SCREENING STUDIES

Affinity Probes for MALDI-MS

MALDI-MS has traditionally been used extensively for the analysis of peptides, proteins, and oligonucleotides. To aid in these assays, a number of researchers, including Hutchins [107–112], Nelson [113,114], Kris [115–117, 118], Weinberger [119], and their co-workers, have developed MALDI probe tips that have been modified by activating the metal surface with a chemical cross-linker that was covalently linked with active biomolecules. These active biomolecules serve as affinity targets for solutions containing ligands to which the surface is exposed. These ligands can then be released and characterized directly by MALDI-MS upon UV-laser irradiation in the presence, or even the absence, of an UV-absorbing matrix. In addition, the probe surface can be subdivided into smaller subregions, each individually activated with a biomolecular target, for high throughput analysis [120]. An extension of this approach is the use of the affinity surface of a surface-plasmon resonance spectroscopy Biacore chip as the probe surface for MALDI-MS [121–123]. Even though the use of affinity MALDI probes and Biacore chips with MALDI-MS detection show great promise as tools for high throughput in vitro bioaffinity drug screening for small molecules, no examples have appeared in the literature at this time.

An example of chemical screening with a modified surface of a MALDI-MS probe was described recently [124]. The gold surfaces of the MALDI plate wells were prepared with a self-assembled monolayer (SAM), formed from a peptide-terminated alkane thiolate. The free peptide terminus, a substrate for anthrax lethal factor, is cleaved by the anthrax enzyme in the absence of a chemical inhibitor. Cocktails consisting of mixtures of eight compounds with anthrax lethal factor were applied to each MALDI plate well containing the SAM. After incubation, the wells were washed, and MALDI mass spectra were obtained. Inhibitors of anthrax lethal factor exhibited spectra in which the peptide-terminated alkane thiolate was intact after the enzymatic reaction. Using this MALDI-MS technique, the rapid screening of chemical libraries for inhibitors of anthrax lethal factor was demonstrated.

Fluorescence-Activated Cell Sorting (Flow Cytometry)/MALDI-MS

Keough and colleagues [125,126] developed a sophisticated method for screening support-bound combinatorial libraries of pharmaceutical interest by using affinity methods coupled with flow cytometry for selecting beads with active components and MALDI-MS for structural identification. By taking advantage of the mass-spectral properties of MALDI samples, support-bound combinatorial libraries were prepared, which, upon MALDI analysis, revealed the chemical structures of the active components. A series of moieties is sequentially linked on a bead by the "randomize and split" method, such that a combinatorial library is generated where each individual bead has one unique structure. The library is designed such that the initial moieties are chemically or photochemically sensitive so that the compound can be easily cleaved from the bead for MALDI analysis. In addition, during each step in the combinatorial synthesis, a small fraction of the growing sequence is capped to produce a mixture of related terminated components on the bead. Capping agents can be used to mass code each step in the synthesis to indicate, for example, the use of D- or L-amino acid residues or to differentiate between structural isomers. Upon cleaving the components from a single bead, the sequence of the compound can be read directly from the mass differences in the molecular ions of the capped components generated in the MALDI mass spectrum. The structure could then be determined unambiguously from the unique masses of the moieties used in the synthesis.

Support-bound combinatorial libraries can be prepared with the chemical properties, as described earlier. Affinity methods were developed where soluble fluorescently labeled proteins were mixed with the library and noncovalently bound to the minority of beads containing active sequences. Using a flow cytometer, the beads with fluorescently bound protein can be sorted from the majority of beads that do not contain the active sequence and the compounds on the active beads sequenced by MALDI-MS. In a model study, an original million-member peptide library was screened against an anti-HIV-1 gp 120 monoclonal antibody. Seven beads with the highest fluorescence were found to have the predicted consensus sequence, as measured by MALDI-MS.

Similar studies with MALDI-MS to sequence members of support-bound combinatorial libraries of peptides [127,128], glycopeptides [129], phosphinic peptides [130], cyclic oligocarbamates [131], and antisense DNA oligonucleotides [132], all *manually* selected after incubation with fluorescently labeled proteins, have been reported. Such studies were also proposed for screening support-bound combinatorial libraries of small molecules with a central scaffold for pharmaceutical applications [125]. In all these screening procedures, fluorescence and MALDI methods are coupled together such that fluorescence methods are used to identify the active components and MALDI-MS essentially plays the role of elucidating the structures of the active components. ESI-MS could also be used to analyze the drugs once cleaved from the beads; however, this may require additional sample manipulation steps to those needed for MALDI-MS.

MALDI-MS Analysis of Biopolymer-Ligand Complexes

Direct analysis of noncovalent biopolymer–ligand complexes by MALDI-MS is difficult [133,134]. The complexes must be prepared in the condensed phase, in a buffer system that maintains the native state, which remains intact upon crystallization and in the gas phase during MALDI analysis. Many factors are involved in preparing the samples, including choice of matrix, matrix-to-analyte concentration, receptor and ligand concentrations, pH, buffers, solvents, and ionic strength. Best results have been obtained with neutral matrixes such as 6-aza-2-thiothymine, 2,6-dihydroxyacetophenone, and neutralized sinapinic acid for UV-MALDI. Good-quality MALDI spectra of noncovalent complexes often are obtained only from the summed first laser shot spectra acquired from regions of the sample not previously irradiated. A number of noncovalent protein–ligand complexes have been studied by UV-MALDI-MS, including RNase S (noncovalent complex of S-protein and S-peptide) [135], myoglobin–heme complex [136], adenylate kinase complexes with AMP, ADP, and ATP [137], Ras-GDP, and Ras-GppNp [guanosine-5′-(β, γ-imido)-triphosphate] [138]. Recently, the noncovalent complex of vancomycin with Ac2-L-Lys-D-Ala-D-Ala was observed in the positive-ion mode by IR-MALDI, but not by UV-MALDI [139]. However, in a control study, where vancomycin was prepared with a 1:1 mixture of Ac_2-L-Lys-D-Ala-D-Ala:$Ac_2(d_6)$-L-Lys-L-Ala-L-Ala, the negative ion IR-MALDI spectrum surprisingly exhibited the vancomycin complexes with an intensity ratio of 1:3, suggesting a considerable difference between the condensed phase and MALDI gas-phase binding affinities, since no complex with the

$Ac_2(d_6)$-L-Lys-L-Ala-L-Ala isomer was expected. At this point, no high throughput direct-screening studies of biopolymers noncovalently bound to drug candidates with UV- or IR-MALDI-MS have been reported. A good starting point for drug screening with UV- and IR-MALDI-MS would be screening for cell-wall inhibitors, as described earlier for the direct analysis of vancomycin–peptide complexes with ESI-MS. In addition, it would be of great interest to compare the MALDI-MS results, where the desorption and ionization steps occur nearly simultaneously, with those obtained with the two-laser mass spectrometry methodology (L2-MS) [140], where the neutral complex is sequentially desorbed and ionized with IR and UV lasers, respectively, prior to mass spectral analysis.

The MALDI-MS method can also be used to analyze the small molecules isolated in the condensed-phase separation techniques just described for resolving biopolymer–drug complexes from unreacted drug or biopolymer. The chromatographic methods, with ESI-MS as a flow-analysis detector, could as well have used MALDI-MS as an off-line detector. However, with the exception of peptides and nucleotides, MALDI-MS has not become a popular detector for screening small molecules, because of the high matrix and chemical background at low mass, the possibility for fragmentation of the small molecule, and the possible reaction of the analyte with the MALDI matrix.

Hydrogen/Deuterium Exchange for Screening Protein–Ligand Complexes with MALDI-MS Detection

Fitzgerald and co-workers [141–145] developed a method for screening the stability of a protein–ligand complex by determining the extent of hydrogen/deuterium (H/D) exchange for the protein in the presence of a ligand as a function of titrated denaturing agent. The mass change of the protein, monitored by MALDI-MS, demonstrated that the lower the extent of H/D exchange, the stronger the ligand bound, corresponding to a lower protein–ligand dissociation constant. This methodology was extended to a single-point measurement, which was proposed as a promising method for high throughput screening for identifying protein–ligand interactions [146].

CONCLUSIONS

The goal of in vitro exploratory and early-discovery drug screening is to develop a universal screening methodology that uses a minimum of precious biochemical resources in an inexpensive high throughput mode where the biochemically active species are easy to characterize, are reacting at the active site of the biomolecule, and bind specifically. In addition, it is desirable to detect both weakly and strongly interacting species with the biomolecules and to discriminate against nonspecific binders. This review has described many exploratory and early-discovery screening modalities, and no doubt newer methods will evolve with time. However, at this point, no one screening method has been shown to solve the ideal universal screening goals described in the preceding sections. Affinity methods show great promise as a means for identifying complexation reactions in a format that minimizes the use of biochemical resources, since the biomolecules often are reusable. Mass spectrometry, especially in the electrospray ionization mode, has been shown to be one of the most promising universal detectors for in vitro screening, since it is very sensitive, readily produces molecular-weight information, and can be set up to be highly specific. In addition, it operates on samples in the condensed phase and can be easily coupled to nearly any chromatographic method. ESI-MS is very efficient since it can be used to deconvolute and analyze complex mixtures as a multiplex detector for molecular weights. Its use becomes very efficient since high throughput analysis is often achieved with mass indexed chemical libraries, whereby the interpretation of the acquired ESI mass spectra can be automated, since the measured molecular weight must correspond only to a specific member of that chemical library. To enhance sample throughput, especially for identifying the tightest binding complexes, prescreening of very large chemical libraries could be useful for reducing the number of drug candidates. Methods useful for prescreening are affinity extraction, affinity chromatography, affinity magnetic beads, gel filtration, flow ultrafiltration, microconcentration, and microdialysis [147,148]. In addition, even higher throughput with ESI-MS systems has been demonstrated with multiple sprayers such that data for multiple infusion or multiple chromatographic samples can be acquired in parallel [50]. Currently, commercial systems are available with up to eight parallel sprayers. Maximum sensitivity and minimum use of sample resources are achieved in ESI-MS by spraying samples in the nanospray mode, which is inefficient in the manual mode, but with recent technological advances can be fully automated [149–153]. Chip-based microfluidic technologies currently are being developed that efficiently incorporate low-flow chromatographic methods (affinity dialysis [64], ultrafiltration [64], capillary electrophoresis [154–158], liquid chromatography [159], affinity chromatography [160]) with low-flow chip-based nanoelectrosprayer devices in very low dead-volume systems. These chip-based technologies are potentially promising methodologies for achieving the idealized goals of in vitro exploratory and early-discovery drug screening by minimizing sample consumption, while using ESI-MS as the optimum universal detector operating with multiple sprayers [161,162] in parallel with duty cycles for the analysis of produced ions that approach 100% [160].

MALDI-MS techniques for high throughput drug screening are not as popular as the ESI-MS techniques. This is due in part to the fact that the *m/z* values of small drug molecules often overlap with the chemical noise of the MALDI matrix and that small drug molecules, unlike peptides or oligonucleotides, often fragment or rearrange. In the near future, these disadvantages may be reduced dramatically by the use of porous silicon chips as MALDI targets [163–165], since no matrix is required, or with sol-gel–derived polymeric matrixes, which produce nearly no chemical background noise [166]. In addition, these laser desorption/ionization mass-spectrometric methods, respectively, referred to as direct ionization off silicon (DIOS) and sol-gel-assisted laser desorption/ionization (SGALDI) may produce molecular ions without any significant fragmentation, and may even be more sensitive than traditional MALDI methods. The future prospects for this technology becomes even more promising when the affinity chromatographic and isolation methods are coupled with the recently demonstrated microfluidic compact disk (CD) technology for MALDI sample preparation [167,168]. A significant breakthrough will also be made when the microfluidic CD technology is coupled to ESI-MS.

Other mass-spectral ionization techniques, such as atmospheric-pressure chemical ionization (APCI) [169,170], atmospheric-pressure photoionization [171], and IR-MALDI [172,173] (vs. UV-MALDI used in nearly all the MALDI applications described in this review) may in the future be found useful for high throughput drug screening. The ultimate screening procedure is to analyze a whole proteome with little or no chromatography, to identify those proteins that react with drug candidates. This may be possible in the not too distant future with an ultrahigh-performance ESI-MS having ultrahigh sensitivity, resolution, and mass accuracy.

The major disadvantage of any mass-spectrometric-based screening program is that mass spectrometry alone cannot routinely reveal structural details of the binding site for the noncovalently bound receptor–ligand complex. In gas-phase electrospray drug-screening studies, strong correlations were shown between observed receptor–ligand complexes and complexes reported with other analytical methods. When large chemical libraries are studied by gas-phase ESI-MS, large numbers of both weak and strong nonspecific complexes are expected to be observed in addition to the specific receptor–ligand complexes. The specific noncovalent binding effects are often verified by mass-spectral studies of the dissociation energetics of the complex and by comparisons of the receptor–ligand complexes produced with native and denatured receptors. Likewise, in both the condensed-phase electrospray and solid-phase MALDI drug-screening studies, many of the noncovalent complexes formed were those predicted for well-understood biomolecule receptors.

Nevertheless, even a larger number of complexes that can be verified with denatured receptor could be nonspecific. A number of methods with mass spectrometry have been evaluated to aid in identifying the structural details of the binding site of specific noncovalent complexes. These techniques include the use of the ligand as a site-specific chemical reagent for cleaving the receptor [174], the use of the ligand as a photoaffinity reagent to label the receptor [175–179], and the use of time-resolved limited proteolysis to distinguish between the relative abundances of digested receptor fragments prepared with and without ligand present [180]. Currently, NMR and X-ray crystallography are the preferred techniques for elucidating the structural details of the binding site and for differentiating between nonspecific and specific receptor–ligand complexes. Ultimately, the lead compounds have to be checked for biological activity in an in vivo biological assay.

ACKNOWLEDGMENTS

The author appreciates the helpful discussions with Jasbir Seehra, Robert Powers, Paul Schnier, John Morin, Dominic Mobilio, Guy Carter, Hanlin Liu and Kenneth Comess during the preparation of this review. He also thanks Hui Tong for preparing the figures.

REFERENCES

1. Loo, J.A. "Electrospray Ionization Mass Spectrometry: A Technology for Studying Noncovalent Macromolecular Complexes," *Int. J. Mass Spectrom.* **200,** 175–186 (2000).
2. Loo, J.A.; Thanabal, V.; Mei, H-Y. "Studying Noncovalent Small Molecule Interactions with Protein and RNA Targets by Mass Spectrometry," pp. 73–90 in Burlingame, A.L.; Carr, S.A.; Baldwin, M.A., eds., *Mass Spectrometry in Biology and Medicine*, Humana Press, Totowa, NJ (2000).
3. Loo, J.A.; Sannes-Lowery, K.A. "Studying Noncovalent Interactions by Electrospray Ionization Mass Spectrometry," pp. 345–367 in Larsen, B.S.; McEwen, C.N., eds., *Mass Spectrometry of Biological Materials*, 2nd ed., Dekker, New York (1998).
4. Loo, J.A. "Studying Noncovalent Protein Complexes by Electrospray Ionization Mass Spectrometry," *Mass Spectrom. Rev.* **16,** 1–23 (1997).
5. Hofstadler, S.A.; Griffey, R.H. "Analysis of Noncovalent Complexes of DNA and RNA by Mass Spectrometry," *Chem. Rev.* **101,** 377–390 (2001).
6. Smith, R.D.; Bruce, J.E.; Wu, Q.; Lei, Q.P. "New Mass Spectrometric Methods for the Study of Noncovalent Associations of Biopolymers," *Chem. Soc. Rev.* **26,** 191–202 (1997).

7. Heck, A.J.R. "Ligand Fishing by Mass Spectrometry," *Spectrosc. Eur.* **11,** 12, 14, 16–17 (1999).
8. Schriemer, D.C.; Hindsgaul, O. "Deconvolution Approaches in Screening Compound Mixtures," *Comb. Chem. High Throughput Screening* **1,** 155–170 (1998).
9. Lim, H-K.; Hsieh, Y.L.; Ganem, B.; Henion, J. "Recognition of Cell-Wall Peptide Ligands by Vancomycin Group Antibiotics: Studies Using Ion Spray Mass Spectrometry," *J. Mass Spectrom.* **30,** 708–714 (1995).
10. Bonnici, P.J.; Damen, M.; Waterval, J.C.M.; Heck, A.J.R. "Formation and Efficacy of Vancomycin Group Glycopeptide Antibiotic Stereoisomers Studied by Capillary Electrophoresis and Bioaffinity Mass Spectrometry," *Anal. Biochem.* **290,** 292–301 (2001).
11. Van de Kerk-Van Hoof, A.; Heck, A.J.R. "Interactions of α- and β-Avoparcin with Bacterial Cell-Wall Receptor-Mimicking Peptides Studied by Electrospray Ionization Mass Spectrometry," *J. Antimicrob. Chemother.* **44,** 593–599 (1999).
12. Van der Kerk-Van Hoof, A.; Heck, A.J.R. "Covalent and Non-Covalent Dissociations of Gas-Phase Complexes of Avoparcin and Bacterial Receptor Mimicking Precursor Peptides Studied by Collisionally Activated Decomposition Mass Spectrometry," *J. Mass Spectrom.* **34,** 813–819 (1999).
13. Heck, A.J.R.; Jorgensen, T.J.D.; O'Sullivan, M.; Von Raumer, M.; Derrick, P.J. "Gas-Phase Noncovalent Interactions Between Vancomycin-Group Antibiotics and Bacterial Cell-Wall Precursor Peptides Probed by Hydrogen/Deuterium Exchange," *J. Am. Soc. Mass Spectrom.* **9,** 1255–1266 (1998).
14. Jorgensen, T.J.D.; Roepstorff, P.; Heck, A.J.R. "Direct Determination of Solution Binding Constants for Noncovalent Complexes Between Bacterial Cell Wall Peptide Analogs and Vancomycin Group Antibiotics by Electrospray Ionization Mass Spectrometry," *Anal. Chem.* **70,** 4427–4432 (1998).
15. Staroske, T.; Heck, A.J.R.; Derrick, P.J.; Williams, D.H. "Interactions Between Vancomycin and Cell-Wall Precursor Analogs Studied by Electrospray Mass Spectrometry," *J. Mass Spectrom. Soc. Jpn.* **46,** 69–73 (1998).
16. Billot-Klein, D.; Gutmann, L.; Bryant, D.; Bell, D.; Van Jeijenort, J.; Grewal, J.; Shlaes, D.M. "Peptidoglycan Synthesis and Structure in *Staphylococcus Hemolyticus* Expressing Increasing Levels of Resistance to Glycopeptide Antibiotics," *J. Bacteriol.* **178,** 4696–4703 (1996).
17. Cheng, X.; Chen, R.; Bruce, J.E.; Schwartz, B.L.; Anderson, A.; Hofstadler, S.A.; Gale, D.C.; Smith, R.D.; Gao, J.; Sigal, G.B.; Mammen, M.; Whitesides, G.M. "Using Electrospray Ionization FTICR Mass Spectrometry to Study Competitive Binding of Inhibitors to Carbonic Anhydrase," *J. Am. Chem. Soc.* **117,** 8859–8860 (1995).
18. Gao, J.; Cheng, X.; Chen, R.; Sigal, G.B.; Bruce, J.E.; Schwartz, B.L.; Hofstadler, S.A.; Anderson, G.A.; Smith, R.D.; Whitesides, G.M. "Screening Derivatized Peptide Libraries for Tight Binding Inhibitors to Carbonic

Anhydrase II by Electrospray Ionization-Mass Spectrometry," *J. Med. Chem.* **39,** 1949–1955 (1996).

19. Bruce, J.E.; Anderson, G.A.; Chen, R.; Cheng, X.; Gale, D.C.; Hofstadler, S.A.; Schwartz, B.L.; Smith, R.D. "Bio-Affinity Characterization Mass Spectrometry," *Rapid Commun. Mass Spectrom.* **9,** 644–650 (1995).
20. Bruce, J.E.; Van Orden, S.L.; Anderson, G.A.; Hofstadler, S.A.; Sherman, M.G.; Rockwood, A.L.; Smith, R.D. "Selected Ion Accumulation of Noncovalent Complexes in a Fourier Transform Ion Cyclotron Resonance Mass Spectrometer," *J. Mass Spectrom.* **30,** 124–133 (1995).
21. Wigger, M.; Eyler, J.R.; Benner, S.A.; Li, W.; Marshall, A.G. "Fourier Transform-Ion Cyclotron Resonance Mass Spectrometric Resolution, Identification, and Screening of Non-Covalent Complexes of Hck Src Homology 2 Domain Receptor and Ligands from a 324-Member Peptide Combinatorial Library," *J. Am. Soc. Mass Spectrom.* **13,** 1162–1169 (2002).
22. Hofstadler, S.A.; Griffey, R.H. "Mass Spectrometry as a Drug Discovery Platform Against RNA Targets," *Curr. Opin. Drug Disc. Dev.* **3,** 423–431 (2000).
23. Griffey, R.H.; Sannes-Lowery, K.A.; Drader, J.J.; Mohan, V.O; Swayze, E.E.; Hofstadler, S.A. "Characterization of Low-Affinity Complexes Between RNA and Small Molecules Using Electrospray Ionization Mass Spectrometry," *J. Am. Chem. Soc.* **122,** 9933–9938 (2000).
24. Sannes-Lowery, K.A.; Griffey, R.H.; Hofstadler, S.A. "Measuring Dissociation Constants of RNA and Aminoglycoside Antibiotics by Electrospray Ionization Mass Spectrometry," *Anal. Biochem.* **280,** 264–271 (2000).
25. Griffey, R.H.; Hofstadler, S.A.; Sannes-Lowery, K.A.; Ecker, D.J.; Crooke, S.T. "Determinants of Aminoglycoside-Binding Specificity for rRNA by Using Mass Spectrometry," *Proc. Natl. Acad. Sci. USA* **96,** 10129–10133 (1999).
26. Hofstadler, S.A.; Sannes-Lowery, K.A.; Crooke, S.T.; Ecker, D.J.; Sasmor, H.; Manalili, S.; Griffey, R.H. "Multiplexed Screening of Neutral Mass-Tagged RNA Targets Against Ligand Libraries with Electrospray Ionization FTICR MS: A Paradigm for High-Throughput Affinity Screening," *Anal. Chem.* **71,** 3436–3440 (1999).
27. Liu, C.; Tolic, L.P.; Hofstadler, S.A.; Harms, A.C.; Smith, R.D.; Kang, C.; Sinha, N. "Probing Rega/RNA Interactions Using Electrospray Ionization-Fourier Transform Ion Cyclotron Resonance-Mass Spectrometry," *Anal. Biochem.* **262,** 67–76 (1998).
28. Gooding, K.B.; Higgs, R.; Hodge, B.; Stauffer, E.; Heinz, B.; McKnight, K.; Phipps, K.; Shapiro, M.; Winkler, M.; Ng, W-L.; Julian, R.K. "High Throughput Screening of Library Compounds Against an Oligonucleotide Substructure of an RNA Target," *J. Am. Soc. Mass Spectrom.* **15,** 884–892 (2004).
29. Borman, S. "Targeting RNA," *C&EN*, October 2, 2000, pp. 54–57.
30. Henry, C.M. "Structure-Based Drug Design," *C&EN*, June 4, 2001, pp. 69–78.
31. Swayze, E.E.; Jefferson, E.A.; Sannes-Lowery, K.A.; Blyn, L.B.; Risen, L.M.; Arakawa, S.; Osgood, S.A.; Hofstadler, S.A.; Griffey, R.H. "SAR by MS: A

Ligand Based Technique for Drug Lead Discovery Against Structured RNA Targets," *J. Med. Chem.* **45,** 3816–3819 (2002).

32. Wan, K.X.; Shibue, T.; Gross, M.L. "Non-Covalent Complexes Between DNA-Binding Drugs and Double-Stranded Oligodeoxynucleotides: A Study by ESI Ion-Trap Mass Spectrometry," *J. Am. Chem. Soc.* **122,** 300–307 (2000).
33. Shibue, T.; Wan, K.X.; Gross, M.L. "A Study of Non-Covalent Interaction Between DNA-Binding Drugs and Double-Stranded Oligodeoxynucleotides by ESI Ion Trap Mass Spectrometry," *J. Mass Spectrom. Soc. Jpn.* **48,** 221–227 (2000).
34. Wan, K.X.; Gross, M.L.; Shibue, T. "Gas-Phase Stability of Double-Stranded Oligodeoxynucleotides and Their Noncovalent Complexes with DNA-Binding Drugs as Revealed by Collisional Activation in an Ion Trap," *J. Am. Soc. Mass Spectrom.* **11,** 450–457 (2000).
35. Gabelica, V.; De Pauw, E.; Rosu, F. "Interaction Between Antitumor Drugs and a Double-Stranded Oligonucleotide Studied by Electrospray Ionization Mass Spectrometry," *J. Mass Spectrom.* **34,** 1328–1337 (1999).
36. Rosu, F.; Gabelica, V.; Houssier, C.; De Pauw, E. "Determination of Affinity, Stoichiometry and Sequence Selectivity of Minor Groove Binder Complexes with Double-Stranded Oligodeoxynucleotides by Electrospray Ionization Mass Spectrometry," *Nucleic Acids Research* **30,** E82/1–E82/9 (2002).
37. Grieg, M.; Robinson, J. "Detection of Oligonucleotide Ligand Complexes by ESI-MS (DOLCE-MS) as a Component of High Throughput Screening," *CPSA 2000—3rd Annual Symposium on Chemical and Pharmaceutical Structure Analysis*, Princeton, NJ, Sept. 26–28, 2000.
38. Greig, M.J.; Robinson, J.M. "Detection of Oligonucleotide-Ligand Complexes by ESI-MS (DOLCE-MS) as a Component of High Throughput Screening," *J. Biomolecular Screening* **5,** 441–454 (2000).
39. Nedved, M.L.; Habibi-Goudarzi, S.; Ganem, B.; Henion, J.D. "Characterization of Benzodiazepine 'Combinatorial' Chemical Libraries by Online Immunoaffinity Extraction, Coupled Column HPLC-Ion Spray Mass Spectrometry-Tandem Mass Spectrometry," *Anal. Chem.* **68,** 4228–4236 (1996).
40. Kelly, M.A.; Liang, H.; Sytwu, I-I.; Vlattas, I.; Lyons, N.L.; Bowen, B.R.; Wennogle, L.P. "Characterization of SH2-Ligand Interactions Via Library Affinity Selection with Mass Spectrometric Detection," *Biochemistry* **35,** 11747–11755 (1996).
41. Schriemer, D.C.; Bundle, D.R.; Li, L.; Hindsgaul, O. "Micro-Scale Frontal Affinity Chromatography with Mass Spectrometric Detection: A New Method for the Screening of Compound Libraries," *Angew. Chem., Int. Ed.* **37,** 3383–3387 (1998).
42. Hindsgaul, O.; Schriemer, D.C. "Methods Using Frontal Chromatography-Mass Spectrometry for Screening Compound Libraries," PCT Int. Appl. (1999), 90 pp., WO 9950669 A1 19991007.
43. Kelly, M.A.; Rosner, P.J.; Zembrowski, W.; Beebe, D.; Mylari, L.; Oates, P.; Chan, N.; Lewis, D.; Schriemer, D. "Evaluation of Frontal Affinity

Chromatography Mass Spectrometry (FAC-MS) as a Universal Binding Assay for Drug Discovery," in *Proceedings of the 49th ASMS Conference on Mass Spectrometry and Allied Topics*, Long Beach, CA, June 11–15, 2000, TOB 10:15.

44. Schriemer, D.C.; Nora Chan, N.; Lewis, D.; Rosner, P.; Hoth, L.; Geoghegan, K.; Kelly, M. "Frontal Affinity Chromatography Mass Spectrometry (FAC-MS): Method Evaluation with a Nuclear Hormone Receptor," in *Proceedings of the 49th ASMS Conference on Mass Spectrometry and Allied Topics*, Long Beach, CA, June 11–15, 2000, TPF 228.
45. Kelly, M.A.; Mclellan, T.J.; Rosner, P.J. "Strategic Use of Affinity-Based Mass Spectrometry Techniques in the Drug Discovery Process," *Anal. Chem.* **74,** 1–9 (2002).
46. Zhang, B.; Palcic, M.M.; Schriemer, D.C.; Alvarez-Manilla, G.; Pierce, M.; Hindsgaul, O. "Frontal Affinity Chromatography Coupled to Mass Spectrometry for Screening Mixtures of Enzyme Inhibitors," *Anal. Biochem.* **299,** 173–182 (2001).
47. Chan, N.W.C.; Lewis, D.F.; Hewko, S.; Hindsgaul, O.; Schriemer, D.C. "Frontal Affinity Chromatography for the Screening of Mixtures," *Comb. Chem. High Throughput Screening* **5,** 395–406 (2002).
48. Wieboldt, R.; Zweigenbaum, J.; Henion, J. "Immunoaffinity Ultrafiltration with Ion Spray HPLC/MS for Screening Small-Molecule Libraries," *Anal. Chem.* **69,** 1683–1691 (1997).
49. Siegel, M.M.; Tabei, K.; Bebernitz, G.A.; Baum, E.Z. "Rapid Methods for Screening Low Molecular Mass Compounds noncovalently Bound to Proteins Using Size Exclusion and Mass Spectrometry Applied to Inhibitors of Human Cytomegalovirus Protease," *J. Mass Spectrom.* **33,** 264–273 (1998).
50. De Biasil, V.; Haskins, N.; Organ, A.; Bateman, R.; Giles, K.; Jarvis, S. "High Throughput Liquid Chromatography/Mass Spectrometric Analyses Using a Novel Multiplexed Electrospray Interface," *Rapid Commun. Mass Spectrom.* **13,** 1165–1168 (1999).
51. Yang, H.; Cheng, X.; Bakhoum, A.; Hajduk, P.; Lico, I.; Voorbach, M.; Dandliker, P.; Schurdak, M.; Gao, L.; Buko, A.; Miesbauer, L.; Schmitt, R.; Martin, Y.; Beutel, B.; Burns, D. "Use of Affinity Selection Mass Spectrometry to Screen Organic Compounds of Diverse Structures for Drug Leads," Abstracts of Papers, American Chemical Society (2001), 221st ANYL-192.
52. Comess, K.M.; Voorbach, M.J.; Coen, M.L.; Tang, H.; Gao, L.; Cheng, X.; Schurdak, M.E.; Beutel, B.A.; Burns, D.J. "Affinity-Based High-Throughput Screening of Orphan Targets: Practical Solutions for Removing Promiscuous Binders," *Abstracts of Papers, 224th ACS National Meeting*, Boston, MA, United States, August 18–22, 2002 (2002).
53. Borman, S. "Mining the Genome," *C&EN*, October 14, 2002, pp. 47–50.
54. Van Breemen, R.B.; Woodbury, C.P.; Venton, D.L. "Screening Molecular Diversity Using Pulsed Ultrafiltration Mass Spectrometry," pp. 99–113 in Larsen, B.S.; McEwen, C.N., eds., *Mass Spectrometry of Biological Materials*, 2nd ed., Dekker, New York (1998).

55. Van Breemen, R.B.; Huang, C-R.; Nikolic, D.; Woodbury, C.P.; Zhao, Y-Z.; Venton, D.L. "Pulsed Ultrafiltration Mass Spectrometry: A New Method for Screening Combinatorial Libraries," *Anal. Chem.* **69,** 2159–2164 (1997).
56. Zhao, Y-Z.; Van Breemen, R.B.; Nikolic, D.; Huang, C-R.; Woodbury, C.P.; Schilling, A.; Venton, D.L. "Screening Solution-Phase Combinatorial Libraries Using Pulsed Ultrafiltration/Electrospray Mass Spectrometry," *J. Med. Chem.* **40,** 4006–4012 (1997).
57. Johnson, B.M.; Nikolic, D.; Van Breemen, R.B. "Application of Pulsed Ultrafiltration-Mass Spectrometry," *Mass Spectrom. Rev.* **21,** 76–86 (2002).
58. Nikolic, D.; Habibi-Goudarzi, S.; Corley, D.G.; Gafner, S.; Pezzuto, J.M.; Van Breemen, R.B. "Evaluation of Cyclooxygenase-2 Inhibitors Using Pulsed Ultrafiltration Mass Spectrometry," *Anal. Chem.* **72,** 3853–3859 (2000).
59. Nikolic, D.; Van Breemen, R.B. "Screening for Inhibitors of Dihydrofolate Reductase Using Pulsed Ultrafiltration Mass Spectrometry," *Comb. Chem. High Throughput Screening* **1,** 47–55 (1998).
60. Gu, C.; Nikolic, D.; Lai, J.; Xu, X.; Van Breemen, R.B. "Assays of Ligand-Human Serum Albumin Binding Using Pulsed Ultrafiltration and Liquid Chromatography-Mass Spectrometry," *Comb. Chem. High Throughput Screening* **2,** 353–359 (1999).
61. Van Breemen, RB.; Nikolic, D.; Bolton, J.L. "Metabolic Screening Using Online Ultrafiltration Mass Spectrometry," *Drug Metab. Dispos.* **26,** 85–90 (1998).
62. Nikolic, D.; Fan, P.W.; Bolton, J.L.; Van Breemen, R.B. "Screening for Xenobiotic Electrophilic Metabolites Using Pulsed Ultrafiltration-Mass Spectrometry," *Comb. Chem. High Throughput Screening* **2,** 165–175 (1999).
63. Jiang, Y.; Lee, C.S. "On-Line Coupling of Hollow Fiber Membranes with Electrospray Ionization Mass Spectrometry for Continuous Affinity Selection, Concentration and Identification of Small-Molecule Libraries,"*J. Mass Spectrom.* **36,** 664–669 (2001).
64. Jiang, Y.; Wang., P-C.; Locascio, L.E.; Lee, C.S. "Integrated Plastic Microfluidic Devices with ESI-MS for Drug Screening and Residue Analysis," *Anal. Chem.* **73,** 2048–2053 (2001).
65. Benkestock, K.; Edlund, P-O.; Roeraade, J. "On-Line Microdialysis for Enhanced Resolution and Sensitivity During Electrospray Mass Spectrometry of Non-Covalent Complexes and Competitive Binding Studies," *Rapid Commun. Mass Spectrom.* **16,** 2054–2059 (2002).
66. Holzman, T.F.; Harlan, J.E.; Egan, D.A.; Buko, A.M.; Solomon, L.R.; Ladror, U.S.; Tang, Q. "Centrifugally-Enhanced Method of Determining Ligand/Target Affinity," PCT Int. Appl. (2000), 57 pp., WO 0055625 A1 20000921.
67. Arkin, M.; Lear, J.D. "A New Data Analysis Method to Determine Binding Constants of Small Molecules to Proteins Using Equilibrium Analytical Ultracentrifugation with Absorption Optics," *Anal. Biochem.* **299,** 98–107 (2001).
68. Dunayevskiy, Y.M.; Lai, J-J.; Quinn, C.; Talley, F.; Vouros, P. "Mass Spectrometric Identification of Ligands Selected from Combinatorial Libraries Using Gel Filtration," *Rapid Commun. Mass Spectrom.* **11,** 1178–1184 (1997).

69. Tong, H.; Tabei, K.; Amin, A.; Olson, M.; Bebernitz, G.; Siegel, M.M. "Identification of Compounds noncovalently Bound to RNA Using GPC Spin Columns with ESI MS Detection," in *Proceedings of the 49th ASMS Conference on Mass Spectrometry and Allied Topics*, Chicago, IL, May 27–31, 2001, Poster Th 274.
70. Kaur, S.; McGuire, L.; Tang, D.; Dollinger, G.; Huebner, V. "Affinity Selection and Mass Spectrometry-Based Strategies to Identify Lead Compounds in Combinatorial Libraries," *J. Protein Chem.* **16,** 505–511 (1997).
71. Davis, R.G.; Anderegg, R.J.; Blanchard, S.G. "Iterative Size-Exclusion Chromatography Coupled with Liquid Chromatographic Mass Spectrometry to Enrich and Identify Tight-Binding Ligands from Complex Mixtures," *Tetrahedron* **55,** 11653–11667 (1999).
72. Wabnitz, P.A.; Loo, J.A. "Drug Screening of Pharmaceutical Discovery Compounds by Micro-Size Exclusion Chromatography/Mass Spectrometry," *Rapid Commun. Mass Spectrom.* **16,** 85–91 (2002).
73. Moy, F.J.; Haraki, K.; Mobilio, D.; Walker, G.; Powers, R.; Tabei, K.; Tong, H.; Siegel, M.M. "MS/NMR: A Structure-Based Approach for Discovering Protein Ligands and for Drug Design by Coupling Size Exclusion Chromatography, Mass Spectrometry, and Nuclear Magnetic Resonance Spectroscopy," *Anal. Chem.* **73,** 571–581 (2001).
74. Tong, H.; Bell, D.; Tabei, K.; Siegel, M.M. "Automated Data Massaging, Interpretation, and E-Mailing Modules for High Throughput Open Access Mass Spectrometry," *J. Am. Soc. Mass Spectrom.* **10,** 1174–1187 (1999).
75. Schnier, P.D.; Deblanc, R.; Woo, G.; Gigante, W.; Cheetham, J. "Ultra-High Throughput Affinity Mass Spectrometry for Screening Protein Receptors," in proceedings of the 51st ASMS Conference on Mass Spectrometry and Allied Topics, Montreal, Canada, June 8–12, 2003, WPC 049.
76. Muckenschnabel, I.; Falchetto, R.; Mayr, L.M.; Filipuzzi, I. "SpeedScreen: Label-Free Liquid Chromatography-Mass Spectrometry-Based High-Throughput Screening for the Discovery of Orphan Protein Ligands," *Anal. Biochem.* **324,** 241–249 (2004).
77. Blom, K.F.; Larsen, B.S.; McEwen, C.N. "Determining Affinity-Selected Ligands and Estimating Binding Affinities by Online Size Exclusion Chromatography/Liquid Chromatography-Mass Spectrometry," *J. Comb. Chem.* **1,** 82–90 (1999).
78. Annis, A. "An Affinity Selection-Mass Spectral Method for the Identification of Small Molecule Ligands from Self-Encoded Combinatorial Libraries," Greater Boston Mass Spectrometry Discussion Group, Dec. 13, 2000, www.Neogenesis.com. Also: Lenz, G.R.; Nash H.M.; Jindal, S. "Chemical Ligands, Genomics and Drug Discovery," *Drug Discovery Today* **5,** 145–156 (2000). Also: Falb, D.; Jindal, S. "Chemical Genomics: Bridging the Gap Between the Proteome and Therapeutics," *Curr. Opinion Drug Disc. Devel.* 2002, **5,** 532–539.
79. Annis, A. "Genome-Scale Drug Discovery by Affinity Selection-Mass Spectrometry-Based Screening of Mass-Encoded Small Molecule Libraries,"

in *Proceedings of the 51st ASMS Conference on Mass Spectrometry and Allied Topics*, Montreal, Canada, June 8–12, 2003, WODam 11:15.

80. Hsieh, Y.F.; Gordon, N.; Regnier, F.; Afeyan, N.; Martin, S.A.; Vella, G. "Multidimensional Chromatography Coupled with Mass Spectrometry for Target-Based Screening," *J. Mol. Diversity* **2,** 189–196 (1997).
81. Hsieh, Y.F. "Multiple Target Screening of Molecular Libraries by Mass Spectrometry," PCT Int. Appl. (1999), 40 pp., WO 9905309 A1 19990204.
82. Van Der Greef, J.; Irth, H. "Identification of Ligands for Orphan Receptors Using On-Line Coupling of Mass Spectrometry to Continuous-Flow Separation Techniques," Eur. Pat. Appl. (2000), 14 pp., EP 1048950 A1 20001102.
83. Van Elswijk, D.A.; Tjaden, U.R.; Van Der Greef, J.; Irth, H. "Mass Spectrometry-Based Bioassay for the Screening of Soluble Orphan Receptors," *Int. J. Mass Spectrom.* **210/211,** 625–636 (2001), and www.kiadis.com.
84. Dunayevskiy, Y.M.; Hughes, D.E. "High Throughput Size-Exclusive Method of Screening Complex Biological Materials for Affinity Ligands," PCT Int. Appl. (2000), 33 pp., WO 0047999 A1 20000817.
85. Chu, Y-H.; Avila, L.Z.; Gao, J.; Whitesides, G.M. "Affinity Capillary Electrophoresis," *Acc. Chem. Res.* **28,** 461–468 (1995).
86. Chu, Y-H.; Kirby, D.P.; Karger, B.L. "Free Solution Identification of Candidate Peptides from Combinatorial Libraries by Affinity Capillary Electrophoresis/ Mass Spectrometry," *J. Am. Chem. Soc.* **117,** 5419–5420 (1995).
87. Chu, Y-H.; Dunayevskiy, Y.M.; Kirby, D.P.; Vouros, P.; Karger, B.L. "Affinity Capillary Electrophoresis-Mass Spectrometry for Screening Combinatorial Libraries," *J. Am. Chem. Soc.* **118,** 7827–7835 (1996).
88. Hughes, D.E.; Karger, B.L. "Method to Detect and Analyze Tight-Binding Ligands in Complex Biological Samples Using Capillary Electrophoresis and Mass Spectrometry," PCT Int. Appl. (2000), 44 pp., WO 0003240 A1 20000120.
89. Rashkovetsky, L.G.; Lyubarskaya, Y.V.; Foret, F.; Hughes, D.E.; Karger, B.L. "Automated Microanalysis Using Magnetic Beads with Commercial Capillary Electrophoretic Instrumentation," *J. Chromatogr., A* **781**(1 + 2), 197–204 (1997).
90. Lyubarskaya, Y.V.; Dunayevskiy, Y.M.; Vouros, P.; Karger, B.L. "Microscale Epitope Mapping by Affinity Capillary Electrophoresis-Mass Spectrometry," *Anal. Chem.* **69,** 3008–3014 (1997).
91. Apruzzese, W.A.; Vouros, P. "Analysis of DNA Adducts by Capillary Methods Coupled to Mass ectrometry: A Perspective," *J. Chromatogr., A* **794,** 97–108 (1998).
92. Hsieh, Y-L.; Cai, J.; Li, Y-T.; Henion, J.D.; Ganem, B. "Detection of Noncovalent FKBP-FK506 and FKBP-Rapamycin Complexes by Capillary Electrophoresis-Mass Spectrometry and Capillary Electrophoresis-Tandem Mass Spectrometry," *J. Am. Soc. Mass Spectrom.* **6,** 85–90 (1995).
93. Dunayevskiy, Y.M.; Lyubarskaya, Y.V.; Chu, Y-H.; Vouros, P.; Karger, B.L. "Simultaneous Measurement of Nineteen Binding Constants of Peptides to Vancomycin Using Affinity Capillary Electrophoresis-Mass Spectrometry," *J. Med. Chem.* **41,** 1201–1204 (1998).

94. Lyubarskaya, Y.V.; Carr, S.A.; Dunnington, D.; Prichett, W.P.; Fisher, S.M.; Appelbaum, E.R.; Jones, C.S.; Karger, B.L. "Screening for High-Affinity Ligands to the Src SH2 Domain Using Capillary Isoelectric Focusing-Electrospray Ionization Ion Trap Mass Spectrometry," *Anal. Chem.* **70,** 4761–4770 (1998).

95. Martinovic, S.; Berger, S.J.; Pasa-Tolic, L.; Smith, R.D. "Separation and Detection of Intact Noncovalent Protein Complexes from Mixtures by On-Line Capillary Isoelectric Focusing-Mass Spectrometry," *Anal. Chem.* **72,** 5356–5360 (2000).

96. Erlanson, D.A.; Braisted, A.C.; Raphael, D.R.; Randal, M.; Stroud, R.M.; Gordon, E.M.; Wells, J.A. "Site-Directed Ligand Discovery," *Proc. Nat. Acad. Sci. USA* **97,** 9367–9372. (2000).

97. Hyde, J.; Braisted, A.; Randal, M.; Arkin, M. "Discovery and Characterization of Cooperative Ligand Binding in the Adaptive Region of Interleukin-2," *Biochemistry* **42,** 6475–6483 (2003).

98. Arkin, M.; Randal, M.; Delano, W.; Hyde, J.; Luong, T.; Oslob, J.; Raphael, D.; Taylor, L.; Wang, J.; Mcdowell, R.; Wells, J.; Braisted, A. "Binding of Small Molecules to an Adaptive Protein-Protein Interface," *Proc. Nat. Acad. Sci. USA* **100,** 1603–1608 (2003).

99. Braisted, A.; Oslob, J.; Delano, W.; Hyde, J.; McDowell, R.; Waal, N.; Yu, C. Arkin, M.; Raimundo, B. "Discovery of a Potent Small Molecule IL-2 Inhibitor through Fragment Assembly," *J. Am. Chem. Soc.* **125,** 3714–3715 (2003).

100. Cancilla, M.T.; Erlanson, D.A.; Braisted, A.C.; Randal, M.; Arkin, M.; Gordon, E.M.; Wells, J.A. "Discovering Antagonists of Enzymes and Protein:Protein Interactions Using a High Throughput Screening Mass Spectrometry Method Termed 'Tethering,'" in *Proceedings of the 50th ASMS Conference on Mass Spectrometry and Allied Topics*, Orlando, FL, June 2–6, 2002, MOEam 10:15, 1107.pdf.

101. Cancilla, M.T.; Evarts, S.L.; Wang, J.; Braisted, A.C.; Jacobs, J.W.; Erlanson, D.A; Wells, J.A. "Development of a True Mass Spectrometry Based High Throughput Screening Method for Discovering Small Molecule Inhibitors for Protein Targets," in *Proceedings of the 51st ASMS Conference on Mass Spectrometry and Allied Topics*, Montreal, Canada, June 8–12, 2003, WODam 10:55.

102. Erlanson, D.; Lam, J.; Wiesmann, C.; Luong, T.; Simmons, R.; DeLano, W.; Choong, I.; Burdett, M.; Flanagan, M., Lee; D.; Gordon, E.; O'Brien, T. "In Situ Assembly of Enzyme Inhibitors Using Extended Tethering," *Nature Biotechnology* **21,** 308–314 (2003).

103. Hoefner, G.; Wanner, K.T. "Competitive Binding Assays Made Easy with a Native Marker and Mass Spectrometric Quantification,“ *Angew. Chem., Int. Ed.* **42,** 5235–5237 (2003).

104. Wanner, K.; Hoefner, G.; Bertling, W.; Kosak, H. "Method for the Determination of Ligand Binding to Target Molecules by Using a Marker Molecule and Mass Spectrometry." PCT Int. Appl. (2002), 44 pp., WO 2002095403 A2 20021128.

105. Wanner, K.; Hofner, G.; Bertling, W. "Method for Determining the Quantity of Ligands that are Bound to Target Molecules," PCT Int. Appl. (2001), 39 pp., WO 2001094943 A2 20011213.

106. Hulme, E.C., ed. *Receptor-Ligand Interactions: A Practical Approach*, Oxford University Press, NY, 1992.

107. Hutchens, T.W.; Yip, T.T. "New Desorption Strategies for the Mass-Spectrometric Analysis of Macromolecules," *Rapid Commun. Mass Spectrom.* **7,** 576–580 (1993).

108. Hutchens, T.W.; Yip, T-T. "Method and Apparatus for Desorption and Ionization of Analytes," PCT Int. Appl. (1994), 146 pp., WO 9428418 A1 19941208.

109. Yip, T-T.; Hutchens, T.W. "Affinity Mass Spectrometry. Probes with Surfaces Enhanced for Affinity Capture (SEAC) of Lactoferrin," *Exp. Biol. Med.* **28,** 39–58 (1997).

110. Hutchens, T.W.; Yip, T-T. "Method and Apparatus for Desorption and Ionization of Analytes," U.S. (1998), 64 pp., US 5719060 A 19980217.

111. Kuwata, H.; Yip, T-T.; Yip, C.L.; Tomita, M.; Hutchens, T.W. "Bactericidal Domain of Lactoferrin: Detection, Quantitation, and Characterization of Lactoferricin in Serum by SELDI Affinity Mass Spectrometry," *Biochem. Biophys. Res. Commun.* **245,** 764–773 (1998).

112. Hutchens, T.W.; Yip, T-T. "Retentate Chromatography and Protein Chip Arrays with Applications in Biology and Medicine," PCT Int. Appl. (1998), 157 pp., WO 9859362 A1 19981230.

113. Nelson, R.W. "The Use of Bioreactive Probes in Protein Characterization," *Mass Spectrom. Rev.* **16,** 353–376 (1997).

114. Nelson, R.W.; Nedelkov, D.; Tubbs, K.A. "Biomolecular Interaction Analysis Mass Spectrometry," *Anal. Chem.* **72,** 404A–411A (2000).

115. Kris, R.M.; Felder, S. "High Throughput Assay System with Quantification of Multiple Array Plate Data," PCT Int. Appl. (2000), 107 pp., WO 0079008 A2 20001228.

116. Kris, R.M.; Felder, S. "High Throughput Assay System Using Multi Array Plate Screen, Nuclease Protection, Oligonucleotide Anchors, Bifunctional Linkers, and Mass Spectrometry," PCT Int. Appl. (2000), 111 pp., WO 0037684 A1 20000629.

117. Felder, S.; Seligmann, B.; Kris, R.M. "High Throughput Assay System for Monitoring Expressed Sequence Tags (Ests) Using Ordered Arrays of Probes," PCT Int. Appl. (2000), 88 pp., WO 0037683 A2 20000629.

118. Felder, S.; Kris, R. "High Throughput Assays Using Microtiter Plates with Individual Wells Carrying Multiple Probes with Defined Locations," PCT Int. Appl. (1999), 93 pp., WO 9932663 A2 19990701.

119. Merchant, M.; Weinberger, S.R. "Recent Advancements in Surface-Enhanced Laser Desorption/Ionization-Time of Flight-Mass Spectrometry," *Electrophoresis* **21,** 1164–1177 (2000).

120. Little, D.P.; Cornish, T.J.; O'Donnell, M.J.; Braun, A.; Cotter, R.J.; Koester, H. "MALDI on a Chip: Analysis of Arrays of Low-Femtomole to Subfemtomole Quantities of Synthetic Oligonucleotides and DNA Diagnostic Products Dispensed by a Piezoelectric Pipet," *Anal. Chem.* **69,** 4540–4546 (1997).

121. Nelson, R.W.; Nedelkov, D.; Tubbs, K.A. "Biosensor Chip Mass Spectrometry: A Chip-Based Proteomics Approach," *Electrophoresis* **21,** 1155–1163 (2000).

122. Nelson, R.W.; Krone, J.R. "Advances in Surface Plasmon Resonance Biomolecular Interaction Analysis Mass Spectrometry (BIA/MS)," *J. Mol. Recognit.* **12,** 77–93 (1999).

123. Nelson, R.W.; Jarvik, J.W.; Taillon, B.E.; Tubbs, K.A. "BIA/MS of Epitope-Tagged Peptides Directly from *E. Coli* Lysate: Multiplex Detection and Protein Identification at Low-Femtomole to Subfemtomole Levels," *Anal. Chem.* **71,** 2858–2865 (1999).

124. Min, D-H.; Tang, W-J.; Mrksich, M. "Chemical Screening by Mass Spectrometry to Identify Inhibitors of Anthrax Lethal Factor," *Nat. Biotechnol.* **22,** 717 (2004).

125. Youngquist, R.S.; Fuentes, G.R.; Miller, C.M.; Ridder, G.M.; Lacey, M.P.; Keough, T. "Use of MALDI Mass Spectrometry for the Identification of Active Compounds Isolated from Support-Bound Combinatorial Libraries," *Adv. Mass Spectrom.* **14**(Chapter 17) 423–448 (1998).

126. Youngquist, R.S.; Fuentes, G.R.; Miller, C.M.; Ridder, G.M.; Lacey, M.P.; Keough, T. "Matrix-Assisted Laser Desorption Ionization for Rapid Determination of the Sequences of Biologically Active Compounds Isolated from Support-Bound Combinatorial Libraries," Pept.: Chem., Struct. Biol., *Proceedings of the 14th American Peptide Symposium*, 1995, pp. 311–312 (1996).

127. Youngquist, R.S.; Fuentes, G.R.; Lacey, M.P.; Keough, T. "Generation and Screening of Combinatorial Peptide Libraries Designed for Rapid Sequencing by Mass Spectrometry," *J. Am. Chem. Soc.* **117,** 3900–3906 (1995).

128. Youngquist, R.S.; Fuentes, G.R.; Lacey, M.P.; Keough, T. "Matrix-Assisted Laser Desorption Ionization for Rapid Determination of the Sequences of Biologically Active Peptides Isolated from Support-Bound Combinatorial Peptide Libraries," *Rapid Commun. Mass Spectrom.* **8,** 77–81 (1994).

129. Hilaire, P.M.St.; Lowary, T.L.; Meldal, M.; Bock, K. "Oligosaccharide Mimetics Obtained by Novel, Rapid Screening of Carboxylic Acid Encoded Glycopeptide Libraries," *J. Am. Chem. Soc.* **120,** 13312–13320 (1998).

130. Buchardt, J.; Schiodt, C.B.; Krog-Jensen, C.; Delaisse, J-M.; Foged, N.T.; Meldal, M. "Solid Phase Combinatorial Library of Phosphinic Peptides for Discovery of Matrix Metalloproteinase Inhibitors," *J. Comb. Chem.* **2,** 624–638 (2000).

131. Cho, C.Y.; Youngquist, R.S.; Paikoff, S.J.; Beresini, M.H.; Hebert, A.R.; Berleau, L.T.; Liu, C.W.; Wemmer, D.E.; Keough, T.; Schultz, P.G. "Synthesis and Screening of Linear and Cyclic Oligocarbamate Libraries. Discovery of High Affinity Ligands for GPIIb/IIIa," *J. Am. Chem. Soc.* **120,** 7706–7718 (1998).

132. Keough, T.; Baker, T.R.; Dobson, R.L.M.; Lacey, M.P.; Riley, T.A.; Hasselfield, J.A.; Hesselberth, P.E. "Antisense DNA Oligonucleotides. II: The Use of Matrix-Assisted Laser Desorption/Ionization Mass Spectrometry for the Sequence Verification of Methylphosphonate Oligodeoxyribonucleotides," *Rapid Commun. Mass Spectrom.* **7,** 195–200 (1993).

133. Farmer, T.B.; Caprioli, R.M. "Determination of Protein–Protein Interactions by Matrix-Assisted Laser Desorption/Ionization Mass Spectrometry," *J. Mass Spectrom.* **33,** 697–704 (1998).

134. Hillenkamp, F. "Matrix-Assisted Laser Desorption/Ionization of Non-Covalent Complexes," *New Methods for the Study of Biomolecular Complexes,* (NATO ASI Ser., Ser. C) **510,** 181–191 (1998).

135. Glocker, M.O.; Bauer, S.H.J.; Kast, J.; Volz, J.; Przybylski, M. "Characterization of Specific Noncovalent Protein Complexes by UV Matrix-Assisted Laser Desorption Ionization Mass Spectrometry," *J. Mass Spectrom.* **31,** 1221–1227 (1996).

136. She, Y-M.; Ji, Y-P. "Studies of Non-Covalent Protein Complexes by MALDI Methods," *Gaodeng Xuexiao Huaxue Xuebao* **20,** 852–855 (1999).

137. Strupat, K.; Sagi, D.; Peter-Katalinic, J.; Bonisch, H.; Schafer, G. "Oligomerization and Substrate Binding Studies of the Adenylate Kinase from Sulfolobus Acidocaldarius by Matrix-Assisted Laser Desorption/Ionization Mass Spectrometry," *Analyst (Cambridge, U. K.)* **125,** 563–567 (2000).

138. Akashi, S.; Shirouzu, M.; Yokoyama, S.; Takio, K. "Detection of Molecular Ions of Non-Covalent Complexes of Ras-GDP and Ras-GppNp by MALDI-TOFMS," *J. Mass Spectrom. Soc. Jpn.* **44,** 269–277 (1996).

139. Budnik, B.A.; Jensen, K.B.; Jorgensen, T.J.D.; Haase, A.; Zubarev, R.A. "Benefits of 2.94 μm Infrared Matrix-Assisted Laser Desorption/Ionization for Analysis of Labile Molecules by Fourier Transform Mass Spectrometry," *Rapid Commun. Mass Spectrom.* **14,** 578–584 (2000).

140. Siegel, M.M.; Tabei, K.; Tsao, R.; Pastel, M.J.; Pandey, R.K.; Berkenkamp, S.; Hillenkamp, F.; De Vries, M.S. "Comparative Mass Spectrometric Analyses of Photofrin Oligomers by Fast Atom Bombardment Mass Spectrometry, UV and IR Matrix-Assisted Laser Desorption/Ionization Mass Spectrometry, Electrospray Ionization Mass Spectrometry and Laser Desorption/Jet-Cooling Photoionization Mass Spectrometry," *J. Mass Spectrom.* **34,** 661–669 (1999).

141. Powell, K.D.; Fitzgerald, M.C. "Accuracy and Precision of a New H/D Exchange- and Mass Spectrometry-Based Technique for Measuring the Thermodynamic Properties of Protein-Peptide Complexes," *Biochemistry* **42,** 4962–4970 (2003).

142. Powell, K.D.; Ghaemmaghami, S.; Wang, M.Z.; Ma, L.; Oas, T.G.; Fitzgerald, M.C. "A General Mass Spectrometry-Based Assay for the Quantitation of Protein-Ligand Binding Interactions in Solution," *J. Am. Chem. Soc.* **124,** 10256–10257 (2002).

143. Powell, K.D.; Wales, T.E.; Fitzgerald, M.C. "Thermodynamic Stability Measurements on Multimeric Proteins Using a New H/D Exchange- and Matrix-Assisted Laser Desorption/Ionization (MALDI) Mass Spectrometry-Based Method," *Protein Science* **11,** 841–851 (2002).

144. Powell, K.D.; Fitzgerald, M.C. "Measurements of Protein Stability by H/D Exchange and Matrix-Assisted Laser Desorption/Ionization Mass Spectrometry Using Picomoles of Material," *Anal. Chem.* **73,** 3300–3304 (2001).

145. Ghaemmaghami, S.; Fitzgerald, M.C.; Oas, T.G. "A Quantitative, High-Throughput Screen for Protein Stability," *Proc. Nat. Acad. Sci. USA* **97,** 8296–8301 (2000).

146. Powell, K.D.; Fitzgerald, M.C. "High-Throughput Analysis of Protein-Ligand Interactions Using a MALDI- and H/D Exchange-Based Technique," in *Proceedings of the 51st ASMS Conference on Mass Spectrometry and Allied Topics*, Montreal, Canada, June 8–12, 2003, WODam 11:15.

147. Smith, R.D.; Liu, C. "Microdialysis Unit for Molecular Weight Separation," U.S. (1999), 24 pp., US 5954959 A 19990921 Application: US 97-855727 19970509.

148. Xiang, F.; Lin, Y.; Wen, J.; Matson, D.W.; Smith, R.D. "An Integrated Microfabricated Device for Dual Microdialysis and Online ESI-Ion Trap Mass Spectrometry for Analysis of Complex Biological Samples," *Anal. Chem.* **71,** 1485–1490 (1999).

149. Corso, T.N.; Schultz, G.A.; Li, J.; Alpha, C.G.; Smith, B.I.; Shinde, N.A.; Ackerman, J.C.; Sheldon, G.S. "A Microchip-Based Multi-Nozzle Nanoelectrospray Device," in *Proceedings of the 50th ASMS Conference on Mass Spectrometry and Allied Topics*, Orlando, Florida, June 2–6, 2002, MPM 390, 1350.pdf, and www.advion.com.

150. Zhang, S.; Van Pelt, C.K.; Schultz, G.A. "Rapid, Fully Automated Nano-ESI/MS/MS Analysis of Excised 2D Gel Spots by the Nanomate Robot and ESI Chip," in *Proceedings of the 50th ASMS Conference on Mass Spectrometry and Allied Topics*, Orlando, Florida, June 2–6, 2002, ThPB 025, 1549.pdf, and www.advion.com.

151. Van Pelt, C.K.; Zhang, S.; Henion, J.D. "Characterizatioin of a Fully Automated Nanospray System with Mass Spectrometric Detection for Proteomic Analyses," *J. Biomolecular Techniques* **13,** 72–84 (2002).

152. Benkestock, K.; Van Pelt, C.K.; Åkerud, T.; Sterling, A.; Edlund, P-O.; Roeraade, J. "Automated Nano-Electrospray Mass Spectrometry for Protein-Ligand Screening by Noncovalent Interaction Applied to Human H-FABP and A-FABP," *J. Biomolecular Screening* **8,** 247–256 (2003).

153. Zhang, S.; Van Pelt, C.K.; Wilson, D.B. "Quantitative Determination of Noncovalent Binding Interactions Using Automated Nanoelectrospray Mass Spectrometry," *Anal. Chem.* **75,** 3010–3018 (2003).

154. Kameoka, J.; Craighead, H.G.; Zhang, H.; Henion, J. "A Polymeric Microfluidic Chip for CE/MS Determination of Small Molecules," *Anal. Chem.* **73,** 1935–1941 (2001).

155. Deng, Y.; Zhang, H.; Henion, J. "Chip-Based Quantitative Capillary Electrophoresis/Mass Spectrometry Determination of Drugs in Human Plasma," *Anal. Chem.* **73,** 1432–1439 (2001).

156. Wachs, T.; Henion, J. "Electrospray Device for Coupling Microscale Separations and Other Miniaturized Devices with Electrospray Mass Spectrometry," *Anal. Chem.* **73,** 632–638 (2001).

157. Deng, Y.; Henion, J.; Li, J.; Thibault, P.; Wang, C.; Harrison, D.J. "Chip-Based Capillary Electrophoresis/Mass Spectrometry Determination of Carnitines in Human Urine," *Anal. Chem.* **73,** 639–646 (2001).

158. Zhang, B.; Liu, H.; Karger, B.L.; Foret, F. "Microfabricated Devices for Capillary Electrophoresis-Electrospray Mass Spectrometry," *Anal. Chem.* **71,** 3258–3264 (1999).

159. Moore, R.E.; Licklider, L.; Schumann, D.; Lee, T.D. "A Microscale Electrospray Interface Incorporating a Monolithic, Poly(Styrene-Divinylbenzene) Support for Online Liquid Chromatography/Tandem Mass Spectrometry Analysis of Peptides and Proteins," *Anal. Chem.* **70,** 4879–4884 (1998).

160. Zhang, S.; Huang, X. "A Microchip Electrospray Device and Column with Affinity Adsorbents and Use of the Same," PCT Int. Appl. (2002), 59 pp., WO 0266135 A1 20020829.

161. Liu, H.; Felten, C.; Xue, Q.; Zhang, B.; Jedrzejewski, P.; Karger, B.L.; Foret, F. "Development of Multichannel Devices with an Array of Elcctrospray Tips for High-Throughput Mass Spectrometry," *Anal. Chem.* **72,** 3303–3310 (2000).

162. Tang, K.; Lin, Y.; Matson, D.W.; Kim, T.; Smith, R.D. "Generation of Multiple Electrosprays Using Microfabricated Emitter Arrays for Improved Mass Spectrometric Sensitivity," *Anal. Chem.* **73,** 1658–1663 (2001).

163. Shen, Z.; Thomas, J.J.; Averbuj, C.; Broo, K.M.; Engelhard, M.; Crowell, J.E.; Finn, M.G.; Siuzdak, G. "Porous Silicon as a Versatile Platform for Laser Desorption/Ionization Mass Spectrometry," *Anal. Chem.* **73,** 612–619 (2001).

164. Siuzdak, G.E.; Buriak, J.; Wei, J. "Improved Desorption/Ionization of Analytes from Porous Light-Absorbing Semiconductor," PCT Int. Appl. (2000), 76 pp., WO0054309 A1 20000914.

165. Wei, J.; Buriak, J.M.; Siuzdak, G. "Desorption-Ionization Mass Spectrometry on Porous Silicon," *Nature (London)* **399,** 243–246 (1999).

166. Lin, Y-S.; Chen, Y-C. "Laser Desorption/Ionization Time-of-Flight Mass Spectrometry on Sol-Gel-Derived 2,5-Dihydroxybenzoic Acid Film," *Anal. Chem.* **74,** 5793–5798 (2002).

167. Holmquist, M.; Palm, A.; Engström, J.; Selditz, U.; Andersson, P. "High-Speed Protein Digestion, Sample Preparation and MALDI-TOF MS Peptide Mapping on a Microfluidic Compact Disc (CD)," in *Proceedings of the 50th ASMS Conference on Mass Spectrometry and Allied Topics*, Orlando, Florida, June 2–6, 2002, WPP 238, 2066.pdf, and www.gyros.com.

168. Gustafsson, M.; Togan-Tekin, E.; Ekstrand, G.; Kånge, R.; Andersson P.; Wallenborg, S. "A CD Microlaboratory for Improved Protein Identification by MALDI MS," in *Proceedings of the 50th ASMS Conference on Mass Spectrometry and Allied Topics*, Orlando, Florida, June 2–6, 2002, ThPA 004, 1558.pdf and www.gyros.com.

169. Thomson, B.A. "Atmospheric Pressure Ionization and Liquid Chromatography/Mass Spectrometry-Together At Last," *J. Am. Soc. Mass Spectrom.* **9,** 187–193 (1998).

170. Doerge, D.R.; Bajic, S. "Analysis of Pesticides Using Liquid Chromatography/Atmospheric-Pressure Chemical Ionization Mass Spectrometry," *Rapid Commun. Mass Spectrom.* **6,** 663–666 (1992).

171. Robb, D.B.; Covey, T.R.; Bruins, A.P. "Atmospheric Pressure Photoionization: An Ionization Method for Liquid Chromatography-Mass Spectrometry," *Anal. Chem.* **72,** 3653–3659 (2000).

172. Berkenkamp, S.; Menzel, C.; Karas, M.; Hillenkamp, F. "Performance of Infrared Matrix-Assisted Laser Desorption/Ionization Mass Spectrometry with Lasers Emitting in the 3 μm Wavelength Range," *Rapid Commun. Mass Spectrom.* **11,** 1399–1406 (1997).

173. Cramer, R.; Burlingame, A.L. "IR-MALDI—Softer Ionization in MALDI-MS for Studies of Labile Macromolecules," *Mass Spectrometry in Biology and Medicines*, pp. 289–307 in Burlingame, A.L.; Carr, S.A.; Baldwin, M.A., eds., Humana Press, Totowa, NJ (2000).

174. Ettner, N.; Hillen, W.; Ellestad, G.A. "Enhanced Site-Specific Cleavage of the Tetracycline Repressor by Tetracycline Complexed with Iron," *J. Am. Chem. Soc.* **115,** 2546–2548 (1993).

175. Jahn, O.; Eckart, K.; Brauns, O.; Tezval, H.; Spiess, J. "The Binding Protein of Corticotropin-Releasing Factor: Ligand-Binding Site and Subunit Structure," *Proc. Nat. Acad. Sci. USA* **99,** 12055–12060 (2002).

176. Borchers, C.; Boer, R.; Klemm, K.; Figala, V.; Denzinger, T.; Ulrich, W-R.; Haas, S.; Ise, W.; Gekeler, V.; Przybylski, M. "Characterization of the Dexniguldipine Binding Site in the Multidrug Resistance-Related Transport Protein P-Glycoprotein by Photoaffinity Labeling and Mass Spectrometry," *Molecular Pharmacol.* **61,** 1366–1376 (2002).

177. Kramer, W.; Sauber, K.; Baringhaus, K-H.; Kurz, M.; Stengelin, S.; Lange, G.; Corsiero, D.; Girbig, F.; Konig, W.; Weyland, C. "Identification of the Bile Acid-Binding Site of the Ileal Lipid-Binding Protein by Photoaffinity Labeling, Matrix-Assisted Laser Desorption Ionization-Mass Spectrometry, and NMR Structure," *J. Biol. Chem.* **276,** 7291–7301 (2001).

178. Webb, Y.; Zhou, X.; Ngo, L.; Cornish, V.; Stahl, O.; Erdjument-Bromage, H.; Tempst, P.; Rifkind, R.A.; Marks, P.A.; Breslow, R.; Richon, V.M. "Photoaffinity Labeling and Mass Spectrometry Identify Ribosomal Protein S3 as a Potential Target for Hybrid Polar Cytodifferentiation Agents," *J. Biol. Chem.* **274,** 14280–14287 (1999).

179. Davidson, W.; McGibbon, G.A.; White, P.W.; Yoakim, C.; Hopkins, J.L.; Guse, I.; Hambly, D.M.; Frego, L.; Ogilvie, W.W.; Lavallee, P.; Archambault, J. "Characterization of the Binding Site for Inhibitors of the HPV11 E1-E2 Protein Interaction on the E2 Transactivation Domain by Photoaffinity Labeling and Mass Spectrometry," *Anal. Chem.* **76,** 2095–2102 (2004).

180. Shields, S.J.; Balhorn, R.L. "Protein-Ligand Complexes: Binding Sites and Protein Conformation," in *Proceedings of the 50th ASMS Conference on Mass Spectrometry and Allied Topics*, Orlando, Florida, June 2–6, 2002, TPG 172, 1589.pdf.

3

ESI-FTICR AS A HIGH THROUGHPUT DRUG DISCOVERY PLATFORM

KRISTIN A. SANNES-LOWERY, LENDELL L. CUMMINS, SHUO CHEN, JARED J. DRADER, AND STEVEN A. HOFSTADLER

INTRODUCTION

With the introduction of electrospray ionization (ESI) [1,2] and matrix-assisted laser-desorption/ionization (MALDI) [3,4] in the mid-1980s, the field of biological mass spectrometry (MS) experienced rapid growth. Both of these ionization methods provide an effective means of promoting intact biological molecules into the gas phase. The combination of electrospray ionization with Fourier transform ion cyclotron resonance (FTICR) mass spectrometers [5–7] provides a powerful platform on which to characterize biological molecules with unprecedented sensitivity, resolution, and mass accuracy [8–14]. The demonstration that labile noncovalent complexes could be analyzed by mass spectrometry showed that mass spectrometry–based analytical methods could be developed to measure binding affinity, stoichiometry, and specificity of macromolecular complexes. Mass spectrometry even plays an important role in the drug-discovery arena, as an integral component of the entire drug-discovery and development process from lead identification to quality control of bulk drug substance.

We have developed a mass spectrometry–based approach for discovery of small-molecule drugs that work by binding to structured regions of ribonucleic acids (RNA). This approach is called *Multitarget Affinity/Specificity Screening*

Integrated Strategies for Drug Discovery Using Mass Spectrometry, Edited by Mike S. Lee

(MASS) and employs electrospray ionization Fourier transform ion cyclotron resonance mass spectrometry (ESI-FTICR-MS) [15–17]. In a single assay, MASS is used to determine the chemical composition of ligands that bind to an RNA target, relative/absolute dissociation constants, and the specificity of binding to one RNA target relative to other RNA targets. Because the ESI-FTICR-MS experiment provides a "snap-shot" of the species present in the solution, solution dissociation constants can be measured from the observed ion abundances of the free and complexed RNA [18]. Ligand–RNA complexes with affinity ranging from 10 nM to at least 1 mM can be observed in the MASS assay [15–17,19].

Structured RNA is an attractive target for therapeutic intervention in a variety of diseases that include bacterial infections, viral infections, and cancer. RNA performs a variety of critical functions in the cell, such as peptide-bond formation, messenger RNA (mRNA) splicing, transfer RNA (tRNA) transport, and the regulation of both transcription and translation. Although RNA has only four building blocks, a plethora of structural diversity, arguably equivalent to that of protein targets, is generated. The three-dimensional (3D) structure of RNA is required for both molecular recognition and functionality [20]. To maintain the functional structures, little or no change can be tolerated at active sites in ribosomal RNAs (rRNA) and at protein-binding regions of mRNAs. Therefore, it is difficult for pathogens to develop resistance to drugs targeted at structured RNAs [21]. Recent advances in the determination of RNA structure and function make targeting unique RNA motifs with small molecules a tractable problem [22–24].

Nature has provided examples of small molecules that exhibit a therapeutic effect by binding to structured RNA. One of the most studied ligand–RNA systems is the binding of aminoglycoside antibiotics to the A-site domain of the Prokaryotic 16S rRNA [25–36]. The 16S rRNA sequence is highly conserved among prokaryotes and is part of the 30S subunit involved in translation. The therapeutic effect of aminoglycoside antibiotics is due to disruption of protein synthesis and RNA splicing. A 27- (*mer*) RNA construct comprising the A-site subdomain has been shown to mimic the behavior of the A-site domain when it is part of the entire 16S rRNA ($\sim$1500 nucleotides) [37]. The aminoglycoside paromomycin binds both the 27-*mer* construct and the 16S rRNA with similar affinities [30]. Additionally, nuclear magnetic resonance (NMR) structures of paromomycin with the 27-*mer* construct have been solved and are similar to the crystal structures of the 30S subunit with paromomycin [25,32]. Since small subdomains can mimic the functional domain of rRNA, it is possible to design RNA constructs that are amenable to analysis by high-performance mass spectrometry [37].

In this chapter, we describe two operating modes of MASS. First, MASS is used as a high throughput affinity screen in which combinatorial libraries (or natural product extracts) are assayed for compounds that bind the RNA target. Second, once lead compounds are identified, MASS is run in a mode that rapidly determines dissociation constants for target–ligand interactions to guide medicinal structure–activity relationship (SAR) efforts to optimize lead ligands. We also show examples from screening natural-product broths and extracts.

MASS AS A HIGH THROUGHPUT SCREEN

The MASS screen uses high-performance ESI-FTICR MS to investigate ligand binding to structured RNA targets. Because of the high mass accuracy of FTICR MS, the exact mass of a small molecule can be used as an "intrinsic mass label" for identification of molecules that bind a target. The intrinsic mass label of the small molecule allows complicated mixtures of ligands to be screened simultaneously. Compounds that bind the RNA target are directly identified without deconvolution of the mixture or rescreening the ligands as individual compounds. Unlike traditional biological assays, neither the target nor the ligands require radiolabeling or fluorescent tagging to be screened. Furthermore, the false-positives that are observed in traditional biological assays due to an aggregate effect of a mixture is avoided in the MASS screen, since each ligand has the opportunity to interact with the RNA target (see below).

Low-affinity ligands (dissociation constants greater than 100 μM), which would be missed with traditional biological assays, will be identified with the MASS assay. Griffey et al. [38] demonstrated that 2-deoxystreptamine (2-DOS) binds to the 27-*mer* 16S A-site construct (**16S**) at multiple locations with dissociation constants (K_d) ranging from 0.6 mM to 15 mM. Additionally, this work demonstrates that the MASS assay can be used to examine concurrent and competitive binding of low-affinity compounds [38]. An example of concurrent binding was shown by the simultaneous binding of 2-DOS and 3,5-diaminotriazole (3,5-DT) to **16S**. From this result, it can be inferred that 2-DOS and 3,5-DT must bind the **16S** in different locations. Alternatively, an example of competitive binding was demonstrated by the absence of 2,4-diaminopyrimidine (DAP) binding to **16S** in the presence of 2-DOS. Since DAP binds to **16S** in the absence of 2-DOS, the competitive binding indicates that 2-DOS and DAP bind to **16S** in the same or overlapping locations.

To multiplex the MASS approach and further increase the throughput, it must be possible to screen multiple targets against multiple ligands in a single well. A multiplexed approach requires that the molecular interaction between any given target–ligand pair is independent of the presence (or absence) of other ligands and targets in solution. In previous work [17], we demonstrated that lividomycin will specifically bind the 27-*mer* 16S A-site construct in the presence of two other RNA targets and 25 compounds, even when the concentration of lividomycin (3 μM) is significantly lower than the total concentration of the other ligands (1.25 mM total). In addition, the original mixture of 25 compounds did not specifically bind the 27-*mer* 16S A-site construct, and the presence of a specific binder (lividomycin) did not cause any of the 25 compounds to bind in a nonspecific manner. Therefore, the MASS technique can characterize the interactions of complex target–ligand mixtures without generating false-positives.

High throughput screening requires that the data acquisition is highly automated [17]. The mass spectrometer used in this work employs an actively shielded 9.4-tesla superconducting magnet. The active shielding constrains the majority of the fringing magnetic field to a relatively small volume and allows robotic components (such as autosamplers) that might be adversely affected by stray magnetic fields to operate in close proximity to the spectrometer. Sample aliquots, typically 12 μL, are extracted directly from 96-well microtiter plates and are injected directly into a sample loop integrated with a custom fluidics handling system. The fluidics module transports sample aliquots to the electrospray source at a high flow rate, and then switches to a lower flow rate for electrospray for improved ESI sensitivity. A room-temperature countercurrent flow of a 50/50 mixture of N_2 and $O_2$is employed to assist in the desolvation process and to stabilize the electrospray plume. Typically, one well is screened every 39 seconds ($\sim$1 plate/hour); 33 seconds of data acquisition (20 co-added scans), and 6 seconds of overhead associated with the autosampler. Currently, three targets at 2.5 μM each are screened against 11 ligands at 25 μM each. The high RNA concentration ensures that there is enough RNA available for all potential binders to interact with the RNA, and the high ligand concentration ensures that even ligands with K_d values of 1mM will be detected. Thus, in a 24-hour period, 22.5K compounds are screened and 67K analyses are performed.

In the high throughput mode, the goal is to identify ligands with K_d values less than 100 μM and with some specificity relative to the other targets. This condition ensures that the ligands bind to a unique structural feature of the target and are not just generic RNA binders. Although the ligands are screened only at a single concentration in the high throughput mode, it is possible to estimate a single-point K_d from the mass spectrometry data. The single-point estimated K_d values can be used to classify compounds as weak, medium, and strong binders, but cannot be used to accurately rank order compounds within the same classification. As illustrated in Figure 3.1A, a ligand binds Target 2 with a single-point estimated K_d of 37 μM. It binds Target 2 with a 3.4 greater specificity than Target 1, and with 1.5 greater specificity than Target 3. This ligand would be a candidate for further SAR by medicinal chemistry to improve both its binding affinity and specificity. An example of a generic

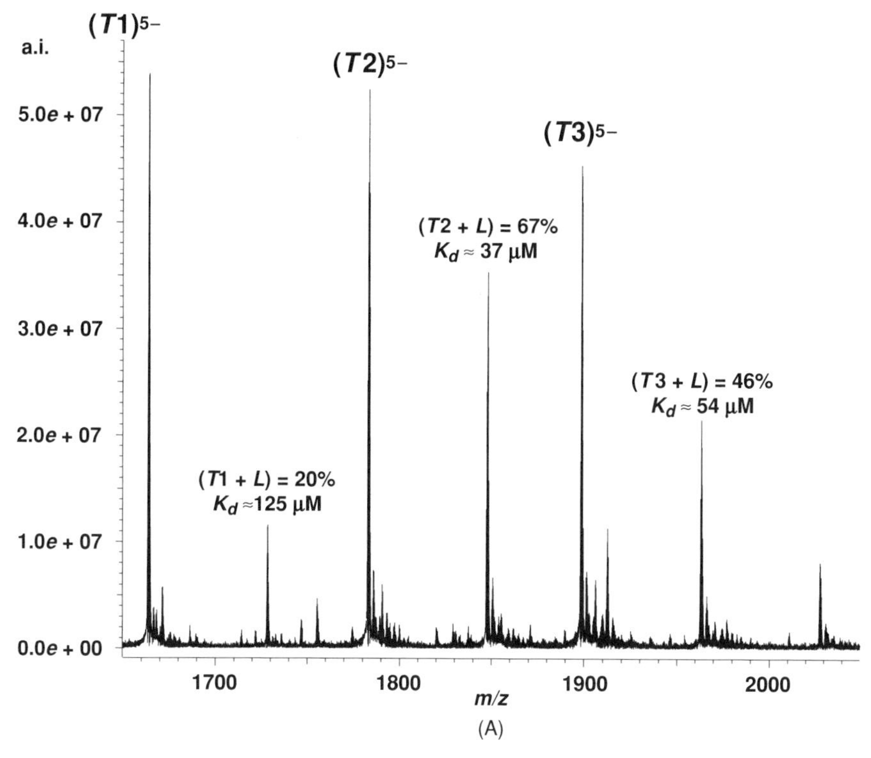

FIGURE 3.1 ESI-FTICR mass spectra of three RNA targets at 2.5 μM, each screened against 11 compounds at 25 μM each. The percent complexes and single-point K_d values are shown for each ligand complex. (A) shows an example of a ligand that specifically binds Target 2. (B) Shows an example of a ligand that nonspecifically binds to all the targets. *Abbreviations.* a.i. = arbitrary intensity; T = target; L = ligand.

RNA binder is shown in Figure 3.1B. The ligand binds Target 1 and Target 2 equally well. In addition, complexes formed by binding two ligands to the target are observed for both Target 1 and Target 2. This result indicates that there are multiple weak binding sites on the target with similar affinities for the ligand. It would likely be difficult to improve the affinity and specificity of this ligand, and therefore, it would probably not be pursued further. Thus, MASS can be used to rapidly identify promising compounds or structural motifs from large chemical libraries.

MASS FOR AUTOMATED K_d DETERMINATION

It has been demonstrated that mass spectrometry can be used to characterize binding properties and stoichiometry for protein–protein interactions, protein–ligand interactions, protein–oligonucleotide interactions,

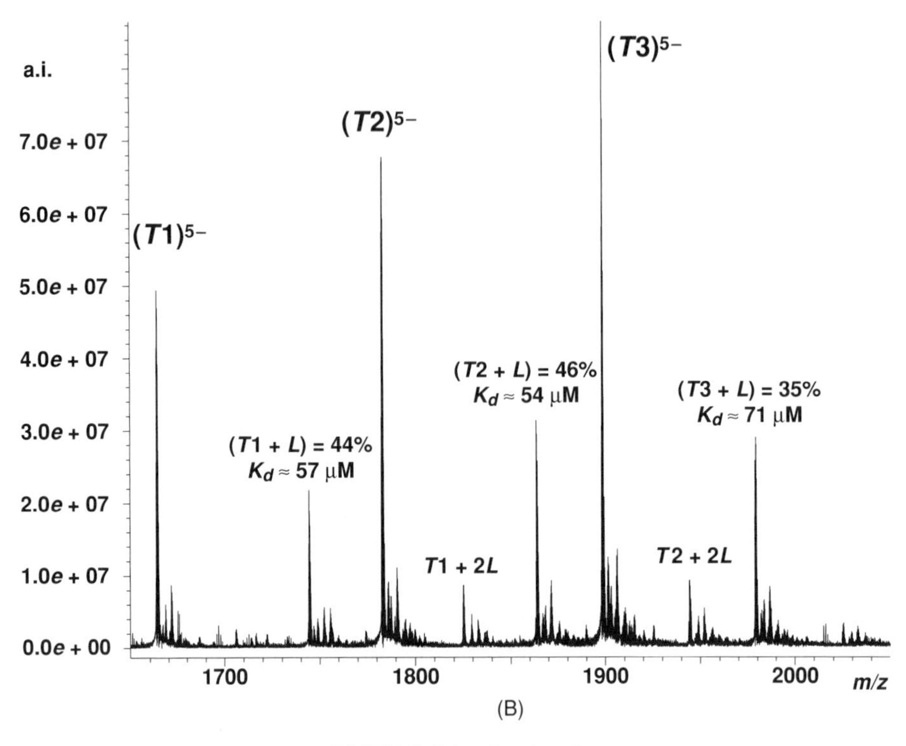

FIGURE 3.1 *Continued*

DNA–ligand interactions, and RNA–ligand interactions [39–53]. Typically, the mass spectrometric results are in good agreement with those derived from more conventional solution-phase techniques. Several groups have shown that dissociation constants can be derived directly from ESI-MS measurements [39–41,46,49,50]. Traditional solution phase methods, such as radioimmunoassays, filter assays, and surface plasmon resonance assays, provide little to no information about the binding stoichiometry and can only measure the equilibrium concentration of one component [54]. In contrast, mass spectrometry can directly determine the binding stoichiometry by measuring the masses of the complexes formed, as well as detect all the components of the equilibrium mixture (i.e., free ligand, free target, and ligand–target complexes). In addition, with appropriate solution conditions, it is possible to simultaneously measure the K_d values for several ligands against a single target [15], and it is possible to simultaneously measure the K_d values for a single ligand to multiple targets (unpublished results). Furthermore, it is possible to easily explore the effect of solution conditions on the measured K_d values using mass spectrometry. This is important because the K_d for ligands with a specific binding site will not be affected by changes in salt concentration, while the

K_d for ligands with nonspecific binding sites can be significantly influenced by the salt concentration [18].

In previous work, we used ESI-MS to measure the K_d values for the aminoglycosides tobramycin and paromomycin binding to a 27-*mer* construct of the 16S A-site [18]. Equations for the determination of a K_d for a ligand with a single binding site (Eq. 3.1) and with two binding sites (Eq. 3.2) were derived since the mass spectrometry data provide information about the binding stoichiometry of the ligands:

$$[RL]/[R] = (1/K_d)([L_i] - [RL]) \tag{3.1}$$

$$([RL] + [RL_2])/[R] = [L]^2/(K_{d1}K_{d2}) + [L]/K_d \tag{3.2}$$

In Eqs. (3.1) and (3.2), $[R]$ is the free RNA target concentration, $[L]$ is the free ligand concentration, $[L_i]$ is the initial ligand concentration, $[RL]$ is the 1:1 ligand–RNA target complex concentration, and $[RL_2]$ is the 2:1 ligand–RNA target complex concentration. In each equation, the mass spectrometrically determined abundances of the appropriate species are used, except for L where L_i is used. It was demonstrated that holding the RNA concentration fixed at or below the expected K_d value and titrating the ligand is the preferred method for determining dissociation constants. To aid our drug-discovery efforts, an automated method for the rapid determination of K_d values for ligands was developed.

Once promising ligands have been identified in the high throughput mode of MASS, medicinal chemistry and organic synthesis are employed for SAR studies of these compounds. These second-generation collections of compounds may contain numerous active compounds with minor differences in affinity. Because the modifications can cause subtle differences in binding affinities and specificities, a multipoint K_d determination is necessary. An accurate K_d measurement requires that the RNA concentration used is below the expected K_d, which, for some classes of compounds, is in the low nM regime. Since K_d determinations are generally done one ligand at a time, well-to-well carryover of both ligand and RNA must be minimized so that the K_d determination of one ligand does not interfere with that of another ligand.

If the estimated K_d values are in the micromolar range, then the RNA is held at 500-nM concentration and the ligand is screened at 750 nM, 2.5 μM, 7.5 μM, and 25 μM. At these concentrations, it is possible to maintain the 39-second/well screening rate used in the high throughput mode. On the other hand, if the estimated K_d values are in the mid-nanomolar range, the RNA concentration is held at 100 nM and the ligand is screened at 250 nM, 750 nM, 2.5 μM, and 7.5 μM. At this RNA concentration, the number of scans collected to maintain the necessary signal-to-noise (S/N) level is increased from 20 scans to 64 scans, and the screening time per well goes from 39 seconds to 95 seconds. For all K_d determinations, one ligand is screened per row of 96-well microtiter-plates, starting with the lowest concentration. Between each ligand concentration, two wells that contain only RNA are run to scavenge any residual ligand in the transfer lines and to minimize ligand carryover within the row. It was found that ligand carryover within the same row interfered with the K_d determination of micromolar binders, but not submicromolar binders. Furthermore, after the highest concentration is screened, two additional rinse steps are done to help minimize the carryover of the ligand to the next row, which would interfere with the next K_d determination.

The K_d values are determined by plotting the fraction of RNA bound by the ligand (fraction bound) versus the total ligand concentration. The fraction bound is determined by dividing the abundances of the ligand–RNA complexes (1:1 and 2:1 ligand–RNA complexes) by the total abundance of the RNA. The fraction bound is related to the K_d with the following equation:

$$Y = [L]_f/(K_d + [L]_f) \tag{3.3}$$

where Y is the fraction bound and $[L]_f$ is the free ligand concentration. The free ligand concentration is calculated from:

$$[L]_f = [L]_i - Y[R]_i \tag{3.4}$$

where $[L]_i$ is the initial ligand concentration and $[R]_i$ is the initial RNA concentration. At a ligand concentration of zero, the fraction bound is zero by definition. Thus, this can be used as an additional data point when the calculations are performed. Nonlinear regression analysis is used to calculate the K_d values, and curve fits with $R^2 > 0.99$ are obtained. The preceding equations only calculate the K_d for the first binding site. This method for K_d determination is reproducible and robust. It provides accurate K_d values to rank ordering ligands and shows the subtle effects of modifications.

The MASS assay was used to determine the K_d values for paromomycin, kanamycin A, and neamine. While the aminoglycoside antibiotics were screened at 250 nM, 750 nM, 2.5 μM, and 7.5 μM, **16S** was held at 100 nM. Figure 3.2 shows a plot of the fraction of **16S** bound by the ligand (fraction bound) versus the total ligand concentration for the three aminogylcosides. The K_d values determined for paromomycin (83.9 nM), kanamycin A (7.8 μM), and neamine (13.1 μM) are consistent with solution phase determinations [26,30] and our previous ESI-MS measurements [18]. Each K_d determination took 380 seconds of data acquisition. Thus, the MASS assay provides a robust way to determine K_d values for 100 compounds a day.

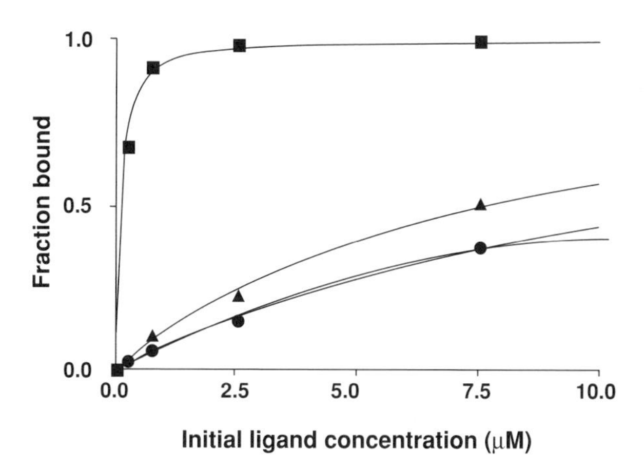

FIGURE 3.2 K_d determination for paromomycin, kanamycin A, and neamine. 16S was held at 100 nM concentration, while the aminoglycosides were screened at 250 nM, 750 nM, 2.5 μM, and 7.5 μM. The fraction bound is plotted versus the initial ligand concentration. Nonlinear regression was used to calculate the K_d and the curves were fit with $R^2 > 0.99$.

MASS FOR SCREENING NATURAL PRODUCTS

For the past twenty years, natural products with their depth and diversity of compounds have been the source of many of the top selling prescription drugs. However, finding the biologically active components from natural product broths and extracts presents many challenges when using traditional biological assays. The issues include detection of active compounds present at low concentrations in a background of other active species and "false" positives that result from the summed activity of many weakly active compounds. In addition, the determination of the active component is often labor-intensive and requires large amounts of the fractionated broth or extract.

The MASS assay is ideally suited for screening the complex mixtures derived from natural products for individual components that bind specifically to biological targets. Because of the small amount of material required for the MASS assay, it is possible to screen the fractionated broths or extracts at multiple concentrations against multiple targets. This protocol ensures that even low-abundance species that bind will be detected in the presence of high-abundance species that bind. In addition, knowledge of the mass of the active component greatly aids the identification process.

We recently extended the MASS assay to the screening of natural-product broths [55]. As a proof of principal concept, fractionated broths from *Streptomyces rimus* sp. *paromomycinus*, which are known to produce the aminoglycoside paromomycin, were screened against a 27-*mer* 16S A-site RNA construct (**16S**) and a 28-*mer* control RNA construct (**16Sc**) (Figure 3.3). The

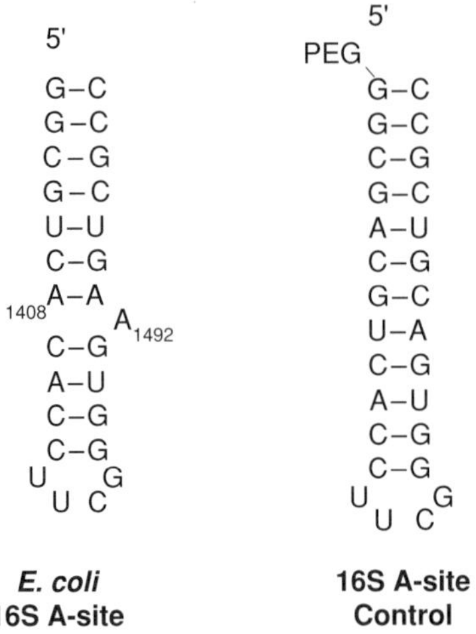

5'
G–C
G–C
C–G
G–C
U–U
C–G
1408 A–A A 1492
C–G
A–U
C–G
C–G
U G
U C

E. coli 16S A-site

5'
PEG
G–C
G–C
C–G
G–C
A–U
C–G
G–C
U–A
C–G
A–U
C–G
C–G
U G
U C

16S A-site Control

FIGURE 3.3 Sequence and secondary structures of the 27-*mer* 16S A-site RNA construct (**16S**) and the 28-*mer* control RNA construct (**16Sc**) in which the internal bulge of **16S** has been replaced with a duplex region. Base numbering is in reference to full length *Escherichia coli* 16S rRNA.

16S (MW_{min} = 8635.1790) contains the essential components of the 16S rRNA A-site, especially the internal A-bulge and the noncanonical U–U base pair. The **16Sc** (MW_{min} = 9300.3388) is a duplexed RNA with the same tetra loop as the **16S.** In addition, **16Sc** contains an 18-atom hexaethylene glycol chain attached to the 5′ terminus of the oligonucleotide that shifts the mass of the construct and minimizes overlaps between ligand complexes with **16S** and **16Sc** [16]. By screening against these two targets simultaneously, ligands that bind specifically to the internal bulge of **16S** can be identified. If a ligand binds equally well to both constructs, then the ligand probably binds to a duplex region or loop region that the two constructs have in common. Thus, the ratio of the abundance of **16Sc** to **16S** can be used to monitor the specificity of ligand binding. If the ligand binds the **16S** specifically, then the 16Sc/16S ratio will increase, since the abundance of **16S** will decrease relative to that of **16Sc**. Conversely, if a ligand binds the two targets equally well, then the 16Sc/16S ratio will not change, since the abundance of both **16S** and **16Sc** will decrease. Similarly, in the absence of ligand binding to either target, the 16Sc/16S ratio will not change.

A bacterial fermentation broth from *S. rimus* sp. *paromomycinus* was fractionated with high-performance liquid chromatography (HPLC), and the individual fractions were screened against **16S** and **16Sc**. Two μL aliquots of each LC fraction was combined with a 15-μL solution containing 2.5 μM each **16S** and **16Sc** in 100 mM ammonium acetate and 33% isopropyl alcohol. This mixture was vortexed and incubated for 60 minutes at room temperature prior

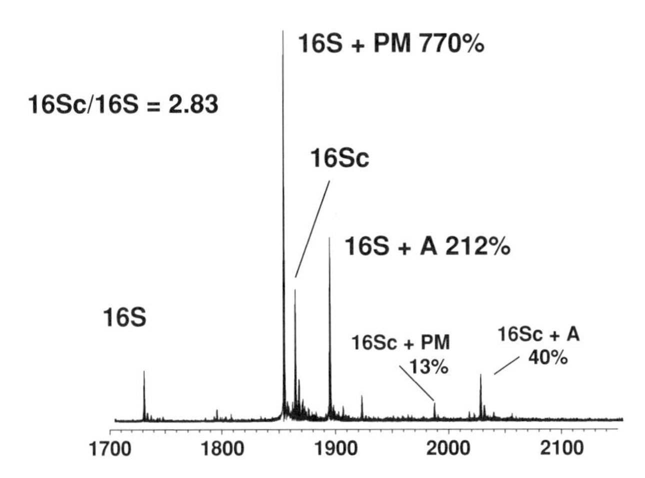

FIGURE 3.4 ESI-FTICR mass spectrum that shows the binding species in fraction 146 (see text). Two species were observed to bind to the 27-*mer* 16S A-site RNA construct (**16S**): (1) paromomycin (**PM**), and (2) **Ligand A** (**A**) that has a mass of 818 Da. The percentages shown are relative to the **16S** or **16Sc** targets, respectively.

to MASS analysis. MASS analysis of fraction 146 (~54.7 minutes) is shown in Figure 3.4. The peak at mass-to-charge ratio (m/z) 1849.90 corresponds to a ligand of mass ~615 Da binding to **16S** and is putatively assigned as the 16S-paromomycin noncovalent complex. In addition, a low-abundance peak that corresponds to paromomycin–16Sc complex is observed at m/z 1982. The peak at m/z 1890.52 represents either a ligand of mass of ~818 Da complexed with **16S** or a ligand of mass ~153 Da complexed with **16Sc**. Further examination of the mass spectrum shows a peak at m/z 2023 that corresponds to a ligand of mass ~818 Da that binds to **16Sc**. Because RNA binding ligands will bind nonspecifically to all RNAs to some degree, the preceding identification of m/z 2023 helps identify the peak at m/z 1890.52 as a ligand of mass ~818 binding to **16S**. The new species with a mass of ~818 will be referred to as **Ligand A** in further discussions. Based on the abundances of the paromomycin–16S complex and the paromomycin–16Sc complex (Figure 3.4), paromomycin binds approximately 59-fold more specifically to the **16S** than it does to **16Sc**. A similar comparison of the **Ligand A** indicates that it binds with approximately a 5-fold specificity to the **16S** over the **16Sc**.

Fraction 146 was further characterized by positive-mode electrospray MS using a quadrupole ion trap mass spectrometer. The components observed in this fraction include the $(M + H^+)$ species of paromomycin (m/z 616), the $(M + H^+)$ species of **Ligand A** (m/z 819), and several other species that do not bind to **16S** or **16Sc** (data not shown). Paromomycin and **Ligand A** were the most abundant peaks detected, and with the assumption of comparable ionization efficiencies, these compounds are likely present at similar concentrations.

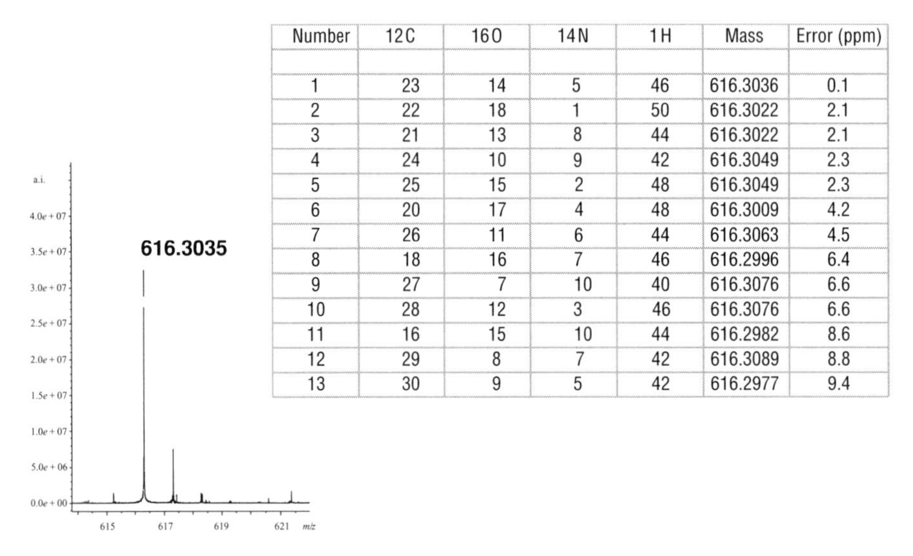

Number	12C	16O	14N	1H	Mass	Error (ppm)
1	23	14	5	46	616.3036	0.1
2	22	18	1	50	616.3022	2.1
3	21	13	8	44	616.3022	2.1
4	24	10	9	42	616.3049	2.3
5	25	15	2	48	616.3049	2.3
6	20	17	4	48	616.3009	4.2
7	26	11	6	44	616.3063	4.5
8	18	16	7	46	616.2996	6.4
9	27	7	10	40	616.3076	6.6
10	28	12	3	46	616.3076	6.6
11	16	15	10	44	616.2982	8.6
12	29	8	7	42	616.3089	8.8
13	30	9	5	42	616.2977	9.4

FIGURE 3.5 High-resolution mass spectrum of the protonated species of paromomycin from fraction 146. The measured mass is 616.3035. The inset is a table of potential elemental compositions with the ppm error between the measured mass and the theoretical mass.

The ~615-Da species putatively identified in fraction 146 as paromomycin was confirmed by accurate mass measurement. The mass spectrometer was operated in the positive-ion mode and angiotensin and bradykinin peptides were used as internal mass standards. The mass accuracy attained using these internal standards was ≤ 1 ppm. The measured mass of the protonated species was 616.3035 with sub-ppm mass measurement error. The elemental composition that is consistent with the observed mass (assuming only C, H, N, and O atoms) is C23 H46 O14 N5, which is the elemental composition of paromomycin (Figure 3.5). The other elemental compositions listed in Figure 3.5 can be eliminated, since the ppm error between these elemental compositions and the observed mass is greater than the experimental mass measurement error. In addition, the MS/MS spectrum of the ~615-Da species was compared with that of commercially available paromomycin. Identical product ions were observed in the MS/MS spectrum for the ~615-Da species and the commercially available paromomycin. Furthermore, the fragmentation patterns observed are consistent with other studies of paromomycin fragmentation pathways [56–58].

To obtain structural information, MS^n was performed on **Ligand A**. MS/MS of **Ligand A** shows a product ion at *m/z* 616 as well as up to three losses of water (*m/z* 801, 783, and 765) (Figure 3.6). Further fragmentation of the ion at *m/z* 616 gave product ions consistent with those found for paromomycin. These results indicate that **Ligand A** is a modified paromomycin derivative. Additional fragmentation experiments were carried out to identify the location of the modification. MS^3 fragmentation of the product ion at *m/z* 801 gave

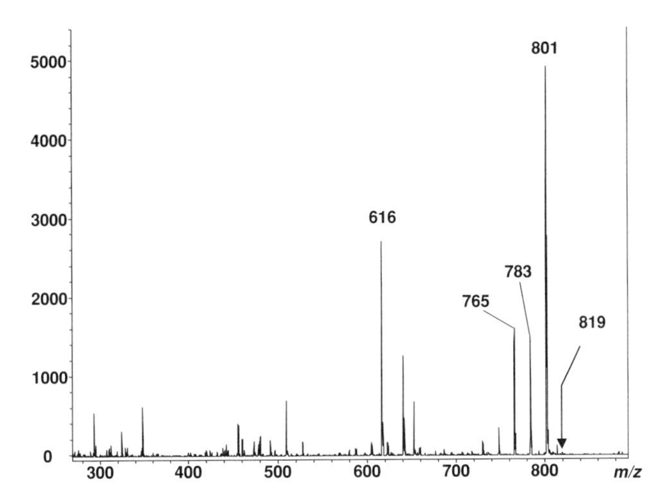

FIGURE 3.6 MS^2 spectrum of **Ligand A** (*m/z* 819) present in fraction 146. The primary fragments are loss of water to produce *m/z* 801 and loss of 203 to produce *m/z* 616, the paromomycin core.

ions consistent with losses of the A ring (*m/z* 640), the D ring (*m/z* 641), and the CD rings (*m/z* 509) of paromomycin (Figure 3.7A). MS^4 fragmentation of the product ion at *m/z* 509 gave ions consistent with the loss of the A ring of paromomycin (*m/z* 348). MS^5 fragmentation of the product ion at *m/z* 348 gave an ion consistent with the B ring of paromomycin (*m/z* 163) (Figure 3.7B). All of the fragmentation patterns are consistent with the modification located on the B ring of paromomycin. All of these MS^n data were collected from the initial fractionation of the natural product broth and required very little material. The initial assignment of the modification being on the B ring will greatly aid further efforts to determine the exact nature of the modification.

In addition to the structural information obtained from the MS^n data, an accurate mass measurement of **Ligand A** was performed to determine the elemental composition of the modification. The mass of this species was measured to be 819.3828 with a sub-ppm mass-measurement error. The elemental composition that is consistent with the observed mass (assuming only C, H, N, and O atoms) is C31 H59 O19 N6 with a mass difference of 0.2 ppm (Figure 3.8). The other elemental compositions listed in Figure 3.8 can be eliminated since the ppm error between these elemental compositions and the observed mass is greater than the experimental mass-measurement error. Since **Ligand A** is consistent with a modified paromomycin, subtraction of the elemental composition of paromomycin (C23 H46 O14 N5) from the determined elemental composition of **Ligand A**, will give the elemental composition of the modification. Thus, the elemental composition of the modification is C8 H13 O5 N1. Scale-up and isolation of **Ligand A** for NMR studies is required to determine the exact structure of the modification.

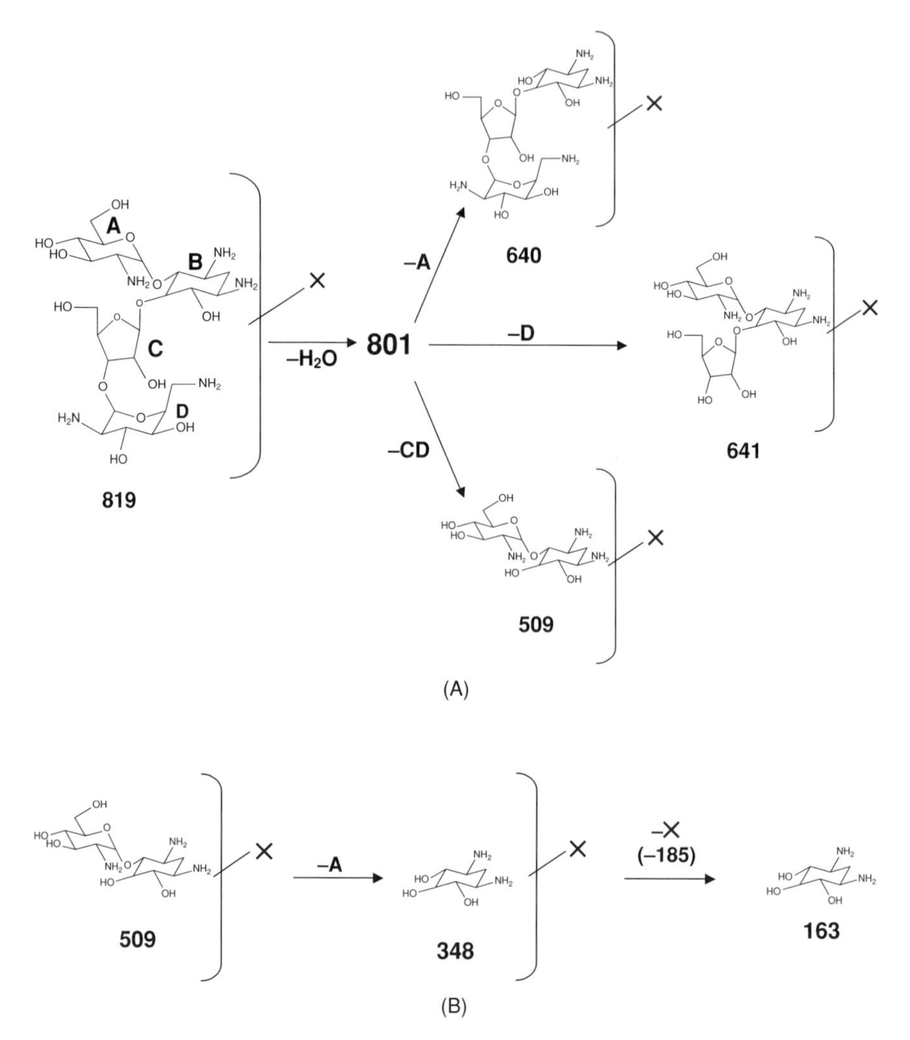

FIGURE 3.7 Fragmentation pathways of **Ligand A**, which is a modified paromomycin. The X represents the modification to paromomycin. (A) shows the MS^2 and MS^3 fragmentation pathways and (B) shows the MS^4 and MS^5 fragmentation pathways of **Ligand A**.

MASS AS A SCREEN FOR ENZYMATIC ACTIVITY

In addition to finding small molecules that bind RNA, MASS can be used to find compounds that exhibit functional activity similar to that of cleavage enzymes. If the functional activity is specific for a structured RNA target, then the compound may exhibit a therapeutic effect by degrading the RNA, and thus, preventing the translation of protein necessary for cell life. Unlike the previous examples of the MASS assay where small molecules are identified

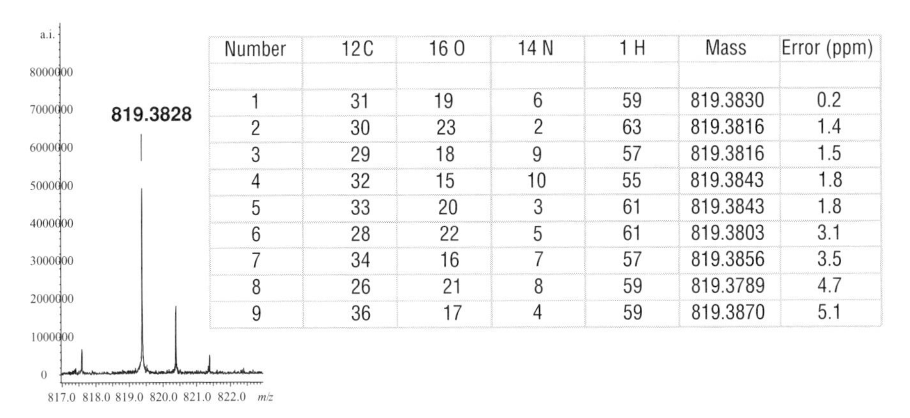

Number	12C	16 O	14 N	1 H	Mass	Error (ppm)
1	31	19	6	59	819.3830	0.2
2	30	23	2	63	819.3816	1.4
3	29	18	9	57	819.3816	1.5
4	32	15	10	55	819.3843	1.8
5	33	20	3	61	819.3843	1.8
6	28	22	5	61	819.3803	3.1
7	34	16	7	57	819.3856	3.5
8	26	21	8	59	819.3789	4.7
9	36	17	4	59	819.3870	5.1

FIGURE 3.8 High-resolution mass spectrum of the protonated species of **Ligand A** from fraction 146. The measured mass is 819.3828. The inset is a table of potential elemental compositions with the ppm error between the measured mass and the theoretical mass.

from the complexes that are formed with the specific RNA target, selective degradation of one RNA target relative to another is evaluated by measuring the abundance of the intact RNA target and concomitantly monitoring the ratio of the abundances of the two targets. For example, an increase in the ratio of 16Sc/16S can also indicate a well where the **16S** has been preferentially degraded relative to the **16Sc**. In this case, the **16Sc** may be more difficult to degrade, since it has a polyethylene glycol (PEG) linker on the 5′ end and a fully Watson–Crick-based pair stem structure. Examples of selective degradation of **16S** were observed in fractionated broths from *S. rimus* sp. *paromomycinus* [55]. Based on the elution profile of the fractions that showed selective degradation, it does not appear that the activity was due to proteinaceous ribonucleases (RNases). Additional examples have been observed from fractionated plant extracts that were treated with proteases as well as from plant extracts from solvents that are not expected to solublize proteins.

For example, Figure 3.9 shows a well that exhibited selective degradation of the **16S** construct. This well contained compounds extracted from a plant with hexane, in which proteins are not expected to be soluble. The natural-product extract was screened at 5×, 10×, 50× dilutions from the stock concentration (25 mg/mL) and selective degradation of **16S** is observed at all concentrations. Since the abundance of **16Sc** only decreases significantly at the highest concentration screened (5× dilution), it indicates that the decrease in the abundance of **16S** is due to cleavage and not overall signal degradation due to a contaminant in the sample. The 50 × dilution gives three cleavage products consistent with **16S** fragments. Based on the accurate mass of the cleavage products, base compositions were derived for the three cleavage products [59]. Using the calculated base compositions, it was determined that a cleavage

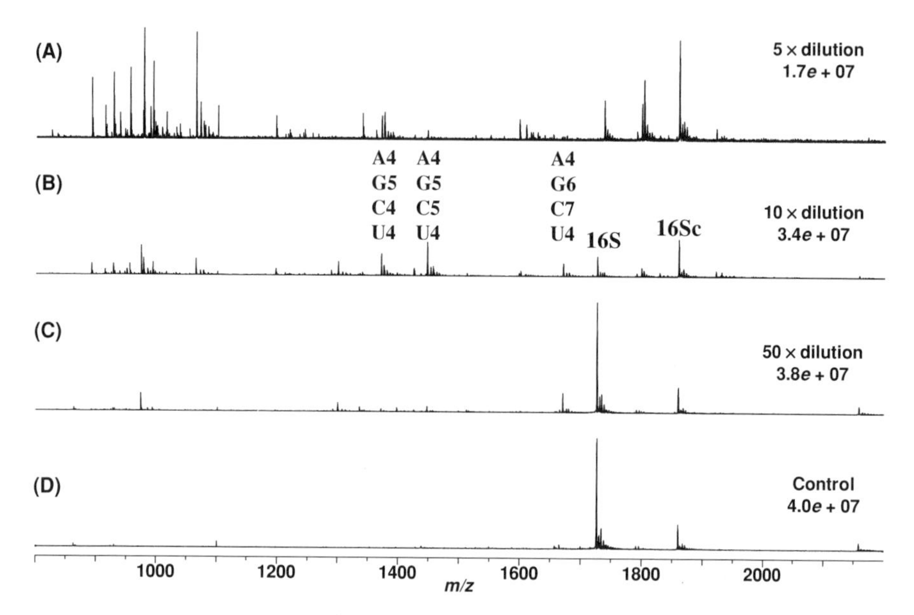

FIGURE 3.9 ESI-FTICR mass spectra of a plant extract that exhibits selective degradation of the 27-*mer* 16S A-site RNA construct (**16S**). The extract was screened at (A) 5×, (B) 10×, and (C) 50× dilutions from the original stock concentration. (D) shows a well in which no extract has been added. The abundances of the 28-*mer* control RNA construct (**16Sc**) are indicated at the right side of the figure. Until the 5× dilution, the abundance of **16Sc** in wells that contain the extract is similar to the well without extract.

occurred at the 3′ side of the 1407C, as indicated by the product with mass 1951, which corresponds to the 5′-GGCGUC-cp and the product with mass 6683 which corresponds to HO-ACACCUUCGGGUGAAGUCGCC-3′ (Figure 3.10). After this initial cleavage, it appears that the 6683 product is further cleaved at position 1496C to give HO-ACACCUUCGGGUGAAGUC-cp. It is also interesting to note that at the highest concentration of the extract, cleavage products that are consistent with cleavage of a C from **16Sc** are observed in addition to the cleavage of **16S**. Further studies are currently underway to identify the agent responsible for the cleavage activity.

AUTOMATED DATA INTERPRETATION AND REPORTING

The ability to screen 22.5K compounds in a 24-hour period necessitates the development of automated data analysis. The high mass-measurement accuracy, combined with a routinely achieved (data-point-limited) mass resolution in excess of 60,000 for species at *m/z* 1725, makes automated data interpretation a tractable undertaking. Consequently, we have developed custom software in

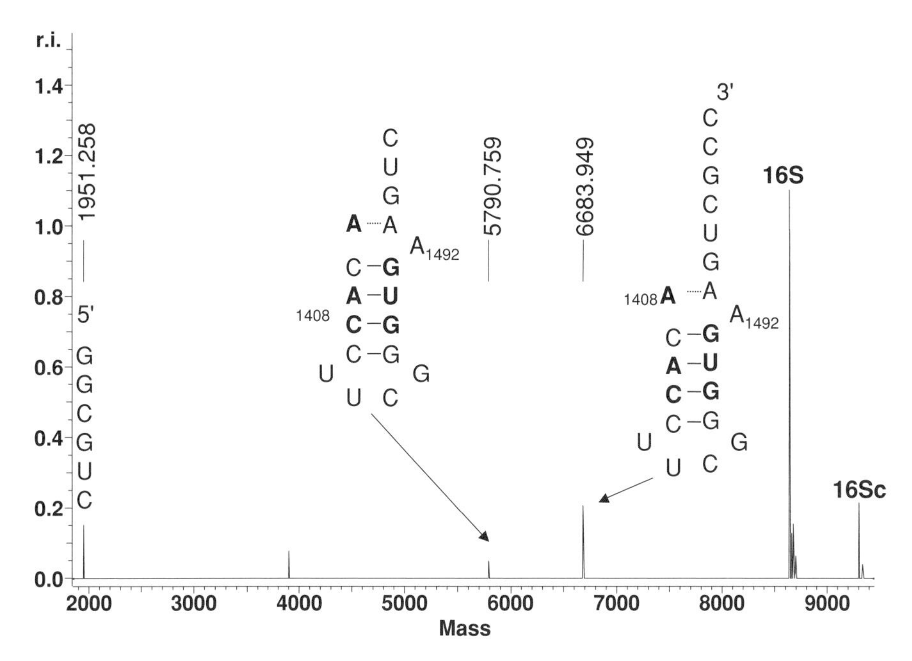

FIGURE 3.10 Deconvoluted spectrum of the 50× dilution of the extract. Based on base composition calculations, the corresponding sequence for the three cleavage products are shown. r.i. = relative intensity.

which raw data files are processed, postcalibrated, interpreted, and tabulated. The results are then exported to a central corporate database and integrated with other high throughput screening (HTS) assay results.

Processing

Once a raw data set (a transient) is stored, it is subjected to a number of standard mathematical manipulations to convert the raw time-domain data into the frequency domain. Following the application of an (optional) apodization function, the transient is zero filled once and a magnitude mode Fourier transform is performed [60,61]. Because the free RNA targets are typically present in high abundance relative to the complexes, and because the exact molecular weights of the RNA targets are, they are used as internal mass standards with which to internally calibrate each mass spectrum. The most abundant charge states of each target are used as reference peaks. The data are typically calibrated to <1 ppm in the region of interest. A peak-picking routine is applied to detect peaks in a given *m/z* range above a pre-specified threshold and store them in a file for subsequent interpretation. In addition, the abundances of the targets are calculated and stored in a table to calculate the target ratios, as described in the section titled “MASS for Screening Natural Products.”

Interpretation

The data interpretation involves two questions: (1) Is a complex observed between compound #1 and target A, and (2) what is the ratio of unbound target A to the target A–compound #1 complex? We ask the same question of each potential target–compound complex. Because the elemental composition of both the molecular targets and the potential ligands are known, and the charge state (z) of the RNA is generally unaffected by ligand binding, an exact m/z value can be calculated for each possible complex. In addition to complexes with 1:1 binding stoichiometry, the m/z values of complexes that arise from the binding of multiple ligands (either a single ligand binding multiple times to a given target or multiple ligands binding to a single target) can be calculated. A simple algorithm goes through each spectrum and looks for the presence of the most abundant isotope peak for each putative complex. If a peak is detected that corresponds to a given complex (within a given mass-measurement uncertainty), then the algorithm looks for isotope peaks on either side of the most abundant isotope peak to both confirm the charge state of the complex and to avoid false-positives due to the occasional spurious noise peak. In addition, a chi-square (χ^2) fit is calculated by comparing the heights of the theoretical isotope pattern to the heights of the experimental isotope pattern. If the χ^2 fit is <0.1, then the peaks are consistent with the putative complex. If the χ^2 is >0.1, then the peaks are not consistent with the putative complex, but are mostly related to another ligand in the well that is close in mass. Once a series of peaks is identified that are consistent with a putative complex, a percent complex is obtained by calculating the ratio of the integrated peak areas of the free and complexed target and multiplying the ratio by 100. A single-point estimated K_d is then calculated by dividing 100 by the percent complex of the ligand and multiplying by the screening concentration. The single-point estimated K_d can be used to rank-order ligands in a given mixture and clearly separates weak binders from strong ones.

In the case of natural products (for which a convenient list of elemental compositions of the components is not available), or for the rare case when an unintentional synthesis product binds to a target, a slightly different protocol is employed. Each unidentified complex consists of an unidentified ligand binding to one of the targets. Since the ligand masses in the mixture are not known *a priori*, an accurate mass measurement alone may not unambiguously reveal to which target the ligand is binding. For example, a complex with a mass greater than that of all targets could be composed of any of the targets binding to a ligand. Thus, if three targets are present, there are three potential target–ligand combinations, and hence three potential ligand masses, that could be consistent with a single observed complex. In most cases, the free ligands will ionize, and examination of the low m/z regime of the mass spectrum will show which of the three putative ligand masses is present in the starting material. Additionally, it is often the case that a ligand with high RNA affinity and low specificity will bind to multiple targets; assignment of a ligand mass is straightforward as redundant (and confirmatory) target–complex mass differences are obtained.

Reporting

Once it is determined which ligand binds to a specific target (and with what relative affinity), the results are tabulated and written to a proprietary corporate database. The rapid turnaround time provides chemists and molecular modelers with immediate feedback as to the affinity of compound collections to the target(s) of interest. Thus, the MASS assay is an invaluable tool in the structure-activity-relationship optimization cycle and represents an exciting new paradigm for lead identification in the drug-discovery arena.

CONCLUSIONS

ESI mass spectrometry is increasingly employed to characterize noncovalent biomolecular complexes formed in solution. A growing body of literature supports the hypothesis that these complexes can be transferred into the gas phase intact, and that mass-spectrometric characterization provides information pertinent to the solution-phase binding affinity, binding specificity, and binding stoichiometry. The MASS assay, which employs ESI-FTICR mass spectrometry to interrogate noncovalent ligand–target complexes, represents an exciting platform for drug discovery. It has been shown that MASS can be used to examine synthetic-compound libraries as well as natural-product libraries for potential drug leads in a quick and robust manner. Although the MASS has been used to study ligand binding to structured RNA drug targets, this technique could easily be extended to protein drug targets.

REFERENCES

1. Yamashita, M.; Fenn, J.B. Electrospray Ion Source. Another Variation on the Free-Jet Theme. *J. Phys. Chem.*, **88**, 4451–4459 (1984).
2. Hofstadler, S.A.; Bakhtiar, R.; Smith, R.D. Electrospray ionization mass spectrometry. 1. Instrumentation and spectral interpretation. *J. Chem. Educ.*, **73**, A82 (1996).

3. Hillenkamp, F.; Karas, M.; Beavis, R.C.; Chait, B.T. Matrix-Assisted Laser Desorption Ionization Mass-Spectrometry of Biopolymers. *Anal. Chem.*, **63**, 1193–1202 (1991).
4. Bakhtiar, R.; Hofstadler, S.; Muddiman, D.; Smith, R. Matrix-assisted laser disorption/ionization mass spectrometry. *J. Chem. Educ.*, **74**, 1288–1292 (1997).
5. Henry, K.D.; Williams, E.R.; Wang, B.H.; McLafferty, F.W.; Shabanowitz, J.; Hunt, D.F. Fourier-Transform Mass Spectrometry of Large Molecules by Electrospray Ionization. *Proc. Natl. Acad. Sci. USA*, **86**, 9075–9078 (1989).
6. Senko, M.W.; Hendrickson, C.L.; Pasatolic, L.; Marto, J.A.; White, F.M.; Guan, S.H.; Marshall, A.G. Electrospray ionization Fourier transform ion cyclotron resonance at 9.4 T. *Rapid Comm. Mass Spectrom.*, **10**, 1824–1828 (1996).
7. Gorshkov, M.V.; Tolic, L.P.; Udseth, H.R.; Anderson, G.A.; Huang, B.M.; Bruce, J.E.; Prior, D.C.; Hofstadler, S.A.; Tang, L.; Chen, L.-Z.; Willett, J.A.; Rockwood, A.L.; Sherman, M.S.; Smith, R.D. Electrospray ionization-Fourier transform ion cyclotron resonance mass spectrometry at 11.5 Tesla: instrument design and initial results. *J. Am. Soc. Mass Spectrom.*, **9**, 692–700 (1998).
8. Fenn, J.B.; Mann, M.; Meng, C.K.; Wong, S.F.; Whitehouse, C.M. Electrospray Ionization for Mass Spectrometry of Large Biomolecules. *Science*, **246**, 64–71 (1989).
9. Smith, R.D.; Loo, J.A.; Loo, R.R.O.; Busman, M.; Udseth, H.R. Principles and Practice of Electrospray Ionization – Mass-Spectrometry for Large Polypeptides and Proteins. *Mass Spectrom. Rev.*, **10**, 359–451 (1991).
10. Tomlinson, A.J.; Benson, L.M.; Naylor, S. Nonaqueous Solvents in the on-line Capillary Electrophoresis Mass Spectrometry Analysis of Drug Metabolites. *HRC – Journal of High Resolution Chromatography*, **17**, 175–177 (1994).
11. Greig, M.; Griffey, R.H. Utility of organic bases for improved electrospray mass spectrometry of oligonucleotides. *Rapid Commun. Mass Spectrometry*, **9**, 97–102 (1995).
12. Limbach, P.A.; Crain, P.F.; McCloskey, J.A. Characterization of oligonucleotides and nucleic acids by mass spectrometry. *Curr. Opin. Biotechnol.*, **6**, 96–102 (1995).
13. Griffey, R.H.; Greig, M.J.; Gaus, H.J.; Liu, K.; Monteith, D.; Winniman, M.; Cummins, L.L. Characterization of oligonucleotide metabolism in vivo via liquid chromatography/electrospray tandem mass spectrometry with a quadrupole ion trap mass spectrometer. *J. Mass Spectrom.*, **32**, 305–313 (1997).
14. Hannis, J.C.; Muddiman, D.C. Accurate characterization of the tyrosine hydroxylase forensic allele 9.3 through development of electrospray ionization fourier transform ion cyclotron resonance mass spectrometry. *Rapid Commun. Mass Spectrom.*, **13**, 954–962 (1999).
15. Griffey, R.H.; Hofstadler, S.A.; Sannes-Lowery, K:A.; Ecker, D.J.; Crooke, S.T. Determinants of Aminoglycoside-Binding Specificity for rRNA by Using Mass Spectrometry. *Proc. Natl Acad. Sci. USA*, **96**, 10129–10133 (1999).

16. Hofstadler, S.A.; Sannes-Lowery, K.A.; Crooke, S.T.; Ecker, D. J.; Sasmor, H.; Manalili, S.; Griffey, R.H. Multiplexed Screening of Neutral Mass-Tagged RNA Targets against Ligand Libraries with Electrospray Ionization FTICR MS: A Paradigm for High-Throughput Affinity Screening. *Anal. Chem.*, **71**, 3436–3440 (1999).
17. Sannes-Lowery, K.A.; Drader, J.J.; Griffey, R.H.; Hofstadler, S.A. Fourier transform ion cyclotron resonance mass spectrometry as a high throughput affinity screen to identify RNA binding ligands. *TrAC, Trends Anal. Chem.*, **19**, 481–491 (2000).
18. Sannes-Lowery, K.A.; Griffey, R.H.; Hofstadler, S.A. Measuring Dissociation Constants of RNA and Aminoglycoside Antibiotics by Electrospray Ionization Mass Spectrometry. *Anal. Biochem.*, **280**, 264–271 (2000).
19. Hofstadler, S.A.; Griffey, R.H. Mass spectrometry as a drug discovery platform against RNA targets. *Curr. Opin. Drug Discovery Dev.*, **3**, 423–431 (2000).
20. Herman, T.; Westhof, E. Rational Drug Design and High-Throughput Techniques for RNA targets. *Combinatorial Chemistry & High-Throughput Screening*, **3**, 219–234 (2000).
21. Sucheck, S.J.; Wong, C.-H. RNA as a target for small molecules. *Curr. Opin. Chem. Biol.*, **4**, 678–686 (2000).
22. Ramos, A.; Gubser, C.C.; Varani, G. Recent solution structures of RNA and its complexes with drugs, peptides and proteins. *Curr Opin Struct Biol*, **7**, 317–323 (1997).
23. Conn, G.L.; Draper, D.E. RNA structure. *Curr. Opin. Struct. Biol.*, **8**, 278–285 (1998).
24. Batey, R.T.; Rambo, R.P.; Doudna, J.A. Tertiary motifs in RNA structure and folding. *Angew. Chem. (Engl)*, **38**, 2326–2343 (1999).
25. Fourmy, D.; Recht, M.I.; Blanchard, S.C.; Puglisi, J.D. Structure of the A site of Escherichia coli 16S ribosomal RNA complexed with an aminoglycoside antibiotic. *Science (Washington, D. C.)*, **274**, 1367–1371 (1996).
26. Recht, M.I.; Fourmy, D.; Blanchard, S.C.; Dahlquist, K.D.; Puglisis, J.D. RNA sequence determinants for aminoglycoside binding to an A-site rRNA model oligonucleotide. *Journal of Molecular Biology*, **262**, 421–436 (1996).
27. Wang, Y.; Hamasaki, K.; Rando, R.R. Specificity of aminoglycoside binding to RNA constructs derived from the 16S rRNA decoding region and the HIV-RRE activator region. *Biochemistry*, **36**, 768–779 (1997).
28. Fourmy, D.; Recht, M.I.; Puglisi, J.D. Binding of Neomycin-class Aminoglycoside antibiotics to the A-site of 16 rRNA. *Journal of Molecular Biology*, **277**, 347–362 (1998).
29. Fourmy, D.; Yoshizawa, S.; Puglisi, J.D. Paromomycin binding induces a local conformational change in the A-site of 16 S rRNA. *Journal of Molecular Biology*, **277**, 333–345 (1998).

30. Wong, C.-H.; Hendrix, M.; Priestley, E.S.; Greenberg, W.A. Specificity of aminoglycoside antibiotics for the A-site of the decoding region of ribosomal RNA. *Chem. Biol.*, **5**, 397–406 (1998).

31. Recht, M.I.; Douthwaite, S.; Dahlquist, K.D.; Puglisi, J.D. Effect of Mutations in the A Site of 16 S rRNA on Aminoglycoside Antibiotic-Ribosome Interaction. *J. Mol. Biol.*, **286**, 33–43 (1999).

32. Carter, A.P.; Clemons, W.M.; Brodersen, D.E.; Morgan-Warren, R.J.; Wimberly, B.T.; Ramakrishnan, V. Functional insights from the structure of the 30S ribosomal subunit and its interactions with antibiotics. *Nature*, **407**, 340–348 (2000).

33. Wimberly, B.T.; Brodersen, D.E.; Clemons, W.M.; Morgan-Warren, R.J.; Carter, A.P.; Vonrhein, C.; Hartsch, T.; Ramakrishnan, V. Structure of the 30S ribosomal subunit. *Nature*, **407**, 327–339 (2000).

34. Hyun Ryu, D.; Rando, R.R. Aminoglycoside binding to human and bacterial A-Site rRNA decoding region constructs. *Bioorg. Med. Chem.*, **9**, 601–2608 (2001).

35. Lynch, S.R.; Puglis, J.D. Structural Origins of Aminoglycoside Specificity for Prokaryotic Ribosomes. *J. Mol. Biol.*, **306**, 1037–1058 (2001).

36. Vicens, Q.; Westhof, E. Crystal Structure of Geneticin Bound to a Bacterial 16 S Ribosomal RNA A Site Oligonucleotide. *Journal of Molecular Biology*, **326**, 1175–1188 (2003).

37. Purohit, P.; Stern, S. Interactions of a small RNA with antibiotic and RNA ligands of the 30S subunit. *Nature*, **370**, 659–662 (1994).

38. Griffey, R.H.; Sannes-Lowery, K.A.; Drader, J.J.; Mohan, V.; Swayze, E.E.; Hofstadler, S.A. Characterization of Low-Affinity Complexes between RNA and Small Molecules Using Electrospray Ionization Mass Spectrometry. *J. Am. Chem. Soc.*, **122**, 9933–9938 (2000).

39. Cheng, X.; Chen, R.; Bruce, J.E.; Schwartz, B.L.; Anderson, G.A.; Hofstadler, S.A.; Gale, D.C.; Smith, RD.; Gao, J.; Sigal, G.B. Using Electrospray Ionization FTICR Mass Spectrometry To Study Competitive Binding of Inhibitors to Carbonic Anhydrase. *J. Am. Chem. Soc.*, **117**, 8859–8860 (1995).

40. Greig, M.J.; Gaus, H.; Cummins, L.L.; Sasmor, H.; Griffey, R.H. Measurement of Macromolecular Binding Using Electrospray Mass Spectrometry. Determination of Dissociation Constants for Oligonucleotide: Serum Albumin Complexes. *J. Am. Chem. Soc.*, **117**, 10765–10766 (1995).

41. Lim, H.K.; Hsieh, Y.L.; Ganem, B.; Henion, J. Recognition of cell-wall peptide ligands by vancomycin group antibiotics: Studies using ion spray mass spectrometry. *J. Mass Spectrom.*, **30**, 708–714 (1995).

42. Gao, J.; Cheng, X.; Chen, R.; Sigal, G.B.; Bruce, J.E.; Schwartz, B.L.; Hofstadler, S.A.; Anderson, G.A.; Smith, R.D.; Whitesides, G.M. Screening derivatized peptide libraries for tight binding inhibitors to carbonic anhydrase II by electrospray ionization-mass spectrometry. *J. Med. Chem.*, **39**, 1949–1955 (1996).

43. Gao, Q.Y.; Cheng, X.H.; Smith, R.D.; Yang, C.F.; Goldberg, I.H. Binding specificity of post-activated neocarzinostatin chromophore drug-bulged DNA complex

studied using electrospray ionization mass spectrometry. *J. Mass Spectrom.*, **31**, 31–36 (1996).

44. Wu, Q.; Cheng, X.; Hofstadler, S.A.; Smith, R.D. Specific metal-oligonucleotide binding studied by high resolution tandem mass spectrometry. *J. Mass Spectrom.*, **31**, 669–675 (1996).
45. Wu, Q.Y.; Gao, J.M.; Joseph-McCarthy, D.; Sigal, G.B.; Bruce, J.E.; Whitesides, G.M.; Smith, R.D. Carbonic anhydrase-inhibitor binding: From solution to the gas phase. *J. Am. Chem. Soc.*, **119**, 1157–1158 (1997).
46. Loo, J.A.; Hu, P.; McConnell, P.; Mueller, W.T. A study of Src SH2 domain protein-phosphopeptide binding interactions by electrospray ionization mass spectrometry. *J. Am. Soc. Mass Spectrom.*, **8**, 234–243 (1997).
47. Smith, R.D.; Bruce, J.E.; Wu, Q.; Lei, Q.P. New mass spectrometric methods for the study of noncovalent associations of biopolymers. *Chem. Soc. Rev.*, **26**, 191–202 (1997).
48. Ayed, A.; Krutchinsky, A.; Chernushevich, I.V.; Ens, W.; Duckworth, H.W.; Standing, K.G. Observation of non-covalent complexes of citrate synthase and NADH by ESI/TOF mass spectrometry. *NATO ASI Ser., Ser. C*, **510**, 135–139 (1998).
49. Ayed, A.; Krutchinsky, A.N.; Ens, W.; Standing, K.G.; Duckworth, H.W. Quantitative evaluation of protein-protein and ligand-protein equilibria of a large allosteric enzyme by electrospray ionization time-of-flight mass spectrometry. *Rapid Commun. Mass Spectrom.*, **12**, 339–344 (1998).
50. Jorgensen, T.J.D.; Roepstorff, P. Direct Determination of Solution Binding Constants for Noncovalent Complexes between Bacterial Cell Wall Peptide Analogues and Vancomycin Group Antibiotics by Electrospray Ionization Mass Spectrometry. *Anal. Chem*, **70**, 4427–4432 (1998).
51. Loo, J.A.; Sannes-Lowery, K.A.; Hu, P.; Mack, D.P.; Mei, H.-Y. Studying noncovalent protein-RNA interactions and drug binding by electrospray ionization mass spectrometry. In *New Methods for the Study of Biomolecular Complexes*; Ens, W., Standing, K. G., Chernushevich, I. V. Eds.; Kluwer: Dordrecht, 1998; pp 83–99.
52. Loo, J.A.; Sannes-Lowery, K.A. Studying noncovalent interactions by electrospray ionization mass spectrometry. In *Mass Spectrom. Biol. Mater. (2nd Ed.)*; Larsen, B. S., McEwen, C. N. Eds.; Marcel Dekker, Inc.: New York, 1998; pp 345–367.
53. Hofstadler, S.A.; Griffey, R.H. Analysis of noncovalent complexes of DNA and RNA by mass spectrometry. *Chemistry Review*, **101**, 377–390 (2001).
54. Winzor, D.J.; Sawyer, W.H. *Quantitative Characterization of Ligand Binding*; Wiley-Liss: New York, 1995; 168 pp.
55. Cummins, L.L.; Chen, S.; Blyn, L.B.; Sonnes-Lowery, K.A.; Drader, J.J.; Griffey, R.H.; Hofstadler, S.A. Multitarget Affinity/Specificity Screening of Natural Products: Finding and Characterizing High Affinity Ligands from Complex Mixtures by using High Performance Mass Spectrometry. *J. Nat. Prod.*, **66**, 1186–1190 (2003).

56. DeJohngh, D.C.; Hribar, J.D.; Hanessian, S.; Woo, P.W.K. Mass Spectrometric Studies on Aminocyclitol Antibiotics. *Journal of the American Chemical Society*, **89**, 3364–3365 (1967).

57. Curcuruto, O.; Kennedy, G.; Hamdan, M. Investigation of several aminoglycoside solutions containing one or more components by positive electrospray mass spectrometry. *Org. Mass Spectrom.*, **29**, 547–552 (1994).

58. Goolsby, B.J.; Brodbelt, J.S. Analysis of protonated and alkali metal cationized aminoglycoside antibiotics by collision-activated dissociation and infrared multiphoton dissociation in the quadrupole ion trap. *J. Mass Spectrom.*, **35**, 1011–1024 (2000).

59. Drader, J.; Anderson, A.; Jiang, Y.; Hannis, J.; Hofstadler, S. *manuscript in preparation*, (2005).

60. Marshall, A.G.; Verdun, F.R. *Fourier Transforms in NMR, Optical, and Mass Spectromery*; Elsevier: New York, 1990.

61. Marshall, A.G.; Grosshans, P.B. Fourier-Transform Ion-Cyclotron Resonance Mass-Spectrometry – The Teenage Years. *Anal. Chem.*, **63**, A215–A229 (1991).

4

MASS SPECTROMETRY–BASED PROTEOMIC APPROACHES FOR DISEASE DIAGNOSIS AND BIOMARKER DISCOVERY

Thomas P. Conrads, Haleem J. Issaq, Josip S. Blonder, and Timothy D. Veenstra

INTRODUCTION

There has been a surge in technology developments over the past decade that has changed the way that many scientists in biomedical-related disciplines design experiments. A vast majority of scientists used to be content with studying a single entity, such as a gene, transcript, or protein in detail; however, there has been an enthusiastic transition to technologies that allow hundreds or thousands of these species to be collectively interrogated. While there are many different reasons to multiplex these analyses, the net result is an increase in the data density and the acceleration of the discovery of important cellular processes. The core technologies that have been responsible for this current and growing trend include the high-speed sequencer for the study of deoxyribonucleic acid (DNA) (i.e., genomics), microarray technologies for the study of messenger ribonucleic acid (mRNA) (i.e., transcriptomics), and mass spectrometry (MS) for the study of proteins (proteomics). The fields of genomics, transcriptomics, and proteomics have in many ways developed independently of one another, however, as within the cell, each field is reliant on the other. For example, global proteomic studies, where the goal is to

Integrated Strategies for Drug Discovery Using Mass Spectrometry, Edited by Mike S. Lee

identify as many proteins as possible within a given system, is only made possible through the generation of large genomic databases to search the data against. The ultimate goal, a complete understanding of cell function, is only possible by the integration of data acquired at the genomic, transcriptonomic, proteomic, as well as the often overlooked metabonomic levels.

This chapter discusses the developing role of proteomics in biomarker discovery and disease diagnostics. More specifically the focus is on the use of MS, as this instrument has arguably been the technology that has driven the field of proteomics. While MS is by no means a new technology, developments made in the past 10–15 years have dramatically increased its capability in the life sciences. The use of proteomic patterns, an emerging technology that holds great promise in the early diagnosis of cancer, are discussed.

DEVELOPMENTS ENABLING MASS SPECTROMETRY AS A TOOL FOR BIOMARKER DISCOVERY

While owing its origins to J.J. Thomson's discovery of electrons and positive rays in a vacuum tube over a century ago [1], MS did not achieve the status it enjoys in the field of biochemistry until the 1990s. Over the past century the primary use of MS has been the characterization of small organic and inorganic molecules; however, several key developments have made MS one of the (if not the) primary analytical instruments in protein chemistry. These developments have enabled MS with the attributes needed to not only characterize proteins, but also in combination have allowed extremely complex mixtures of proteins to be characterized in an automated, robust, and high throughput fashion. One of the first developments was tandem MS (MS/MS), introduced in 1968 by professors Keith R. Jennings of the University of Warwick, England, and Fred McLafferty, who was then at Purdue University [2]. In MS/MS, a precursor ion is mass-selected and typically fragmented by collision-induced dissociation (CID), followed by mass analysis of the resulting product ions. MS/MS provides structural information concerning the precursor ion, thereby allowing its identification through the measurement of its product ions. Another significant step was Professor Klaus Biemann's application of fast atom-bombardment methods to peptide sequencing [3]. His group was able to show for the first time the usefulness of MS/MS to elucidate the fragmentation behavior of peptides in the MS. This seminal work paved the way to the high throughput identification of peptides in complex mixtures using MS.

Electrospray ionization (ESI) and matrix-assisted laser-desorption/ ionization (MALDI) are two recently developed MS techniques that have had a major impact on the ability to use MS for the study of biomolecules such as proteins and peptides [4,5]. In ESI-MS, highly charged droplets dispersed from a capillary in an electric field are evaporated, and the resulting ions are drawn into an MS inlet and measured. Chemistry professor Malcolm Dole of Northwestern University, Evanston, Illinois, first conceived the technique in the 1960s; however, it was put into practice in the early 1980s by Professor John B. Fenn of Yale University [6]. Dr. Fenn was awarded the 2002 Nobel Prize for his development of ESI and its use in the MS analysis of biological macromolecules. MALDI-MS, a form of laser desorption MS, was developed in 1987 at the University of Frankfurt, Germany, by professor of biophysics Franz Hillenkamp and Michael Karas [5], and independently by Koichi Tanaka and co-workers at Shimadzu Corp., Kyoto, Japan [7]. In MALDI, sample molecules are laser-desorbed from a solid or liquid matrix containing a highly UV-absorbing substance. Both ESI-MS and MALDI-MS have made MS increasingly useful for sophisticated biomedical analysis. While MS used to be restricted to the analysis of small organic and inorganic molecules, the development of ESI-MS and MALDI-MS have allowed for such applications as the sequencing and analysis of peptides and proteins, studies of noncovalent complexes and immunological molecules, DNA sequencing, and the analysis of intact viruses.

While the preceding developments allowed for the proteins to be identified and characterized, the mass spectrometer is still limited in the number of species that can be identified during a single experiment. Since the discovery of biomarkers typically requires the analysis of very complex mixtures containing tens, and possibly hundreds, of thousands of different species, effective fractionation of these species is necessary prior to MS analysis. The two most popular means of fractionating complex biological mixtures prior to MS analysis are two-dimensional polyacrylamide gel electrophoresis (2D-PAGE) [8] and reversed-phase liquid chromatography (LC) [9], which has the capability of being coupled directly to the spectrometer so that species separated on the column can be characterized as they elute (i.e., LC/MS). While both means are used to fractionate complex mixtures of proteins, 2D-PAGE and LC/MS can also be used to quantitate the relative abundances of proteins within two distinct samples. The difference between the two methods of comparison, however, are that when using 2D-PAGE fractionation, the relative abundances of proteins in the samples are compared prior to MS analysis, while in LC/MS, the two samples are compared only after the MS data have been acquired.

CONVENTIONAL MS-BASED PROTEOMICS TECHNOLOGIES

The discovery, identification, and validation of proteins associated with a particular disease state is a difficult and laborious task. This process often requires hundreds, if not thousands, of samples to be analyzed. As mentioned earlier, the predominant proteomic method of disease biomarker discovery in use today is a combination of 2D-PAGE and MS [10]. In this method, proteins from two distinct samples (i.e., diseased vs. normal, treated vs. control, etc.) are analyzed via 2D-PAGE and their protein expression patterns compared. Protein spots of interest are excised from the gel, proteolytically or chemically digested, and the resultant peptides analyzed by MS to identify the protein. As a separation technique, 2D-PAGE provides excellent resolution of complex protein mixtures. The method, however, is limited by its laboriousness, the low sensitivity of conventional stains, and its inability to resolve proteins with extreme molecular weights, hydrophobicity, and isoelectric points. Regardless of the limitations of 2D-PAGE, it has proven to be a vital tool in proteomics.

An excellent example of the use of 2D-PAGE MS analysis for the discovery of potential cancer-related biomarkers is illustrated in the study by Chen et al. [11], who analyzed 93 lung adenocarcinomas and 10 uninvolved lung samples (Figure 4.1). After measuring and comparing the relative abundances of the individual protein spots, those that showed a difference in relative intensity were identified via peptide mapping using MALDI-MS or ESI-MS/MS.

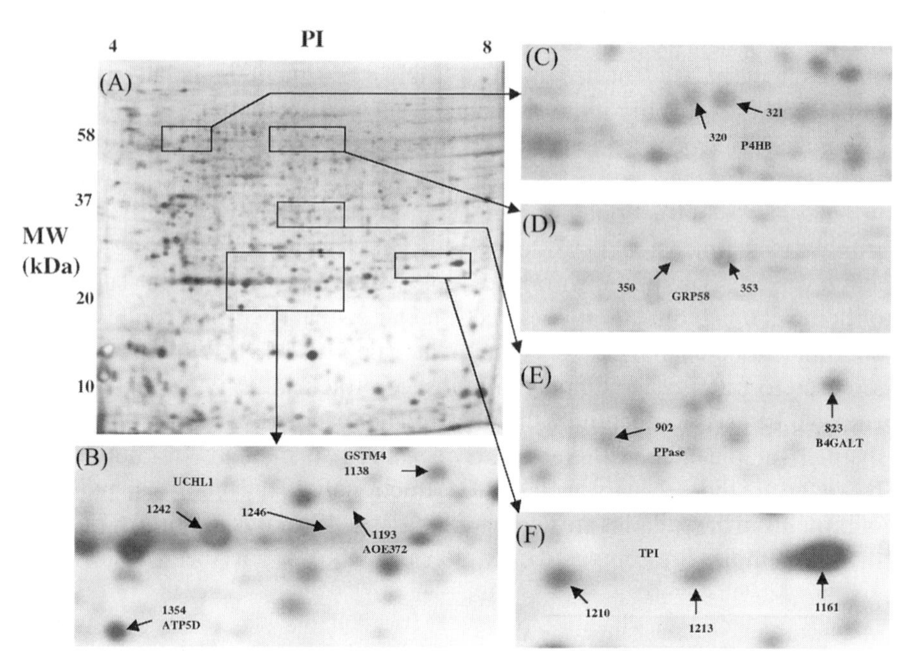

FIGURE 4.1 (A) 2D-PAGE gel separation of proteins identified with silver staining from a stage I lung adenocarcinoma. The proteins are separated by isoelectric point (pI) in the first dimension and by molecular weight (MW) in the second dimension. (B–F) The outlined areas of part A showing proteins significantly increased in lung adenocarcinoma. (Reprinted from Chen et al. [11], used with permission from the American Association of Cancer Research.)

Several candidate tumor markers were identified to be upregulated (1.4- to 10.6-fold) in lung adenocarcinomas when compared with normal lung tissue. Among these, the antioxidant enzyme AOE372, cytosolic inorganic pyrophosphatase (PPase), mu-class glutathione transferase 4 (GSTM4), and ubiquitin carboxy–terminal hydrolase L1 (UCHL1) were increased 10.6-, 7.6-, 4.0-, and 3.5-fold, respectively. The frequency of elevated expression of these proteins in lung adenocarcinomas was found to range from 35.5% to 96.8% among the 93 tumors examined. GSTM4 was the most consistently overexpressed protein, being upregulated in 96.8% of the tumors. Correlations were observed between overexpression of some proteins and specific clinicopathological variables, including tumor differentiation (AOE372), tumor subhistology (PPase), and a positive smoking history (PPase and UCHL1). In addition, the increased abundance of both AOE372 and UCHL1 correlated with the upregulation of these genes at the mRNA level.

Identification of Clinically Useful Biomarkers

In the design of studies aimed to discover and ultimately use a clinically useful biomarker, a set of characteristics that define the acquisition and measurement of the biomarker needs to be established. For a biomarker to have the greatest impact it should be present within an easily obtainable sample, such as urine or blood. If a biomarker is only assayable within a sample that requires a biopsy to be recovered, it is likely to only be interrogated as a secondary screen to confirm the original diagnosis. Second, the assay to clinically measure and validate the overall positive predictive value (PPV) of the biomarker must be capable of screening thousands of samples in a high throughput manner. In an attempt to fulfill the first criterium, a major focus in proteomics is in the discovery of biomarkers within serum. The second criterium is absolutely necessary for the analysis of serum due to the wide variability in the protein content of serum from different individuals. Thousands of samples need to be assayed to ensure that the potential biomarker is indeed related to the pathophysiology or pathohistology of the disease state and is not simply a function of the variability within the serum of patients due to differences in diet, genetic background, lifestyle, etc. At first glance, serum presents many beneficial attributes for proteomic investigation, since it has a high protein content (i.e., 60–80 mg/mL), with many of these proteins being secreted and shed from cells and tissues [12]. Serum, however, is one of the most difficult proteome samples to characterize adequately. Unfortunately, serum proteins are present across an extraordinary dynamic range of concentration that is likely to span more than 10 orders of magnitude [13]. This large dynamic range exceeds the analytical capabilities of traditional proteomic methods, making the detection of lower abundance serum proteins extremely challenging. The protein

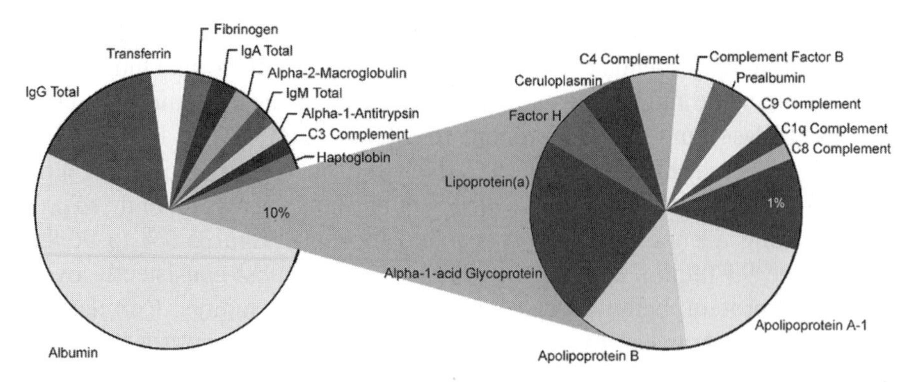

FIGURE 4.2 Pie chart representing the relative contribution of proteins within plasma. Twenty-two proteins constitute approximately 99% of the protein content of plasma.

content of serum is dominated by a handful of proteins, such as albumin, transferrin, haptoglobulin, immunoglobulins, and lipoproteins [14]. Indeed, only 10 proteins constitute 90% of the protein content of serum. Of the remaining 10%, 12 proteins make up 90% of this remaining total. In fact, only 1% of the entire protein content of serum is made up of proteins that are considered to be in low abundance and of great interest in proteomic studies in search of potential biomarkers (Figure 4.2).

SURFACE-ENHANCED LASER-DESORPTION/IONIZATION TIME-OF-FLIGHT MASS SPECTROMETRY

A major development in the use of MS as a diagnostic instrument has been fueled by the development of surface-enhanced laser-desorption/ionization time-of-flight MS (SELDI-TOF) [15]. Whereas HPLC-MS combines elution chromatography with MS, SELDI-TOF MS uses retention chromatography. Chromatographic surfaces arrayed on the surface of protein chips are used to retain proteins from complex mixtures according to some physicochemical property of the protein, such as hydrophobicity, charge, specific affinity, etc. After the protein mixture has been deposited on the protein-chip surface, it is washed and an energy-absorbing molecule (i.e., matrix) is applied. The proteins are then ionized and desorbed from the protein-chip surface using a nitrogen laser, and their molecular masses are measured by TOF MS.

The SELDI-TOF MS format enables proteins from a variety of complex biological specimens, such as serum, plasma, intestinal fluid, urine, cell lysates, and cellular secretion products, to be profiled in a high throughput manner [16]. In most cases biofluids can be added directly to the protein chip without the need for extensive sample processing steps prior to analysis. The sensitivity of the TOF MS enables protein profiles to be generated from as little as a single microliter of serum, and have been generated from as few as 25–50 cells [17]. SELDI-TOF MS has proved to be useful in the discovery of potential diagnostic markers and patterns for prostate [18], bladder [19], breast [20], and ovarian cancers [21]. While SELDI-TOF MS has been used primarily for the analysis of crude biological fluids, and its greatest potential is within the field of diagnostic medicine, it can also be used for more targeted studies, such as the characterization of phosphoproteins [22], glycoproteins [23], and protein-DNA interactions [24].

The SELDI-TOF MS Components

The most popular MS instrument to perform SELDI-TOF MS analysis is the PBS-II, manufactured by Ciphergen Biosystems Inc. [16]. The SELDI-TOF MS instrument is composed of three major components: the ProteinChip® arrays, the mass analyzer, and the data-analysis software.

Protein Chip Arrays

The ProteinChip arrays are the unique components that distinguish SELDI-TOF MS technology from other MS-based systems. The protein-chip arrays are composed of different chromatographic surfaces spotted onto a 10-mm-wide × 80-mm-long protein chip that is made of aluminum. Each chromatographic spot is 2 mm in diameter, and each protein chip contains either 8 or 16 of these spots. The chromatographic surfaces are designed to capture proteins based on either a chemical (anionic, cationic, hydrophobic, hydrophilic, metal ion, etc.) or biochemical (immobilized antibody, receptor, DNA, enzyme, etc.) interaction, as shown in Figure 4.3. Typically, chemically active surfaces retain whole classes of proteins, whereas biochemical surfaces are used to target a single protein of interest through a specific interaction with an affinity reagent, such as an antibody. The selectivity of the chemically active surfaces is illustrated in Figure 4.3B, which shows the protein profiles of a cell lysate using the different surfaces. Different surfaces retain different proteins, and these differences depend, for example, on the pH of the sample, when anion and cation exchange surfaces are used. While the chemically treated ProteinChip arrays are commercially available, the biochemical surfaces are generally custom-made by using an open preactivated platform, to which a bait molecule of choice, such as an antibody, is immobilized. Any crude extract or sample can be applied to the surface thus retaining those target proteins that interact with the bait molecule.

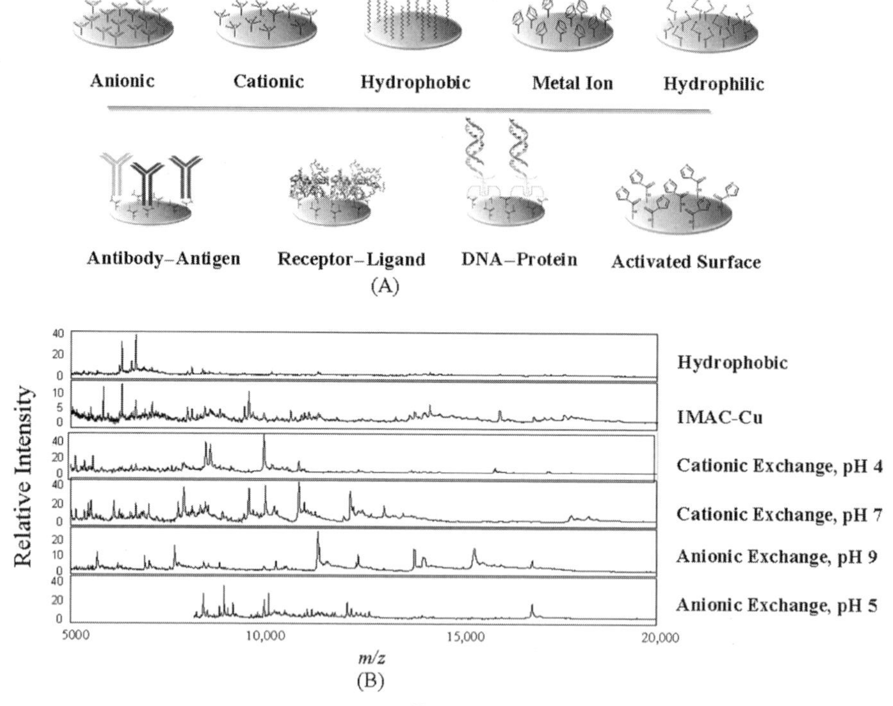

FIGURE 4.3 The variety of ProteinChip® arrays available for sample preparation. (A) The upper arrays represent chemically modified chromatographic surfaces, while the bottom arrays are biochemically modified surfaces. Chemically modified surfaces are used to retain a group of proteins, while biochemically modified surfaces are typically used to isolate a specific protein or functional class of proteins. (B) Protein profile of a cell lysate on different ProteinChip surfaces. As shown in the figure for a selection of protein chips, the individual surfaces retain different groups of proteins, depending on their physiochemical properties. The proteins retained are also dependent on the pH of the sample for the cation and anion exchange surfaces.

The Mass Analyzer

The mass analyzer is a relatively simple TOF MS equipped with a pulsed UV nitrogen laser, as shown in Figure 4.4. While the acronym SELDI has become almost synonymous with this technology, in principle it is really a form of MALDI. Upon laser activation, samples become irradiated, causing the sample to be desorbed and ionized, analogous to the MALDI process. The ionized molecules are accelerated through the spectrometer under vacuum (the so-called TOF tube) toward an ion detector. The mass-to-charge ratio (m/z) of each species is recorded based on the time each species requires to pass through the TOF tube (i.e., its time-of-flight). The design of the mass analyzer is somewhat rudimentary when compared to higher-performance mass spectrometers, but the SELDI-TOF MS does have reasonably high sensitivity. While time-lag focusing is used to increase data resolution and mass accuracy,

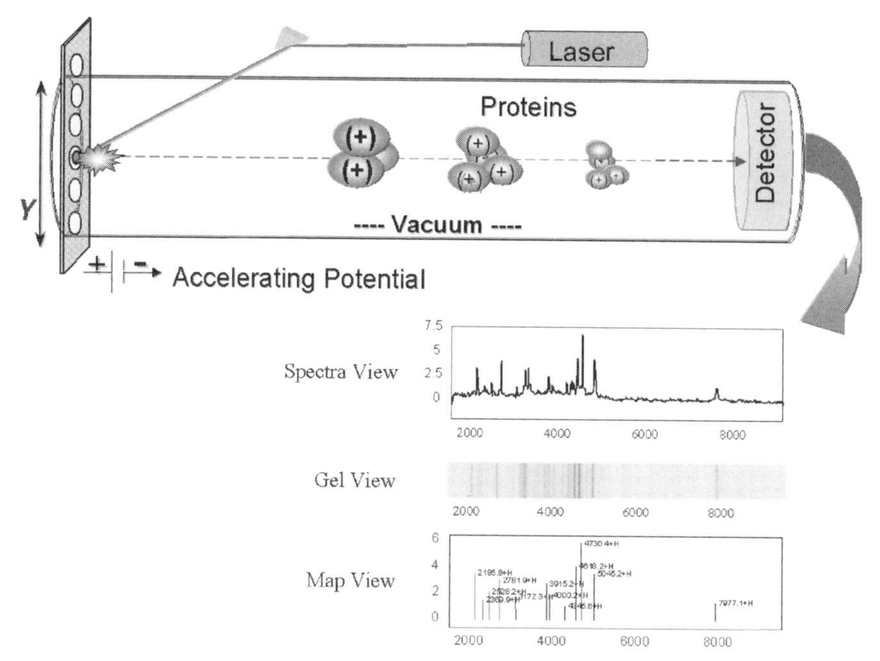

FIGURE 4.4 Schematic diagram of the SELDI Ciphergen mass spectrometer. After sample preparation, the ProteinChip arrays are analyzed by a laser-desorption/ionization (LDI) time-of-flight mass spectrometer (TOF MS). The TOF MS measures the molecular weights of the various proteins that are retained on the array. For comparison purposes, the software associated with the SELDI Ciphergen instrument is capable of displaying the resultant data as either a spectral, map, or gel view.

the achievable mass accuracy is much less than that afforded using more conventional, high-resolution TOF MS instrumentation [25].

The Software

One of the original intents was to use SELDI-TOF MS to identify differences in the protein expression profiles of two or more distinct biological samples, such as diseased vs. healthy, treated vs. control, differentiated vs. immature, etc. [16,25]. Accordingly, the samples being analyzed are often quite complex, particularly in the field of biomarker discovery, in which clinical samples such as serum are measured. The resulting MS spectrum shows the relative abundance versus the *m/z* ratios of the detected proteins, as shown in Figure 4.5. To simplify the visual output, the spectrum can be displayed as a trace view, gel view, or map view. The recognition of peaks that show differences in intensity is simplified by converting the MS peak trace into a simulated 1D gel electrophoresis display (i.e., the gel view). The software also can be used to compare the displays and identify unique mass-spectral peaks or those that show a significant abundance difference in one of the samples via cluster

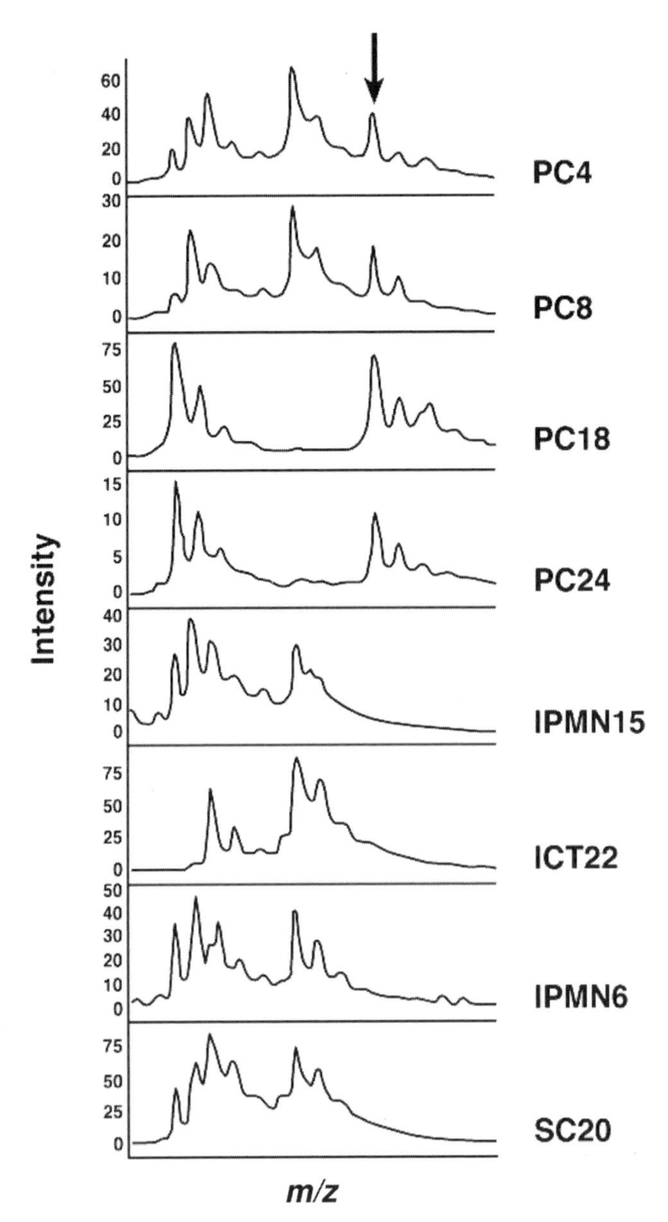

FIGURE 4.5 Representative spectrum examples of SELDI analysis of pancreatic juice samples bound to IMAC-3 cupper ProteinChip array. A peak of ~ 16,570 Da (arrow) was present in the four pancreatic juice samples from patients with pancreatic adenocarcinoma (PC4, PC8, PC18, PC24) but absent in the four patients with other pancreatic diseases (IPMN; islet cell tumor (ICT); serous cystadenoma (SC)). (Reprinted from Rosty et al. [26], used with permission from the American Association of Cancer Research.)

analysis. As described later in this chapter, significant software developments, both by the manufacture and independent laboratories, have enabled SELDI-TOF MS spectral data to be analyzed as a pattern rather than as individual peaks. It has been the analysis of the proteomic patterns that has propelled the use of SELDI-TOF MS as a potentially revolutionary diagnostic tool.

Protein Identification by SELDI-TOF MS

Essentially the SELDI-TOF MS spectrum is a low-resolution profile of molecular species that were retained on the protein-chip surface. The resolution, mass accuracy, and lack of MS/MS capabilities of the mass analyzer makes direct protein identification from complex mixtures extremely challenging, unless a protein of interest is selectively targeted using an affinity-based surface. So what is the value of the results? The value lies in the ability to obtain and compare spectra from a significant number of samples in a relatively short time period, with very little sample preparation or sophisticated chromatography. For example, a single operator can acquire mass spectra of >250 different samples in a single day. Ideally the ability to compare this number of samples will reveal a reliable biomarker signal that is unique to a particular disease state.

Once a signal that is consistently unique to a particular disease state has been recognized in a statistically empowered sample set, the next phase is to identify the potential biomarker. Due to the low-mass accuracy of the TOF MS used to acquire the proteomic spectrum, direct identification is generally impossible; however, invaluable information can be garnered from the SELDI analysis to design the necessary purification strategy. Matching its m/z recorded by SELDI-TOF MS with that measured during the purification process provides an assay to follow the purification of the potential biomarker. The purification strategy can be based on any type of chromatography; however, the type of protein chip that the protein(s) of interest binds to provides a useful starting point in the design of a purification scheme. Once isolated to a reasonable level of purity, high-resolution MS with tandem MS capabilities can then be used to identify the potential biomarker. A simpler approach would be to identify the differentially expressed proteins that are captured directly on the chip. In this procedure, the peak of interest is selected for tandem MS analysis directly off the protein chip without the need to isolate it from its complex matrix (i.e., serum). This procedure would be much easier, faster, and less costly. Fortunately, new ion sources are available that allow the protein chips to be analyzed using a hybrid triple quadrupole TOF MS (Qq-TOF MS). The aim is to use this high-resolution MS/MS capable instrument to directly identify peptides desorbed and ionized directly off the protein chip surface. While in theory this seems to be a reasonable approach, in practice this method is still unproved as being routine with clinical samples.

In some cases, however, investigators have been fortuitous in the identification of potential biomarkers using the SELDI-TOF MS. A good example of this is the recognition and eventual identification of a biomarker for pancreatic cancer by Rosty et al. [26]. This group's focus was on the identification of a reliable biomarker for pancreatic cancer, the fourth leading cause of cancer death in both men and women in the United States [27]. Few if any patients with pancreatic cancer are cured without resection, and unfortunately only 10–15% of patients are resectable at the time of diagnosis [28]. Current methods for diagnosing pancreatic cancer at an early stage are relatively ineffective. Their study used the SELDI-TOF MS method essentially as described earlier to analyze and compare the proteomic profiles of 15 pancreatic-juice samples from patients with pancreatic adenocarcinoma to those obtained from seven fluid samples acquired from patients with other pancreatic diseases. No samples were obtained from normal control patients, since all of the samples were collected from patients undergoing pancreatectomy. A peak with a molecular weight of ~16,570 Da was present in higher intensity in the pancreatic juice samples obtained from patients with pancreatic adenocarcinoma, as shown in Figure 4.6.

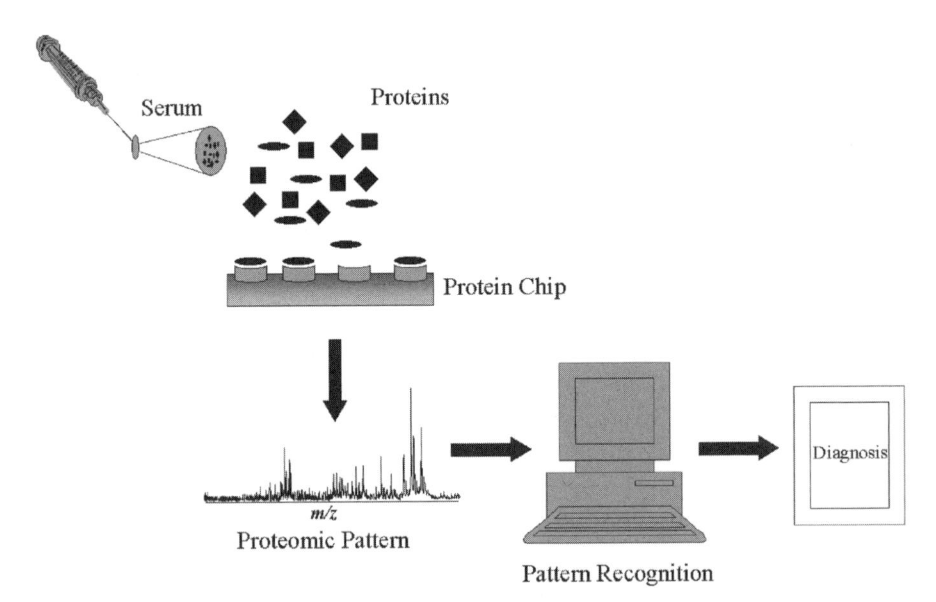

FIGURE 4.6 Disease diagnostics using proteomic patterns. The sample drawn from the patient is applied to a protein chip, which is made up of a specific chromatographic surface. After several washing steps and the application of an energy-absorbing molecule, the species that are retained on the surface of the chip are analyzed via mass spectrometry. The pattern of peaks within the spectrum is analyzed using sophisticated bioinformatic software to diagnose the source of the biological sample.

To identify this 16,572.9-Da protein, the investigators searched this mass against the masses of the proteins listed within the SWISS-PROT and TrEMBL protein databases. The mass approximately matched that of the secreted form of the human pancreatic–associated protein 1 (PAP-I), which has a calculated mass of 16,566.5 Da, and is known to originate within the pancreas. The match between the calculated mass of PAP-I and the experimentally determined mass was within the margin of error of the SELDI-TOF MS mass-measurement accuracy (i.e., ~1000 ppm). Due to the degeneracy of protein masses within the human databases, the identity of the peak at 16,572.9 Da required confirmation. To confirm the identification of this peak, the investigators performed a SELDI immunoassay by coupling an anti-hepatocarcinoma–intestine–pancreas (HIP)/PAP-I polyclonal antibody to a biochemically activated protein-chip surface and applying 12 of the different pancreatic-juice samples used in the original screening. Six of the samples were those in which the 16,572.9-Da peak was present, and six were those in which it was absent. The specific peak at mass 16,569.2 was present in all of the six samples that showed this peak in the original screening and not detected in the six samples that did not display the peak in the original screening. This peak was not detected in the control spots using an irrelevant antibody. Using this information, the investigators proceeded to develop an enzyme-linked immunosorbant assay (ELISA) to measure the levels of HIP and PAP-I in the pancreatic juice and serum of additional patients, providing a potentially simple, noninvasive diagnostic assay for early-stage pancreatic cancer. While this group was somewhat fortuitous, in that the mass of their proposed biomarker obtained from the initial screening matched that of a pancreatic-related protein, it still shows a useful example of the need of classic biochemistry techniques to identify the peaks that are seen within a SELDI-TOF MS profile.

DISEASE DIAGNOSTICS VIA PROTEOMIC PATTERN ANALYSIS

The diagnosis of cancer through the analysis of proteomic patterns is an area that is drawing much interest. Dr. Lance Liotta of the National Cancer Institute and Dr. Emmanuel Petricoin of the Food and Drug Administration were the first to report on this revolutionary concept in a seminal paper describing its use in the diagnosis of women suffering from ovarian cancer [21]. Ovarian cancer is the leading cause of gynecological malignancy, which resulted in approximately 13,900 deaths in 2002 [29]. Unfortunately, almost 80% of women with common ovarian cancer are not diagnosed until the disease is advanced in stage, and has spread to the upper abdomen (stage III) or beyond (stage IV) [29]. The 5-year survival rate for these women is only 15 to 20%. The 5-year survival rate for ovarian cancer patients that are diagnosed at stage I, however, approaches 95% with surgical intervention. Early diagnosis, therefore, could dramatically decrease the number of deaths from ovarian cancer.

Cancer Antigen 125 (CA 125) is currently the most widely used diagnostic biomarker for ovarian cancer [30]. Unfortunately, CA 125 is elevated in only 50–60% of patients with stage I ovarian cancer, lending it a PPV of 10% [29]. A combined strategy of CA 125 determination with ultrasonography increases the PPV to approximately 20% [31]. Moreover, CA 125 can be elevated in other nongynecologic and benign conditions, resulting in a significant number of false-positives for ovarian cancer [29]. Obviously a biomarker with a higher PPV for early-stage ovarian cancer would have saved thousands of lives every year.

The procedure using proteomic patterns to diagnose ovarian cancer is illustrated in Figure 4.6 [21]. Serum is obtained from patients and approximately 5 μL is applied directly to the surface of a protein chip. After processing the sample as described previously, the protein chip is placed within the interface of a PBS-II TOF MS and interrogated with a laser to desorb and ionize the retained proteins and peptides from the chip surface. The measured m/z values of these species are then used to provide the proteomic pattern of that particular serum sample. After acquiring the proteomic patterns of serum samples drawn from a number of healthy and diseased patients, bioinformatic tools are used to identify key features that enable the patterns from healthy and diseased serum to be segregated. The use of proteomic patterns as a diagnostic tool represents a contrast to how the discovery of biomarkers has been carried out in MS-based proteomic approaches. As described earlier most MS-based approaches are designed to separate and compare species across different complex samples with the goal to identify the specific entity and its changes between the sample sets. Proteomic pattern analysis does not rely on the identification of any of the species observed during the analysis, instead relying solely on the changes in relative abundance of a number of different points within the mass spectrum. A key aspect to their study was the application of a high-order self-organizing cluster-analysis approach based on a genetic algorithm that was "trained" on SELDI-TOF MS spectra from serum derived from either affected women or women with ovarian cancer, which is described later in the chapter. The "trained" algorithm was applied to a masked set of samples and resulted in a sensitivity of 100%, a specificity of 95%, and an overall PPV for ovarian cancer of 94% [21]. The success of the use of proteomic patterns for the diagnosis of stage I ovarian cancer suggests that patterns generated from other biomolecules within biofluids may also provide a useful indicator of the early onset of a particular disease state.

The success of proteomic pattern analysis is highly dependent on the use of bioinformatic tools to analyze the spectrum. It is virtually impossible to correctly identify the source of a serum sample based on a visual analysis of the spectra, unless there is a prominent signal that is unique to either the healthy or diseased state. Typically, there is significant variability between sera obtained from the two sample sets (i.e., healthy and diseased); however, there also exists variability within a single sample set. This inherent variability makes it virtually impossible to identify signals that are consistently unique to either the control or disease samples. In addition this variability makes the identification of a disease-specific biomarker difficult in the background of changes related to genetic differences, diet, lifestyle, etc.

While different types of bioinformatics tools can be used to discriminate samples based on their source, the initial paper by Petricoin and Liotta used ProteomeQuest^{TM}, a software tool developed by Correlogics of Bethesda, Maryland. This software combines elements from genetic algorithm methods and cluster analysis. For the analysis of mass-spectral data, each of the input files is composed of m/z values on the x axis along with their corresponding amplitudes on the y axis. The output of the algorithm is the most robust subset of amplitudes at defined frequency values that best separates the preliminary data acquired from the samples obtained from either healthy or diseased patients [21].

The analysis was divided into two phases: a pattern-discovery phase and a pattern-matching phase, as shown in Figure 4.7. In the pattern-discovery phase, a set of mass spectra of serum from both healthy and ovarian cancer–affected individuals were analyzed to identify a subset of key frequency values and their related amplitudes that are able to completely segregate the data

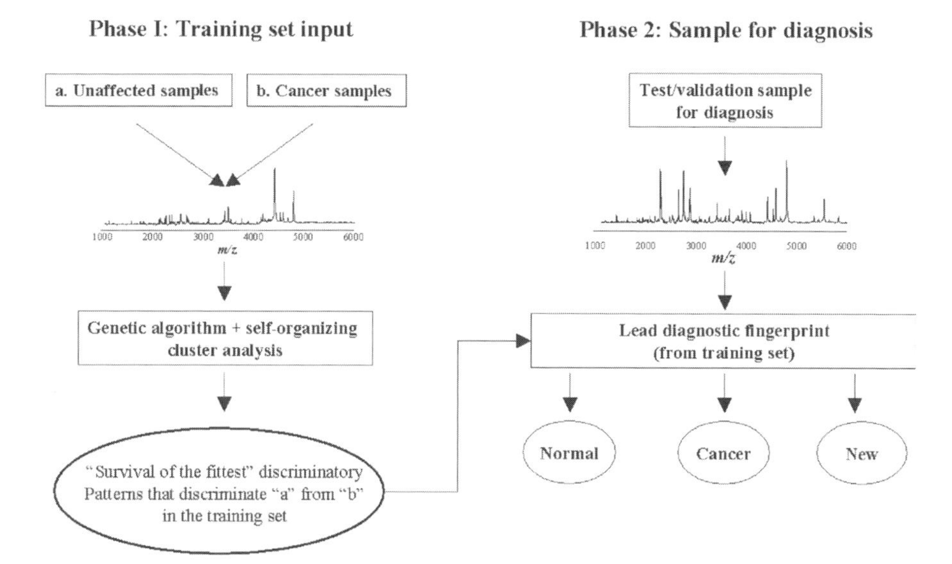

FIGURE 4.7 Bioinformatic analysis of proteomic pattern spectra for the determination of the discriminatory patterns in the training and diagnostic (i.e., testing and blind validation) phases.

acquired using serum samples from patients with ovarian cancer and unaffected individuals. This set of spectra is known as the training set. In the pattern-discovery phase, the source of the serum (i.e., from healthy or ovarian cancer–affected individuals) is known and is included as part of the data that are provided to the algorithm. The process starts with hundreds of arbitrary choices of small sets of exact m/z values selected along the x axis of the mass spectra. Each candidate subset contains five to 20 of the potential x axis values contained within the spectra. The diagnostic pattern is formed by plotting the combined y-axis amplitudes of the candidate set of the key m/z values in N-dimensional space, where N is the number of m/z values found within the training set of spectra. The pattern that is formed by the relative amplitude of the spectrum data for this set of chosen m/z values is rated for its ability to distinguish the urine samples from the healthy and cancer-affected individual sets. The frequency values within the highest-rated sets are reshuffled to form new subset candidates, and the resultant defined amplitude values are rated iteratively until the set that fully discriminates the preliminary sample sets is revealed.

Once the algorithm recognizes key m/z values, the model is tested using a set of masked test spectra in the pattern-matching phase. In the pattern-matching phase, the source of the sample used to acquire the inputted data is not provided to the algorithm. As opposed to the pattern discovery phase, which uses all of the frequency values within the entire spectral data set, in the pattern-matching phase only the key subset of the frequency values identified in that is used to classify the unknown samples as being from healthy or cancer-affected individuals. The pattern formed by the relative amplitudes of the key frequency values in each unknown is then matched to the optimum pattern defined in the pattern-matching phase. Each of the unknown samples is classified into a cluster diagnostic of either an unaffected or cancer-affected individual or will generate a new cluster if it is found not to match any of the patterns defined in the pattern-discovery phase. If the sample generates a new cluster, then it means that the point in N-space of the unknown sample is outside the defined likeness boundaries of the ovarian cancer and unaffected clusters.

HIGH-RESOLUTION SELDI PROTEOMIC PATTERNS

Since this original study, several laboratories have validated the use of serum proteomic pattern analysis for the diagnosis of breast [20] and prostate cancer [18]. The most common analytical platform used to acquire the proteomic patterns is the PBS-II, a low-resolution TOF MS as described previously. Many of these studies have been able to correctly diagnose serum samples with sensitivities and specificities in the high 90% range. Even with this high PPV, this technology as it exists would not be useful as a clinical screening tool for low-prevalence diseases such as ovarian cancer. While a PPV of 94% is extremely high, when extrapolated over a large population, in which very few patients would actually have ovarian cancer, the net result would be six out of every 100 patients diagnosed with ovarian cancer would be sent for an unnecessary biopsy or progress to an advanced disease stage when false-negatives are encountered. This situation would overwhelm the available medical resources and would likely not be looked upon favorably by health management organizations (HMOs), which would not pay for an expensive procedure based on a test with a 6% failure rate. An effective screening tool for a low-prevalence disease such as ovarian cancer requires a PPV of 99.6% [32].

Our laboratory, in collaboration with Drs. Liotta and Petricoin, conducted a study to determine if a higher-resolution MS analysis of a well-controlled serum-sample set would result in a diagnostic for ovarian cancer with a higher sensitivity, specificity, and overall PPV. The high-resolution instrument used in this study is a hybrid Qq-TOF MS manufactured by Applied Biosystems Incorporated (ABI), fitted with a Ciphergen SELDI source. An example of a serum sample processed on a weak cation-exchange (WCX2) protein chip and analyzed using the PBS-II TOF MS and the Qq-TOF MS instruments is shown in Figure 4.8. While the spectra are qualitatively similar, the resolution

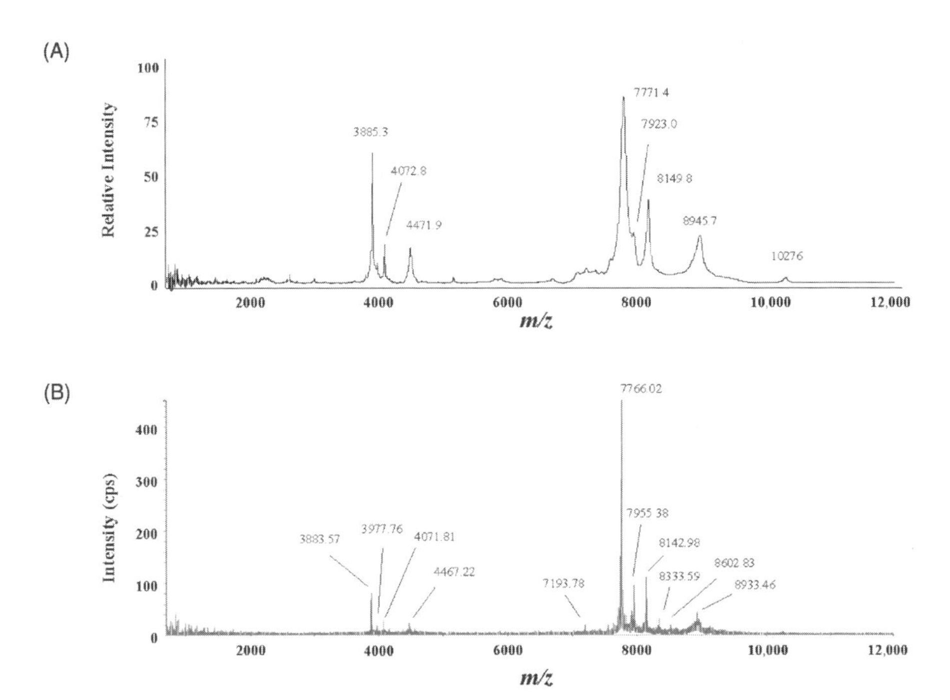

FIGURE 4.8 Comparison of mass spectra acquired using a SELDI-TOF MS and a Qq-TOF MS equipped with a SELDI source.

obtainable with the Qq-TOF MS is 50–60-fold higher than that obtainable with the PBS-II TOF MS. The hope is that the increased resolution will allow a greater number of features to be detected in the spectra acquired using the Qq-TOF MS, thereby increasing the opportunity for finding unique features and patterns within serum proteome profiles that are diagnostic for ovarian cancer.

Analysis of Serum Samples

Two hundred and forty-eight serum samples provided from the National Ovarian Cancer Early Detection Program clinic at Northwestern University Hospital (Chicago, Illinios) were analyzed on the same ProteinChip arrays using both a PBS-II and a Qq-TOF MS fitted with a SELDI interface. The proteomic patterns of the serum samples were acquired on the PBS-II TOF MS, immediately followed by their acquisition on the Qq-TOF MS. The key to this study is that the identical set of serum samples was analyzed on the exact same protein-chip surface, eliminating all experimental variability other than the use of two different instruments.

Data Analysis

The proteomic patterns acquired on both MS instruments were analyzed using the ProteomeQuest bioinformatics tool employing ASCII files consisting of m/z and intensity values [21]. As described earlier, the data were analyzed by the algorithm in an attempt to find a set of features at precise m/z values whose combined, normalized relative intensity values in n-space best segregate the data derived from the training set of spectra. The entire set of spectra acquired from the serum samples was divided into three data sets: (1) a training set that was used to discover the hidden diagnostics patterns; (2) a testing set; and (3) a validation set. The key subset of m/z values identified using the training set were used to classify, in a blinded fashion, the testing and validation sets, hence the algorithm had no prior knowledge of the spectra in the testing and validation sets.

The diagnostic models derived from the training sets were tested using blind serum-sample spectra obtained from 31 unaffected women and 63 women with ovarian cancer. The ovarian cancer diagnostic models were further validated using blind serum-sample spectra obtained from 37 unaffected women and 40 women with ovarian cancer. The results (Figure 4.9) clearly showed the ability of the bioinformatic algorithm to recognize feature sets from mass spectra acquired using the higher-resolution Qq-TOF MS, which result in statistically superior models over those generated using the lower-resolution PBS-II TOF MS. These models were statistically superior, not only in testing

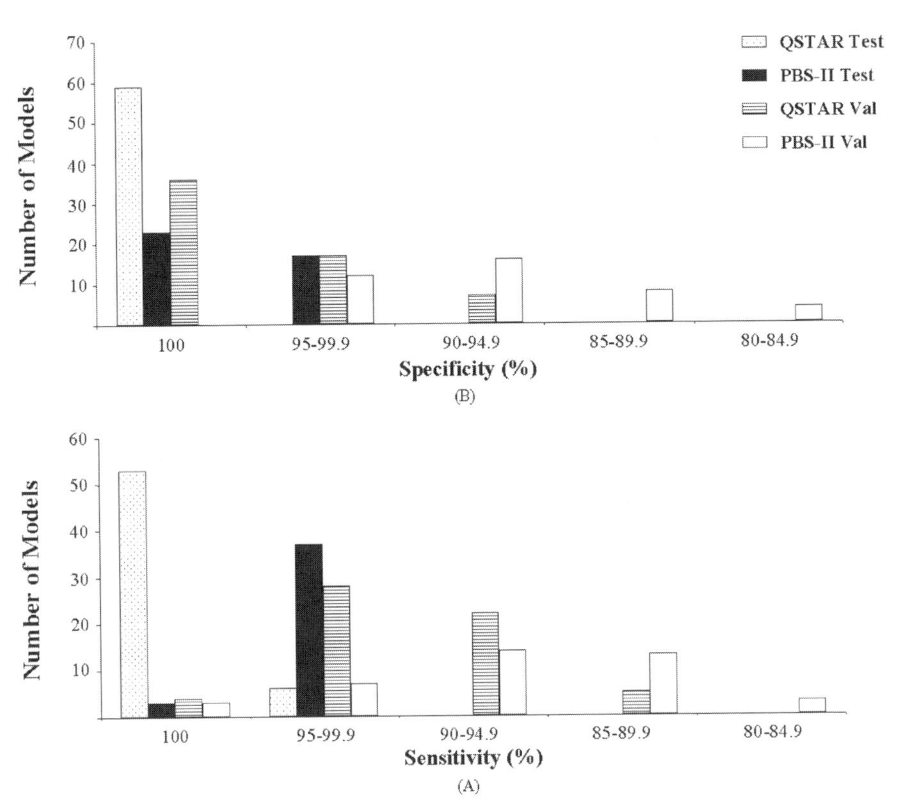

FIGURE 4.9 Histograms representing the testing and blinded validation results of (A) sensitivity and (B) specificity of the diagnostic models for MS data acquired on either a Qq-TOF or a PBS-II TOF mass spectrometer.

(sensitivity, $P_2 < .0001$; specificity, $P_2 < 3 \times 10^{-19}$), but also in validation (sensitivity, $P_2 < 9 \times 10^{-9}$; specificity, $P_2 < 6 \times 10^{-6}$).

Four of the diagnostic models were found to be both 100% sensitive and specific in their ability to correctly discriminate serum samples taken from unaffected women from those taken from suffering from ovarian cancer. These models were generated from data acquired using the Qq-TOF MS. Just as importantly, and key if this technology is to become a viable screening tool, no false-positive or false-negative classifications occurred using these models, giving each of these models a PPV of 100% using the patient cohort employed in this study. No models generated from the PBS-II TOF MS data were both 100% sensitive and specific.

Another key aspect to this study is found in the examination of the key m/z features that comprise the four best performing patterns. One criticism of the use of proteomic patterns for diagnostic purposes is that the identity of the key m/z features is not known. At this point, it is debatable as to whether it is worth

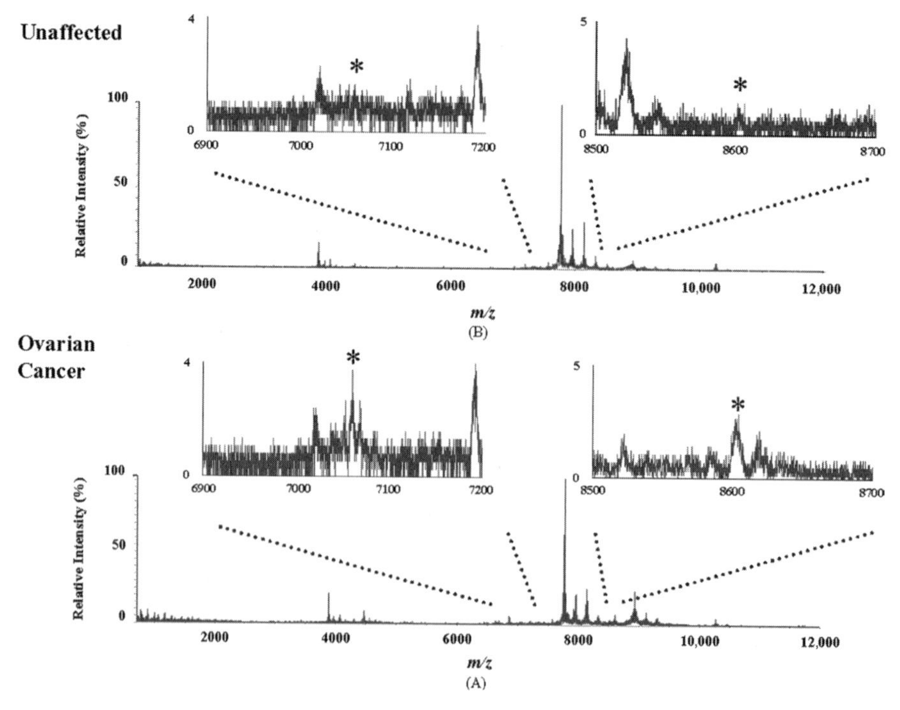

FIGURE 4.10 Comparison of SELDI Qq-TOF mass spectra of serum from an ovarian cancer patient (panel A) and an unaffected individual (panel B). Insets show expanded m/z regions highlighting significant intensity differences of the peaks in the m/z bins 7060.121 and 8605.678, identified by the algorithm as belonging to the optimum discriminatory pattern.

the effort to identify these features, as they may provide little aid in developing an alternative diagnostic platform. There is also the likelihood that these key values may represent proteins that provide exciting insights to the manifestation and progression of cancer, and therefore identifying them is most likely a worthwhile effort. Examination of the four models that had 100% PPV for ovarian cancer reveals certain consistent features. Although the proteomic patterns generated from both healthy and cancer patients using the Qq-TOF MS are quite similar (Figure 4.10), careful inspection of the raw mass spectra reveals that key features within the binned m/z values 7060.121 and 8605.678 are indeed differentially abundant in a selection of the serum samples obtained from ovarian cancer patients as compared to unaffected individuals (Figure 4.10, insets). The results indicate that these MS peaks originate from species that may be consistent indicators of the presence of ovarian cancer and represent good candidates that may be key disease-progression indicators. The consistency of these peaks within spectra of serum acquired from patients with ovarian cancer, provide an excellent target for identification. Efforts at identifying these low molecular-weight components in serum are ongoing in our laboratory. Indeed, one of the overlooked powers of the proteomic pattern approach is its ability to screen hundreds of serum samples in a high throughput manner, and therefore quickly determine targets for further investigation. It must be reiterated that the ability to distinguish sera from an unaffected individual or an individual with ovarian cancer based on a single serum proteomic *m/z* feature alone, however, is not possible across the entire serum study set. Correct diagnosis is only possible when the key *m/z* features and their intensities are analyzed as a whole.

While the sensitivity and specificity of a previous report using a lower-resolution PBS-II TOF MS to diagnose ovarian cancer was impressive, to screen for diseases of relatively low prevalence, such as ovarian cancer, a diagnostic test must exceed 99.6% sensitivity and specificity to minimize false-positives, while correctly detecting early-stage disease when it is present [32]. In blinded testing and validation studies any one of the four best models generated using Qq-TOF MS data were able to correctly classify 22/22 women with stage I ovarian cancer, 81/81 women with stage II, III, and IV ovarian cancer, and 68/68 benign disease controls. It can be envisioned that in the near future a clinical test would simultaneously employ several combinations of highly accurate diagnostic proteomic patterns. Taken together, these patterns can achieve an even higher degree of accuracy in screening a large population heterogeneity and potential variability in sample quality and handling. Hence, a high-resolution system, such as the Qq-TOF MS employed in this study, is preferred based on the present results that serve as a platform for clinical trials of serum proteomic patterns.

DISCUSSION

Proteomics has been ushered in with the hope that the use of high throughput technologies to characterize complex biological matrices will lead to the discovery of a plethora of novel disease-specific biomarkers. Unfortunately, the lack of true success stories of biomarker discovery, despite the considerable intellectual and financial resources currently invested in the use of conventional proteomic technologies, gives one pause. It is probable that a vast majority of disease states do not manifest themselves in such a manner that a single recognizable change in a protein can be used to diagnose it with high accuracy. For example, both CA 125 for ovarian cancer, and PSA for prostate cancer, have low PPV over a large population. When the complexity of an individual cell and the aberrations caused by such disease states as cancer are considered, a vast number of differences between the protein character of healthy and diseased tissues would be expected. Obviously one of the main reasons that the discovery of disease-specific biomarkers has been so elusive is that for a diagnostic marker to be clinically relevant it should be assayed from a sample that can be relatively noninvasively obtained in sufficient quantity from patients, therefore, the search for biomarkers has largely focused on plasma and serum. While serum constantly perfuses tissues, hence potentially endowing an archive of disease relevant information, this information is comprised not only of the expected circulatory proteins in serum such as immunoglobulins but also of peptides and proteins that are secreted into the blood and species shed from diseased, dying, or dead cells present throughout the body [12]. This background matrix within biofluids, such as serum, represents a complex milieu in which, low-abundance, disease-specific biomarkers may be discovered. While the identification of a reliable biomarker relies on the comparison of samples from thousands of healthy and disease-stricken individuals, the comparison of even two distinct serum samples is incredibly laborious using conventional proteomic technologies. More to the point, in the comparison of just two serum samples, a multitude of changes in protein abundances are observed due simply to differences in age, gender, lifestyle, etc., making the assumption that a particular difference is a result of a specific disease state tenuous at best.

Disease diagnosis using proteomic patterns has emerged as a potentially revolutionary new method to use proteomic technology for the early detection of diseases such as cancer. A major criticism of the disease diagnosis using proteomic patterns is that the identity of the proteins or peptides giving rise to the diagnostic features within each spectrum is not known. It is debatable as to whether it is worth the effort to identify these features, since their identity may provide little novel information regarding the mechanism of cancer progression or aid in the development of an alternative diagnostic platform. Most of the key features within the proteomic patterns are of low m/z (i.e., $<10{,}000$ Da), therefore it is likely that these could be from fragment species generated from larger proteins that are proteolyzed either within the circulatory system or in the tumor/host microenvironment. It would be difficult to generate an affinity reagent with specificity to a peptide fragment without considerable cross-reactivity to its parent protein. In addition, there are many tools in medicine today in which physicians rely solely on a pattern to base their diagnosis, such as an electroencephalogram. Even the identification of a specific biomarker may not provide any direct insight into how a disease may arise or progress. Take, for instance, PSA, which is used to indicate the possible presence of a prostatic tumor. Its role in cancer development, however, remains unclear. Conversely, the likelihood that these key features may represent proteins that provide exciting insights to the manifestation and progression of cancer still exists. Therefore, their identification is most likely a worthwhile effort, although the advancement of disease diagnostics using proteomic patterns should not be hindered by this exercise.

Disease diagnostics using proteomic patterns has rapidly emerged as a potentially revolutionary tool to detect and monitor disease progression or therapeutic response. It represents a complete about face in proteomic analysis. While the trend in proteomic technology has been to identify and characterize an increasing number of proteins from a particular clinical sample in order to find a disease-specific biomarker, proteomic patterns rely simply on a crude proteomic survey that provides all of the necessary diagnostic information. Although the potential is great, much still needs to be learned. The concept of using a proteomic pattern as a diagnostic tool is in its infancy, therefore every step in this analytical process requires optimization. This optimization process will include such aspects as sample acquisition and processing, pattern acquisition, and data analysis. Since the diagnostic power of proteomic patterns relies heavily on the use of bioinformatics, it is important to discover the biological basis behind the mathematical solution. While the identification of key peaks that are called out by the bioinformatic analysis may not provide any clues as to the manifestation or progression of the disease, the hope is they can at least validate the results being provided. While many critics still abound, one simple fact cannot be ignored: the diagnostic models generated from proteomic patterns continue to provide highly sensitive and specific results in testing and blind validation studies, even as the number of samples being analyzed continues to increase.

The next few years will be critical in the validation of the use of proteomic patterns in disease detection. While currently the information present in proteomic patterns may provide an extremely powerful complementary tool to assist physicians in disease diagnosis, the impact of proteomics in disease diagnosis is even greater. The niche that proteomics will fill within the field of diagnostic medicine remains to be determined. The most obvious benefit of using proteomic patterns to diagnose disease states is in screening large populations to detect diseases, such as cancer, at earlier stages, to enable more effective medical intervention. The ease by which proteomic patterns can be acquired makes it feasible to screen populations at high risk for a variety of different cancers. If the sensitivity and specificity of diagnosing cancer using proteomic patterns can approach 100%, its use may revolutionize diagnostic medicine. Even if this level of sensitivity and specificity is not achieved, proteomic patterns will still provide an invaluable complement to determine the need for a patient biopsy or response to therapy.

ACKNOWLEDGEMENTS

The authors acknowledge the laboratories of Dr. Emmanuel Petricoin of the Food and Drug Administration and Dr. Lance Liotta of the National Cancer

Institute, as well as Correlogics Systems, for their collaboration on using high-resolution mass spectrometry to diagnose ovarian cancer.

This project was funded in whole or in part with federal funds from the National Cancer Institute, National Institutes of Health, under Contract No. NO1-CO-12400.

REFERENCES

1. Thomson, J.J. "Discharge of Electricity through Gases," *Philos. Mag. V.* **44,** 293 (1897).
2. McLafferty, F.W., ed. *Tandem Mass Spectrometry*, Wiley, New York (1983).
3. Martin, S.A.; Rosenthal, R.S.; Biemann, K. "Fast Atom Bombardment Mass Spectrometry and Tandem Mass Spectrometry of Biologically Active Peptidoglycan Monomers from *Neisseria gonorrhoeae*," *J. Biol. Chem.* **262,** 7514–7522 (1987).
4. Dole, M.; Ferguson, L.D.; Hines, R.L.; Mobley, R.C.; Alice, M.B. "Molecular Beams of Macroions," *J. Chem. Phys.* **49,** 2240–2249 (1968).
5. Karas, M.; Hillenkamp, F. "Laser Desorption Ionization of Proteins with Molecular Masses Exceeding 10,000 Daltons," *Anal. Chem.* **78,** 53–68 (1987).
6. Fenn, J.B.; Mann, M.; Meng, C.K.; Wong, S.F.; Whitehouse, C.M. "Electrospray Ionization for Mass Spectrometry of Large Biomolecules," *Science* **246,** 64–71 (1989).
7. Tanaka, K.; Waki, H.; Ido, Y.; Akita, S.; Yoshida, Y.; Yoshida, T. "Protein and Polymer Analyses up to *m/z* 100 000 by Laser Ionization Time-of-flight Mass Spectrometry," *Rapid Commun. Mass Spectrom.* **2,** 151–153 (1988).
8. O'Farrell, P.H.; "High Resolution Two-dimensional Electrophoresis of Proteins," *J. Biol. Chem.* **250,** 4007–4021 (1975).
9. McDonald, W.H.; Yates, J.R., III "Shotgun Proteomics and Biomarker Discovery," *Dis Markers* **18**(2), 99–105 (2002).
10. Cash, P.; Kroll, J.S.; "Protein Characterization by Two-dimensional Gel Electrophoresis," *Methods Mol. Med.* **71,** 101–118 (2003).
11. Chen, G.; Gharib, T.G.; Huang, C.C.; Thomas, D.G.; Shedden, K.A.; Taylor, J.M.; Kardia, S.L.; Misek, D.E.; Giordano, T.J.; Iannettoni, M.D.; Orringer, M.B.; Hanash, S.M.; Beer, D.G. "Proteomic Analysis of Lung Adenocarcinoma: Identification of a Highly Expressed Set of Proteins in Tumors," *Clin. Cancer Res.* **8**(7), 2298–2305 (2002).

12. Adkins, J.N.; Varnum, S.M.; Auberry, K.J. et al. "Toward a Human Blood Serum Proteome: Analysis by Multidimensional Separation Coupled with Mass Spectrometry," *Mol. Cell. Proteomics* **1**(12), 947–955 (2002).

13. Pieper, R.; Su, Q.; Gatlin, C.L.; Huang, S.T.; Anderson, N.L.; Steiner, S. Multi-Component Immunoaffinity Subtraction Chromatography: An Innovative Step Towards a Comprehensive Survey of the Human Plasma Proteome," *Proteomics* **3**(4), 422–432 (2003).

14. Anderson, N.L.; Anderson, N.G. The Human Plasma Proteome: History, Character, and Diagnostic Prospects, *Mol. Cell. Proteomics* **1,** 845–867 (2002).

15. Hutchens, T.W.; Yip, T.T. "New Desorption Strategies for the Mass Spectrometric Analysis of Macromolecules," *Rapid. Commun. Mass Spectrom.* **7,** 576–580 (1993).

16. Issaq, H.J.; Conrads, T.P.; Prieto, D.A.; Tirumalai, R.; Veenstra, T.D. "SELDI-TOF MS for Diagnostic Proteomics," *Anal. Chem.* **75**(7), 148A–155A (2003).

17. Paweletz, C.P.; Liotta, L.A.; Petricoin, E.F. "New Technologies for Biomarker Analysis of Prostate Cancer Progression: Laser Capture Microdissection and Tissue Proteomics," *Urology* **57,** 160–163 (2001).

18. Wright, G.L.; Cazares, L.H.; Leung, S.M.; et al. "ProteinChip Surface Enhanced Laser Desorption/ionization (SELDI) Mass Spectrometry: A Novel Protein Biochip Technology for Detection of Prostate Cancer Biomarkers in Complex Protein Mixtures, *Prostate Cancer and Prostatic Dis.*, **2,** 264–276 (2000).

19. Vlahou, A.; Schellhammer, P.F.; Medrinos, S.; et al. "Development of a Novel Approach for the Detection of Transitional Cell Carcinoma of the Bladder in Urine," *Am. J. Pathology* **158,** 1491–1501 (2001).

20. Wulfkuhle, J.D.; McLean, K.C.; Pawaletz, C.P.; Sgroi, D.C.; Trock, B.J.; Steeg, P.S.; Petricoin, E.F. "New Approaches to Proteomic Analysis of Breast Cancer," *Proteomics* **1,** 1205–1215 (2001).

21. Petricoin, E.F.; Ardekani, A.M.; Hitt, B.A.; Levine, P.J.; Fusaro, V.A.; Steinberg, S.M.; Mills, G.B.; Simone, C.; Fishman, D.A.; Kohn, E.C.; Liotta, L.A. "Use of Proteomic Patterns in Serum to Identify Ovarian Cancer," *Lancet*, **359,** 572–577 (2002).

22. Cardone, M.H.; Roy, N.; Stennicke, H.R.; Salvesen, G.S.; Franke, T.F.; Stanbridge, E.; Frisch, S.; Reed, J.C. "Regulation of Cell Death Protease Caspase-9 by Phosphorylation," *Science* **282,** 1318–1321 (1998).

23. Chernyak, A.; Karavanov, A.; Ogawa, Y.; Kovac, P. "Conjugating Oligosaccharides to Proteins by Squaric Acid Diester Chemistry: Rapid Monitoring of the Progress of Conjugation, and Recovery of the Unused Ligand," *Carbohydrate Res.* **330,** 479–486 (2001).

24. Forde, C.E.; Gonzales, A.D.; Smessaert, J.M.; Murphy, G.A.; Shields, S.J.; Fitch, J.P.; McCutchen-Maloney, S.L. "A Rapid Method to Capture and Screen for Transcription Factors by SELDI Mass Spectrometry," *Biochem. Biophys. Res. Commun.* **290,** 1328–1335 (2002).

25. Merchant, M.; Weinberger, S.R. "Recent Advancements in Surface-enhanced Laser Desorption/Ionization-Time of Flight-Mass Spectrometry," *Electrophoresis* **21**(6), 1164–1177 (2000).
26. Rosty, C.; Christa, L.; Kuzdzal, S.; Baldwin, W.M.; Zahurak, M.L.; Carnot, F.; Chan, D.W.; Canto, M.; Lillemoe, K.D.; Cameron, J.L.; Yeo, C.J.; Hruban, R.H.; Goggins, M. "Identification of Hepatocarcinoma-Intestine-Pancreas/Pancreatitis-Associated Protein I as a Biomarker for Pancreatic Ductal Adenocarcinoma by Protein Biochip Technology," *Cancer Res* **62**(6), 1868–1875 (2002).
27. Greenlee, R.T.; Murray, T.; Bolden, S.; Wingo, P.A. "Cancer Statistics, 2000," *CA Cancer J. Clin.*, **50,** 7–33 (2000).
28. Yeo, C.J.; Cameron, J.L.; Lillemoe, K.D.; Sitzmann, J.V.; Hruban, R.H.; Goodman, S.N.; Dooley, W.C.; Coleman, J.; Pitt, H.A. "Pancreaticoduodenectomy for Cancer of the Head of the Pancreas, 201 patients," *Ann. Surg.* **221,** 721–731 (1995).
29. Jacobs, I.J.; Skates, S.J.; MacDonald, N. et al. "Screening for Ovarian Cancer: A Pilot Randomised Controlled Trial," *Lancet* **353,** 1207–1210 (1999).
30. Menon, U.; Jacobs, I. Recent Developments in Ovarian Cancer Screening. *Curr. Opin. Obstet. Gynaecol.* **12,** 39–42 (2000).
31. Cohen, L.S.; Escobar, P.F.; Scharm, C.; Glimco, B.; Fishman, D.A. "Three-dimensional Power Doppler Ultrasound Improves the Diagnostic Accuracy for Ovarian Cancer Prediction." *Gynecol. Oncol.* **82,** 40–48 (2001).
32. Kainz, C. "Early Detection and Preoperative Diagnosis of Ovarian Carcinoma," *Wien. Med. Wochenschr.* **146**(1–2) 2–7 (1996).

5

METABONOMIC APPLICATIONS IN TOXICITY SCREENING AND DISEASE DIAGNOSIS

JOHN P. SHOCKCOR AND ELAINE HOLMES

INTRODUCTION

History of Metabonomics

In the current climate within the pharmaceutical industry, there is increased pressure to optimize the efficiency of high throughput toxicity screening for lead compound selection. This focus has resulted in a concerted effort to evaluate the role of technologies such as genomics, proteomics, and metabonomics. These technologies have the potential to generate megavariate data sets that contain information that relates to an organism's response to drugs at the gene expression, cellular protein, or metabolic level. The data-rich matrices generated can then be interrogated with appropriate data mining and multivariate analytical tools [1–4]. This chapter addresses the contribution and potential value of nuclear magnetic resonance (NMR)-based metabonomic approaches to toxicity screening and disease diagnosis.

Metabonomic analysis involves the quantitation of the dynamic multivariate metabolic response of an organism to a pathological event or genetic modification [1]. The concept of metabonomics has evolved over two decades of ^{1}H NMR spectroscopic analysis of the multicomponent metabolic composition of biofluids, cells, and tissues under different physiological and

Integrated Strategies for Drug Discovery Using Mass Spectrometry, Edited by Mike S. Lee

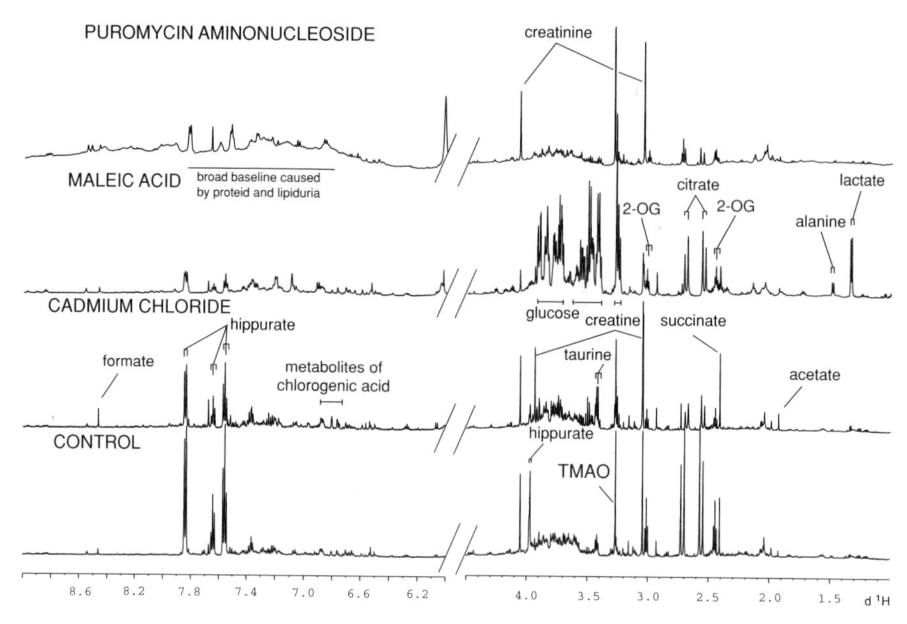

FIGURE 5.1 600 MHz ^{1}H NMR spectra of urine samples obtained from a control rat and rats treated with cadmium chloride (testicular toxin), maleic acid (proximal tubular renal toxin) and puromycin aminonucleoside (renal glomerular toxin).

pathophysiological conditions [5–14]. To date, high field NMR spectroscopy coupled with advanced chemometric data analysis has been the dominant analytical platform for acquiring metabonomic data. However, high-performance liquid chromatography [HPLC], mass spectrometry [MS], gas chromatography MS [GC-MS] and near infrared [NIR] spectroscopy have also been used to generate metabonomic data [15–21]. High-resolution ^{1}H NMR spectra of biological matrices, such as urine, blood plasma, or tissue samples, generate complex spectral profiles (Figure 5.1) that contain a wealth of metabolic information that relates to the physiological or pathophysiological status of the organism [1,13,22–26]. ^{1}H NMR spectroscopic analysis is nondestructive, cost effective, and typically takes only a few minutes per sample. Little or no sample pretreatment or reagents is required, and is therefore bioanalytically more efficient than the methods used to characterize either the genetic or proteomic composition of samples. Typically, 600–800 MHz ^{1}H NMR spectra of biofluids such as urine and plasma contain thousands of signals arising from hundreds of endogenous molecules that represent many biochemical pathways. Toxin- or disease-induced changes in the biochemical composition of these biofluids are reflected by modulations in the pattern of the ^{1}H NMR spectral profiles (Figure 5.2). Application of automated data-reduction algorithms and chemometric analysis to spectral biofluid databases facilitates

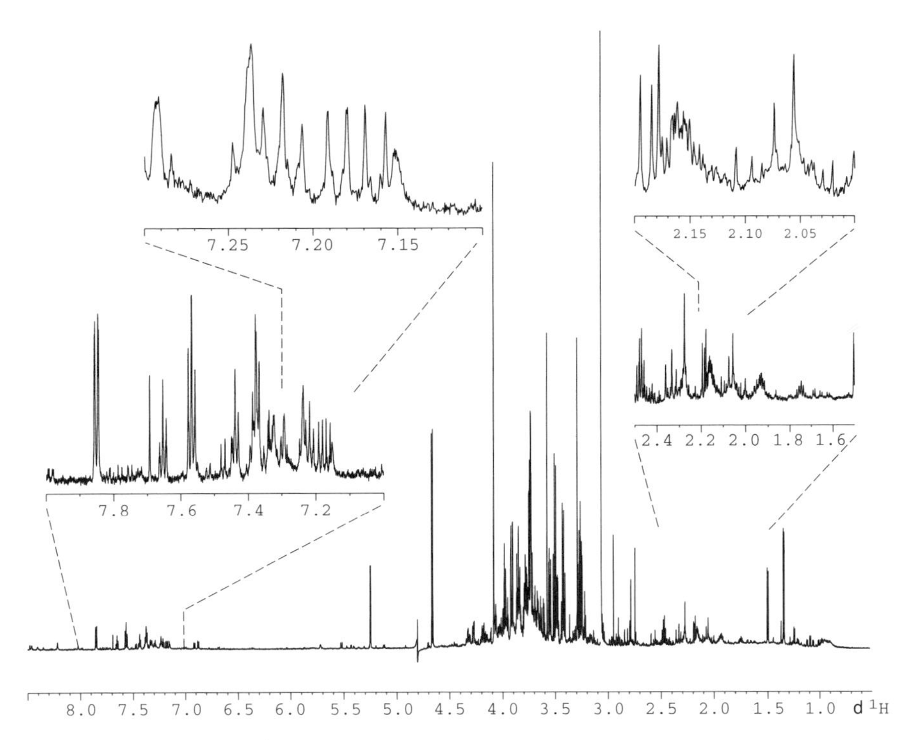

FIGURE 5.2 600 MHz ^{1}H NMR urine spectrum of control urine with expansions of selected spectral regions.

the interpretation of these complex modulations in spectral profiles, and can enable the classification of samples according to their metabolic status based on similarity of biochemical composition. Pattern recognition (PR) and related chemometric approaches can be used to discern significant patterns in complex data sets [27–30], and are particularly appropriate in situations where there are more variables than samples in the data set, such as is the case with spectral data. The general aim of PR is to classify objects (in this case ^{1}H NMR spectra of biofluid or tissue samples) or to predict the origin of objects (the class of toxicity or disease in the current example), based on identification of inherent patterns in a set of indirect measurements. Moreover, PR can be used to reduce the dimensionality of complex data sets via two-dimensional 2D or 3D mapping procedures, thereby facilitating the visualization of inherent patterns in the data set. Both the theory and the application of the basic mathematical models used in PR have been well documented [27–30]. Typically, 1D biofluid spectra are digitized into a series of spectral descriptors that are scaled to provide input variables to map samples and construct mathematical models for specific types of physiological or pathological condition (Figure 5.3). Metabonomic technology has proved a powerful toxicological

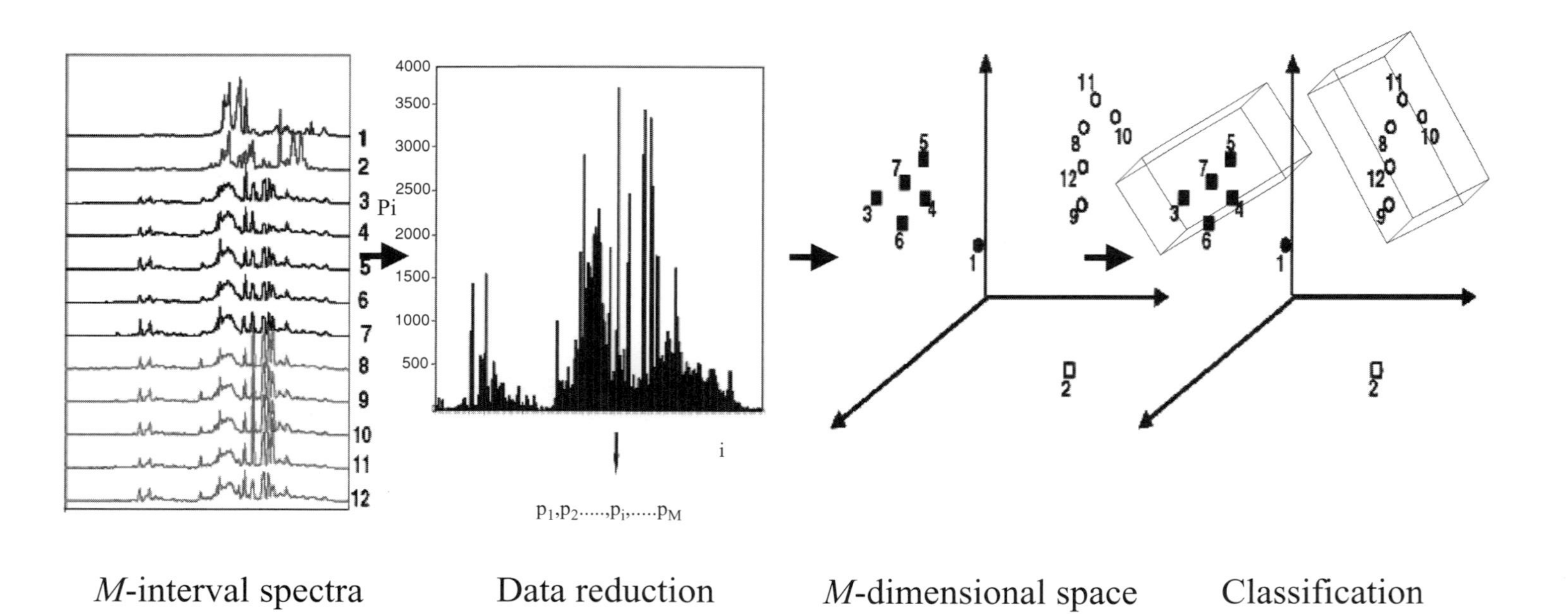

FIGURE 5.3 Schematic diagram of the sequence of spectral data reduction and analysis.

probe, both for the characterization of site- or mechanism-specific toxicity and for identification of toxicological biomarkers in vivo [3,7,12,31,32]. However, although metabonomic analysis has proved extremely successful in terms of toxixty profiling and screening, it is by no means limited to this field. Indeed, metabonomic analysis has a wide range of applications, including quality control of food and chemical products [33–37], evaluation of environmental contamination [38–40], diagnosis of metabolic diseases [41–43], classification of tumors [44–48], evaluation of drug efficacy [49], and monitoring physiological variation [24,50,51].

TOXICOLOGICAL APPLICATIONS OF METABONOMICS

High-resolution ^{1}H NMR spectroscopic analysis of biofluids has proved to be one of the most powerful techniques for the investigation organism response to xenobiotics. Exposure of an organism to a xenobiotic results in subtle modifications in the biochemical composition of intra- and extracellular fluids as the organism attempts to maintain homeostasis (constancy of internal environment). This metabolic adjustment involves altering the composition of body fluids such as urine and plasma, and this change in biochemical composition can be rapidly profiled with ^{1}H NMR spectroscopic analysis. ^{1}H NMR spectral profiles of biofluids provide a "unique" fingerprint of the metabolic state of an organism, and can provide information on the nature of a drug or toxin to which an animal has been exposed [1,5–7,52–55]. The site or basic mechanism of toxicity can often be determined from characteristic changes in the concentrations and patterns of endogenous metabolites in biofluids (Figure 5.2). For example, increased urinary excretion of taurine and creatine together with bile aciduria generally reflects a cholestatic lesion [5]. While each toxin will give rise to a unique spectral profile, compounds that target the same tissue will possess similar metabolic profiles. ^{1}H NMR urine spectra obtained from $HgCl_2$ and hexachlorobutadiene (HCBD), both of which target the renal proximal tubules, contain elevated resonances that pertain to glucose, amino acids, and organic acids, together with a decrease in signals from citrate and succinate [31]. The biochemical consequences of over 100 drugs and model toxins have been characterized metabonomically via ^{1}H NMR spectroscopy of biofluids such as urine, plasma, and bile, and large spectral databases, defining toxicological events, have been constructed. Selected toxins analyzed with metabonomic methods have been listed in Table 5.1 together with the major associated biomarkers of toxicity.

TABLE 5.1 Metabolic Markers Associated with Various Toxins as Determined via Metabonomic Analysis (as illustrated in published literature [5–7, 11–14, 75–79])

Toxin and Reference	Target Organ/Toxicity Type	Associated Biomarkers
Adriamycin	Heart and kidney (glomerulus)	↓citrate, ↑creatine, ↑taurine, ↑↓2-OG
Allyl alcohol	Liver (periportal)	↑creatine, ↓citrate, ↓2-OG, ↑lactate, ↑phenylacetylglycine, ↑NMN, ↑taurine
Amiodarone	Phospholipidosis (lung)	↑phenylacetylglycine, ↑DMG
α-Naphthylisothiocyanate (ANIT)	Liver (cholestasis)	↑acetate, ↑bile acids, ↓citrate, ↑glucose, ↓hippurate, ↓2-OG, ↓succinate,
2-Bromoethanamine	Kidney (papilla) and mitochondrial dysfunction	↑adipic acid, ↑DMG, ↑glutaric acid, ↑ *N*-acgly, ↓↑succinate, ↓↑TMAO
Butylated hydroxytoluene	Liver	↑glucose, ↑taurine
Cadmium chloride	Testicular	↑creatine, ↑glucose, ↓citrate
Carbon tetrachloride	Liver	↑taurine, ↑creatine, ↓citrate, ↓2-OG, ↓succinate
Cefoperazone[a]		↑glucose, ↓hippurate
Cephaloridine[a]		↑alanine, ↑citrate, ↑glucose, ↑glutamine ↑glycine, ↓hippurate, ↑lactate,
Cephalothin[a] Z3		↑glucose, ↓hippurate
2-Chloroethanamine	Kidney (papilla) and mitochondrial dysfunction	↑adipic acid, ↑DMG, ↑glutaric acid, ↑*N*-acgly, ↓↑succinate, ↓↑TMAO
Chloroquine	Phospholipidosis and liver necrosis	↑phenylacetylglycine, ↑DMG
S-(1,2-dichlorovinyl)-L-cysteine (DCVC) Z4	Kidney (S2/S3 proximal tubular)	↑citrate ↑succinate
S-(1,2-dichlorovinyl)-L-homocysteine (DCVHC)	Kidney (S2/S3 proximal tubular)	↑acetate, ↑amino acids, ↓citrate ↓creatinine, ↑glucose, ↓hippurate, ↑organic acids, ↓2-OG, ↓succinate
Ethionine	Liver	↑glucose, ↓2-OG, ↑taurine
Galactosamine	Liver (hepatitis-like lesion)	↑acetate, ↑betaine, ↑bile acides, ↑creatine, ↓hippurate, ↑organic acids, ↓2-OG, ↓succinate, ↑taurine, ↑urocanic acid
Hexachlorobutadiene	Kidney (S3 proximal tubular)	↑acetate, ↑amino acids, ↓citrate ↓creatinine, ↑glucose, ↓hippurate, ↑organic acids, ↓2-OG, ↓succinate

Hydrazine	Liver (steatosis)	↑2-amino adipate, ↑β-alanine, ↑creatine, ↓creatinine, ↓fumarate, ↓hippurate, ↑NAC, ↓TMAO
Ifosfamide Z2		↑glycine, ↑glucose, ↓hippurate, ↑histidine, ↑lactate, ↑TMAO
Imipenem[b] Z5	Kidney	β-hydroxybutyrate and other ketone boides
Lead acetate	Liver/lung/kidney	↑acetate, ↑creatine, ↓citrate, ↑glucose, ↓hippurate, ↑lactate, ↓2-OG, ↑*N*-acetyls, ↓succinate, ↑taurine,
Mercuric chloride	Kidney (S3 proximal tubular)	↑acetate, ↑amino acids, ↓citrate ↓creatinine, ↑glucose, ↓hippurate, ↑organic acids, ↓2-OG, ↓succinate
para-Aminophenol	Kidney (S3 proximal tubular)	↑acetate, ↑amino acids, ↓citrate ↓creatinine, ↑glucose, ↓hippurate, ↑organic acids, ↓2-OG, ↓succinate
Paraquat	Kidney and lung	↑amino acids, ↓citrate ↓creatinine, ↑glucose, ↓hippurate, ↑lactate and organic acids, ↓valine
Propyleneimine	Kidney (papilla)	↑*N*-acgly, ↓↑2-OG, ↓↑TMAO
Puromycin aminonucleoside	Kidney (glomerulus and proximal tubular)	↑acetate, ↑alanine, ↓citrate, ↑creatine, ↑formate, ↑glucose, ↑macromolecules (proteins and lipids), ↓2-OG, ↑taurine, ↑TMAO
Sodium chromate	Kidney (S1 proximal tubular)	↓citrate, ↑glucose, ↓hippurate, ↓2-OG
Sodium fluoride	Kidney (proximal tubular)	↑acetate, ↑amino acids, ↓citrate ↓creatinine, ↑glucose, ↓hippurate, ↑organic acids, ↓2-OG, ↓succinate, ↑threonine
1,1,2-Trichloro-3,3,3-trifluoro-1-propene (TCTFP)	Kidney (S3 proximal tubular)	↑acetate, ↑amino acids, ↓citrate ↓creatinine, ↑glucose, ↓hippurate, ↑organic acids, ↓2-OG, ↓succinate
Thioacetamide	Liver and kidney	↑acetate, ↑amino acids, ↓citrate ↓creatine, ↑glucose, ↓hippurate, ↑organic acids, ↓2-OG, ↓succinate, ↑taurine, ↑TMAO
Urany1 nitrate	Kidney (S3 proximal tubular)	↑acetate, ↑amino acids, ↓citrate ↓creatinine, ↑glucose, ↓hippurate, ↑organic acids, ↓2-OG, ↓succinate, ↑urea

[a] New Zealand white rabbit.
[b] Cynomolgus monkey.
[c] human.

Abbreviations: DMG—dimehtylglycine; NAC—*N*-acetyl-citrulline; NMN—*N*-methyl nicotinamide; TMAO–trimethylamine-*N*-oxide.

Overt toxicity, such as is detected following maleic acid or puromycin toxicity (Figure 5.2), is relatively easy to characterize with principal-components analysis (PCA). In the simplest case, scored or quantitated measurements of selected metabolite signals that indicate an elevation or depletion in the levels of selected urinary metabolites after toxic insult, followed by appropriate mathematical analysis such as PCA can be used to characterize toxic lesions [6,26,31]. Despite the rudimentary nature of the scoring systems, clear relationships between metabolic composition and the dominant site of toxicity could be established for toxins that targeted the renal cortex, renal medulla, liver, and testes [26]. More recent methods of selecting spectral descriptors have involved automated approaches that incorporate the whole NMR spectrum, either with computer points or integrated spectral regions (as shown in Figure 5.3), thereby any degree of subjectivity in metabolite selection is removed and the speed of analysis is increased.

For the purpose of screening for overt toxicity, PCA is one of the most useful and easily applied PR techniques that require no a priori knowledge as to the class of the samples. Principal components (PCs) are linear combinations of the original variables and are calculated such that (1) each PC is orthogonal (uncorrelated) with all other PCs, and (2) the first PC contains the largest part of the variance of the data set (information content) with subsequent PCs that contain correspondingly smaller amounts of variance. Thus, the "best" representation given in two or three dimensions, in terms of biochemical variation in the data set, can be obtained by a plot of the first two or three PCs. However, for more subtle or complex toxin or disease models, unsupervised chemometric methods such as PCA, have limited capabilities of classification, particularly where large numbers of classes exist within a data set. Therefore, when the existence of clustering behavior (relating to the metabolic status of an organism) is established, supervised methods of analysis can be used to maximize the separation between two or more sample classes and to define features (i.e., biochemical markers) that distinguish each class of toxin-treated urine samples from control. These supervised methods include soft independent modeling of class analogy (SIMCA), K nearest-neighbor (KNN), and neural-network analysis [27–30].

Although many of the differences between ^{1}H NMR spectra obtained from control and toxin-treated rats can be readily observed without detailed mathematical analysis [54,55], the biomarker information within NMR spectra of biofluids is much more subtle and rich than a small set of biochemicals that define a single metabolic event. Hundreds of compounds that represent many pathways can often be measured simultaneously. The overall metabonomic response to toxic insult (occurring over time) that so well characterizes a lesion [5,56] and perturbations in the levels of metabolites present at low concentration may be equally diagnostic and important to understand the biochemical sequelae that follow a toxic insult. For example, several bile acids are excreted in elevated amounts in the urine of rats treated with the hepatotoxin galactosamine [5]. Although these bile acids are present in relatively low concentrations, the pattern of bile acids can be indicative of the site of toxicity within the liver and have more diagnostic potential than other more dominant spectral changes, such as the marked reduction in urinary citrate [5]. The use of supervised mathematical modeling techniques such as partial least square (PLS) discriminant analysis allows an unbiased assessment of the response of metabolites, regardless of the extent of their contribution to the overall composition of a biofluid.

For the purposes of this chapter, two compounds that are known to induce phospholipidosis, chloroquine, and amioderone [57], have been selected as models for illustrating NMR-based metabonomic analysis. Urine samples obtained from Fischer rats treated with chloroquine (at 60 mg/kg (i.p.)) or amiderone (at 80 mg/kg i.p.) for a period of 21 consecutive days analyzed via ^{1}H NMR spectroscopy at 600.13 MHz [58]. Using PCA to map these samples along with control samples, it can be seen that three clusters of data relating to treatment can be obtained (Figure 5.4A). Analysis of these data indicated that both chloroquine and amioderone generated phospholipidosis, but in addition, chloroquine exhibited metabolic effects consistent with hepatotoxicity, causing the chloroquine-derived samples to map separately from the amioderone samples. Data-filtering methods can be used to deconvolve biological matrices such that any biochemical effect not related to the class of interest is removed [59]. We show the application of data-filtering methods to remove the influence of chloroquine-induced hepatotoxicity from the data set to focus solely on the biochemical consequences of phospholipidosis (Figure 5.4B). The PC plot of the filtered data now shows only two groups that relate to control and treated samples, which are clearly separated in the first principal component. The loadings plot generated from the first principal component indicated that the

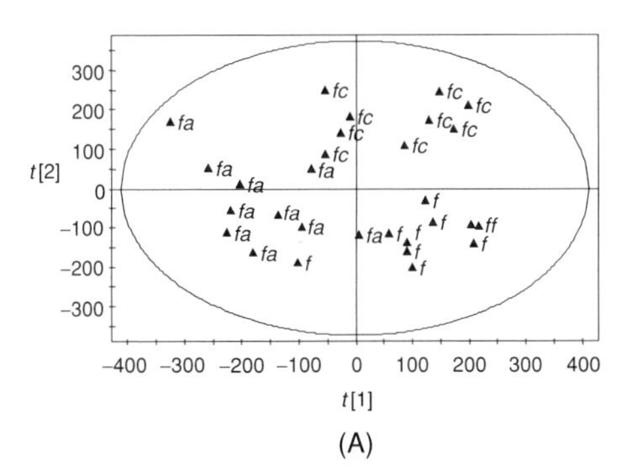

FIGURE 5.4A PCA scores plot from analysis of Fischer Rat urines. Vehicle control rats (f), rats dosed with amiodarone (fa) and rats dosed with chloroquine (fc).

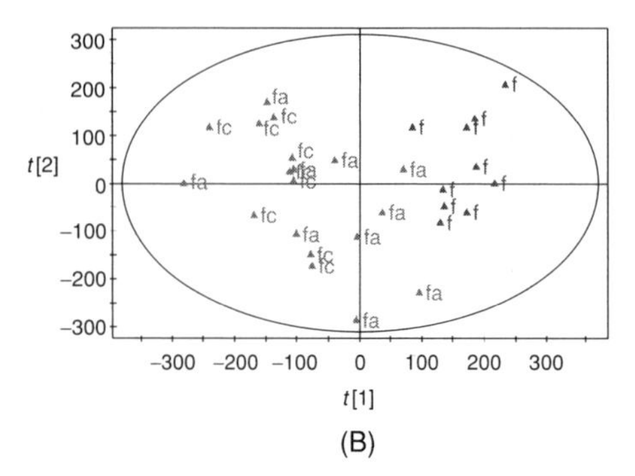

FIGURE 5.4B PCA scores plot from analysis of Fischer Rat urines after application of data filtering methods to remove the influence of chloroquine-induced hepatotoxicity from the data set to focus on the biochemical consequences of phospholipidosis. Vehicle control rats (f), rats dosed with amiodarone (fa) and rats dosed with chloroquine (fc).

spectral regions responsible for differentiation between the treated and control populations were strongly influenced by phenylacetyl glycine, 2-oxoglutarate, and dimethylglycine (Figure 5.4C) [58,60]. Having identified the metabolic effects of phospholipidosis, mathematical models were constructed for both the control group and the phospholipidosis group with SIMCA (Figure 5.4D). In this way, a series of multivariate models can be built to define individual classes that relate to specific types or mechanisms of toxicity or disease.

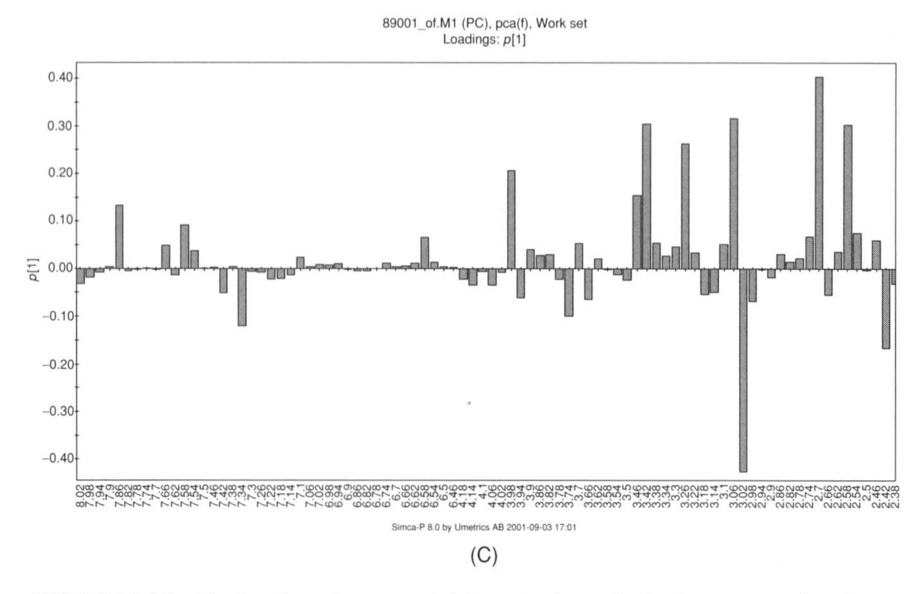

FIGURE 5.4C The loadings plot generated from the first principal component that shows the spectral regions responsible for differentiation between the treated and control populations.

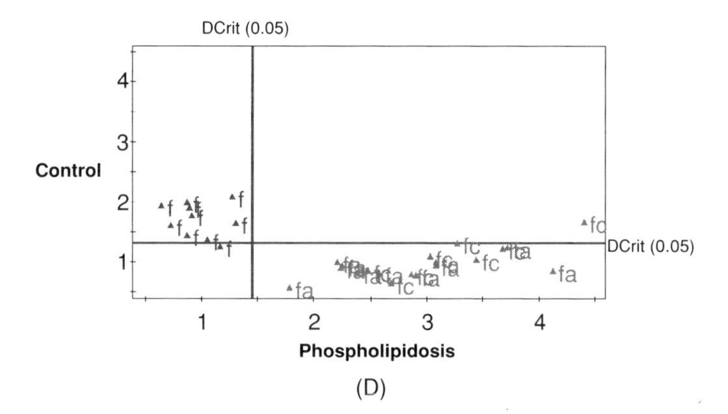

FIGURE 5.4D Cooman's plot that shows the results from SIMCA modeling. Rats treated with chloroquine (fc) and amiodarone (fa) are classified in the phospholipidosis region with only one rat outside the 95% confidence limit.

Evolution of Toxic Lesions

Toxicological data are seldom straightforward with drugs that target multiple tissues rather than single organs and induce a time series of interrelated biochemical events. Since lesions develop and resolve in real time, time-related changes in NMR-detected metabolic profiles for each toxin must be taken into account, and indeed the time profile of altered biochemical composition is in itself a feature of the toxicity [5,56] (Figure 5.5A). Therefore, ^{1}H NMR spectra of biofluids represent complex indices of the metabolic response of an organism to xenobiotic exposure. The evolution of a toxin-induced lesion can be mapped with biochemical trajectory plots (as shown in Figure 5.5B), where the mean response of a group of animals to a particular xenobiotic at a given timepoint is expressed as a single coordinate in the PC map, and the coordinates are connected in chronological order to generate a trajectory that corresponds to the evolving metabolic status of the population [5]. The direction and magnitude of the trajectory's deviation from the predose or control state contains information as to the nature and extent of the lesion, and the behavior of the trajectory can be related to specific phases of the biochemical lesion, such as the development of secondary toxicity or recovery. In addition, a comparison of the geometry of toxin trajectories provides an efficient method of assessing the similarity of metabolic response for two or more toxins. More sophisticated means of monitoring the evolution of a toxic lesion, such as partial least square–based multivariate batch-process control, can be used to generate a better hierarchy of data and to assign biomarkers to specific windows of time in relation to the onset of toxicity.

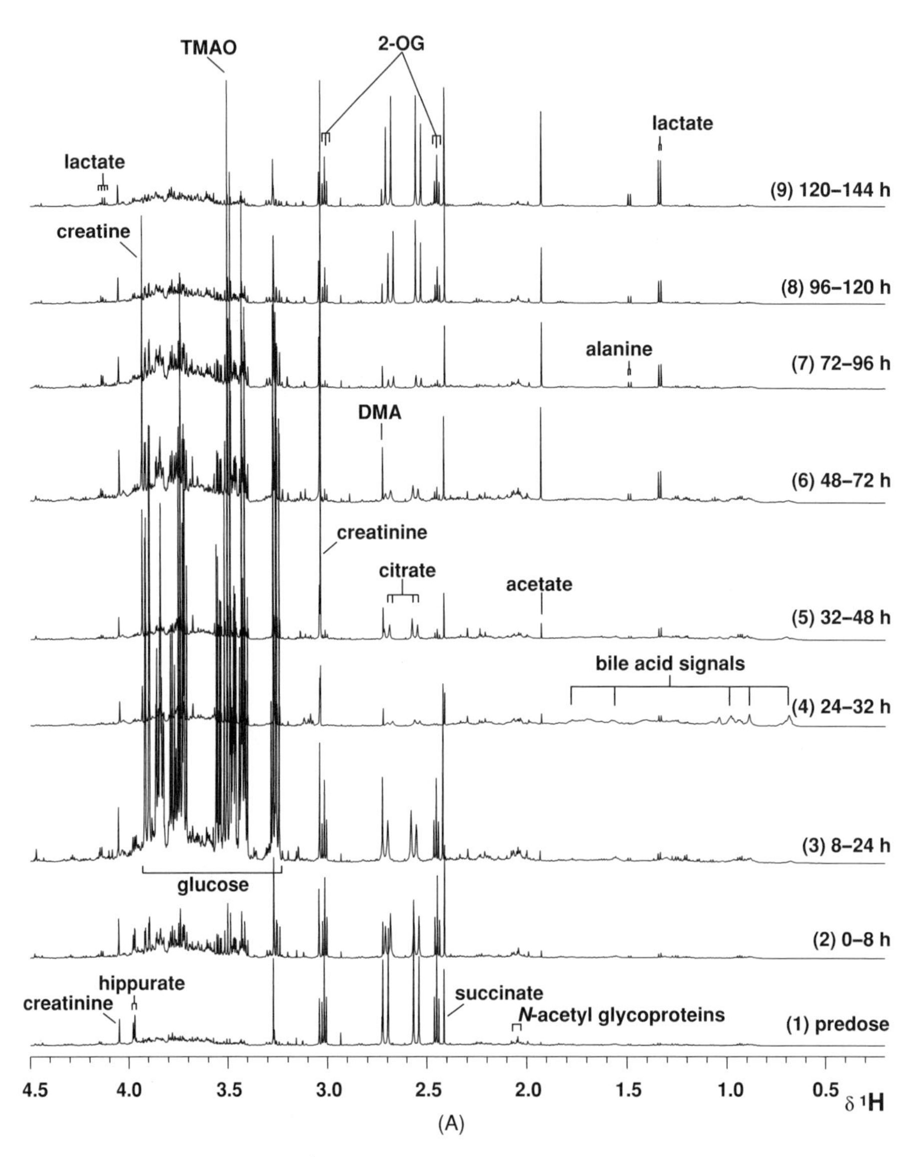

FIGURE 5.5 (A) 600 MHz ^{1}H NMR spectra of sequential urine samples obtained from a rat treated with ANIT, B) PCA trajectory that shows evolution of ANIT induced lesion in time together with trajectories for galactosamine and butylated hydroxytoluene.

Development of NMR-PR–Based "Expert Systems" for Toxicological Screening

For efficient toxicity screening, rapid diagnosis of samples obtained from animals treated with potential drug candidates is necessary. Metabonomic expert systems that comprise a series of mathematical models derived from a training database have been built where the class of toxicity for all samples

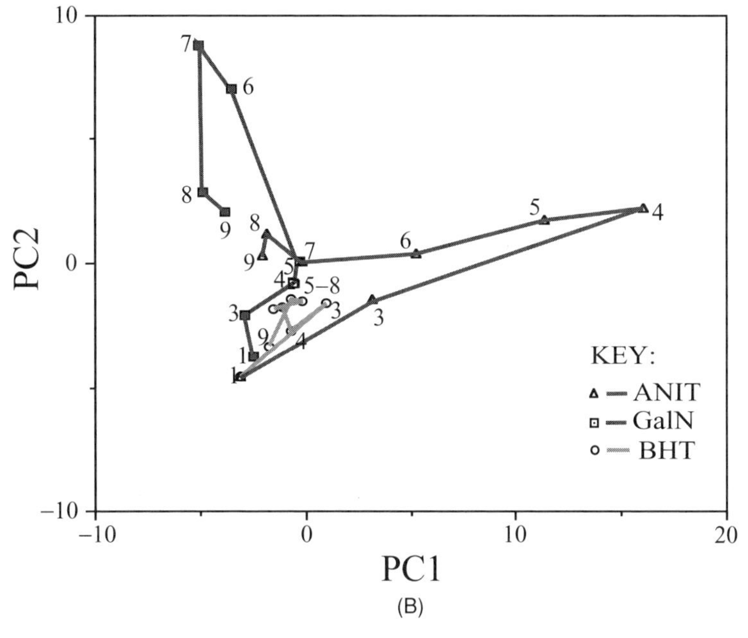

(B)

FIGURE 5.5 (*Continued*)

in the database is known. Prediction of toxicity of novel compounds can then be achieved by fitting test compounds to the models in the database to determine if test samples fall within any of the boundary limits calculated for each model of toxixity. These multivariate models can be derived from one or more of a range of multivariate statistical methods, including PCA, neural-network analysis, SIMCA, rule induction, PLS analysis, and discriminant analysis [5]. The statistical models are then validated with an independent or "test" set of samples where the outcome of toxicity is known but not used in the mathematical algorithm. Once the robustness of the models is checked with a test set, the system can then be used to assess and predict the toxicity of novel xenobiotics. Expert systems can operate at three separate levels.

Level 1: Classification of a Sample or Organism as "Normal" or "Abnormal" For this level of classification it is necessary to build a well-defined model that contains samples that cover all aspects of normality. Subtle variations in control spectral profiles that relate to sampling time, genetic strain, hormonal status, level of activity, and diet have been identified with NMR-based metabonomic technology [24,61,62]. An example of separation of a control population based on diurnal differences in urine composition is given in Figure 5.6. Only when the class boundaries for "normal" biofluid composition have been adequately defined can the "truly abnormal" samples that relate to pathological dysfunction be identified. Using

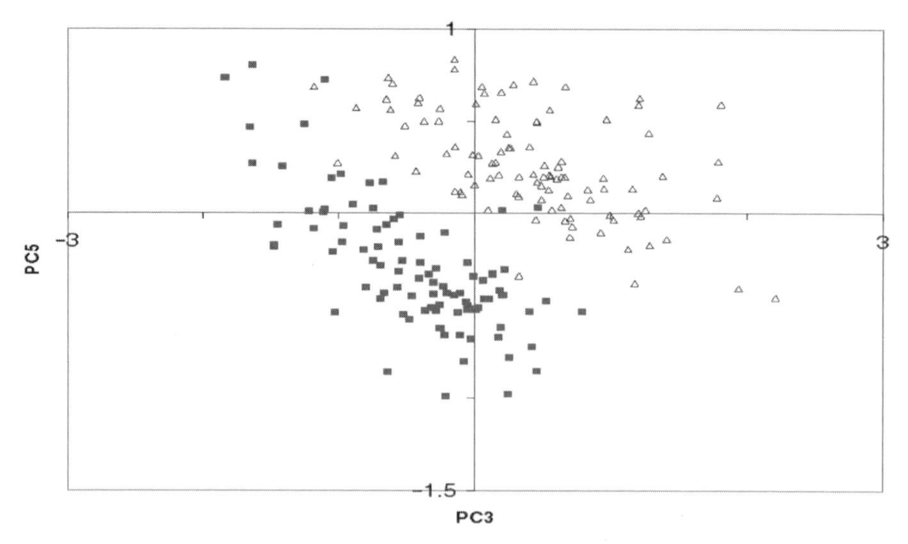

FIGURE 5.6 A PCA plot that shows separation of a control population of rats based on diurnal differences in urine composition as detected by ^{1}H NMR. PC3 versus PC5 scores plot of mean-centred data from NMR spectra of female rat urine collected during the day and night. ■ = night, △ = day.

appropriate software, selection of abnormal samples can be achieved automatically on-line, with any sample defined as abnormal undergoing further NMR measurements or multivariate statistical analysis to ascertain the nature of the abnormality.

Level 2: Classification of Toxicity Samples identified as dissimilar to matched control samples can be fitted to a series of mathematical models that define the multivariate boundaries for known classes of toxicity (Figures 5.4 and 5.5) [6,7,12,63]. Therefore, biofluid or tissue samples from experimental animals treated with novel drugs can be tested to ascertain if the drug induces biochemical effects that would infer a particular site or mechanism of toxicity.

Level 3: Identification of the Biomarkers Once a given compound is established to reproducibly alter the metabolic state of an organism, the metabolites that differentiate between biofluid samples obtained from drug-treated and control rats can be elucidated to give insight into possible mechanisms of toxicity or dysfunction. For example, renal papillary necrosis was a condition for which no early biochemical markers of damage previously existed. However, following ^{1}H NMR spectroscopic analysis of urine obtained from rats treated with model renal papillary toxins, perturbations in the levels of trimethylamine-*N*-oxide, *N*, *N*-dimethylglycine, dimethylamine, and succinate were found to be indicative of damage to the renal papilla [64,65]. Identification of biomarkers can be achieved by a variety of means. For many compounds, chemical structure can be identified by reference to databases containing information relating to chemical shift, spin-spin coupling, and relative intensity of resonances and the structures verified by spiking biofluids with authentic solutions of the proposed metabolites. For more challenging cases, then, it is necessary to disperse the ^{1}H resonances either with selected 2D NMR experiments such as total correlation spectroscopy (TOCSY) or with chromatographic techniques to isolate the metabolite of interest prior to NMR or MS analysis. This procedure is illustrated in a study of the systemic biochemical effects of oral hydrazine-administration dosed in male Han Wistar rats, using metabonomic analysis of ^{1}H NMR spectra of urine and plasma, HPLC-NMR (Figure 5.7), conventional clinical chemistry, and liver histopathology [66]. ^{1}H NMR spectra of the biofluids were analyzed visually and via pattern recognition with PCA. PCA showed that there was a dose-dependent biochemical effect of hydrazine treatment on the levels of a range of low molecular-weight compounds in urine and plasma, which was correlated with the severity of the hydrazine-induced liver lesions determined by histopathology. In plasma, increases in the levels of free glycine, alanine, isoleucine, valine, lysine, arginine, tyrosine, citrulline, 3-D-hydroxybutyrate, creatine,

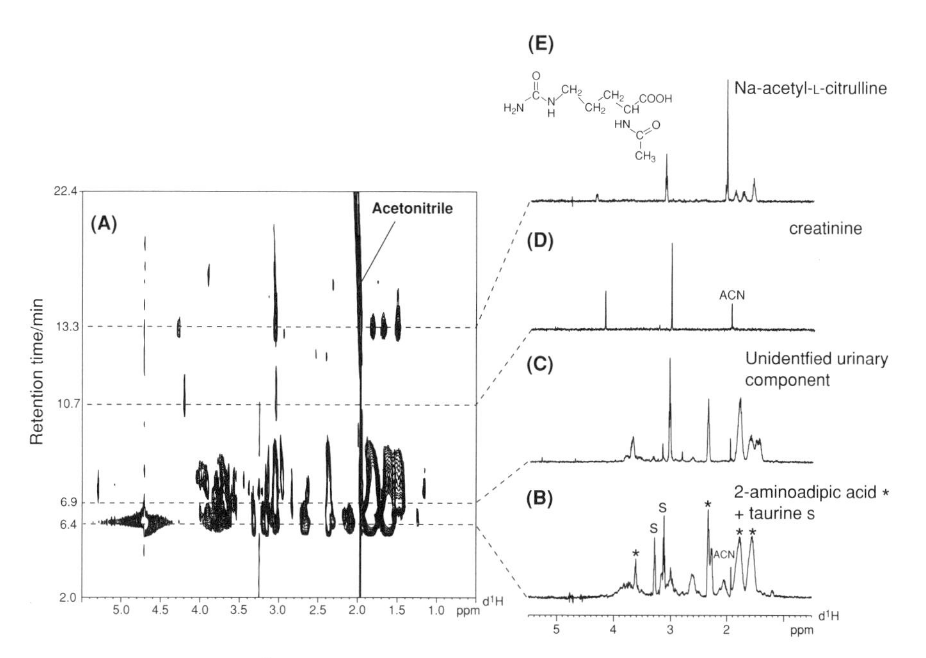

FIGURE 5.7 500 MHz ^{1}H NMR spectra of endogenous metabolites separated from whole rat urine 56h post-dose hydrazine (120 mg/kg) with on-flow LC-NMR spectroscopy: (A) on-flow HPLC-NMR pseudo-2D NMR spectrum, (B) 2-aminoadipic acid and unidentified co-eluting metabolites, (C) Unidentified urinary component, (D) creatinine (E) N-acetyl-citrulline.

histidine, and threonine were observed. Urinary excretion of hippurate, citrate, succinate, 2-oxoglutarate, trimethylamine-*N*-oxide, fumarate, and creatinine were decreased following hydrazine dosing, whereas taurine, creatine, threonine, *N*-methylnicotinic acid, tyrosine, β-alanine, citrulline, Na-acetyl-L-citrulline, and argininosuccinate levels were increased. In addition, several other previously unassigned resonances were detected in urine and plasma. Using HPLC-NMR spectroscopy, these were assigned to 2-aminoadipate, which has previously been shown to lead to neurological effects in rats (Figure 5.7). High urinary levels of 2-aminoadipate may explain the poorly understood neurological effects of hydrazine.

Metabonomics in Disease Diagnosis

^{1}H NMR spectroscopy of biofluid and tissue samples has been applied to the investigation of many diseases and disease models that include inborn errors of metabolism, classification of tumors, evaluation of transplant patients, detection of markers for neurodegenerative diseases, and monitoring drug-overdose cases [41–48]. As with episodes of toxicity, the temporal progression of a disease is represented by a complex and dynamic metabolic profile that may be best evaluated with sequential samples obtained over an appropriate time course. The evaluation of human biofluid samples is further complicated by a high degree of normal physiological variation caused by lifestyle differences. We have shown that alterations in a range of physiological conditions such as diurnal, hormonal, and dietary variation can be easily detected in control populations of experimental animals with NMR spectroscopy. Since the diets and lifestyles of humans are generally less well controlled than those of laboratory animals, the range of metabolic variation (particularly in urine composition) is greater [62]. However, we have also shown that the use of data-filtering techniques, such as orthogonal signal correction, can be used to deconvolve much of the biological variation that is not directly related to disease class.

In addition to characterization of disease states, NMR-based metabonomic analysis offers an efficient means to monitor the response of patients to drug therapy or other therapeutic interventions. For example, in a study of patients with end-stage renal failure, the response of patients to hemodialysis was monitored. Plasma samples were obtained from healthy subjects and from patients with renal failure immediately preceding and following hemodialysis. Samples were analyzed by NMR spectroscopy and mapped with pattern-recognition methods. Samples obtained from the majority of patients following hemodialysis were observed to map more closely to the cluster of samples obtained from healthy subjects than those samples obtained prior to dialysis therapy, with the exception of one patient who responded badly to the therapy and mapped separately to all other samples [13]. Thus, this methodology can be used to select appropriate therapies for patients.

Recent Advances in Metabonomic Technology

The development of automated flow probe technology has had a substantial impact on sampling speed, making this technology viable as a high throughput metabolic screen. Using such technology, it is possible to acquire >300 samples per day. Moreover, recent advances in spectrometer technology, such as the development of cryoprobes and microprobes, have resulted in an increase in the sensitivity of the technique and improved signal dispersion, which results in the amount of latent metabolic information contained in the biofluid spectra being increased. Moreover, the potential of ^{1}H NMR spectroscopy for classification of toxin- or disease-induced lesions, and for elucidation of markers of toxicity, increases with field strength. However, analysis and interpretation of these spectra presents an analytical challenge, since there are thousands of partially overlapped signals. Thus toxicity profiling and disease diagnosis based on large databases of biofluid spectra generally requires the use of automated data-reduction and PR techniques to access the latent biochemical information present in the spectra. Software packages capable of performing multivariate statistical analysis on spectral data have recently become commonly available and provide a user friendly means of data reduction and visualization to explore intrinsic clustering behavior of samples, e.g., Pirouette (infometrix, Seattle, Washington) and SIMCA P (Umetrics, Umea, Sweden), making metabonomic technology generally more accessible.

The Role of Magic-Angle Spinning Spectroscopy in Metabonomic Analysis

In the last decade, the introduction of high resolution magic angle spinning (HRMAS) NMR spectroscopy to the analysis of intact biological tissues has made a major contribution to the ability to directly correlate NMR-detected biomarkers in biofluids with tissue damage. Although ^{1}H NMR-detected perturbation in biofluids such as urine and plasma can give rise to surrogate markers of tissue-specific toxicity and can lend insight into the mechanism of toxicity, it cannot give unequivocal evidence for damage to specific tissues per se. Urine data contains components from metabolic processes throughout the body, and therefore it is often necessary to analyze tissues directly to provide a direct link between the histopathology of a lesion and biofluid NMR spectroscopic data. In vivo NMR spectroscopy has been used to investigate abnormal tissue biochemistry, but spectral quality is always severely compromised by the high heterogeneity in the sample. These sample conditions cause magnetic-field inhomogeneity, and the constrained molecular motions of molecules in some tissue compartments lead to poor resolution. Therefore, NMR spectral analysis of tissues has largely relied upon tissue-extraction methods. However, extraction processes result in the

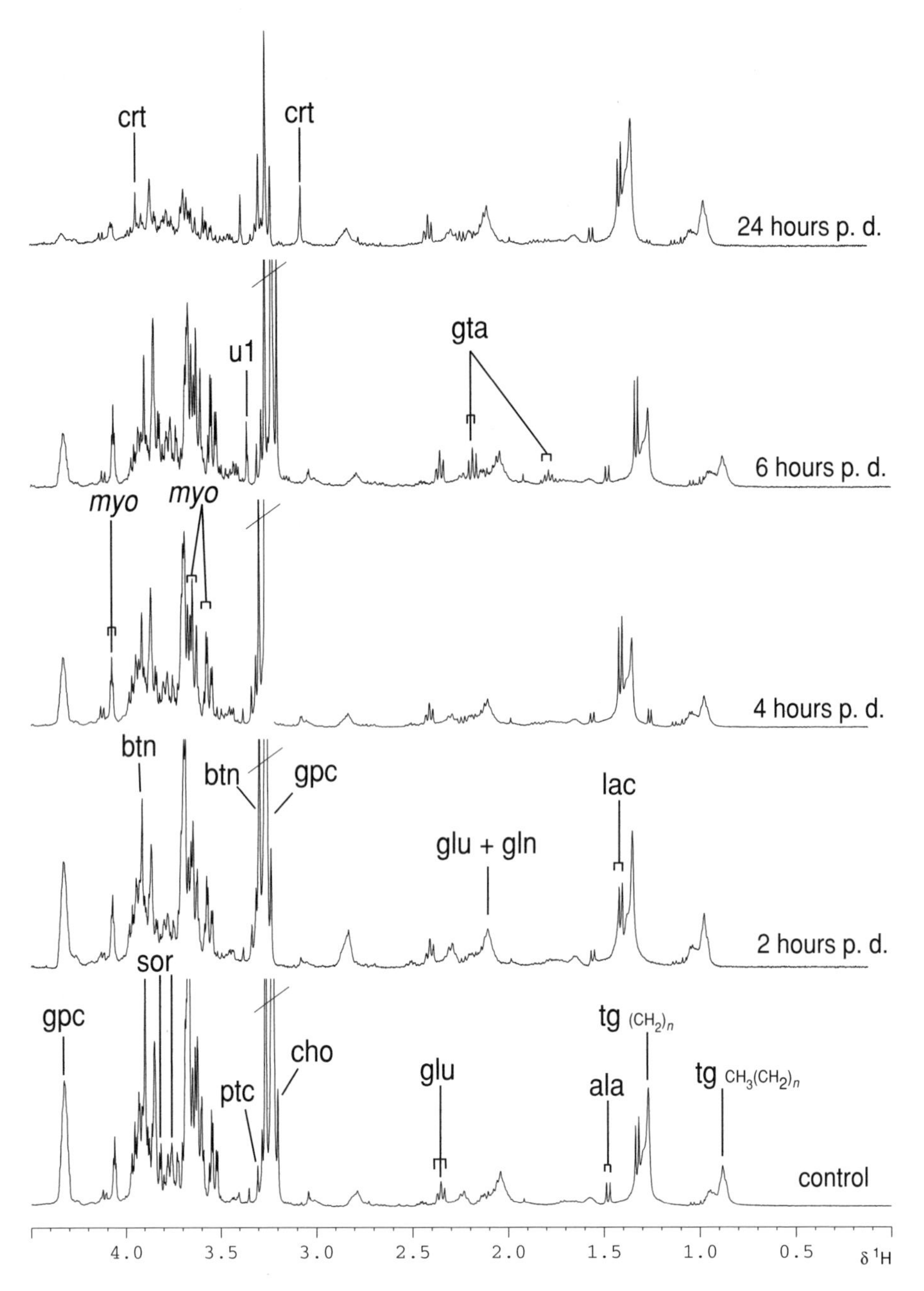

FIGURE 5.8 A series of ^{1}H MAS-NMR spectra of intact renal papilla obtained from rats 24 h after dosing with the model papillotoxin 2-bromoethanamine at 150 mg/kg.

loss of tissue components, such as proteins and lipids. By spinning solid or semisolid samples, such as biological tissues, at the magic angle (54.7° relative to the applied magnetic field), several important line-broadening effects are reduced, and it is possible to obtain very high-quality NMR spectra of whole tissue samples with no sample pretreatment [67]. At this angle, line-broadening effects due to sample heterogeneity and inherent magnetic-field inhomogeneity, residual dipolar couplings, and chemical shift anisotropy are reduced by scaling the free induction decay (FID) by $(3\cos^2\theta - 1)/2$. High-resolution MAS NMR spectroscopy has been used to characterize the low MW composition of a range of biochemical tissues and organelles, including liver, kidney, brain, heart, adipose, and mitochondria, and to evaluate the biochemical consequences of several toxins and disease processes [68–75]. For example, MAS-NMR analysis of intact renal papilla obtained from rats 24 hours after dosing with the model papillotoxin 2-bromoethanamine showed a decrease in tissue concentrations of renal osmolytes such as glycerophosphocholine, myo-inositol, and betaine (Figure 5.8). In addition to bridging the gap between

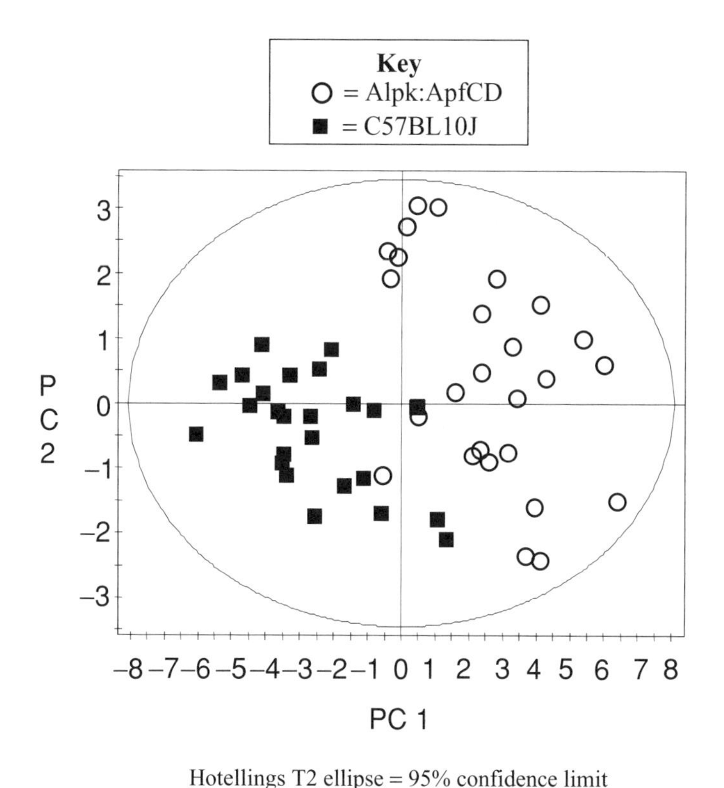

FIGURE 5.9A A PCA plot that differentiates C57BL10J (black mice) and Alpk:ApfCD (white mice) on the basis of biochemical differences observed in ^{1}H NMR spectra.

hisopathology and biofluid analysis, MAS spectroscopy can be used to visualize dynamic processes and to gain insight into the compartmentalization of metabolites within cellular environments [71].

Functional Genomics

The explosion of information in gene expression data has resulted in the production of megavariate data sets. However, the pattern of changes in gene profiles requires interpretation as to the time effect of induction on the metabolic status of the organism. NMR-based metabonomic analysis offers a means to evaluate the direct effect of altered gene expression on the metabolic status of an organism. NMR studies have shown that it is possible to detect biochemical differences in urine samples obtained from different strains of rat or mouse. For example, it is possible to differentiate between Han Wistar and Sprague-Dawley rats or between C57BL10J and Alpk:ApfCD mice[23](Figure 5.9A). Moreover the key differences in intermediary biochemical pathways between strain can be determined (Figure 5.9B). This technology has obvious potential

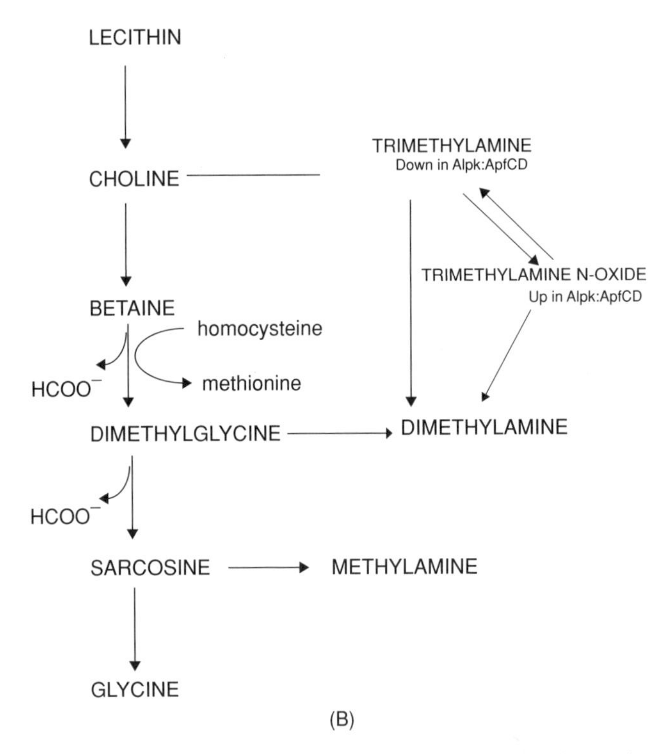

FIGURE 5.9B The key differences in intermediary biochemical pathways between C57B-L10J (black mice) and Alpk:ApfCD (white mice).

for the characterization of transgenic models, assessment of the quality of such models, and to monitor the response to therapy.

CONCLUSIONS

NMR-based metabonomics can be used to address a large range of toxicological, clinical, and environmental problems. In combination with pattern-recognition techniques, ^{1}H NMR spectroscopy has been used to identify changes in biofluid metabolite concentrations to reflect site- and mechanism-specific toxicity, to define novel indices of toxic insult, to evaluate control data, to monitor disease progression and response to therapeutic intervention, and to track progression and regression of toxin-induced lesions over a time period. Furthermore, this metabonomic approach has been shown to be sensitive enough to characterize biochemical differences in urine composition in closely related strains of rat (Han Wistar and Sprague-Dawley), and therefore has potential in the evaluation of genetically modified animals.

Current technology enables the generation of substantial amounts of metabolic data from even simple ^{1}H NMR experiments on whole biofluids to give a comprehensive representation of the biochemical processes that occur in whole organisms under different physiological and pathophysiological conditions. Metabonomics already has been accepted as a valuable part of toxicological assessment in the pharmaceutical industry. Ongoing developments in instrumentation, multivariate statistical techniques, and user-friendly software should serve to make metabonomics an integral component of toxicological screening and lead compound selection in the pharmaceutical industry.

REFERENCES

1. Nicholson, J.K.; Lindon, J.C.; Holmes, E. " 'Metabonomics':Understanding the Metabolic Responses of Living Systems to Pathophysiological Stimuli via Multivariate Statistical Analysis of Biological NMR Spectroscopic Data," *Xenobiotica* **11,** 1181–1189 (1999).
2. Sinclair, B. "Everything's Great When It Sits on a Chip: A bright future for DNA arrays," *The Scientist* **13,** 11, 18–20 (1999).
3. Geisow, M.J. "Proteomics: One Small Step for a Digital Computer, One Giant Leap for Humankind," *Nature Biotechnol.* **16,** 206 (1998).
4. Smith, L.L. "Key Challenges for Toxicologists in the 21st Century," *Trends Pharm. Sci.* **22**(6), 281–285 (2001).
5. Beckwith-Hall, B.M.; Nicholson, J.K.; Nicholls, A.; Foxall, P.J.D.; Lindon, J.C.; Connor, S.C.; Abdi, M.; Connelly, J.; Holmes, E. "Nuclear Magnetic Resonance Spectroscopic and Principal Components Analysis Investigations into

Biochemical Effects of Three Model Hepatotoxins," *Chem. Res. Toxicol.* **11,** 260–272 (1998).

6. Anthony, M.L.; Beddell, C.R.; Lindon, J.C.; Nicholson, J.K. "Studies on the Comparative Toxicity of S-(1,2-dichlorovinyl)-L-homocysteine (DCVHC), S-(1,2-dichlorovinyl)-L-cysteine (DCVC) and 1,1,2-trichloro-3,3,3-trifluoro-1-propene (TCTFP) in the Fischer 344 rat," *Arch. Toxicol.* **69**(2), 99–110 (1994).
7. Holmes, E.; Nicholls, A.W.; Lindon, J.C.; Ramos, S.; Spraul, M.; Neidig, P.; Connor, S.C.; Connelly, J.; Damment, S.J.P.; Haselden, J.N.; Nicholson, J.K. "Development of a Model for Classification of Toxin-induced Lesions with ^{1}H NMR Spectroscopy of Urine Combined with Pattern Recognition," *NMR Biomed.* **11,** 1–10 (1998).
8. Maxwell, R.J.; Martinez-Perez, I.; Cerdan, S.; Cabanas, M.E.; Arus, C.; Moreno, A.; Capdevila, A.; Ferrer, E.; Bartomeus, F.; Aparicio, A.; Conesa, G.; Roda, J.M.; Carcellar, F.; Pascual, J.M.; Howells, S.L.; Mazucco, R.; Griffiths, J.R. "Pattern Recognition Analysis of ^{1}H NMR Spectra from Perchloric Acid Extracts of Human Brain Tumour Biopsies," *Magn. Res. Med.* **39,** 869–877 (1998).
9. El-Deredy, W. "Pattern Recognition Approaches in Biomedical and Clinical Magnetic Resonance Spectroscopy: A Review," *NMR Biomed.* **10,** 99–124 (1997).
10. Howells, S.L.; Maxwell, R.J.; Peet, A.C.; Griffiths, J.R. "An Investigation of Tumour ^{1}H Nuclear Magnetic Resonance Spectra by the Application of Chemometric Techniques," *Mag. Res. Med.* **28,** 214–236 (1992).
11. Nicholson, J.K.; Wilson, I.D. "High Resolution Proton NMR Spectroscopy of Biological Fluids," *Prog. NMR Spectrosc.* **21,** 444–501 (1989).
12. Robertson, D.G.; Reily, M.D.; Sigler, R.E.; Wells, D.F.; Paterson, D.A.; Braden, T.K. "Metabonomics: Evaluation of Nuclear Magnetic Resonance (NMR) and Pattern Recognition Technology for Rapid in vivo Screening of Liver and Kidney Toxicants," *Toxicol. Sci.* **57**(2), 326–337 (2000).
13. Lindon, J.C.; Holmes, E.; Nicholson, J.K. "Pattern Recognition Methods and Applications in Biomedical Magnetic Resonance," *Prog. NMR Spectros.* **39**(1), 1–40 (2001).
14. Lindon, J.C.; Nicholson, J.K.; Holmes, E.; Everett, J.R. "Metabonomics: Metabolic Processes Studied by NMR Spectroscopy of Biofluids," *Concepts Mag. Res.* **12**(5), 289–320 (2000).
15. Moreda-Pineiro, A.; Marcos, A.; Fisher, A.; Hill, S.J. "Evaluation of the Effect of Data Pretreatment Procedures on Classical Pattern Recognition and Principal Components Analysis: A Case Study for the Geographical Classification of Tea," *J. Environ. Monit.* **3**(4), 352–360 (2001).
16. Queralto, J.M.; Torres, J.; Guinot, M. "Neural Networks for the Biochemical Prediction of Bone Mass Loss." *Clin. Chem. Lab. Med.* **37**(8), 831–838 (1999).
17. Halket, J.M.; Przyborowska, A.; Stein, S.E.; Mallard, W.G.; Down, S.; Chalmers, R.A. "Deconvolution of Gas Chromatography/Mass Spectrometry of Urinary

Organic Acids—Potential for Pattern Recognition and Automated Identification of Metabolic Disorders," *Rapid Commun. Mass Spectrom.* **13**(4), 279–284 (1999).

18. Goodacre, R.; Rooney, P.J.; Kell, D.B. "Discrimination Between Methicillin-Resistant and Methicillin-Susceptible *Staphylococcus Aureus* Using Pyrolysis Mass Spectrometry and Artificial Neural Networks," *J. Antimicrob. Chemother.* **41**(1), 27–34 (1998).
19. Kim, K.R.; Kim, J.H.; Cheong, E,; Jeong, C. "Gas Chromatographic Amino acid Profiling of Wine Samples for Pattern Recognition," *J. Chromatogr. A* **722**(1–2), 303–309 (1996).
20. Freeman, R.; Goodacre, R.; Sisson, P.R.; Magee, J.G.; Ward, A.C.; Lightfoot, N.F. "Rapid Identification of Species within the Mycobacterium Tuberculosis Complex by Artificial Neural Network Analysis of Pyrolysis Mass Spectra," *J. Med. Microbiol.* **40**(3), 170–173 (1994).
21. Ramos L.S. "Characterisation of Mycobacteria species by HPLC and Pattern Recognition," *J. Chromatogr. Sci.* **32**(6), 219–227 (1994).
22. Gartland, K.P.R.; Bonner, F.W.; Nicholson, J.K. "Investigations into the Biochemical Effects of Region-specific Nephrotoxins," *Mol. Pharmacol.* **35,** 242–250 (1989).
23. Gavaghan, C.L.; Holmes, E.; Lenz, E.; Wilson, I.D.; Nicholson, J.K. "An NMR-based Metabonomic Approach to Investigate the Biochemical Consequences of Genetic Strain Differences: Application to the C57BL10J and Alpk:ApfCD Mouse," *Febs. Lett.* **484,** 169–174 (2000).
24. Gavaghan, C.L.; Nicholson, J.K.; Connor, S.C.; Wilson, I.D.; Wright, B.; Holmes, E. "HPLC-NMR Spectroscopic and Chemometric Studies on Metabolic Variation in Sprague Dawley Rats," *Anal. Biochem.* **291** (2), 245–252 (2001).
25. Phipps, A.N.; Stewart, J.; Wright, B.; Wilson, I.D. "Effect of Diet on the Urinary Excretion of Hippuric Acid and Other Dietary-derived Aromatics in Rat," *Xenobiotica*, **28**(5), 527–537 (1998).
26. Anthony, M.L.; Sweatman, B.C.; Beddell, C.R.; Lindon, J.C.; Nicholson, J.K. "Pattern Recognition Classification of the Site of Nephrotoxicity Based on Metabolic Data Derived from High Resolution Proton Nuclear Magnetic Resonance Spectra of Urine," *Mol. Pharmacol.* **46,** 199–211 (1994).
27. Manley, B.F.J. *Multivariate Statistical Methods: A Primer*, Chapman & Hall, London (1986).
28. Beebe, K.R.; Pell, R.J.; Seahsholt, M.B. *Chemometrics: A Practical Guide*, Wiley, New York (1998).
29. Jurs, P.C. "Pattern Recognition Used to Investigate Multivariate Data in Analytical Chemistry," *Science* **232,** 1219–1224 (1986).
30. Eriksson, L.; Johansson, E.; Kettanah-Wold, N.; Wold, S. "Introduction to Multi and Megavariate Data Analysis Using Projection Methods (PCA and PLS) UMETRICS ACADEMY (1999).

31. Holmes, E.; Nicholson, J.K.; Tranter, G. "Metabonomic Characterization of Genetic Variations in Toxicological and Metabolic Responses Using Probabilistic Neural Networks," *Chem. Res. Toxicol.* **14**(2), 182–191 (2001).
32. Gartland, K.P.R.; Beddell, C.R.; Lindon, J.C.; Nicholson, J.K. "Application of Pattern-Recognition Methods to the Analysis and Classification of Toxicological Data Derived from Proton Nuclear-Magnetic-Resonance Spectroscopy of Urine." *Mol. Pharmacol.* **39**(5), 629–642 (1991).
33. Kosir, I.J.; Kocjancic, M.; Ogrinc, N.; Kidric, J. "Use of SNIF-NMR and IRMS in Combination with Chemometric Methods for the Determination of Chaptalisation and Geographical Origin of Wines (the Example of Slovenian wines)," *Anal. Chimica Acta* **429**(2), 195–206 (2001).
34. Zamora, R.; Alba, V.; Hidalgo, F.J. "Use of high-resolution C-13 nuclear magnetic resonance spectroscopy for the screening of virgin olive oils," *J. Am. Oil Chem. Soc.* **78**(1), 89–94 (2001).
35. "Characterization of the Geographic Origin of Bordeaux Wines by a Combined Use of Isotopic and Trace Element Measurements," *Am. J. Enology Viticulture* **50**(4), 409–417 (1999).
36. Belton, P.S.; Colquhoun, I.J.; Kemsley, E.K.; Delgadillo, I.; Roma, P.; Dennis, M.J.; Sharman, M.; Holmes, E.; Nicholson, J.K.; Spraul, M. "Application of Chemometrics to the H-1 NMR Spectra of Apple Juices: Discrimination between Apple Varieties," *Food Chem.* **61**(1–2), 207–213 (1998).
37. Vogels, J.T.W.E.; Terwel, L.; Tas, A.C.; vandenBerg, F.; Dukel, F.; VanderGreef, J. "Detection of Adulteration in Orange Juices by a New Screening Method Using Proton NMR Spectroscopy in Combination with Pattern Recognition Techniques," *J. Agric. Food Chem.* **44**(1), 175–180 (1996).
38. Bundy, J.G.; Osborn, D.; Weeks, J.M.; Lindon, J.C.; Nicholson, J.K. "An NMR-based Metabonomic Approach to the Investigation of Coelomic Fluid Biochemistry in Earthworms under Toxic Stress," *FEBS Letters* **500**(1–2), 31–35 (2001).
39. Warne, M.A.; Lenz, E.M.; Osborn, D.; Weeks, J.M.; Nicholson, J.K. "An NMR-based metabonomic investigation of the toxic effects of 3-trifluoromethyl-aniline on the earthworm *Eisenia veneta*," *Biomarkers* **5**(1), 56–72 (2000).
40. Gibb, J.O.T.; Holmes, E.; Nicholson, J.K.; Weeks, J.M. "Proton NMR Spectroscopic Studies on Tissue Extracts of Invertebrate Species with Pollution Indicator Potential," *Comp. Biochem. Physiol.* B-*Biochem. Mol. Biol.* **118**(3), 587–598 (1997).
41. Griffin, J.L.; Williams, H.J.; Sang, E.; Clarke, K.; Rae, C.; Nicholson, J.K. "Metabolic Profiling of Genetic Disorders: A Multitissue H-1 Nuclear Magnetic Resonance Spectroscopic and Pattern Recognition Study into Dystrophic Tissue," *Anal. Biochem.* **293**(1), 16–21 (2001).
42. Holmes, E.; Foxall, P.J.D.; Spraul, M.; Farrant, R.D.; Nicholson, J.K.; Lindon, J.C. "750 MHz H-1 NMR Spectroscopy Characterisation of the Complex Metabolic Pattern of Urine from Patients with Inborn Errors of Metabolism: 2-Hydroxyglutaric Aciduria and Maple Syrup Urine Disease," *J. Pharm. Biomed. Anal.* **15**(11), 1647–1659 (1997).

43. Alanen, A.; Komu, M.; Penttinen, M.; Leino, R. "Magnetic Resonance Imaging and Proton MR Spectroscopy in Wilson's disease." *Br. J. Radiol.* **72**(860), 749–56 (1999).
44. Burtscher, I.M.; Holtas, S. "Proton Magnetic Resonance Spectroscopy in Brain Tumours: Clinical Applications," *Neuroradiology* **43**(5), 345–352 (2001).
45. Roda, J.M.; Pascual, J.M.; Carceller, F.; Gonzalez-Llanos, F.; Perez-Higueras, A.; Solivera, J.; Barrios, L.; Cerdan, S. "Nonhistological Diagnosis of Human Cerebral Tumors by H-1 Magnetic Resonance Spectroscopy and Amino Acid Analysis," *Clin. Cancer Res.* **6**(10), 3983–3993 (2000).
46. Tate, A.R.; Foxall, P.J.D.; Holmes, E.; Moka, D.; Spraul, M.; Nicholson, J.K.; Lindon, J.C. "Distinction between Normal and Renal Cell Carcinoma Kidney Cortical Biopsy Samples Using Pattern Recognition of H-1 Magic Angle Spinning (MAS) NMR Spectra," *Nmr Biomed.* **13**(2), 64–71 (2000).
47. Morales, F.; Ballesteros, P.; Cerdan, S. "Neural Networks in Automatic Diagnosis Malignant Brain Tumors," *Eng. Appl. Bio-Inspired Artificial Neural Networks* **607,** 778–787 (1999).
48. Hagberg, G. "From Magnetic Resonance Spectroscopy to Classification of Tumors. A Review of Pattern Recognition Methods," *NMR Biomed.* **11**(4–5), 148–156 (1998).
49. Connor, S.C.; Hughes, M.G.; Moore, G.; Lister, C.A.; Smith, S.A. Antidiabetic Efficacy of BRL49653 a Potent Orally Active Insulin Sensitising Agent, Assessed in the C57BL/KsJ db/db Diabetic Mouse by Non Invasive Proton NMR Studies in Urine," *J. Pharm. Pharmacol.* **49,** 336–344 (1997).
50. Warne, M.A.; Lenz, E.M.; Osborn, D.; Weeks, J.M.; Nicholson, J.K. "Comparative Biochemistry and Short-term Starvation Effects on the Earthworms *Eisenia veneta* and *Lumbricus terrestris* Studied by H-1 NMR Spectroscopy and Pattern Recognition," *Soil Biol. Biochem.* **33**(9), 1171–1180 (2001).
51. Holmes, E.; Foxall, P.J.D.; Nicholson, J.K., Neild, G.H.; Brown, S.M.; Beddell, C.R.; Sweatman, B.C., Rahr, E.; Lindon, J.C.; Spraul, M.; Neidig, P. "Automatic Data Reduction and Pattern-Recognition Methods for Analysis Of H-1 Nuclear-Magnetic-Resonance Spectra of Human Urine from Normal and Pathological States." *Anal. Biochem.* **220**(2), 284–296 (1994).
52. Nicholson, J.K.; Timbrell, J.A.; Sadler, P.J. Proton NMR Spectra of Urine as Indicators of Renal Damage: Mercury Nephrotoxicity in Rats. *Mol. Pharmacol.* **27,** 644–651 (1985).
53. Nicholson, J.K.; O'Flynn, N.; Sadler, P.J.; Mcleod, P.; Juul, S.M.; Sonksen, P.H. "Proton-Nuclear-Magnetic-Resonance Studies of Serum, Plasma and Urine from Fasting Normal and Diabetic Subjects." *Biochem. J.* **217,** 365–375 (1984).
54. Nicholson, J.K.; Higham, D.; Timbrell, J.A.; Sadler, P.J; Quantitative ^{1}H NMR Urinalysis Studies on the Biochemical Effects of Acute Cadmium Exposure in the Rat." *Mol. Pharmacol.* **36,** 398–404 (1989).
55. Sanins, S.M.; Timbrell, J.A.; Elcombe, C.R.; Nicholson, J.K. "Hepatotoxin-induced Hypertaurinuria: A Proton NMR Study," *Arch. Toxicol.* **64,** 407–411 (1990).

56. Holmes, E.; Bonner, F.W.; Sweatman, B.C.; Lindon, J.C.; Beddell, C.R.; Rahr, E.; Nicholson, J.K. "NMR Spectroscopy and Pattern Recognition Analysis of the Biochemical Processes Associated with the Progression and Recovery from Nephrotoxic Lesions in the Rat Induced by Mercury (II) Chloride and 2-Bromoethanamine," *Mol. Pharmacol.* **42,** 922–930 (1992).

57. Tjiong, H.B.; Debuch H. "Lysosomal bis (monoacylglycerol)phosphate of Rat Liver, Its Induction by Chloroquine and Its Structure. *Hoppe-Seyler's Physil. Chem.* **359,** 71–79 (1978).

58. Espina, J.R.; Herron, W.J.; Shockcor, J.P.; Car, B.D.; Contel, N.R.; Ciaccio, P.J.; Lindon, J.C.; Holmes. E.; Nicholson, J.K. "Detection of *in vivo* Biomarkers of Phospholipidosis Using NMR-based Metabonomic Approaches," *J. Mag. Res. Chem.* **39,** 559–565 (2001).

59. Wold, S.; Antti, H.; Lindgren, F.; Öhman, J. "Orthogonal Signal Correction of Near-Infrared Spectra, "*Chemomet. Intelligent Lab. Systems* **44,** 175–185 (1998).

60. Nicholls, A.W.; Nicholson, J.K.; Haselden, J.H.; Waterfield, C.J. "Metabonomic Investigations into Hydrazine Toxicity in the Rat." *Biomarkers* **5,** 410–423 (2000).

61. Holmes, E.; Foxall, P.J.D.; Nicholson, J.K.; Neild, G.H.; Brown, S.M.; Beddell, C.; Sweatman, B.C.; Rahr, E.; Lindon, J.C,; Spraul, M.; Neidig, P. "Automatic Data Reduction and Pattern Recognition Methods for Analysis of ^{1}H Nuclear Magnetic Resonance Spectra of Human Urine form Normal and Pathological States," *Anal. Biochem.* **220,** 284–296 (1994).

62. Bollard, M.E.; Holmes, E.; Lindon, J.C.; Mitchell, S.C.; Branstetter, D.; Zhang, W.; Nicholson, J.K. "Investigations into Biochemical Changes Due to Diurnal Variation and Estrus Cycle in Female Rats Using High Resolution ^{1}H NMR Spectroscopy of Urine and Pattern Recognition," *Anal. Biochem.* **295,** 194–202 (2001).

63. Nicholson, J.K.; Wilson I.D. "High Resolution Proton NMR Spectroscopy of Biological Fluids," *Prog. NMR Spectrosc.* **21,** 444–501 (1989).

64. Holmes, E.; Bonner, F.W.; Nicholson, J.K. ^{1}H NMR Spectroscopic and Histopathological Studies on Propyleneimine-induced Renal Papillary Necrosis in the Fischer 344 rat and the Multimammate Desert Mouse (*Mastomys natalensis*). *Compar. Biochem. Pharmacol. (C)* **116**(2), 125–134 (1997).

65. Holmes, E.; Bonner, F.W.; Nicholson, J.K. "Comparative Studies on the Nephrotoxicity of 2-bromoethanamine in the Fischer 344 and Multimammate Desert Rats (*Mastomys natalensis*)" *Arch. Toxicol.* **70,** 89–95 (1995).

66. Nicholls, A.W.; Holmes, E.; Lindon, J.C.; Shockcor, J.P.; Farrant, R.D.; Haselden, J.N.; Damment, S.J.P.; Waterfield, C.J.; Nicholson, J.K. "Metabonomic Investigations into Hydrazine Toxicity in the Rat," *Chemical Research in Tox.* **14,** 975–987.

67. Andrew, E.R.; Eades, R.G. "Removal of Dipolar Broadening of NMR Spectra of Solids by Specimen Rotation," *Nature* **183,** 1802 (1959).

68. Moka, D.; Vorreuther, R.; Schicha, H.; Spraul, M.; Humpfer, E.; Lipinski, M.; Foxall, P.J.D.; Nicholson, J.K.; Lindon, J.C. "Magic Angle Spinning Proton Nuclear Magnetic Resonance Spectroscopic Analysis of Intact Kidney Tissue Samples." *Anal. Comm.* **34,** 107–109 (1997).

69. Moka, D.; Vorreuther, R.; Schicha, H.; Spraul, M.; Humpfer, E.; Lipinski, M.; Foxall, P.J.D.; Nicholson, J.K.; Lindon, J.C. "Biochemical Classification of Kidney Carcinoma Biopsy Samples Using Magic-Angle-Spinning ^{1}H Nuclear Magnetic Resonance Spectroscopy," *J. Pharm. Biomed. Anal.* May; **17**(1), 125–32 (1998).
70. Millis, K.; Maas, E.; Cory, D.G.; Singer, S. "Gradient High Resolution Magic Angle Spinning Nuclear Magnetic Resonance Spectroscopy of Human Adipocyte Tissue," *Mag. Res. Med.* **38,** 399–403 (1997).
71. Humpfer, E.; Spraul, M.; Nicholls, A.W.; Nicholson, J.K.; Lindon, J.C. "Direct Observation of Resolved Intracellular and Extracellular Water Signals in Intact Human Red Blood Cells Using ^{1}H MAS NMR Spectroscopy," *Mag. Res. Med.* **38,** 334–336 (1997).
72. Tomlins, A.; Foxall, P.J.D.; Lindon, J.C.; Lynch, M.J.; Spraul, M.; Everett, J.; Nicholson, J.K. "High Resolution Magic Angle Spinning ^{1}H Nuclear Magnetic Resonance Analysis of Intact Prostatic Hyperplastic and Tumour Tissues. *Anal. Comm.* **35,** 113–115 (1998).
73. Cheng, L.L.; Ma, M.J.; Becerra, L.; Hale, T.; Tracey, I.; Lackner, A.; Gonzalez, R.G. "Quantitative Neuropathology by High Resolution Magic Angle Spinning Proton Magnetic Resonance Spectroscopy," *Proc. Natl. Acad. Sci. USA* **94,** 6408–6413 (1997).
74. Cheng, L.L.; Lean, C.L.; Bogdanova, A.; Wright, Jr., S.C.; Ackerman, J.L.; Brady, T.J.; Garrido, L. "Enhanced Resolution of Proton NMR Spectra of Malignant Lymph Nodes Using Magic Angle Spinning." *Mag. Res. Med.* **36,** 653–658 (1996).
75. Garrod, S.; Humpher, E.; Connor, S.C.; Connelly, J.C.; Spraul, M.; Nicholson, J.K.; Holmes, E. "High Resolution ^{1}H NMR and Magic Angle Spinning NMR Spectroscopic Investigation of the Biochemical Effects of 2-Bromoethanamine in Intact Renal and Hepatic Tissue," *Mag. Res. Med.* **45**(5), 781–790 (2001).
76. Foxall, P.J.; Singer, J.M.; Hartley, J.M.; Neild, G.H.; Lapsley, M.; Nicholson, J.K. "Urinary Proton Magnetic Resonance Studies of Early Ifosfamide-induced Nephrotoxicity and Encephalopathy," *Clin. Cancer Res.* **3**(9), 1507–1518 (1997).
77. Halligan, S.; Byard, S.J.; Spencer, A.J.; Gray, T.J.; Harpur, E.S.; Bonner, F.W. "A study of the Nephrotoxicity of Three Cephalosporins in Rabbits Using ^{1}H NMR Spectroscopy," *Toxicol. Lett.* **81**(1), 15–21 (1995).
78. Harrison, M.P.; Jones, D.V.; Pickford, R.J.; Wilson, I.D. "Beta-Hydroxybutyrate: A Urinary Marker of Imipenem Induced Nephrotoxicity in the Cynomolgus Monkey Detected by High Field NMR Spectroscopy," *Biochem. Pharmacol.* **41**(12), 2045–2049 (1991).
79. Bairaktari, E.; Katopodis, K.; Siamopoulos, K.C.; Tsolas, O. "Paraquat-Induced Renal Injury Studied by 1H NMR Spectroscopy of Urine," *Clin. Chem.* **44**(6), 1256–1261 (1998).

6

THE CENTRAL ROLE OF MASS SPECTROMETRY IN NATURAL PRODUCTS DISCOVERY

JEFFREY R. GILBERT, PAUL LEWER, DENNIS O. DUEBELBEIS, ANDREW W. CARR

INTRODUCTION

The demand for new chemical entities (NCEs) in lead discovery is greater now than at any time in the history of the pharmaceutical industry. Natural products have traditionally provided an excellent source of NCEs, often contributing structures that fall outside the diversity space encompassed by synthetic approaches [1]. Natural products represent a diverse range of chemistry classes, including secondary metabolites, peptides, proteins, polysaccharides, and oligonucleotides. Many commercial drugs are either natural products, or have their roots in natural product chemistry [2], and at present over 60% of small-molecule anticancer leads are natural product–derived [3]. A recent review [3] summarizing the contribution that natural products have made to drug discovery concluded that "natural products play a dominant role in the discovery of leads for the development of drugs." The reemphasis of natural products in drug discovery was the subject of a recent series of articles in *Chemical and Engineering News* [4–6], which stated that "the alarming decline in the number of new chemical entities in the past decade, from an average of 30 or so to as few as 17, has correlated with decreased interest in natural product discovery."

Integrated Strategies for Drug Discovery Using Mass Spectrometry, Edited by Mike S. Lee

One key factor in the success of any natural product discovery effort is minimizing the time required to identify active compounds from complex natural sources. The use of high throughput and parallel chromatographic techniques has considerably reduced this cycle time. Concomitantly, the broad use of liquid chromatography/mass spectrometry (LC-MS) has significantly improved the efficiency of natural product screening and structure elucidation, which will be the focus of this chapter. Indeed, mass spectrometry is now a tool that can impact many, if not all, of the stages of lead generation from natural products.

The discovery of novel biologically active compounds derived from nature has evolved from a relatively simple, linear "bioassay guided" process to one that encompasses several different approaches, as shown in Figure 6.1. The traditional approach to natural product screening involves selection of source material, extraction and assay of that material, and separation of components using bioassay-guided fractionation, culminating in the isolation and identification of a purified active. Conversely, in the natural product library approach, extracts are semipurified and concentrated prior to screening for activity—almost the reverse of the traditional process. Both of these avenues rely on a variety of mass spectrometry techniques, which can have a significant impact on their success.

This chapter explores in more detail the role of mass spectrometry in several phases of the natural product discovery process, including (1) selection of source material, (2) screening, (3) dereplication of known compounds, and (4) identification of unknowns. It is clear that over the last 10 years mass spectrometry has evolved into an indispensable tool in the generation of lead candidates from nature.

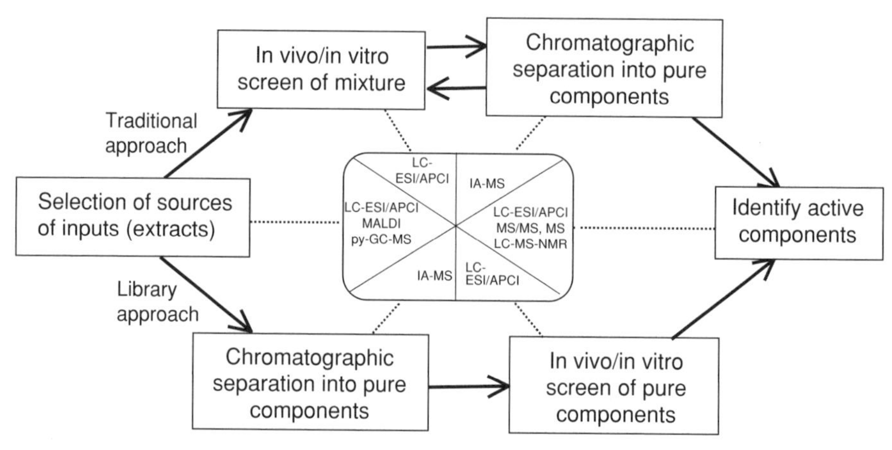

FIGURE 6.1 The central role of mass spectrometry in natural products lead discovery.

SELECTION OF SOURCE MATERIAL

Chemical Fingerprinting and Diversity Assessment

Traditionally, the first step in natural product lead discovery is the selection of source material. The search for novel organisms, and their associated novel chemistries, has taken many forms that include collection from unique locales and microclimates as well as the use of ethnobotany and traditional medicine. Once collected, several methods have been developed to prioritize these library candidates for further study. One of the more traditional methods applied to this process is chemical fingerprinting [7–10].

Chemical fingerprinting can take several forms, but has for many years involved the detection of the molecules of interest by thin-layer chromatography (TLC) separation with staining [11], or by liquid chromatographic (LC) separation with diode-array detection. In such cases, the molecules of interest were often differentiated from inactive molecules by their in vivo (e.g., antibiotic) or in vitro (e.g., receptor binding inhibition) activity. Consequently, the ability to find novel compounds with this approach depended on the combination of novel source material with a unique detection assay. There are clearly many advantages to these approaches that directly detect the molecules of interest. For example, once new molecules are observed and their novelty is recognized, one method for their future detection (and, possibly, optimization) is already established. However, when faced with a large (tens to hundreds of thousands) number of sources to screen, this method can become inefficient due to replicate detection of actives. Methods to prequalify samples according to their likelihood to contain novel chemistry have thus become attractive.

A somewhat less direct method for seeking molecular novelty lies in the detection of "talented organisms." This approach is based on the observation that organisms which produce greater numbers of secondary metabolites at higher concentrations more frequently contain rare or novel metabolites [12]. In this scenario, the search for novel metabolites is initiated by searching for prolific metabolite producers, and LC with ultraviolet (LC-UV) or mass spectral (LC-MS) detection has been commonly used to achieve this objective. In one regard, this approach makes the overall task of source selection somewhat easier, since it is not necessary to determine the novelty of individual components in the initial step. Rather, an overall assessment of each organism's biosynthetic potential is made, and those organisms that appear to have the best biosynthetic potential are given a higher priority for further study.

Efforts to accelerate the discovery of organisms that are likely to produce novel compounds have recently included significant shortening, or eliminating, the chromatography step. Smedsgard et al. [9] acquired MS profiles of crude fungal extracts in a flow injection analysis electrospray system to create "mass profile" libraries. Higgs et al. took the evaluation phase one step further by developing a "metabolite productivity index" computed from similarly acquired data using custom-written software [13]. This surrogate measure of productivity was shown to be a viable method to survey large sets of bacteria and identify highly productive strains for future study. Zahn et al. [14] showed that this method could also be used to identify growth conditions that enhance the expression of secondary metabolites from actinomycetes. Cremin et al. [15] developed an alternative method to increase the speed of the analysis step using an eight-channel parallel LC-MS. Their system used fast parallel LC separations that fed into an eight-channel multiplexed (MUX) interfaced MS for the analysis of prefractionated plant extracts [16]. Custom software was then used to extract characteristic parent ions, fragment ions, and retention times for all detected compounds. These were compared to an existing database to highlight infrequently occurring compounds. The frequency of occurrence of these rare compounds over a given library provided a measurement of chemodiversity.

Chemotaxonomy: Identification of Organisms

While the methods discussed previously produce a significant amount of physical data on the secondary metabolites, they provide little or no direct data on the producing organisms themselves. To address this need, an entirely different approach must be used, that of first searching for novel organisms or production systems. Methods to unequivocally identify novel strains within a bacterial collection, would enhance the opportunity for discovery of novel secondary metabolites. While it is well known that taxonomically distinct organisms can produce similar metabolites, several MS techniques have been successfully applied to the process of bacterial identification. Older chemotaxonomic methods often targeted specific compound classes, for example, fatty acid methyl ester (FAME) profiles, which were determined using gas chromatography with mass spectral detection (GC-MS) [17]. The development of methods for oligonucleotide sequencing led to what is often regarded as the current "gold standard" for bacterial identification: *16S* ribosomal-RNA (rRNA) sequencing [18,19]. While this method will frequently give unambiguous genus-level strain identification based on comparison with a very large body of *16S*-sequence data, it requires significant wet chemical skills, which are not necessary with MS-based techniques. Bacterial identification by pyrolysis-MS, for example, can be rapidly and simply performed, and has been used by several research groups [20–23]. However, pyrolysis-MS has largely been supplanted by the recent development of intact cell analysis using matrix-assisted laser-desorption/ionization time-of-flight mass spectrometry (MALDI-TOF), discussed in the following paragraphs [24–30].

Intact-cell MALDI-TOF analysis offers several attractive features for rapid screening of bacterial collections. Analysis is performed directly on the cells after minimal sample preparation, and data acquisition is complete in only a matter of minutes. Intact "biomarkers" are introduced into the MALDI-TOF instrument under these conditions. Whether the observed biomarker molecules are desorbed directly from the surface of the cell wall or are extracted from the cells and co-crystallized with the matrix is currently unresolved, but MALDI spectra of intact bacteria generally contain a large number of peaks in the mass range 1–20 kDa [31]. For bacterial cells, proteins are the most often observed biomarkers. While this approach samples only a small percentage of the total proteins produced in the cells, these profiles have been reported by many groups to be suitable for taxonomic identification, down to at least the strain level. The wide availability of the MALDI-TOF instrumentation and its relative ease of use, coupled with relatively simple sample preparation procedures, have been key features in the rapid advancement of this approach.

While the spectral complexity of intact-cell MALDI-TOF can provide good bacterial differentiation, relatively sophisticated data-handling methods are required to successfully implement this approach. A key challenge for this method is reproducibility. Meaningful comparisons of spectra are highly dependent on reproducible spectra. This difficult issue has been the subject of much research [28]. Various labs have developed their own protocols to minimize spectral variability that include attention to factors such as culture growth time, cell-handling and storage, temporal changes in protein production in the cells, sample/matrix preparation procedures (i.e., pH, MALDI laser type), use of internal standards, and sample concentration on the MALDI probe. Several investigators have reported that while some peaks in the mass spectrum are dependent on conditions and may not always be detected, a "core" of observable peaks exist that can form the basis for a usable bacterial differentiation method [32].

Related to the difficulties associated with reproducibility, is the challenge presented by comparison of many—perhaps hundreds to tens of thousands of—complex spectra and the resulting data interpretation. Several groups have developed their own analytical and software solutions to this problem. Jarman et al. [33] and Bright et al. [29], for example, developed methods to automatically compare spectra of unknown bacteria with those of library entries to accomplish high-quality identifications. The recent emergence of bacteria as potential "terror weapons" will likely spur further efforts in the development of data-evaluation tools. Achievements to date show that this approach is non-trivial, and while significant progress has been made, the methods have been used primarily for rapid species-level identification of strains, often responsible for food-borne illnesses, from relatively small databases. Nonetheless, the published data indicate that this approach might be successfully applied to new strain discovery and lead to novel natural products. From the standpoint of novel natural products discovery with industrial potential, the importance of 100% accuracy for the identification of an individual bacterium contained in a library is clearly less critical than for public health/forensic/weapons applications. The ability to differentiate the bacteria with corresponding mass spectra and rank the spectra according to their degree of similarity or difference with the others in the library may be all that is needed.

Further technical advances which could contribute to the successful application of MALDI-TOF to bacterial discovery include the ability to analyze simple mixtures, as reported by Wahl et al. [34]. While cultures analyzed in a MALDI-based program would be ideally pure, in reality this may not be the case. Indeed, these authors were led by their MALDI data evaluation to the conclusion that one of their cultures had become contaminated during the study, demonstrating the value of their analytical method. In an interesting extension of this work, the same group recently reported a first attempt at what amounts to a biological analog of LC-MS [35]. By coupling flow field-flow fractionation with MALDI, they were able to demonstrate both the separation of cell types in a mixed bacterial population, and its interface with a MALDI instrument. Due to the difficulties inherent with an on-line coupling of these two techniques, the authors reported development of an off-line mode. However, the researchers were able to investigate these difficulties with a view to future improvements. The first application of Fourier transform mass spectrometry (FTMS) for the characterization of MALDI-produced ions up to approximately 10 kDa directly from whole cells was recently reported [32]. While the mass resolution produced by this technique may not be routinely needed in a search for novel bacteria, the extra resolution may prove to be a valuable asset for the differentiation of strains that appear similar with nominal mass MALDI data.

To date, the vast majority of applications of intact cell MALDI-TOF analyses have been to bacteria. However, preliminary extensions of the approaches used in bacteria have been applied to fungi. From the standpoint of the discovery of novel chemotypes via natural products, this is obviously a desirable step, but it is also clearly in its infancy. Two groups reported [36,37] the observation of relatively simple mass spectral fingerprints from the intact fungal tissue they analyzed. This application might have the advantage of simplified spectral comparisons, relative to those for bacteria, but also means that such comparisons would have fewer data points to support them.

In summary, there are various MS-based approaches available that can be considered in design of a novelty-based source evaluation method for natural product discovery. While each approach has technical challenges to overcome, a key feature in their successful implimentation is the development of software to allow appropriate evaluation and value extraction from the large databases produced. Overcoming these hurdles would allow natural products to be used more effectively as part of the suite of lead-generating technologies employed by a research group.

Purification of Natural Product Libraries

As discussed earlier, natural product discovery has traditionally been performed using the process of bioassay-guided fractionation [38,39]. In this approach, each sample is extracted, typically with an organic solvent, to produce a small number of extracts. These extracts contain mixtures of compounds that are subsequently submitted for biological screening. Extracts that show activity against the target of interest are then advanced for extensive fractionation and bioassay to determine the identity of the active component(s).

Although bioassay-guided fractionation has been successfully applied for many years, this approach has several limitations. Because natural product extracts often contain complex mixtures, they can create several challenges when submitted for high throughput screening. For example, many extracts contain ubiquitous compounds that may mask or interfere with the biological assays [40–42]. In addition, the extracts often contain compounds covering a wide range of concentrations. As a result, only a limited number of the components may be present at the appropriate concentration for the screen of interest. In these cases, minor components with interesting biological activity may be missed.

Modern pharmaceutical screening campaigns can currently screen more than 200,000 compounds from an internal library in a few weeks [43]. These screening efforts are generally optimized to screen relatively pure compounds produced using traditional or combinatorial synthesis. Recent advances in genomics and proteomics have allowed the rapid development of new targets, which has emphasized the need to maintain libraries of diverse, well-characterized, and relatively pure chemistries that can be submitted for high throughput biological screening (HTS). Although the bioassay-guided fractionation process can ultimately generate isolated compounds compatible with HTS, the cost and resources required to generate these libraries is viewed by many as not competitive with approaches such as combinatorial chemistry.

Recently, several approaches have been developed to address these challenges via automated separation of natural product extracts into purified or semipurified fractions. The purpose of these systems is to improve the compatibility of natural product libraries with HTS campaigns by (1) reducing of the number of components in each sample, (2) adjusting the concentrations to levels appropriate for HTS, (3) and eliminating nontarget (e.g., high molecular-weight) compounds that may produce false positives or negatives. Mass spectrometry has played an important role in the development and implementation of several of these approaches.

Significant effort has been made to automate the process of natural-product fractionation prior to screening. Sepiatec GmbH has recently developed several systems based on the SepBox® technology [44]. These systems can perform automated two-dimensional (2D) preparative LC directly coupled with solid-phase extraction (SPE) under high pressure. A schematic diagram of a SepBox 2D-5000 system is shown in Figure 6.2. Using these systems, large quantities (1–5 g) of crude extract can be intelligently fractionated into several hundred fractions in less than 24 hours. This fractionation is accomplished with preparative LC, followed by polarity adjustment, SPE trapping, washing, and elution. Fraction collection can be triggered with time,

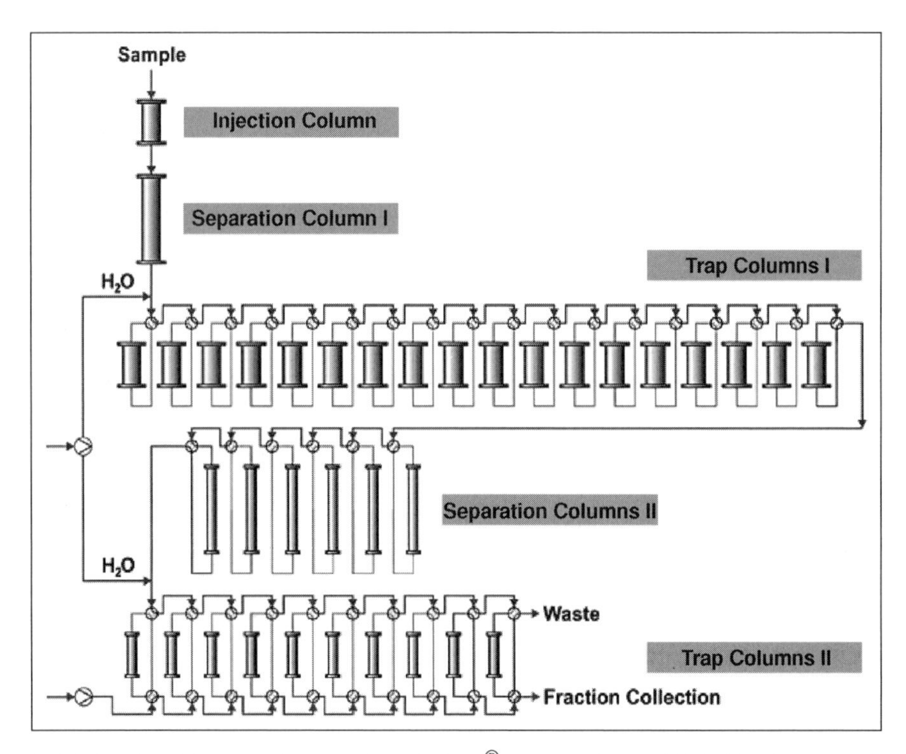

FIGURE 6.2 Schematic diagram of the SepBox® automated HPLC/SPE/HPLC/SPE purification system. (Reprinted with permission.)

UV, or evaporative light-scattering detection (ELSD). In a related approach, Schmid et al. described an automated robotics system based on SPE [45]. This system uses 1–3 fractionation steps, followed by a concentration step to generate semipurified samples. MALDI-TOF was used to assess the purity and identity of the resulting fractions. The complexity of the final fractions was dramatically reduced, resulting in better compatibility with their biological screens. These developments in automated fractionation have been used to generate a purified central natural product pool, as described by Koch et al. [46].

Another successful approach to improve the quality of libraries prior to screening has been mass-triggered fractionation. This approach uses MS-triggered LC fraction collection for sample purification and was first introduced by Kassel et al. [47,48], who demonstrated the purification of a 192 compound library to a purity sufficient for biological correlation (80–90% purity) in a 24-hour period. Mass-triggered fractionation has seen broad application in the purification of combinatorial libraries [49,50] as well as natural product samples [51]. Several commercial systems now provide this capability [52,53]. Mass-triggered fractionation has also been adapted to supercritical fluid chromatography (SFC) formats for the purification of compound libraries and natural product extracts [54–58].

Both the automated fractionation and mass-triggered fractionation approaches suffer the throughput limitations of a serial technique. Significant advances have also been made in the development of parallel separations for both purification and analysis. LC-MS has proved to be a critical component to this approach, particularly in the area of library quality assessment. The first parallel LC multiple-sprayer system for high throughput sample characterization was developed by Zeng and Kassel [59], who were able to adapt a commercial electrospray source to allow the simultaneous monitoring of the effluent from two LC columns.The molecular weight and purity of up to 384 compounds (four 96-well plates) were characterized per day with this system. This approach was subsequently adapted to a four-channel MUX design with a fast-scanning quadrupole instrument by Xu et al. [60] for the high-throughput parallel purification of combinatorial samples.More recently, eight-channel MUX systems, similar to that shown in Figure 6.3, have been developed, which allows for the analysis of a 96-well plate in under 40 minutes [61]. Isbell et al. recently demonstrated the ability to purify an average of 500 samples/day with a four-channel MUX fast-scanning quadrupole-based purification system and less than three full-time employees [62]. Parallel MUX LC-MS has also been used with TOF detection to provide accurate mass data. Nogle and Mallis recently described the use of a four-channel MUX system with a Micromass LCT instrument to provide accurate mass data for high throughput library assessment [63]. The LC separation was shortened to 2.5 minutes/

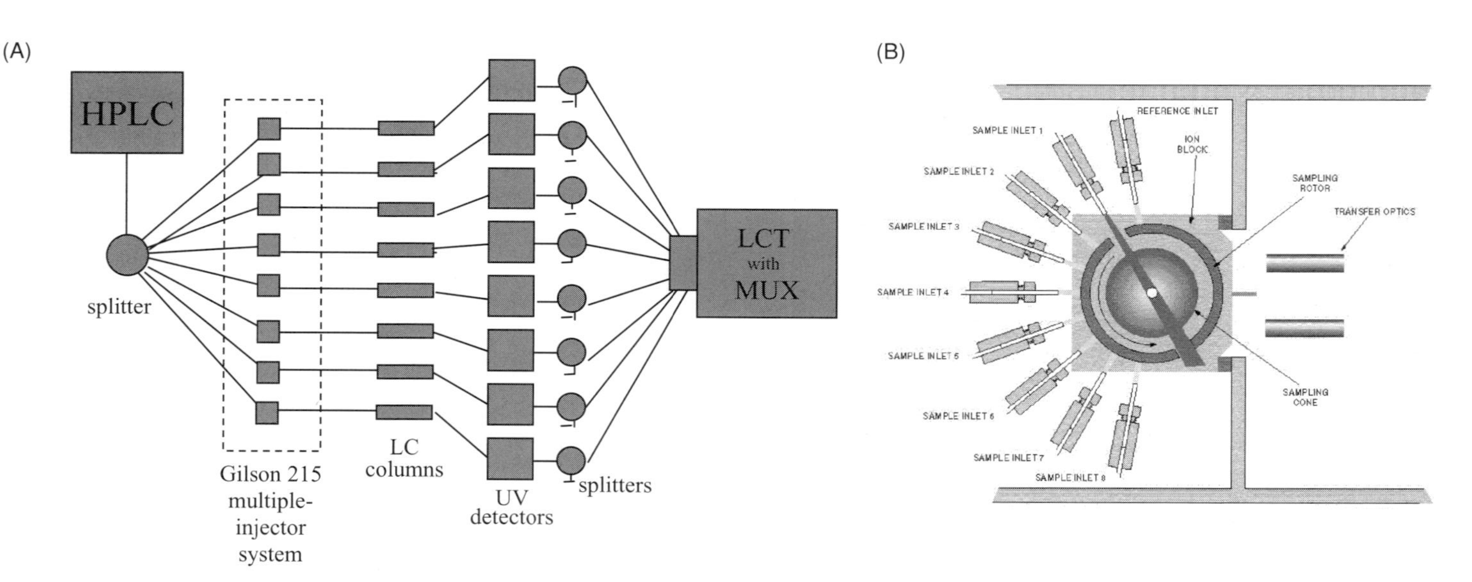

FIGURE 6.3 Schematic diagram of (A) a complete eight-channel parallel LC-MS system, and (B) the index system used in the eight-channel MUX interface.

sample, and both purity and accurate mass data were provided on compound libraries at a rate of 1000 samples/day.

Improvements in the accurate mass capabilities of MUX systems were recently reported by Cremin and Zeng [64], who compared the use of a dedicated internal standard (IS) channel versus single-time-point correlation for accurate mass measurement of purified natural product libraries. Using the single-point calibration method on an eight-channel MUX system, the authors reported 10-ppm mass accuracy without sacrificing a channel for IS infusion. Additionally, a new nine-channel MUX interface was recently demonstrated that provides a dedicated IS channel, while still allowing the system to monitor eight parallel LC streams [65].

Several of these techniques have been integrated into a standardized approach for the purification of natural product samples described by Eldridge et al. [66,67]. Their goal was to harness the biodiversity of plants with the high throughput technologies developed in combinatorial chemistry to produce an ordered collection of relatively pure compounds compatible with HTS. Libraries of purified fractions containing known amounts of 5–7 components were generated from plant tissue using a combination of automated flash chromatography with parallel four-channel preparative LC. A schematic representation of this purification process is shown in Figure 6.4. Prior to screening these libraries were analyzed with a parallel eight-channel LC/ELSD/MS (MUX) system to determine the number, molecular weight, and relative quantity of the components in each fraction. In one example, this approach was used to isolate 147 purified compounds from an extract of *Taxus brevifolia*. The crude extract and its fractions were screened at the National Cancer Institute (NCI) for anticancer activity. Although the initial crude extract did not show activity, several of the purified fractions were found to contain actives, including paclitaxel and several minor paclitaxel analogs. The lack of activity of the crude taxane-containing extracts demonstrated the value of the purification and titre enhancement provided by this approach.

In conclusion, there have been several successful approaches to implementing a "purified library" strategy to maximize the value of natural product sources. Each of these approaches has been enhanced by the integration of mass spectrometry. The further evolution of this strategy is still in progress.

SCREENING: LIGAND–RECEPTOR BINDING IN THE DETECTION OF NOVEL NATURAL PRODUCTS

Technologies that focus on the direct interaction of molecules such as natural products with molecular targets can serve to reduce the cycle time inherent in bioassay-directed fractionation and identification. The use of affinity-based

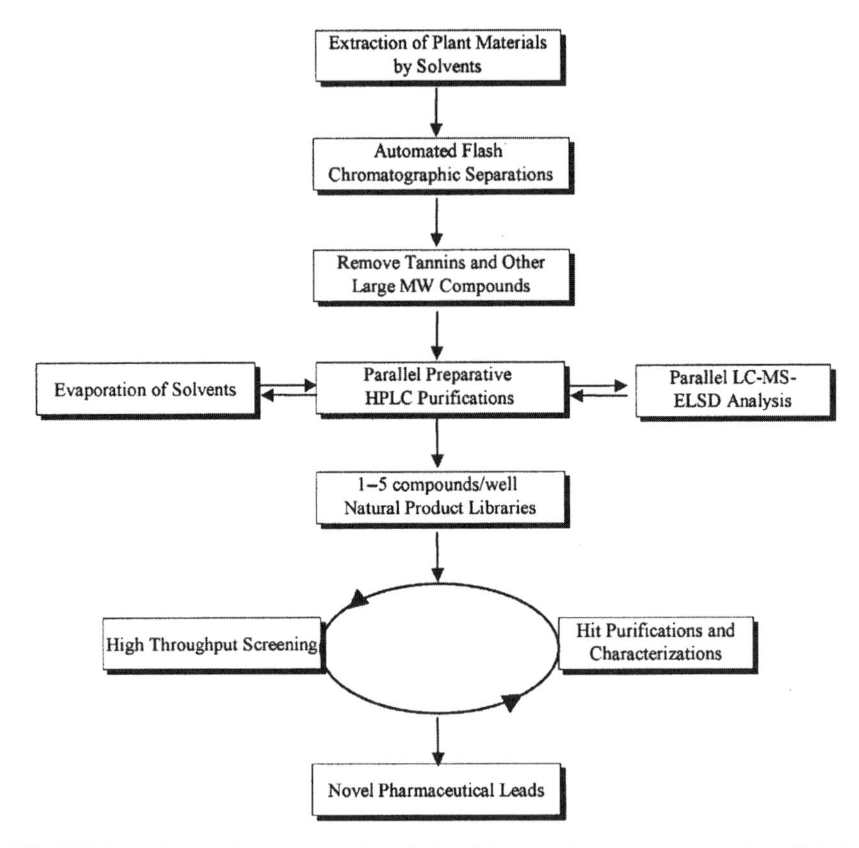

FIGURE 6.4 Schematic representation of the high-throughput methods used by Eldridge et al. (Reprinted with permission from Anal. Chem (see ref 66))

techniques coupled with mass-spectral detection for screening is an area of active research and has been the subject of recent review [68]. One such technique is the "high resolution screening" (HRS) MS-based technique described by Schobel et al. [69]. In this example, plant extracts were initially screened in an estrogen receptor–binding fluorescence-enhancement assay that used the fluorescent ligand coumesterol. Extracts that contained ligands which bind to the estrogen receptor were detected by a reduction in coumesterol fluorescence. Binding activity was then confirmed with competitive fluorescence polarization, functional activity, and transcriptional activation assays. Natural product extracts that were confirmed to be active in all these assays were next processed in the HRS MS mode. After injection and chromatographic separation of the extract, the coumesterol + estrogen receptor solution was infused into the effluent stream of the LC. A decrease in fluorescence from the coumesterol/estrogen receptor was used to indicate activity. Active peaks were directed to the mass spectrometer where MS and MS/MS spectra were acquired in data-dependent scan mode. The LC retention time, molecular weight, and MS/MS spectra were then used to perform database searches to elucidate the structures of the bioactive compounds. In this fashion, molecular weight information on bioactive compounds as well as MS/MS spectra to support structure elucidation was obtained in a single analysis. Limitations of this technique include various matrix effects (i.e., fluorescence quenching, absorbance, light scattering, nonspecific binding between matrix compounds and target), and nontrivial statistical analysis. Nonetheless, reduced cycle time was demonstrated through this on-line ligand–receptor binding MS-based approach.

Another direct method for the detection of active natural products has centered on the binding of ligands to ribonucleic acid (RNA) targets [70]. One such technique is referred to as multitarget affinity/specificity screening (MASS). This approach is based on the identification of library members bound to a macromolecular target with the direct measurement of the accurate mass of the target–ligand complex. The technique uses the resolving power and mass accuracy of FTMS, allowing the simultaneous screening of multiple targets against multiple ligands. High-resolution accurate mass scans of RNA targets are first acquired without the addition of a screening mixture (extract). The extract is then added, and the mass spectrum of the mixture is evaluated for mass shifts that indicate binding of the ligand to a target. The number of targets screened is limited by the complexity of the target spectrum, but can utilize as many as 10 targets simultaneously, or fewer targets with increasingly complex ligand mixtures. Interestingly, the screening mixture concentration can be modified to select for weak (μM) to strong (nM) binders. Not surprisingly, with this technique data interpretation is complex and presents a significant challenge.

DEREPLICATION OF KNOWN COMPOUNDS

The Need for Dereplication

Considerable time and effort are required to isolate and characterize NCEs from natural product sources. It has been estimated that, on average, it takes $50,000 and three months to isolate and characterize an active compound from a single natural source [71]. To date, more than 150,000 known compounds have been isolated from nature, and this list expands at a rate of approximately 10,000 per year [72]. As a result, the probability of rediscovering previously isolated compounds steadily increases, and the potential cost associated with reisolating known compounds is high. Consequently, significant effort has been invested to avoid the replicate identification of previously identified compounds from natural product extracts.

This process, commonly termed *dereplication*, has been defined as "the attempt to determine the occurrence of previously known active compounds in crude extracts as early as possible to minimize the effort lost in their isolation" [43]. Traditionally, dereplication was performed using TLC with bioautographic detection. In this approach, antimicrobial extracts were separated using TLC, and monitored for active compounds with bioautographic detection (i.e., growth inhibition of an organism of interest). The advent of LC with UV/visible (VIS) diode-array detection significantly improved dereplication by providing a combination of high resolution separation with spectroscopic detection. These systems allowed the development of in-house libraries of LC retention times and UV spectra for known actives that could be readily compared to unknowns.

The next step in dereplication is to determine whether actives that do not match internal libraries are, in fact, novel. Several commercial databases have been developed that allow researchers to search for previously discovered natural products using criteria including: taxonomy of the producing organism, bioactivity, molecular formula, molecular weight, substructures, and UV maxima [73]. These databases include: Chapman and Hall's *Dictionary of Natural Products, Bioactive Natural Products Database* (Berdy), DEREP, Chemical Abstracts Services Registry File (CAS/STN), MARINLIT, and the *Marine Natural Products Database* (MNP Database), and can provide powerful tools to evaluate the novelty of compounds of interest. Unfortunately, at the dereplication stage, many of the search criteria may not be known. Prior to isolation, generally only the source family or genus and the UV maxima are available for database searching, and often produce long lists of potential matches when used for database searches. One of the most powerful means to narrow these search results is to use molecular weight or molecular formula limits, both of which can be provided by mass spectrometry.

LC/MS-Based Dereplication

Prior to the advent of electrospray ionization (ESI), the primary use of mass spectrometry in natural product discovery was the structural elucidation of compounds that had been isolated with bioassay-guided fractionation. With the advent of commercial ESI and atmospheric-pressure chemical ionization (APCI) sources in the early 1990s [74,75], researchers gained access to LC-MS instrumentation that could be used to directly analyze natural product mixtures. This allowed the integration of mass spectrometry into the earliest stages of natural product discovery. The impact of ESI and APCI on natural products discovery has been the subject of recent reviews [76,77].

Most modern LC-MS dereplication systems use the basic configuration introduced by Constant et al. [78], which is shown schematically in Figure 6.5A.

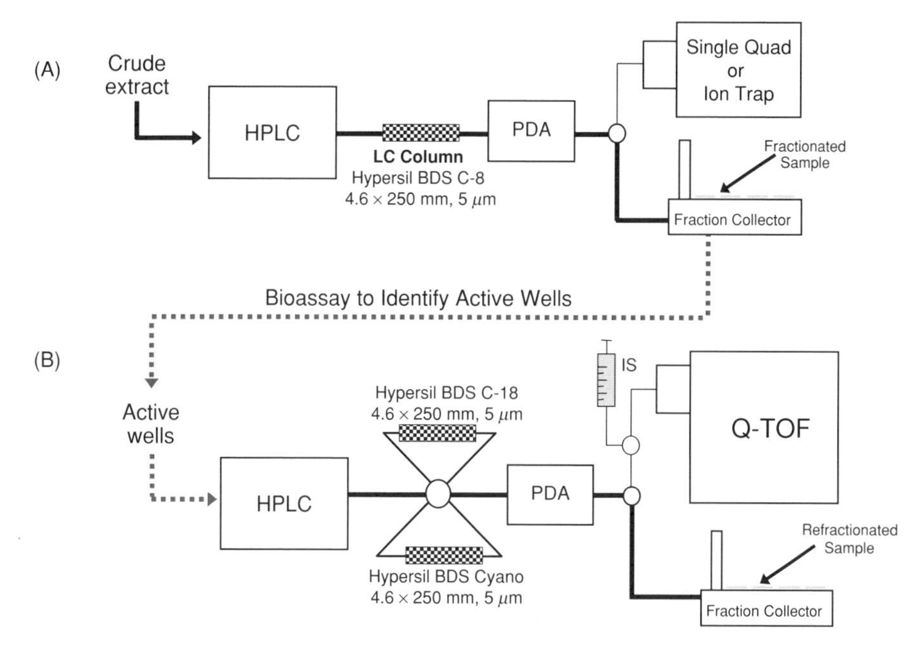

FIGURE 6.5 General schematic diagram for (A) an LC-MS dereplication system, and (B) a quadrupole TOF based LC-MS for identification.

Crude extracts are separated with reverse-phase LC, and the column effluent is directed to a UV/VIS diode-array detector. The effluent is then split between a fraction collector and the ESI source of an LC-MS. The majority of the flow is directed toward the fraction collector and collected in a 96-well or 384-well plate, while a small percentage is ionized and detected with positive or negative ESI. This approach takes advantage of the concentration dependence of electrospray [79] to provide molecular-weight information, while the majority of the sample is retained for subsequent bioassay. The molecular weight and UV/VIS spectral data produced by these systems, combined with any known taxonomic information on the source, can then be used for database searching. Often these searches produce high-confidence matches. If such a match is made, the extract is often deprioritized, allowing the researcher to focus their activities on higher-value samples. Ackermann et al. [80,81] improved this approach by combining LC-MS with MALDI-TOF MS detection to confirm the molecular weights of unknowns. This approach was used in the screening of antibiotic-containing mixtures from fermentation broths. Julian [82] extended this approach with the addition of MS/MS and MS^n spectra obtained with an ion trap, and compared the resulting spectral data to the proposed matches from natural product database searches. This approach has also been used by other research groups [83].

Frequently, organisms will produce several members (factors) within a family of compounds. Product-ion MS/MS spectra can often be used to propose new structures within these series of natural product factors. For example, LC-MS/MS can provide both the molecular weight and product-ion MS/MS spectra of the unknowns. Database searches using these spectra may produce a match to a known natural product in the sample, which can be confirmed as previously described. In addition, the correlation of product ions and neutral losses from the MS/MS spectra with the structure of these known compounds can often be used to propose structures for the other related factors in the sample. An example of this correlation was shown by Janota et al. [84], who used this approach to propose several structures within the roridin family.

Data-Dependent MS/MS and MS^n

One of the major advances in dereplication was the advent of instrumentation capable of data-dependent MS/MS and MS^n scanning. Although these scan functions had been available for several years on triple quadrupole instruments the introduction of LC-MS ion trap instruments made data-dependent MS/MS practical for routine dereplication [85]. The ability of ion traps to intelligently switch between MS, MS/MS, or MS^n in real time proved extremely useful for screening the complex mixtures present in natural product extracts. In addition, scientists could design experiments to acquire MS and MS/MS spectra of suspect compounds from a dereplication "inclusion list," as well as MS/MS and MS^n spectra for unknowns. The latter often proved particularly useful for structure elucidation. An example of the application of data-dependent MS^n scans in natural-product structure elucidation is shown in Figure 6.6. The MS^n fragmentation pathway of homolelobanidine, a novel compound found in an extract from *Polygonum flacidum*, was obtained using

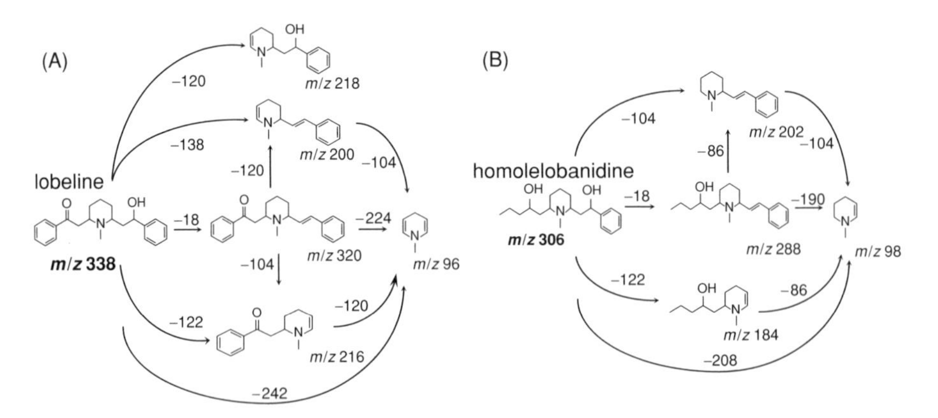

FIGURE 6.6 The MS^n fragmentation pathway of (A) lobeline, and (B) homolelobanidine.

an ion trap configured for data-dependent MS/MS and MS^n scans [86]. Several known members of the lobeline family were found to be present in the sample. Using the MS^n (n = 1–3) fragmentation pathway observed for one known member of this family (lobeline), a structure was proposed for the novel compound homolelobanidine. The MS^n fragmentation pathways for both lobeline and homolelobanidine are shown in Figure 6.6. The proposed structure for homolelobanidine was subsequently confirmed via isolation and NMR.

Data-dependent scanning can also provide highly specific MS/MS spectra of known compounds that can be used for mass-spectral library searching in dereplication [87]. Using automated library searches, the MS/MS spectra of unknowns can be readily compared against MS/MS spectra of known compounds that were analyzed on the same instrument. By combining the retention time, UV, MS, MS/MS, and/or MS^n spectra of unknowns to those acquired for purified standards, dereplication of natural products can be readily performed [88]. In addition, the use of MS, MS/MS, and MS^n data provided by ion-trap instruments for structure elucidation of natural products has been demonstrated by several groups [89–92]. These approaches allow the rapid screening of extracts for known nuisance compounds with high specificity, and can often be performed *prior* to bioassay to save time and conserve bioassay assets.

One limitation in the use of ion-trap instruments for dereplication and structure elucidation is their tendency to produce simple product-ion spectra with low specificity, such as those resulting from neutral losses of water. The quality of MS/MS library spectra was improved through the use of several methods to generate more extensive fragmentation. Sanders et al. [87] described the use of a modified parent-ion excitation method termed *wide-band excitation*, which was capable of exciting both the parent and water-loss ions, improving the richness of the resulting product-ion spectra. This procedure was combined with normalization of collision energies to produce information-rich product-ion MS/MS spectra over a wide mass range that could then be used to search MS/MS spectral libraries [93].

One important aspect in designing an effective dereplication system is to employ ionization techniques that can successfully detect the broad range of chemistries encountered in nature. Positive ESI has generally been found to ionize the broadest range of compounds from nature, with estimates of approximate coverage at $>70\%$ [94]. Unfortunately, not all compounds are ionized in this mode, and of those that are ionized, not all produce molecular adduct ions that are readily converted to the molecular weight data required for searching external databases [95]. For example, the mycotoxin verruculogen (molecular weight-511), a secondary metabolite of *Aspergillus fumigatus*, does not produce a molecular adduct ion with positive electrospray, as shown in Figure 6.7A. Instead, an ion at m/z 494 resulting from water loss from the

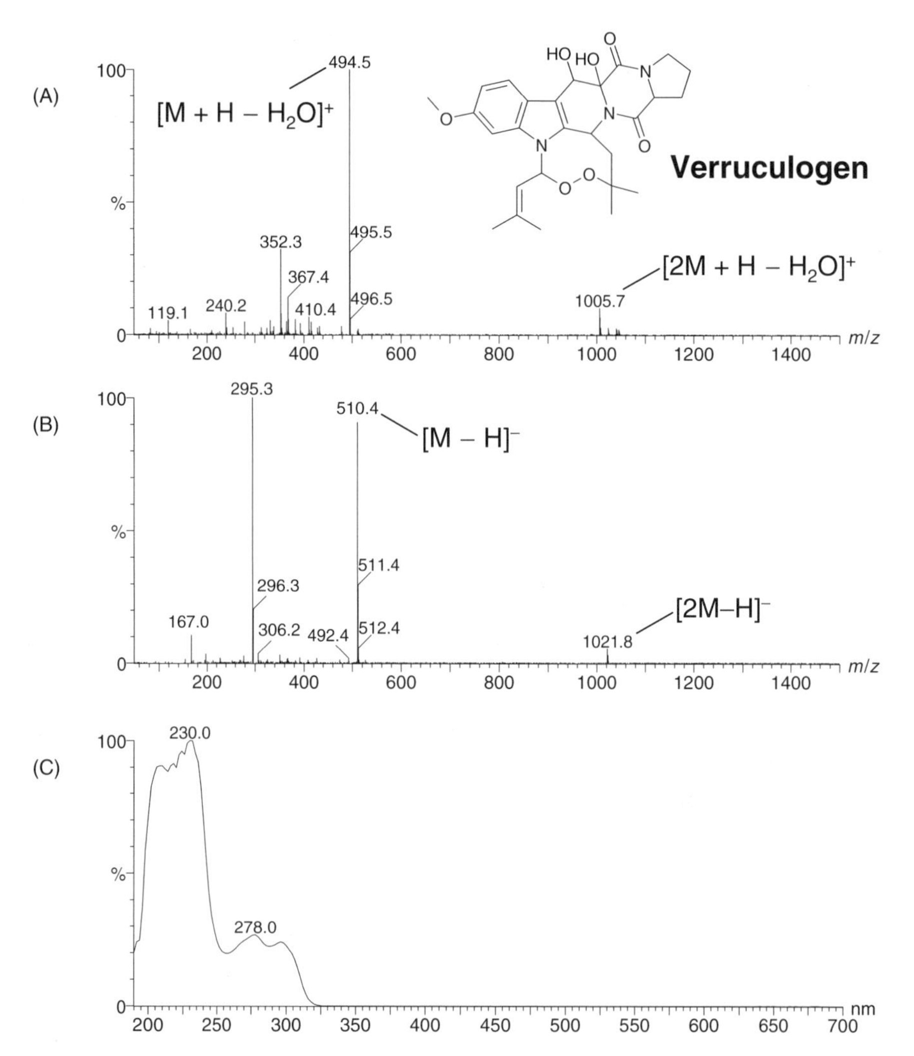

FIGURE 6.7 The (A) positive ESI-MS, (B) negative ESI-MS, and (C) UV spectra of verruculogen.

protonated molecule, $[M + H - H_2O]^+$, is observed. When verruculogen is ionized using negative ESI, both the deprotonated parent and deprotonated dimer of the molecule are observed. As a result, the molecular weight of the compound can be determined. Thus, the value of a second ionization mode, in this case negative ESI, is clearly demonstrated.

Advances in both LC separations and MS ionization modes have also increased the range of natural product chemistries accessible to dereplication. Reverse-phase LC continues to be the most widely used separation approach, and works well for molecules with moderately polar to nonpolar chemistries [77]. However, this approach has proved inadequate for more highly polar compounds, where dereplication becomes considerably more difficult. Highly polar compounds have, in some cases, been successfully addressed through use of ion-exchange (acids/bases) and graphite stationary phases. The disadvantage to ion-exchange methods is that they require separate approaches for acids and bases, and do not retain neutral chemistries. Graphite columns can suffer from irreversible retention of unknowns, making their use problematic. The introduction of hydrophilic interaction liquid chromatography (HILIC) offers a single-column approach applicable to a broad range of polar chemistries [96]. In the author's laboratory, HILIC separation has proved useful for dereplication for a variety of highly polar molecules [97]. Finally, GC-MS dereplication methods, typically requiring derivatization, can be applicable in special cases where LC-MS methods lack the sensitivity or resolution to detect the molecules of interest [98].

The Advent of the Quadrupole TOF—Accurate MS and MS/MS

With the introduction of commercial quadrupole time-of-flight (Q-TOF) instruments in the late 1990s [99–101], natural product researchers were, for the first time, able to obtain both accurate MS and accurate MS/MS data directly from crude natural product mixtures. This provided a significant advancement in the dereplication and identification process, often allowing the unambiguous assessment of compound novelty in a single LC-MS analysis. Where searches of external databases using nominal molecular weights generally produced several possible matches, accurate mass-based searches often resulted in a single molecular formula match [102,103], even when searching large databases such as Chapman and Hall. Accurate MS/MS data also provided a useful tool for confirming the resulting matches, allowing the direct comparison of the proposed structures to the observed accurate fragment masses and neutral losses of the unknown [104,105].

One limitation of Q-TOF instruments was their inability to perform pulsed positive and negative ionization on a scan-to-scan basis. As a result, a two-tiered LC-MS screening approach to dereplication and identification was often used, as shown in Figure 6.5A and 6.5B [106]. In this approach initial dereplication of crude extracts was performed with a UV detector–equipped single quadrupole mass spectrometer system and fraction collector. This system used pulsed positive and negative electrospray with and without source-induced dissociation to provide retention time, UV, positive ESI, negative ESI, and fragmentation data from a single injection. These data were compared to a spectral library of known bioactive compounds, and the collected effluent was bioassayed to localize the region of activity.

A second injection was made on a Q-TOF system operated in data-dependent MS/MS mode with both UV detection and fraction collection. The accurate MS and MS/MS data from this system were used to search commercial databases using a combination of taxonomic, UV maxima, and accurate mass data. This approach often produced a single formula match, which could be confirmed by comparing the structural features of each proposal to the observed accurate MS/MS fragments. Additional MS^n data were also acquired, when necessary, by infusion of the contents of the fractionated sample from the second 96-well plate into an ion trap configured with a nanospray source.

One example illustrating this approach is shown in Figure 6.8. The LC/UV/MS chromatogram of a marine algal methanol extract is shown in Figure 6.8A. The (RT), UV, and MS data from the LC/UV/MS analysis were

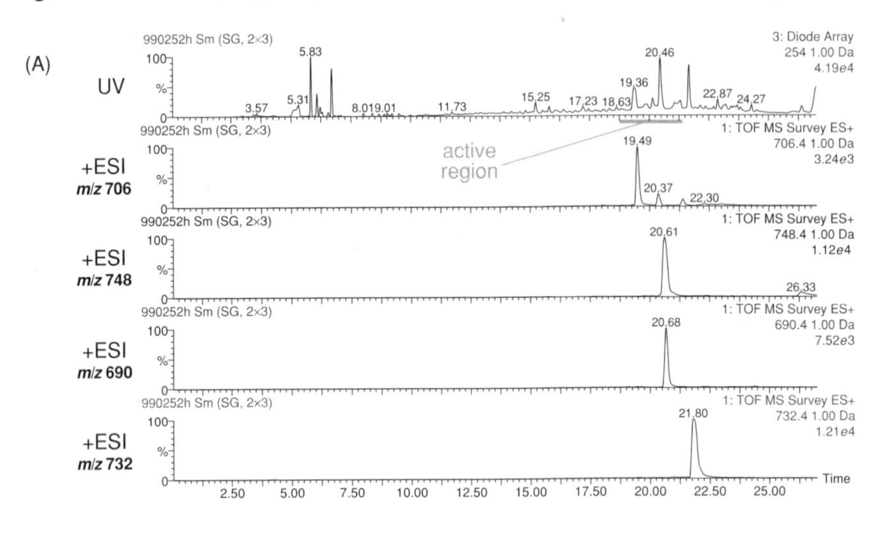

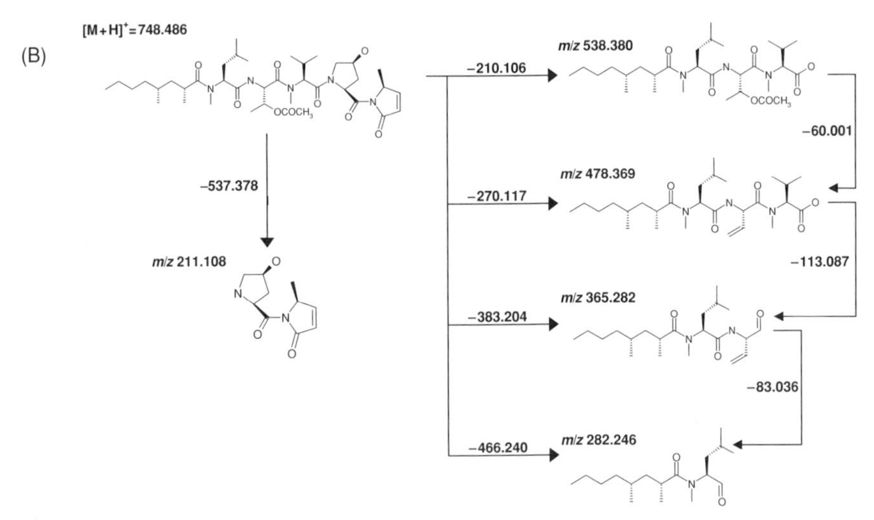

FIGURE 6.8 The (A) LC-MS chromatogram of a marine algal extract, and (B) the MS^n fragmentation pathway of the peak detected at 20.6 minutes, microcolin A.

searched against an internal mass-spectral library and did not produce a match with any previously isolated compounds. The biological activity was then localized to the 19–21-minute region of the chromatogram through bioassay. This region was examined in a second analysis performed on a Micromass Q-TOF to provide both accurate MS and MS/MS spectra of the components in the active region. In addition, the MS^n product-ion fragmentation pathway ($n = 1 - 4$) was mapped using an ion trap. A complete picture of the molecular composition of the parent ion, fragment ions, and neutral losses was developed from these experiments. As a result, one of the compounds in the active region was identified as microcolin A, whose structure and MS/MS fragmentation are shown in Figure 6.8B. This approach was applied to the other biologically active components in the mixture, resulting in the identification of three additional known microcolins as well as the proposal of six new members of the microcolin family. This combination of accurate MS, accurate MS/MS, and MS^n data has been used by many groups for the dereplication and structural elucidation of several classes of natural products, including the flavonoids [107,108], rapamycin [109], and humic substances [110].

One of the operational challenges with the Q-TOF instrument, as configured in Figure 6.5B, is that an internal standard (IS) solution must be introduced into the column effluent prior to the ESI source. Due to limitations in the dynamic range of these instruments, the intensity of the IS peak must be closely controlled throughout the chromatographic run. Often the ionization efficiency of the IS changes under gradient conditions, making it difficult to maintain constant IS peak intensity. This issue was recently addressed with the introduction of the "lock spray" source design [111,112]. This source uses two independent ESI sprayers to decouple the ionization of the sample stream from the IS stream. A mechanical indexing system is used to switch between the two sprayers, allowing both streams to be alternately sampled. The result is that the ionization of the IS stream is independent of changes in the effluent-stream composition. This allows the IS peak intensity to be readily controlled throughout the chromatographic run. This source has been used by several groups for the analysis of natural products, and mass accuracies between 2 and 10 ppm have been reported [113]. It was recently used to develop a database of 474 fungal mycotoxins and metabolites, which were indexed by name, formula, retention index, peak symmetry, UV maxima, accurate mass adduct ions, and accurate fragment ions [95]. These data were then used for rapid dereplication of extracts for these compounds, and provided a framework for the proposal of several novel analogs of known compounds.

Data Reduction Tools

A significant challenge in the process of dereplication is the extraction of useful data from the large and complex data sets generated with hyphenated techniques such as LC/UV/MS. To accommodate high throughput sample processing, this extraction should occur in an automated fashion where data files are automatically searched for known components. Most modern LC-MS systems have some capability to perform this operation. For compounds that have been analyzed previously on a similar system, data files can generally be searched within known retention-time windows for the presence of specific mass-spectral or UV peaks. Matches are then compared (often manually) to the MS or UV library entries from in-house libraries. One major limitation to this approach is the need to generate RT, UV, and MS data on each standard of interest. This restricts the utility of these comparative searches to compounds that are available as standards, or that were previously isolated in-house, *and* have been previously run on a similar system. Obviously, this case will only represent a small fraction of the compounds that are encountered in nature.

Rather than rely solely on comparisons with standards, significant research has been performed to develop tools for the automated extraction of peaks from chromatographic data. Many of these are based on component-detection algorithm (CODA)-based peak detection [114], or the Automated Mass Spectral Deconvolution and Identification System (AMDIS) software developed by the National Institute of Standards and Technology [115]. Application of the AMDIS software for component detection and searching was recently described by Zink et al. [116]. In this work, a single quadrupole LC-MS was operated in the pulsed positive and negative ESI mode with concurrent UV detection. AMDIS was used to extract mass spectral peaks from both the positive and negative ESI mass spectra. These MS spectra were combined with UV spectra and retention-time data to search against an internal compound library to produce a composite matching score. Wang et al. [94] described the development of an automated deconvolution system based on an ion trap that used fast positive/negative switching routines in combination with the acquisition of data-dependent MS/MS and MS^n spectra. This system was used to generate a "spectral fingerprint" of each unknown, which included its retention time, MS polarity, as well as MS and MS/MS spectra. These data were then searched against a database of several thousand known natural products.

Some of the most elegant work in data reduction for natural products screening has been reported in the area of diversity assessment, as discussed earlier [10]. Several of these examples involve LC/UV/MS analysis of natural product extracts on an extremely large scale. The resulting data were converted to the netCDF format and evaluated with customized software. Three-dimensional images were generated for the visualization of sample component differences, and a measure of similarity was calculated to allow for diversity assessment. This approach was also applied to the automatic deconvolution of the mass spectra of secondary metabolites from LC-MS data with factor analysis [117].

Diversity assessment has also been combined with the accurate mass capabilities offered by TOF instruments. He et al. [118] described the use of an LCT instrument with UV detection to rank crude natural product extracts. These data were converted to netCDF format along with the accurate mass data, and processed using the CODA algorithm to extract peaks from the raw data. Accurate mass, RT, UV, and MS data were then combined to generate over 4000 unique "signatures," which were used to prioritize new samples according to their diversity. This approach was later applied to an LCT, incorporating a lock-spray inlet, resulting in improved mass accuracy and system stability [119].

IDENTIFICATION OF UNKNOWNS

Once the novelty of an active component is established, the next step in the lead generation cycle is its identification. Often this process starts with a significant amount of existing data, including accurate mass, accurate MS/MS, and UV–visible spectra. These data provide an excellent framework for structure elucidation that can be augmented with other spectroscopic tools.

Fourier Transform Mass Spectrometry

Although significant data are available from the dereplication process, the mass accuracy of spectra produced by TOF instruments (± 5 ppm) is often insufficient to allow for the proposal of a single molecular formula for an unknown. This case is especially true for the higher molecular-weight molecules that are often encountered in nature. In these instances, the increased mass accuracy of the MS, MS/MS, and MS^n data provided by FTMS systems can prove critical to structure elucidation.

The extremely high-resolution ($>30{,}000$ full width half maximum (FWHM)) and mass accuracy (<1 ppm) that can be achieved with FTMS systems can have significant impact on the structural elucidation of unknowns [120]. FTMS has been used for the detailed structural elucidation of several classes of natural products that include the erythromycins [121,122], muramycins [123], saccharomicins [124], and other polyketides [125,126]. The ability of FTMS to provide MS^n fragmentation combined with extremely high mass accuracy makes it an exceptional tool for structure elucidation. In one example, the use of this type of data was demonstrated by Kearney et al. [127]. In this study, MS^n data provided by ion trap was insufficient to unambiguously assign the fragmentation pathway of erythromycin due to the redundant nature of the molecule. When this system was studied with FTMS, the complete fragmentation pathway for erythromycin was readily elucidated [122,128].

A systematic strategy for elucidation of fragmentation pathways with FTMS called the *Top-down/bottom-up approach* was described by McDonald et al. [129] and applied to the structure elucidation of Muraymycins A1 and B1. A multi correlated harmonic excitation fields excitation signal was used to isolate multiple ions in the FTMS analyzer cell. Only the target precursor ion was activated by sustained off-resonance irradiation CID, allowing the observation of IS ions in the product-ion spectrum of the unknown. These IS ions were then used for internal calibration to improve the mass accuracy of the product-ion spectrum. In the top-down/bottom up approach, the elemental compositions ($C_aH_bN_cO_dS_e$, etc.) of neutral losses were first calculated from the measured accurate mass differences between sequential fragment ions. These losses were generally small enough to allow the proposal of single molecular formulas per neutral loss. The accurate mass of the lowest mass fragment was then combined with these neutral loss formulas, allowing the *calculation* of a more accurate parent-ion mass than could be determined experimentally.

LC-MS/NMR

Ultimately, the definitive structure elucidation of unknown molecules is most often accomplished via NMR. NMR has traditionally been performed on the purified natural product isolated using bioassay-guided fractionation. With the advent of hyphenated techniques such as LC/NMR, these data can now be obtained prior to purification [130,131]. LC/NMR can prove useful even in the dereplication phase, particularly when LC/UV/MS data are insufficient for unambiguous peak identification. LC/NMR has played an important role in natural products structure elucidation, where several related compounds (factors) are often encountered in a single sample. For example, isobaric or isomeric mixtures that may prove difficult or impossible to differentiate by MS, can often be readily distinguished by NMR. Several thorough reviews of LC/NMR in natural products discovery and phytochemical analysis have recently appeared [132,133].

LC/NMR spectra are generally acquired in one of two modes, either with on-flow (real-time) acquisition during the LC run, or off-flow (trapping, stopped-flow) operation. LC/NMR was first demonstrated in the characterization of sesquiterpene lactones from the Mexican plant *Azlusania grayana* with an on-flow probe [134], and the method became more widely used in the late 1990s. On-flow methods are the simplest conceptually, but even with high-field instruments, the sensitivity of on-flow LC/NMR generally limits these experiments to the acquisition of 1H-NMR spectra of major components. Greater than 10 micrograms per analyte is typically required to obtain useful 500-MHz 1H-NMR spectra in on-flow LC/NMR [135].

Thus, the main challenge in the application of LC/NMR to natural products discovery has been improving sensitivity. Instrument developments, including higher-field (>500 MHz) instruments, digital signal processing, advanced probe designs, and gradient pulse sequences have helped to improve the sensitivity of LC/NMR. Sensitivity has also been improved by combining solvent suppression and use of deuterated LC solvents with methods to increase the dwell time of the sample within the flow cell [136]. In the loop-storage approach, the column effluent is trapped in sample loops with an automated peak-sampling system. The contents of the loops can be subsequently directed to the NMR probe for signal averaging. An alternative to this approach is the use of stopped-flow LC, in which the LC flow is either stopped at fixed intervals or stopped based upon the observation of a UV or MS signal [137]. This approach can improve the NMR sensitivity and allows for the acquisition of 1H-NMR spectra at the low microgram range [138], but can adversely affect chromatographic resolution. Modern limits of detection (LODs) are approximately 100 nanograms (MW = 500) on a 600-MHz NMR with stopped-flow or loop-storage methods.

Recently, combined LC/NMR/MS was applied to the identification of an ecdysteroid in an extract of *Sliene otides* [139]. In this example, identification could only be achieved through the combination of both NMR and MS data. In another example, LC/UV/MS/MS analysis on a Q-TOF combined with LC/1H-NMR was used to identify the components of a crude extract of *Erythrina vogelii*. These results are shown in Figure 6.9 [140]. Accurate MS and MS/MS spectra were acquired for a total of 12 peaks, and the molecular weight of each unknown was determined within 5 ppm. LC/1H-NMR was conducted by loading 10 mg of the extract onto a high-capacity C-18 radial compression column. This system was operated at low flow rates (0.1 mL/min) and resulted in the observation of 10 LC peaks during a 19-hour overnight run. Ten-minute fractions from this run were collected and bioassayed for antifungal activity. A UV-shift reagent was used to allow localization of hydroxyl groups, resulting in the partial or total identification of eight polyphenols in the extract.

In a third example, LC/UV/MS^n/NMR was used for the identification of asterosaponins (with differing oligosaccharide side chains) isolated from the Baltic starfish *Asterias rubens* [141]. The isobaric compounds solasteroside A and ruberoside E produced common MS/MS fragment ions. However, different fragment-ion intensities were observed in the MS^n spectra obtained on the ion trap, indicative of structural differences. On-flow NMR was used to propose oligosaccharide chain structures for each isomer that was subsequently isolated and confirmed by 2D NMR. Similar results were reported by Bobzin et al. [133] for the identification of the marine alkaloid aaptamine from an *Aaptos* species sponge. In their application, LC-MS did not allow conclusive identification of the active compounds responsible for the observed

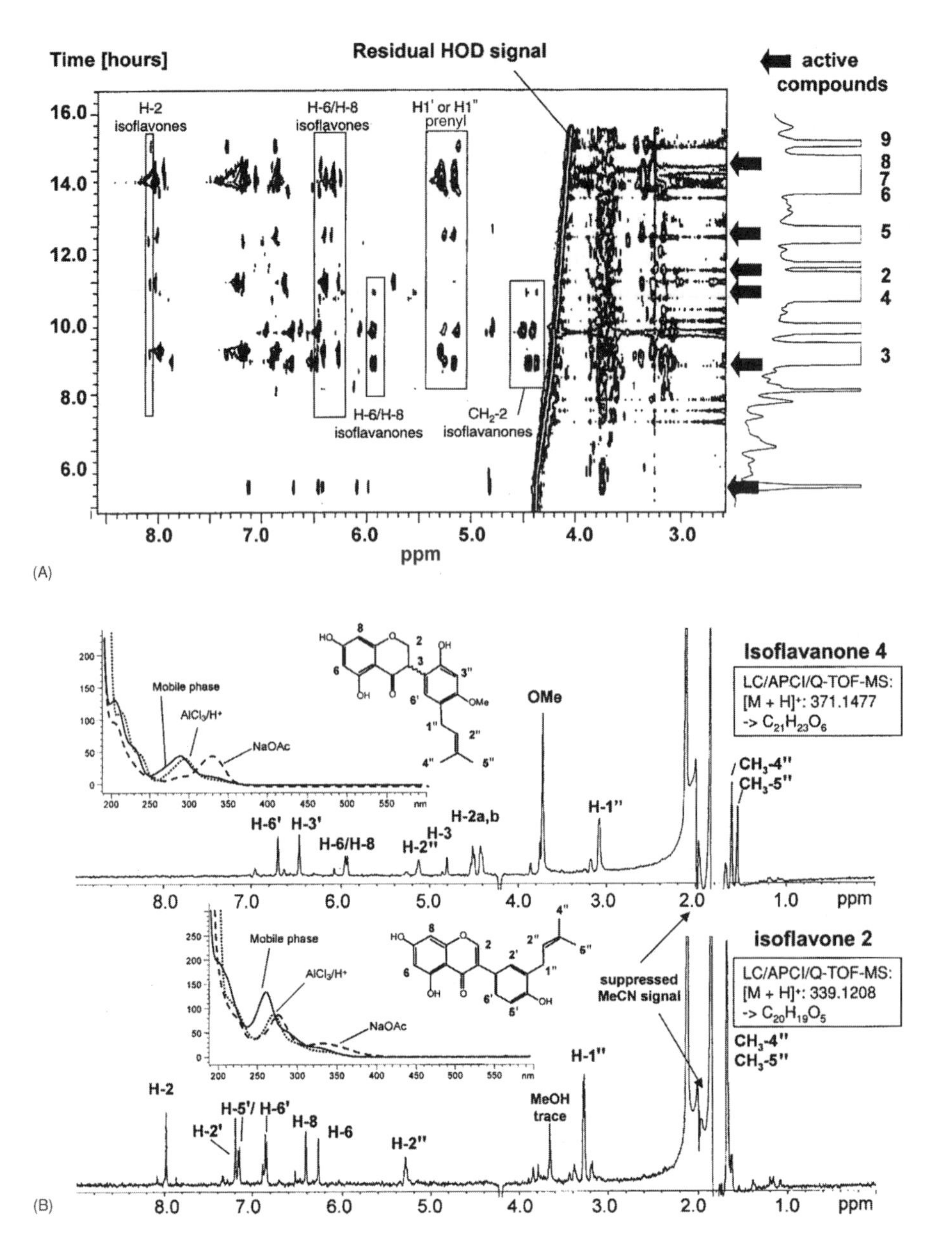

FIGURE 6.9 The (A) 2D LC/1H-NMR chromatogram, and (B) on-flow 1H-NMR spectra of isoflavanones detected in a crude extract of *E. Vogelii*. (Reprinted with permission by John Wiley & Sons.)

enzyme-inhibition activity. In this case, NMR again proved especially valuable for the differentiation of isobaric or isomeric compounds. More recently, an integrated LC/UV/FTIR/NMR/MS system was also developed [142], which has potential to further increase the amount of spectral data available from a single LC run.

CONCLUSIONS

The continued advancement of chromatographic techniques and mass-spectral instrumentation will undoubtedly open up new applications for mass spectrometry in natural products discovery. The development of new chromatographic approaches, such as turbulent-flow chromatography [143], monolithic column supports [144], and ultraperformance liquid chromatography (UPLC) [145] promise to improve the speed and efficiency of natural product screening and purification. New mass spectrometers including ion-trap–FTMS systems promise to combine pulsed positive/negative ionization, sub-100 ppb mass accuracy [146] with MS, MS/MS, and MS^n spectra to provide an extremely powerful set of tools for the structural elucidation of natural products. The recent development of mixed-mode sources combining ESI with APCI, APCI with atmospheric-pressure photoionzation (APPI), or ESI with APPI should expand the range of chemistries accessible to LC-MS [147–150]. The combination of ESI with APCI for small neutral chemistries, or APPI for enhanced ionization of aromatic neutral molecules [151], has potential application in both the dereplication and identification of natural products using LC-MS.

Mass spectrometry has been, and will continue to be, a critical tool for the discovery of new lead chemistries from nature. It continues to play a central role in several phases of the natural products discovery processes that include: source selection, screening, dereplication, and identification. Mass spectrometry will likely remain a key technology that should contribute to the success of natural product lead generation programs in both the pharmaceutical and agrochemical industries.

REFERENCES

1. Henkel, T.; Brunne, R.M.; Muller, H.; Reichel, F. "Statistical Investigation into the Structural Complementarity of Natural Products and Synthetic Compounds," *Angew Chem. Int. Ed.* **38,** 643–647 (1999).
2. Shu, Y-Z. "Recent Natural Products Based Drug Development: A Pharmaceutical Industry Perspective," *J. Nat. Prod.* **61,** 1053–1071 (1998).
3. Newman, D.J.; Cragg, G.M.; Snader, K.M. "Natural Products as Sources of New Drugs over the Period 1981–2002," *J. Nat. Prod.* **66,** 1022–1037 (2003).

4. Rouhi, A.M. "Rediscovering Natural Products," *Chem. Eng. News*, **81**(41), 77–78, 82–83, 86, 88–91 (2003).
5. Rouhi, A.M. "Betting on Natural Products for Cures," *Chem. and Eng. News*, **81**(41), 93–98, 100–103 (2003).
6. Rouhi, A.M. "Moving Beyond Natural Products," *Chem. Eng. News*, **81**(41), 104–107 (2003).
7. Zaehner, H.; Drautz, H.; Weber, W. "Novel Approaches to Metabolite Screening," pp. 51–70 in Bu'Lock, J.D.; Nisbet, L.J.; Winstanley, D.J., eds., *Bioactive Microbial Products: Search and Discovery*, Academic Press, London, (1982).
8. Feistner, G.J. "Profiling of Bacterial Metabolites by Liquid Chromatography-Electrospray Mass Spectrometry: A Perspective," *Am. Lab.* **32L–32Q** (Sept. 1994).
9. Smedsgard, J.; Frisvad, J.C. "Using Direct Electrospray Mass Spectrometry in Taxonomy and Secondary Metabolite Profiling of Crude Fungal Extracts," *J. Microbiol. Methods* **25,** 5–17 (1996).
10. Julian, R.K., Jr.; Higgs, R.E.; Gygi, J.D.; Hilton, M.D. "A Method for Quantitatively Differentiating Crude Natural Product Extracts Using High Performance Liquid Chromatography-Electrospray Mass Spectrometry," *Anal. Chem.* **70,** 3249–3254 (1998).
11. Burkhardt, K.; Fiedler, H.P.; Grabley S.; Thierick, e.R.; Zeeck, A. "New Cineromycins and Musacins Obtained by Metabolite Pattern Analysis of Sterptomyces griseoviridis (FH-S 1832)—I. Taxonomy, Fermentation, Isolation and Biological Activity," *J. Antibiot.* **49**(5), 432–437 (1996).
12. Monaghan, R.L.; Polishook, J.D.; Pecore, V.J.; Bills, G.F.; Nallin-Omstead, M.; Streicher, S.L. "Discovery of Novel Secondary Metabolites from Fungi—Is It Really a Random Walk Through a Random Forest?" *Can. J. Bot.* **73**(1), S925–S931 (1995).
13. Higgs, R.E.; Zahn, J.A.; Gygi, J.D.; Hilton, M.D. "Rapid Method to Estimate the Presence of Secondary Metabolites in Microbial Extracts," *Appl. Environ. Microbiol.* **67,** 371–376 (2001).
14. Zahn, J.A.; Higgs, R.E.; Hilton, M.D. "Use of Direct-Infusion Electrospray Mass Spectrometry To Guide Empirical Development of Improved Conditions for Expression of Secondary Metabolites from Actinomycetes," *Appl. Environ. Microbiol.* **67,** 377–386 (2001).
15. Cremin, P.A.; Zeng, L.; Hart, S. "High Throughput Parallel LC-MS for the Estimation of Natural Product Library Chemodiversity," in *Proceedings of the 51st ASMS Conference on Mass Spectrometry and Allied Topics*, Quebec, Canada, June 8–12, 2003.
16. Cremin, P.A.; Zeng, L. "High-Throughput Analysis of Natural Product Compound Libraries by Parallel LC-MS Evaporative Light Scattering Detection," *Anal. Chem.* **74,** 5492–5500 (2002).

17. Odham, G.; Larsson, L.; Mardh, P.-A., eds. *Gas Chromatography-Mass Spectrometry: Applications in Microbiology*, Plenum Press, New York and London (1984).
18. Lane, D.J.; Pace, B.; Olsen, G.J.; Stahl, D.A.; Sogin, M.L.; Pace, N.R. "Rapid Determination of 16S Ribosomal RNA Sequences for Phylogenetic Analyses," *Proc. Nat. Acad. Sci. USA* **82**(20), 6955–6959 (1985).
19. Weisburg, W.G.; Barns, S.M.; Pelletier, D.A.; Lane, D.J. "16S Ribosomal DNA Amplification for Phylogentic Study," *J. Bacteriol.* **173**(2), 697–703 (1991).
20. Pragai, Z.; Ward, A.C.; Harwood, C.R. "Whole Cell Fingerprinting—Pyrolysis Mass Spectrometry," pp.75–76 in Schumann, W.; Erlich, S.D.; Ogasawara, N., eds., *Functional Analysis of Bacterial Genes*, Wiley (2001).
21. Wilkes, J.G. "Defining and Using Microbial Spectral Databases," *J. Am. Soc. Mass. Spectrom.* **13**(7), 875–887 (2002).
22. Heller, D.N.; Murphy, C.M.; Cotter, R.J.; Fenselau, C.; Uy, O.M. "Constant Neutral Loss Scanning for the Characterization of Bacterial Phospholipids Desorbed by Fast Atom Bombardment," *Anal. Chem.* **60**(24), 2787–2791 (1988).
23. Cole, M.J.; Enke, C.G. "Direct Determination of Phospholipid Structures in Microorganisms by Fast Atom Bombardment Triple Quadrupole Mass Spectrometry," *Anal. Chem.* **63**(10), 1032–1038 (1991).
24. Holland, R.D.; Wilkes, J.G.; Rafii, F.; Sutherland, J.B.; Persons, C.C.; Voorhees, K.J.; Lay, J.O., Jr. "Rapid Identification of Intact Whole Bacteria Based on Spectral Patterns Using Matrix-Assisted Laser Desorption/Ionization With Time-of-Flight Mass Spectrometry," *Rapid Commun. Mass Spectrom.* **10**(10), 1227–1232 (1996).
25. Claydon, M.A.; Davey, S.N.; Edwards-Jones, V.; Gordon, D.B. "The Rapid Identification of Intact Microorganisms Using Mass Spectrometry," *Nature Biotechnol.* **14**(11), 1584–1586 (1996).
26. Krishnamurthy, T.; Ross, P.K. "Rapid Identification of Bacteria By Direct Matrix-Assisted Laser Desorption/Ionization Mass Spectrometric Analysis of Whole Cells" *Rapid Commun. Mass Spectrom.* **10,** 1992–1996 (1996).
27. Fenselau, C.; Demirev, P.A. "Characterization of Intact Microorganisms by MALDI Mass Spectrometry," *Mass Spectrom. Rev.* **20,** 157–171 (2001).
28. Lay, J.O. "MALDI-TOF Mass Spectrometry of Bacteria," *Mass Spectrom. Rev.* **20,** 172–194 (2001).
29. Bright, J.J.; Claydon, M.A.; Soufian, M.; Gordon, D.B. "Rapid Typing of Bacteria Using Matrix-assisted Laser Desorption Ionisation Time-of-flight Mass Spectrometry and Pattern Recognition Software," *J. Microbiol. Methods* **48**(2–3), 127–138 (2002).
30. Evason, D.J.; Claydon, M.A.; Gordon, D.B. "Exploring the Limits of Bacterial Identification by Intact Cell-mass Spectrometry," *J. Am. Soc. Mass Spectrom.* **12**(1), 49–54 (2001).

31. Williams, T.L., Andrzejewski, D., Lay, J.O.; Musser, S.M. "Experimental Factors Affecting the Quality and Reproducibility of MALDI-TOF Mass Spectra Obtained from Whole Bacteria Cells," *J. Am. Soc. Mass Spectrom.* **14**(4), 342–351 (2003).
32. Jones, J.J.; Stump, M.J.; Fleming, R.C.; Lay, J.O.; Wilkins, C.L. "Investigation of MALDI-TOF and FT-MS Techniques for Analysis of *Escherichia coli* Whole Cells," *Anal. Chem.* **75**(6), 1340–1347 (2003).
33. Jarman K.H.; Cebula S.T.; Saenz A.J.; Petersen C.E.; Valentine N.B.; Kingsley M.T.; Wahl K.L. "An Algorithm for Automated Bacterial Identification Using Matrix-Assisted Laser Desorption/Ionization Mass Spectrometry," *Anal. Chem.* **72**(6), 1217–1223 (2000).
34. Wahl, K.L.; Wunschel, S.C.; Jarman, K.H.; Valentine, N.B.; Petersen, C.E.; Kingsley, M.T.; Zartolas, K.A.; Saenz, A.J. "Analysis of Microbial Mixtures by Matrix-Assisted Laser Desorption/Ionizaion Time-of-Flight Mass Spectrometry," *Anal. Chem.* **74**(24), 6191–6199 (2002).
35. Lee, H.; Williams, S.K.R.; Wahl, K.L.; Valentine, N.B. "Analysis of Whole Bacterial Cells by Flow Field-Flow Fractionation and Matrix-Assisted Desorption/Ionization Time-of-Flight Mass Spectrometry," *Anal. Chem.* **75**(11), 2746–2752 (2003).
36. Welham, K.J.; Domin, M.A.; Johnson, K.; Jones, L.; Ashton, D.S. "Characterization of Fungal Spores by Laser Desorption/Ionization Time-of-Flight Mass Spectrometry," *Rapid Commun. Mass Spectrom.* **14,** 307–310 (2000).
37. Valentine, N.B.; Wahl, J.H.; Kingsley, M.T.; Wahl, K.L. "Direct Surface Analysis of Fungal Spores by Matrix-Assisted Laser Desorption/Ionizaion Time-of-Flight Mass Spectrometry," *Rapid Commun. Mass Spectrom.* **16**(14), 1352–1357 (2002).
38. Wall, M.E.; Wani, M.C. "Camptothecin and Taxol: From Discovery to Clinic," *J. Ethnopharmacol.* **51**(1–3), 239–254 (1996).
39. Atta-ur-Rahman, C.; Iqbal, M. "Bioactive Natural Products as a Potential Source of New Pharmacophores, A Theory of Memory," *Pure Appl. Chem.* **73**(3), 555–560 (2001).
40. Ingkaninan, K.; Von Kuenzel, J. K.; Ijerman, A.P.; Verpoorte, R. "Interference of Linoleic Acid Fraction in Some Receptor Binding Assays," *J. Nat. Prod.* **62**(6), 912–914 (1999).
41. Haberlein H.; Tschiersch K.P.; Boonen G.; Hiller K. "Chelidonium majus L.: Components With in vitro Affinity for the GABAA Receptor. Positive Cooperation of Alkaloids," *Planta Med.* **62**(3), 227–231 (1996).
42. Menzies, J.R.W.; Paterson, S.J.; Duwiejua, M.; Corbett, A.D. "Opioid Activity of Alkaloids Extracted from Picralima Nitida (fam. Apocynaceae)," *Eur. J. Pharmacol.* **350**(1), 101–108 (1998).
43. Hook, D.J.; Pack, E.J.; Yacobucci, J.J.; Guss, J. "Approaches to Automating the Dereplication of Bioactive Natural Products—The Key Step in High Throughput

Screening of Bioactive Materials From Natural Sources," *J. Biomol. Screening*, **2**(3), 145–152 (1997).

44. God, R.; Gumm, H.; Heuer, C.; Juschka, M. "Online Coupling of HPLC and Solid-Phase Extraction for Preparative Chromatography," *GIT Lab. J.* **3/99**, Darmstadt, G.I.T. VERLAG PUBLISHING LTD, (1999).
45. Schmid, I.; Sattler, I.; Grabley, S.; Thiericke, R. "Natural Products in High Throughput Screening: Automated High-Quality Sample Preparation," *J. Biomol. Screening* **4**(1), 15–25 (1999).
46. Koch, C.; Neumann, T.; Thiericke, R.; Grabley, S. "A Central Natural Product Pool—New Approach in Drug Discovery Strategies," *Drug Discovery from Nature*, Chapter 3, 51–55, Susanne Grabley and Ralf Thiericke, eds, Berlin, Springer (1999).
47. Zeng, L.; Wang, X.; Wang, T.; Kassel, D.B. "New Developments in Automated PrepLCMS Extends the Robustness and Utility of the Method for Compound Library Analysis and Purification," *Comb. Chem. High Throughput Screening* **1**(2), 101–111 (1998).
48. Zeng, L.; Burton, L.; Yung, K.; Shushan, B.; Kassel, D.B. "Automated Analytical/Preparative High-Performance Liquid Chromatography-Mass Spectrometry System for the Rapid Characterization and Purification of Compound Libraries," *J. Chromatogr. A* **794**(1 + 2), 3–13 (1998).
49. Nemeth, G.A.; Kassel, D.B. "Existing and Emerging Strategies for the Analytical Characterization and Profiling of Compound Libraries," *Annu. Rep. Med. Chem.* **36,** 277–292 (2001).
50. Kassel, D.B. "Combinatorial Chemistry and Mass Spectrometry in the 21st Century Drug Discovery Laboratory," *Chem. Rev.* (Washington, D.C.), **101**(2), 255–267 (2001).
51. Yu, K.; Balogh, M. "Characterization, Fractionation, and Purification of Isoflavonoids from Novel Plant Extracts by LC-MS," in *Proceedings of the 49th ASMS Conference on Mass Spectrometry and Allied Topics*, Chicago, Illinois, May 27–31, 2001.
52. Smith, B. "FractionLynx for CombiChem Purification," Chemical and Pharmaceutical Structure Analysis (CPSA) symposium, Princeton, New Jersey, September 26–28, 2000.
53. Zang, L.; Chen, R.; Deguchi, K. "High Throughput Analysis for Synthetic Compounds Using Automated MS- and UV-Trigged High Throughput Purification System," Chemical and Pharmaceutical Structure Analysis (CPSA) symposium, Princeton, New Jersey, September 26–28, 2000.
54. Zeng, L.; Xu, R.; Aparicio, A.; Manuel, M.; Kassel, D. "Mass-Directed Purification of Drug Metabolites, by SFC/MS," in *Proceedings of the 51st ASMS Conference on Mass Spectrometry and Allied Topics*, Quebec, Canada, June 8–12, 2003.
55. Xu, R.; Cai, Z.; Fogelman, K.; Wilfors, R.; Worley, V.; Stublen, N.; Kassel, D. "Mass-Directed Purification of Compound Libraries by Automated

Semi-Preparative SFC/MS," in *Proceedings of the 50th ASMS Conference on Mass Spectrometry and Allied Topics*, Orlando, Florida, June 2–6, 2002.

56. Wang, T.; Barber, M.; Hardt, I.; Kassel, D. "Mass Directed Fractionation and Isolation of Pharmaceutical Compounds by Packed-Column Supercritical Fluid Chromatography/Mass Spectrometry," *Rapid Commun. Mass Spectrom.* **15,** 2067–2075 (2001).
57. Wang, T; Hardt, I.; Kassek, D.; Zeng, L. "Characterization and Semi-Preparative Purification of Natural Product Ginsenosides by SFC/ELSD/UV/MS and prepSFC/MS," in *Proceedings of the 49th ASMS Conference on Mass Spectrometry and Allied Topics*, Chicago, Illinois, May 27–31, 2001.
58. Wang, T.; Barber, M.; Kassel, D.B. "Purification of Pharmaceutical Compounds Using Packed Column Supercritical Fluid Chromatography Based on Target Mass Triggered Fraction Collection," Chemical and Pharmaceutical Structure Analysis (CPSA) symposium, Princeton, New Jersey, September 26–28, 2000.
59. Zeng, L.; Kassel, D.B. "Developments of a Fully Automated Parallel HPLC/Mass Spectrometry System for the Analytical Characterization and Preparative Purification of Combinatorial Libraries," *Anal. Chem.*, **70**(20), 4380–4388 (1998).
60. Xu, R.; Wang, T.; Isbell, J.; Cai, Z.; Sykes, C.; Brailsford, A.; Kassel, D.B. "High-Throughput Mass-Directed Parallel Purification Incorporating a Multiplexed Single Quadrupole Mass Spectrometer," *Anal. Chem.* **74**(13), 3055–3062 (2002).
61. Organ, A. "MUX for Parallel LC/MS," Chemical and Pharmaceutical Structure Analysis (CPSA) symposium, Princeton, New Jersey, October 9–11, 2001.
62. Isbell, J.; Zhou, Y.; Backes, B.; Weslak, M.; Rund, M.; Chang, J.; Jiang, S.; Ek, J.; Brailsford, A.; Shave, D. "Purifying 10,000 Samples Per Month Using a MUX Purification System," in *Proceedings of the 51st ASMS Conference on Mass Spectrometry and Allied Topics*, Quebec, Canada, June 8–12, 2003.
63. Nogle, L.M.; Mallis, L.M. "Utilization of Multiplexed Liquid Chromatography/Mass Spectrometry in the Purity and Accurate Mass Determination of Pharmaceutical Compound Libraries," in *Proceedings of the 51st ASMS Conference on Mass Spectrometry and Allied Topics*, Quebec, Canada, June 8–12, 2003.
64. Cremin, P.; Zeng, L.; "High Throughput Parallel LC-MS Accurate Mass Measurements for the Dereplication of Natural Products," in *Proceedings of the 50th ASMS Conference on Mass Spectrometry and Allied Topics* Orlando, Florida, June 2–6, 2002.
65. Fang, L.; Demee, M.; Cournoyer, J.; Sierra, T.; Young, C.; Yan, B. "Parallel High-Throughput Accurate Mass Measurement Using a Nine-Channel Multiplexed Electrospray Liquid Chromatography Ultraviolet Time-of-Flight Mass Spectrometry," *Rapid Commun. Mass Spectrom.* **17**(13), 1425–1432 (2003).
66. Eldridge, G.R.; Vervoort, H.C.; Lee, C.M.; Cremin, P.A.; Williams, C.T.; Hart, S.M.; Goering, M.G.; O'Neil-Johnson, M.; Zeng, L. "High-Throughput Method for the Production and Analysis of Large Natural Product Libraries for Drug Discovery," *Anal. Chem.* **74**(16), 3963–3971 (2002).

67. Eldridge, G.; Zeng, L.; Lee, C.; Cremin, P.; Vervoort, H.C.; Ghanem, M. "Chemical Compound Libraries Prepared from Biological Sources for Use in Screening and Methods of Preparing Such Libraries," U.S. Pat. Appl. Publ. (2003), 48 pp., Cont.-in-part of WO 2001 33,193. CODEN: USXXCO US 2003044846 A1 20030306 CAN 138:180677 AN 2003:174345 CAPLUS.
68. Siegel, M.M. "Early Discovery Drug Screening Using Mass Spectrometry," *Curr. Top. Med. Chem.* **2**(1), 13–33 (2002).
69. Schobel, U.; Frenay, M.; van Elswijk, D.A.; McAndrews, J.M.; Long, K.R.; Olson, L.M.; Bobzin, S.C.; Irth, H. "High Resolution Screening of Plant Natural Product Extracts for Estrogen Receptor Alpha and Beta Binding Activity Using an Online HPLC-MS Biochemical Detection System," *J. Biomol. Screen.* **6**(5), 291–303 (2001).
70. Hofstadler, S.A.; Griffey, R.H. "Mass Spectrometry as a Drug Discovery Platform Against RNA Targets," *Curr. Opin. Drug Disc. Dev.* **3**(4), 423–431 (2000).
71. Corley, D.G.; Miller-Wideman, M.; Durley, R.C. "Isolation and Structure of Harzianum A: A New Trichothecene from *Trichoderma harzianum*," *J. Nat. Prod.* **57**(3), 422–425 (1994).
72. Bradshaw, J.; Butina, D.; Dunn, A.J.; Green, R.H.; Hajek, M.; Jones, M.M.; Lindon, J.C.; Sidebottom, P.J. "A Rapid and Facile Method for the Dereplication of Purified Natural Products," *J. Nat. Prod.* **64**(12), 1541–1544 (2001).
73. Corley, D.G.; Durley, R.C. "Strategies for Database Dereplication of Natural Products" *J. Nat. Prod.* **57**(11), 1484–1490 (1994).
74. Balogh, M.P. "The Commercialization of LC-MS During 1987–1997: A Review of Ten Successful Years," *LC-GC*, **16**(2), 135–136, 138, 140, 142, 144 (1998).
75. Niessen, W.M.A. "Advances in Instrumentation in Liquid Chromatography-Mass Spectrometry and Related Liquid-Introduction Techniques," *J. Chromatogr. A* **794**(1 + 2), 407–435 (1998).
76. Ganguly, A.K.; Pramanik, B.N.; Chen, G.; Shipkova, P.A. "Characterization of Pharmaceuticals and Natural Products by Electrospray Ionization Mass Spectrometry," *Pract. Spectr.* **32,** 149–185 (2002).
77. Strege, M.A. "High-Performance Liquid Chromatographic-Electrospray Ionization Mass Spectrometric Analyses for the Integration of Natural Products with Modern High-Throughput Screening," *J. Chromatogr. B* **725,** 67–78 (1999).
78. Constant, H.L.; Beecher, C.W. "A Method for the Dereplication of Natural Product Extracts Using Electrospray HPLC/MS," *Nat. Prod. Lett.* **6**(3), 193–196 (1995).
79. Kebarle, P.; Tang, L. "From Ions in Solution to Ions in the Gas Phase—The Mechanism of Electrospray Mass Spectrometry," *Anal. Chem.* **65**(22), 972A–986A (1993).
80. Ackermann, B.L.; Regg, B.T.; Colombo, L.; Stella, S.; Coutant, J.E. "Rapid Analysis of Antibiotic-Containing Mixtures from Fermentation Broths by Using Liquid Chromatography-Electrospray Ionization-Mass Spectrometry and Matrix-Assisted Laser Desorption Ionization-Time-of-Flight-Mass Spectrometry," *J. Am. Soc. Mass Spectrom.* **7**(12), 1227–1237 (1996).

81. Colombo, L.; Stalla, S.; Regg, B.T.; Ackermann, B.J. "Rapid Analysis of Crude Mixtures from Fermantation Broths Using LC-UV-ESI/MS and MALDI-TOF," in *Proceedings of the 42nd ASMS Conference on Mass Spectrometry and Allied Topics*, Chicago, Illinois, May 29–June 3, 1994.
82. Julian, R.K. Jr. "Tandem Mass Spectometry in Natural Products Discovery," in *Proceedings of the 44th ASMS Conference on Mass Spectrometry and Allied Topics*, Portland, Oregon, May 12–16, 1996.
83. Cordell, G.A.; Shin, Y.G. "Finding the Needle in the Haystack. The Dereplication of Natural Product Extracts," *Pure Appl. Chem.* **71**(6), 1089–1094 (1999).
84. Janota, K.; Carter, G.T. "Natural Products Library Screening Using LC/MS and LC/MS/MS," in *Proceedings of the 46th ASMS Conference on Mass Spectrometry and Allied Topics*, Orlando, Florida, May 31–June 4, 1998.
85. Lopez, L.L.; Yu, X.; Cui, D.; Davis, M.R. "Identification of Drug Metabolites in Biological Matrixes by Intelligent Automated Liquid Chromatography/Tandem Mass Spectrometry," *Rapid Commun. Mass Spectrom.* **12**(22), 1756–1760 (1998).
86. Gilbert, J.R.; Lewer, P. "LC/MSn as a Tool for the Identification of Biologically Active Compounds from Natural Product Mixtures," in *Proceedings of the 46th ASMS Conference on Mass Spectrometry and Allied Topics*, Orlando, Florida, May 31–June 4, 1998.
87. Sanders, M.; Josephs, J., Schwartz, J.; Tymiak, A.; DiDonato, G. "Rapid Identification of Natural Products Using a Modified Ion Trap Mass Spectrometer and MS/MS Spectral Library Searching," in *Proceedings of the 47th ASMS Conference on Mass Spectrometry and Allied Topics*, Dallas, Texas, June 13–17, 1999.
88. Sander, P.; Wang, Y.; Baessmann, C.; Schneider, B.; Zurek, G.; Wunderlich, D. "High Throughput MSn Library Search in Natural Product Research," in *Proceedings of the 51st ASMS Conference on Mass Spectrometry and Allied Topics*, Quebec, Canada, June 8–12, 2003.
89. Goodley, P.; Moore, S.; Sadilek, M. "The Use of API-Ion Trap Mass Spectrometry to Characterize Biosynthesis Products," in *Proceedings of the 47th ASMS Conference on Mass Spectrometry and Allied Topics*, Dallas, Texas, June 13–17, 1999.
90. Estes-Sealey, R.; Morrison, R.; Meadows, L.; Price, W. "The Isolation, Identification, and Analysis of an Analgesic from *Clibadium asperum*," in *Proceedings of the 47th ASMS Conference on Mass Spectrometry and Allied Topics*, Dallas, Texas, June 13–17, 1999.
91. Stacey, C.; Phillips, R. "Specific Compound Identification Using Multiple Stages of Fragmentation With and ESI-Ion Trap," in *Proceedings of the 45th ASMS Conference on Mass Spectrometry and Allied Topics*, Palm Springs, California, June 1–5, 1997.
92. Kearney, G.; Gates, P.; Boehm, G.; Staunton, J.; Leadlay, P.; Jones, R. "Fragmentation Studies on Erythromycin by Electrospray MS/MS," in *Proceedings*

of the 46th ASMS Conference on Mass Spectrometry and Allied Topics, Orlando, Florida, May 31–June 4, 1998.

93. Lopez, L.L.; Tiller, P.R.; Senko, M.W.; Schwartz, J.C. "Automated Strategies for Obtaining Standardized Collisionally Induced Dissociation Spectra on a Benchtop Ion Trap Mass Spectrometer," *Rapid Commun. Mass Spectrom.* **13**(8), 663–668 (1999).
94. Wang, Y.; Rudolph, S.; Di-Leonardo, K.; Trentani, A.; Peterson, F.; Sander, P.; Baessmann, C.; Schneider, B. "Automated Deconvolution in Natural Product Screening," in *Proceedings of the 51st ASMS Conference on Mass Spectrometry and Allied Topics*, Quebec, Canada, June 8–12, 2003.
95. Nielsen, K.F.; Smedsgaard, J. "Fungal Metabolite Screening: Database of 474 Mycotoxins and Fungal Metabolites for Dereplication by Standardized Liquid Chromatography-UV-Mass Spectrometry Methodology," *J. Chromatogr. A* **1002**(1–2), 111–136 (2003).
96. Strege, M.A.; Sevenson, S.; Lawrence, S.M. "Mixed-Mode Anion-Cation Exchange/Hydrophilic Interaction Liquid Chromatography-Electrospray Mass Spectrometry as an Alternative to Reversed Phase for Small Molecule Drug Discovery," *Anal. Chem.* **72,** 4629–4633 (2000).
97. Duebelbeis, D.O.; Snipes, C.E.; Werk, T.L.; Gilbert, J.R.; Martin, L.J. "Dereplication of Gougerotin in Raw Fermentation Extracts: A Problem Solved by HILIC LC/MS," in *Proceedings of the 50th ASMS Conference on Mass Spectrometry and Allied Topics*, Orlando, Florida, June 2–6, 2002.
98. Nielsen, K.F.; Thrane, U. "Fast Methods for Screening of Trichothecenes in Fungal Cultures Using Gas Chromatography-Tandem Mass Spectrometry," *J. Chromatogr. A* **929,** 75–87 (2001).
99. Morris, H.R.; Paxton, T.; Dell, A.; Langhorne, J.; Berg, M.; Bordoli, R.S.; Hoyes, J.; Bateman R.H. "High Sensitivity Collisionally-Activated Decomposition Tandem Mass Spectrometry on a Novel Quadrupole/Orthogonal-Acceleration Time-of-Flight Mass Spectrometer," *Rapid Commun. Mass Spectrom.* **10**(8), 889–896 (1996).
100. Shevchenko, A.; Chernushevich, I.; Ens, W.; Standing, K.G.; Thomson, B.; Wilm, M.; Mann, M. "Rapid 'de Novo' Peptide Sequencing by a Combination of Nanoelectrospray, Isotopic Labeling and a Quadrupole/Time-of-Flight Mass Spectrometer," *Rapid Commun. Mass Spectrom.* **11**(9), 1015–1024 (1997).
101. Eckers, C.; Haskins, N.; Langridge, J. "The Use of Liquid Chromatography Combined With a Quadrupole Time-of-Flight Analyzer for the Identification of Trace Impurities in Drug Substance," *Rapid Commun. Mass Spectrom.* **11**(17), 1916–1922 (1997).
102. Gilbert, J.R.; Lewer, P.; Carr, A.W.; Snipes, C.E.; Balcer, J.L., Gerwick, W. "Natural Product Dereplication and Structural Elucidation Using LC/MSn Combined With Accurate Mass LC/MS and LC/MS/MS," in *Proceedings of the 47th ASMS Conference on Mass Spectrometry and Allied Topics*, Dallas, Texas, June 13–17, 1999.

103. Potterat, O.; Wagner, K.; Haag, H. "Liquid Chromatography-Electrospray Time-of-Flight Mass Spectrometry for On-Line Accurate Mass Determination and Identification of Cyclodepsipeptides in a Crude Extract of the Fungus Metarrhizium anisopliae," *J. Chromatogr. A* **872**(1+2), 85–90 (2000).

104. Janota, K. "Session 6: Impurities, Degradants, and Natural Products," Chemical and Pharmaceutical Structure Analysis (CPSA) symposium, Princeton, New Jersey, 1998.

105. Gilbert, J.R. "LC/MS Strategies for Screening and Identifying Biologically Active Compounds from Natural Product Mixtures," Chemical and Pharmaceutical Structure Analysis (CPSA) symposium, Princeton, New Jersey, 1998.

106. Gilbert, J.R.; Lewer, P.; Duebelbeis, D.O.; Carr, A.W.; Snipes, C.E.; Williamson, R.T. "Identification of Biologically-Active Compounds from Nature Using LC-MS," in *Mass Spectrometry, LC/MS/MS and TOF/MS: Analysis of Emerging Contaminants*, ACS Symposium vol. 80, I. Ferrer; E.M. Thurman, eds., Oxford University Press and the American Chemical Society, 52–65 (2003).

107. Wolfender, J.L.; Waridel, P.; Ndjoko, K.; Hobby, K.R.; Major, H.J.; Hostettmann, K. "Evaluation of Q-TOF-MS/MS and Multiple Stage IT-MSn for the Dereplication of Flavonoids and Related Compounds in Crude Plant Extracts," *Analusis* **28**(10), 895–906 (2001).

108. Waridel, P.; Wolfender, J.-L.; Ndjoko, K.; Hobby, K.R.; Major, H.J.; Hostettmann, K. "Evaluation of Quadrupole Time-of-Flight Tandem Mass Spectrometry and Ion-Trap Multiple-Stage Mass Spectrometry for the Differentiation of C-glycosidic Flavonoid Isomers," *J. Chromatogr. A* **926**(1), 29–41 (2001).

109. Reather, J.; Bohm, G.; Staunton, J.; Leadlay, P.F. "Fragmentation Patterns of Rapamycin by Electrospray Ion-Trap MS/MS," in *Proceedings of the 46th ASMS Conference on Mass Spectrometry and Allied Topics*, Orlando, Florida, May 31–June 4, 1998.

110. Kramer, R.W.; Kujawinski, E.B.; Zang, X.; Green-Church, K.B.; Jones, B.; Freitas, M.A.; Hatcher, P.G. "Studies of the Structure of Humic Substances by Electrospray Ionization Coupled to a Quadrupole-Time of Flight (QQ-TOF) Mass Spectrometer," *R. Soc. Chem. Spec. Publ.* **273,** 95–107 (2001).

111. Eckers, C.; Wolff, J.C.; Haskins, N.J.; Sage, A.B.; Giles, K.; Bateman, R. "Accurate Mass Liquid Chromatography/Mass Spectrometry on Orthogonal Acceleration Time-of-Flight Mass Analyzers Using Switching Between Separate Sample and Reference Sprays. 1. Proof of Concept," *Anal. Chem.* **72**(16), 3683–3688 (2000).

112. Wolff, J.C.; Eckers, C.; Sage, A.B.; Giles, K.; Bateman, R. "Accurate Mass Liquid Chromatography/Mass Spectrometry on Quadrupole Orthogonal Acceleration Time-of-Flight Mass Analyzers Using Switching Between Separate Sample and Reference Sprays. 2. Applications Using the Dual-Electrospray Ion Source," *Anal. Chem.* **73**(11), 2605–2612 (2001).

113. Gao, J.; He, L.; Chu, M.; Mierzwa, R.; Mahesh, P. "Optimizing Mass Measurement for Natural Product Fraction Library Via the Lock Spray Source Interfaced

to LC/MS System," in *Proceedings of the 50th ASMS Conference on Mass Spectrometry and Allied Topics*, Orlando, Florida, June 2–6, 2002.

114. Windig, W.; Phalp, J.M.; Payne, A.W. "A Noise and Background Reduction Method for Component Detection in Liquid Chromatography/Mass Spectrometry," *Anal. Chem.* **68**(20), 3602–3606 (1996).
115. Stein, S.E. "An Integrated Method for Spectrum Extraction and Compound Identification from Gas Chromatography/Mass Spectrometry Data," *J. Am. Soc. Mass Spectrom.* **10**(8), 770–781(1999).
116. Zink, D.; Dufresne, C.; Liesch, J.; Martin, J. "Automated LC-MS Analysis of Natural Products: Extraction of UV, MS, and Retention Time Data for Component Identification and Characterization," in *Proceedings of the 50th ASMS Conference on Mass Spectrometry and Allied Topics*, Orlando, Florida, June 2–6, 2002.
117. Higgs, R.E.; Julian, R.K.; Hilton, M.D. "Advances in the Application of LC-MS to Mixture Analysis in Natural Products Discovery," in *Proceedings of the 48th ASMS Conference on Mass Spectrometry and Allied Topics*, Long Beach, California, June 11–15, 2000.
118. He, L.; Chu, M.; Mierzwa, R.; Gao, J.; Xu, L.; King, A.; Patel, M. "Development of a LC/MS Database for Automatic Natural Products Dereplication," in *Proceedings of the 49th ASMS Conference on Mass Spectrometry and Allied Topics*, Chicago, Illinois, May 27–31, 2001.
119. He, L.; Chu, M.; Mierzwa, R.; Gao, J.; Xu, L.; Kink, A.; Patel, M. "Automatic Natural Products Crude Extracts De-convolution Using LC/MS Technology," in *Proceedings of the 50th ASMS Conference on Mass Spectrometry and Allied Topics*, Orlando, Florida, June 2–6, 2002.
120. Marshall, A.G.; Hendrickson, C.L.; Jackson, G.S. "Fourier Transform Ion Cyclotron Resonance Mass Spectrometry: A Primer," *Mass Spectrom. Rev.* **17**(1), 1–35 (1998).
121. Kearney, G.C.; Gates, P.J.; Leadlay, P.F.; Staunton, J.; Jones R. "Structural Elucidation Studies of Erythromycins by Electrospray Tandem Mass Spectrometry II," *Rapid Commun. Mass Spectrom.* **13**(16), 1650–1656 (1999).
122. Gates, P.J.; Kearney, G.C.; Jones, R.; Leadlay, P.F.; Staunton, J. "Structural Elucidation Studies of Erythromycins by Electrospray Tandem Mass Spectrometry," *Rapid Commun. Mass Spectrom.* **13**(4), 242–246 (1999).
123. McDonald, L.A.; Barbeiri, L.R.; Carter, G.T.; Kruppa, G.; Siegel, M.M. "Structure Elucidation of Muraymycins A1 and B1 Using ESI Multi-CHEF FTMSn," in *Proceedings of the 50th ASMS Conference on Mass Spectrometry and Allied Topics*, Orlando, Florida, June 2–6, 2002.
124. Shi, S.D.; Hendrickson, C.L.; Marshall, A.G.; Siegel, M.M.; Kong, F.; Carter, G.T. "Structural Validation of Saccharomicins by High Resolution and High Mass Accuracy Fourier Transform–Ion Cyclotron Resonance–Mass Spectrometry and Infrared Multiphoton Dissociation Tandem Mass Spectrometry," *J. Am. Soc. Mass Spectrom.* **10**(12), 1285–1290 (1999).

125. Reather, J.; Gates, P.; Boehm, G. "Structural Elucidation Studies on Rapamycin," in *Proceedings of the 47th ASMS Conference on Mass Spectrometry and Allied Topics*, Dallas, Texas, June 13–17, 1999.

126. Gates, P.; Reather, J.; Staunton, J. "Accurate-Mass MSn Studies of Natural Products (Polyketides) by ESI-FT-ICR Mass Spectometry," in *Proceedings of the 47th ASMS Conference on Mass Spectrometry and Allied Topics*, Dallas, Texas, June 13–17, 1999.

127. Kearney, G.; Gates, P.; Boehm, G.; Staunton, J.; Leadley, P.; Jones, R. "Fragmentation Studies on Erythromycins by Electrospray MS/MS," in *Proceedings of the 46th ASMS Conference on Mass Spectrometry and Allied Topics*, Orlando, Florida, May 31–June 4, 1998.

128. Gates, P.J.; Kearney, G.C.; Skelton, P.C.; Leadley, P.F.; Staunton, J. "Structural Elucidation Studies on Polyketides by Electrospray FT-ICR-MSn," in *Proceedings of the 46th ASMS Conference on Mass Spectrometry and Allied Topics*, Orlando, Florida, May 31–June 4, 1998.

129. McDonald, L.A.; Barbieri, L.R.; Carter, G.T.; Kruppa, G.; Feng, X.; Lotvin, J.A.; Siegel, M.M. "FTMS Structure Elucidation of Natural Products: Application to Muraymycin Antibiotics Using ESI Multi-CHEF SORI-CID FTMSn, the Top-Down/Bottom-Up Approach, and HPLC ESI Capillary-Skimmer CID FTMS," *Anal. Chem.* **75**(11), 2730–2739 (2003).

130. Hostettmann, K.; Potterat, O.; Wolfender, J.L. "Strategy in the Search for New Bioactive Plant Constituents," *Pharmazeutische Industrie*, **59**(4), 339–347 (1997).

131. Acker, P.; Guenat, C.; Moss, S.; Ramseier, U. "Approaches to Micro-HPLV/MS/IR Coupling," in *Proceedings of the 47th ASMS Conference on Mass Spectrometry and Allied Topics*, Dallas, Texas, June 13–17, (1999).

132. Wolfender, J.L.; Ndjoko, K.; Hostettmann, K. "The Potential of LC-NMR in Phytochemical Analysis," *Phytochem. Anal.* **12**(1), 2–22 (2001).

133. Bobzin, S.C.; Yang, S.; Kasten, T.P. "LC-NMR: A New Tool to Expedite the Dereplication and Identification of Natural Products," *J. Ind. Microbiol. Biotechnol.* **25**(6), 342–345 (2000).

134. Spring, O.; Buschmann, H.; Vogler, B.; Schilling, E.E.; Spraul, M.; Hoffmann, M. "Sesquiterpene Lactone Chemistry of *Zaluzania grayana* from Online LC-NMR Measurements," *Phytochemistry* **39**(3), 609–612 (1995).

135. Corcoran, O.; Spraul, M. "LC-NMR-MS in Drug Discovery," *Drug Discov. Today* **8**(14), 624–631 (2003).

136. Wolfender, J.L.; Ndjoko, K.; Hostettmann, K. "LC/NMR in Natural Products Chemistry," *Curr. Organic Chem.* **2**(6), 575–596 (1998).

137. Holt, R.M.; Newman, M.J.; Pullen, F.S.; Richards, D.S.; Swanson, A.G., "High-Performance Liquid Chromatography/NMR Spectrometry/Mass Spectrometry: Further Advances in Hyphenated Technology," *J. Mass Spectrom.* **32**(1), 64–70 (1997).

138. Wolfender, J.L.; Terreaux, C.; Hostettmann, K. "The Importance of LC-MS and LC-NMR in the Discovery of New Lead Compounds from Plants," *Pharm. Biol.* **38,** 41–54 (2000).
139. Wilson, I.D.; Morgan, E.D.; Lafont, R.; Shockcor, J.P.; Lindon, J.C.; Nicholson, J.K.; Wright, B. "High-Performance Liquid Chromatography Coupled to Nuclear Magnetic Resonance Spectroscopy and Mass Spectrometry Applied to Plant Products. Identification of Ecdysteroids from Silene Otites," *Chromatographia*, **49**(7/8), 374–378 (1999).
140. Wolfender, J.L.; Ndjoko, K.; Hostettmann, K. "Liquid Chromatography with Ultraviolet Absorbance-Mass Spectrometric Detection and with Nuclear Magnetic Resonance Spectrometry: A Powerful Combination for the On-Line Structural Investigation of Plant Metabolites," *J. Chromatogr A.* **1000**(1–2), 437–455 (2003).
141. Sanvoss, M. "Application of LC-NMR and LC-NMR-MS Hyphenation to Natural Products Analysis," *On-Line LC-NMR and Related Techniques* 111–128, John Wiley & Sons, Ltd (2002).
142. Louden, D.; Handley, A.; Taylor, S.; Lenz, E.; Miller, S.; Wilson, I.D.; Sage, A.; Lafont, R. "Spectroscopic Characterization and Identification of Ecdysteroids Using High-Performance Liquid Chromatography Combined With On-line UV-Diode Array, FT-Infrared and 1H-Nuclear Magnetic Resonance Spectroscopy and Time of Flight Mass Spectrometry" *J. Chromatogr. A* **910**(2), 237–246 (2001).
143. Ynddal, L.; Hansen, S.H. "On-Line Turbulent-Flow Chromatography-High-Performance Liquid Chromatography-Mass Spectrometry for Fast Sample Preparation and Quantitation," *J. Chromatogr. A* **1020**(1), 59–67 (2003).
144. Tolstikov, V.V.; Lommen, A.; Nakanishi, K.; Tanaka, N.; Fiehn, O. "Monolithic Silica-Based Capillary Reversed-Phase Liquid Chromatography/Electrospray Mass Spectrometry for Plant Metabolomics," *Anal. Chem.* **75**(23), 6737–6740 (2003).
145. Castro-Perez, J., Preece, S.; Joncour, K.; Beattie, I.; Wright, A.; Plumb, R.; Granger, J. "A Novel Approach to Metabolite ID by High Resolution Chromatography Coupled to High Resolution TOFMS," presented at the 52nd session of the American Society for Mass Spectrometry, Nashville, Tennessee, May 23–27, 2004.
146. Senko, M.J.; Zabrouskov, V.; Lange, O.; Weighaus, A.; Griep-Raming, J.; Horning, S. "LC/MS with External Calibration Accuracies Approaching 100 ppb," presented at the 52nd session of the American Society for Mass Spectrometry, Nashville, Tennessee, May 23–27, 2004.
147. Siegel, M.M.; Tabei, K.; Lambert, F.; Candela, L.; Zoltan, B. "Evaluation of a Dual Electrospray Ionization/Atmospheric Pressure Chemical Ionization Source at Low Flow Rates for the Analysis of Both Highly and Weakly Polar Compounds," *J. Am. Soc. Mass Spectrom.* **9**(11), 1196–1203 (1998).

148. Gallagher, R.T.; Balogh, M.P.; Davey, P.; Jackson, M.R.; Sinclair, I.; Southern, L.J. "Combined Electrospray Ionization-Atmospheric Pressure Chemical Ionization Source for Use in High-Throughput LC-MS Applications," *Anal. Chem.* **75,** 973–977 (2003).

149. Hanold, K.A.; Horner, J.; Thakur, R.; Miller, C. "Dual APPI/APCI Source for LC/MS," presented at the 50th session of the American Society for Mass Spectrometry, Orlando, Florida, June 2–6, 2002.

150. Hanold, K.A. "Evaluation of a Combined Electrospray and Photoionization Source," presented at the 1st session of the Conference on Small Molecule Science, Bristol, Rhode Island, August 9–12, 2004.

151. Robb, D.B.; Covey, T.R.; Bruins, A.P. "Atmospheric Pressure Photoionization: An Ionization Method for Liquid Chromatography-Mass Spectrometry," *Anal. Chem.* **72**(15), 3653–3659 (2000).

7

APPLICATION OF MASS SPECTROMETRY TO COMPOUND LIBRARY GENERATION, ANALYSIS, AND MANAGEMENT

XUEHENG CHENG AND JILL HOCHLOWSKI

INTRODUCTION

Modern drug discovery is a highly integrated process that contains various interactive and interdependent components. Mass spectrometry plays an important role in every step of this process. Several other chapters of this book illustrate the application of mass spectrometry in drug discovery, such as target identification and validation (genomics, proteomics, and metabonomics), high throughput screening (HTS), and drug metabolism and pharmacokinetics (DMPK). In this chapter, we discuss the application of mass spectrometry in the generation of compound libraries for HTS and for lead optimization, and in the analysis and management of these compound libraries.

In a small-molecule-based drug-discovery process, the creation of large numbers of well-designed compounds is a critical step to ensure the success of drug lead discovery and optimization. Combinatorial chemistry or its more common form, high throughput parallel organic synthesis (HTOS), is often used to generate these compound libraries [1–4]. Mass spectrometry is used throughout the compound generation process from design and optimization of libraries, to synthesis monitoring, and finally, to high throughput purification (HTP) and characterization of the library compounds. In addition,

Integrated Strategies for Drug Discovery Using Mass Spectrometry, Edited by Mike S. Lee

the high throughput, high sensitivity, and structural information provided by mass spectrometry facilitate profiling of physiochmical and pharmaceutical properties of the library compounds. Finally, as pharmaceutical companies increasingly rely on HTS for lead discovery, the establishment and maintenance of a high-quality repository compound collection have become an important issue. Mass spectrometry has been used in the study of quality and stability of the repository compound collection.

Small-molecule libraries synthesized in the parallel format may be those directed toward either lead discovery or lead optimization, and may be synthesized either on a solid-support or in solution. The large numbers of discrete compounds generated by these parallel formats require analysis, and most frequently, purification. This chapter will illustrate how mass spectrometry assists these efforts.

With the success of combinatorial chemistry and HTS, more and more new structural entities are created or acquired and lead compounds discovered. As a consequence, there is an increased demand for high throughput measurement of the quality, physiochemical, and pharmaceutical properties of absorption, distribution, metabolism, and excretion (ADME) of these compounds. This increased demand has created opportunities for the application of mass spectrometry to these research activities.

Finally, the maintenance and efficient management of the increased number of compounds in the chemical repository of drug-discovery operations pose new challenges. The need for the investigation of large numbers of compounds with diverse properties and the various conditions and formats for compound storage and handling require improved analytical methods. Mass spectrometry has been the workhorse in these investigations.

The various analyses and measurements of the large numbers of organic compounds just described have several common requirements: (1) high throughput, (2) general applicability to compounds of diverse structures and properties, (3) selectivity and often separation of complex components, and (4) identification as well as quantitative analysis. Mass spectrometry, especially in combination with high-performance liquid chromatography (HPLC), is ideally suited for such applications. A number of reviews on the application of mass spectrometry to compound library synthesis and analysis have been published [5–20]. Good reviews on HPLC as well as specific equipment such as columns and detectors are also available [21,22].

MASS SPECTROMETRY IN LIBRARY DESIGN, OPTIMIZATION, AND SYNTHESIS

Analytical techniques are important in product characterization and process monitoring of compound synthesis. These techniques are used extensively in the design and optimization of compound libraries. Due to the high speed, specificity, and ability to provide compound identification, mass spectrometry is widely used in these applications. Atmospheric-pressure ionization (API) techniques, which include electrospray ionization (ESI), atmospheric-pressure chemical ionization (APCI), and atmospheric pressure photo-ionization (APPI), are commonly used to interface with liquid-sample introduction and separation techniques.

Reaction Monitoring and Optimization

Flow injection mass spectrometry analysis (FIA-MS) has been used extensively in reaction monitoring and optimization due to its simplicity and easy of use. Detection of enantiomers in stereospecific reactions has been reported. Reetz et al. [23] prepared isotopically labeled pseudoenantiomers and pseudoprochiral compounds that behave chemically as a racemate or as a meso compound. Kinetic resolution of a racemate gives two products with different molecular weights, which enables the ratio to be determined with mass spectrometry. The authors used FIA-MS with a microplate autosampler to allow analysis of up to 1000 reactions/day. Guo et al. [24] reported an alternative mass-tagging approach to the measurement of enantiomeric excess by FIA-MS. In their method, an equimolar mixture of pseudoenantiomeric mass-tagged pairs of reagents was prepared that differ in a substituent remote to the chiral center. Szewczyk et al. developed a quantitative method to monitor the progress of the tagging experiment and to optimize the reaction by conversion of the product ketones into ESI-active derivatives [25]. The method was used to monitor three experiments involving 33 different substrates with more than 170 determinations of yield. McKeown et al. incorporated mass spectrometry–sensitive linkers into the synthesis of combinatorial libraries [26]. Upon photochemical cleavage after synthesis, these linkers can be monitored by high throughput FIA-MS to provide information about the yield of the synthesis and the photocleavage processes. A similar strategy that incorporates analytical constructs into the library synthesis for analysis and quality control was used by Lorthioir et al. [27]. These researchers selected an analytical fragment that is highly sensitive in ESI-MS for this purpose. The usefulness of the method was demonstrated by assessment of synthesis of library compounds that are otherwise not detected in ESI-MS analysis due to low sensitivity. Congreve and Jamieson have recently reviewed strategies with high throughput analytical techniques for reaction optimization [28].

In library synthesis that involves a solid-state polymer-bead support, matrix-assisted laser-desorption/ionization time-of-flight mass spectrometry (MALDI-TOF-MS) and secondary ion mass spectrometry (SIMS) have been used to monitor products of reaction either through direct analysis of bead–bound compounds or in situ analysis of the compounds after cleavage. These approaches have been reviewed extensively in the literature [5–20]. A recent development in the use of MALDI-TOF and SIMS in reaction monitoring and optimization was reported by Enjalbal et al. [29]

Open-Access Operation and Automation of Compound Analysis

As FIA-MS and liquid chromatography mass spectrometry (LC-MS) become more pervasive in the analysis of compound libraries, open-access instrumentation is increasingly used in HTOS laboratories as well as in support of general medicinal chemistry. These open-access systems are most often used for reaction monitoring and optimization, and in some cases, for library quality control and synthesis product purification.

The first open-access systems were described by Pullen et al. [30] and Taylor et al. [31]. These systems were based on FIA-MS or generic LC-MS analysis with single quadrupole mass spectrometers to ensure the ease of use and ruggedness of the system. Spreen et al. [32] also described an open-access facility. The use of the open-access facility has increased sample throughput by 60% and has allowed the mass spectrometrists to spend more time on nonroutine problems. Mallis et al. [33] described an open-access facility in a drug-discovery environment that was comprised of FIA-MS and LC-MS units to monitor synthetic chemistry reactions as well as the integrity and purity of new chemical entities. Greaves [34] described an open-access mass spectrometry in an academic environment, primarily in support of organic synthesis but also for the use of biological scientists. The open-access methods featured FIA-MS, gas chromatography mass spectography (GC-MS), and MALDI-TOF-MS systems, as well as autosamplers with plate-handling interfaces. This level of automation allows access to the instruments by a user community of more than 100 users, day or night.

Automation of data acquisition, processing, interpretation, and reporting is another trend in the effort to improve the throughput of compound library analysis. Daley et al. [35] reported automated high throughput LC-MS methods for verification of library compound structure from 96 wells. Greig [36] reported an automated procedure for the calculation and storage of sample purity information based on the LC-MS results. The combination of a fast LC-MS method and automated data-processing techniques enabled high throughput analysis of combinatorial library samples at the rate of up to 2300 per week per instrument. Tong et al. [37] developed hardware components and software modules to enhance the automation, efficiency, and reliability of a commercial open-access FIA-MS system. The software modules include utilities for data manipulation/reduction, data interpretation, data transmission, and reporting to the desktop computer of the submitter. Choi et al. [38,39] reported an approach to apply intelligent automation for high throughput LC-MS analysis of compound libraries with Microsoft Visual Basic software. Compounds were analyzed by a generic primary HPLC method. Those that failed, in the initial analysis, were reanalyzed automatically with secondary analytical methods based on the information derived from the target analyte structure. Examples were described where a secondary method with a longer column and slower gradient was selected for target compounds that failed in primary analysis with ClogP values less than one. Williams et al. [40] reported an automated molecular-weight assignment method for positive-ion ESI-MS spectra. The software application (MassAssign) differentiates $[M + H]^+$ ions from other signals in a complex mass spectrum and reports assignments in the form of either a single component that has the displayed molecular weight, multiple components, or an undetermined molecular weight. Initial tests with the program yielded a 90% success rate compared with manual interpretation for 55 samples [40]. Yates et al. [41] reported a method of rapid characterization of multi- or single-component libraries. The methodology compares LC-MS analysis results with predicted ESI mass spectra to confirm library products, identify chemical-synthesis errors, and assess overall library integrity. In general, equal signal intensities were observed for most combinatorial mixture components, and indicated that differences in ESI efficiency was not a major limitation to this approach. High throughput data-processing programs and informatics tools were used to speed data analysis and simplify the presentation of the library characterization results. Potential limitations of the method include unequal sensitivity of library members and multiply charged-ion formation for large compounds. Klagkou et al. reported an initial study on the determination of the fragmentation rules for some classes of compounds in ESI tandem mass spectrometry (MS/MS). Based on studies carried out on several combinatorial libraries, it was established that different classes of drug molecules follow unique fragmentation pathways. The results were incorporated into an artificial intelligence (AI) software package with the goal of addressing the issue of high throughput, automated MS/MS data interpretation [42].

Automated data processing, data reduction, spectral interpretation, and reporting are also integral parts of open-access sample analysis. Mallis et al. [33] described the use of an open-access system with e-mail capabilities to distribute processed data reports to the chemist. Several methods have been developed, including structure elucidation with in-source collision-induced dissociation (CID).

LIBRARY PURIFICATION

Whether synthesized on a solid support or in the solution phase, most HTOS libraries require some degree of postsynthesis purification prior to biological evaluation. Purification is necessitated by the fact that impurities, of either inactive contaminants or related components, can cause inaccuracy in the determination of IC_{50} results or toxicity evaluations, and lead to incorrect structure–activity relationship (SAR) conclusions. Impurities in a reaction mixture can cause false-positive and false-negative HTS assay results through the effects of either synergy or antagonism. Purification can sometimes be accomplished via structurally specific techniques, such as covalent scavenging or solid-phase extraction–ionic extraction, or via low-resolution techniques such as liquid–liquid partitioning or flash chromatography. High-performance chromatographic techniques, either HPLC or supercritical fluid chromatography (SFC), are generally employed to accomplish the purification of compounds of broad structural scope. Mass spectrometry is routinely used in conjunction with these purification methodologies as either a structural validation tool postpurification or as the system detector by which the correct reaction component is collected.

Weller et al. [43] described an early high throughput UV-triggered HPLC purification system to support the parallel synthesis efforts at Bristol-Myers Squibb. This open-access instrument used fast flow rates and rapid universal reverse-phase gradient elution methods that enabled the purification of up to 200 samples per day at weights up to 200 mg/sample in an unattended mode. Customized software and hardware were developed for this system to optimize efficiency and throughput.

A UV-triggered purification system was described by Kibby [44] in support of the purification of combinatorial libraries generated at Parke-Davis. This system is operated in either reverse-phase or normal-phase mode, and is employed as well for chiral separations. Multiple column sizes allow the system to accommodate the purification of samples in weight up to 50 mg. The operational protocol involves an initial "scouting run" by analytical HPLC with APCI-MS detector. The conditions that are selected are based on structural information. Fraction collection is controlled by customized software, and sample identity, UV, MS data along with chromatographic data are imported from the analytical LC-MS. Peaks are collected only when the UV threshold is met within an appropriate collection window; thus, the number of fractions obtained is limited. Postpurification loop injection mass spectra are collected on these fractions to determine the desired component from each sample.

Abbott Laboratories [45,46] developed a high throughput HPLC system capable of UV-triggered or evaporative light-scattering detector (ELSD) triggered fraction collection in support of the combinatorial chemistry libraries. Column switching and delay loops allow the accommodation of samples sized from 10 mg to 300 mg. All members of each combinatorial library synthesized are triaged by the acquisition of an analytical, reverse-phase HPLC with UV, ELSD, and MS detectors. The analytical LC-MS systems are operated on an open-access basis, and the preparative HPLC as a service to the chemists. Multiple standard-gradient methods are available on both the analytical and preparative HPLC systems, and the purification chemist selects optimal-gradient and fraction-collection trigger (either ELSD, or a specific UV wavelength) conditions for the latter based upon elution times and detector peak intensities of the former. To select the desired component from each sample processed, collected fractions are fed into a loop-injection mass spectrometer that is operated in either ESI or APCI mode, positive or negative ion, dependant upon structure. Four preparative HPLC instruments are thus accommodated by a single mass spectrometer. Customized software provides; intranet based sample log-in, sample and fraction tracking, labeling, and report generation.

A parallel-format UV-triggered HPLC system, commercially available from Biotage [47] as Parallax®, was co-developed with workers at and MDS Panlabs [48]. This open-access system comprises four injectors—accessing samples from 48-well microtiter plates—and four parallel HPLC columns developed with identical gradients. Column eluants flow through a custom-designed parallel flow-path UV detector and allow fraction collection from each column to be triggered by two simultaneous wavelengths. This system is employed for the purification of Panlabs' solution-phase parallel syntheses and accommodates the purification of diverse libraries synthesized at the one millimolar scale. Researchers at GlaxoSmithKline [49] evaluated the high throughput capacity of the Parallax system to purification of their solid-phase combinatorial libraries. The system is robust and capable for the purification of up to 200 samples in 10 hours with a 93% success rate, along with excellent purity and acceptable recovery.

All of the aforementioned HPLC purification systems employ detectors other than mass spectrometers as the selection criteria by which peaks are collected. With these systems, however, mass spectrometry is generally employed as a primary structural validation tool. More recently, systems have been described whereby a mass spectrometer has been added to preparative HPLC format to detect compounds of interest for collection. A recent review by Kassel [9] discusses the relative merits of employing MW-triggered versus UV-triggered fraction collection in various environments for the purification of combinatorial libraries.

The development of the first HPLC system with MW-triggered fraction collection was described by Zeng et al. [50] at CombiChem. This system was a dual analytical–preparative instrument with parallel-column format, termed "parallel Analyt/PrepLCMS," developed by the modification of commercially available instrumentation. This system had software-controlled valves that applied sample to each path from a single autosampler and had the capacity to purify and analyze more than 100 samples per day. Initial analytical LC-MS data acquired by the system allowed the identification of samples that require purification. UV detection is incorporated into the system and the percentage of desired compound is estimated by a postprocessing script that averages the intensity of UV peaks at both 220 nm and 254 nm. Samples that are estimated by this technique to be <85–90% pure are processed by the same parallel Analyt/PrepLCMS system with MW-triggered fraction collection. Fraction collection is thus initiated only when the desired component is detected above a preset threshold level. Therefore, only compounds of interest are collected. Sample validation of identity is accomplished in concert with the purification step. The preparative HPLC employs short HPLC columns and high flow rates. Various-sized columns can accommodate samples sized from 1 mg to 100 mg. The system is made available to the synthetic chemists on an open-access format. Later modifications to this same system describe [51] an improvement in throughput through the addition of two *separate* parallel analytical and preparative HPLC paths. This modification expands the capacity of the instrument to accommodate both the analysis and the purification of 200 samples per day on a single instrument. Isbell et al. [52] reported the development of a validated, streamlined high throughput process for the purification of parallel-synthesis-derived combinatorial libraries. The steps involved in this library purification process include dissolution of dry films of crude synthetic material, dual-column LC-MS purification, dual-column postpurification analysis, quantitation, reformatting, and submission of pure compounds for registration.

Kiplinger, in collaboration with Gilson Instruments and Micromass, developed a fraction-collection system [53] in either the UV or MW modality. This system is applied to the purification of parallel synthesis libraries synthesized in the 10–20-mg weight range at Pfizer. A mass spectrometer–controlled program provides peak trigger signal based upon output from either the diode array UV detector or the mass spectrometer. Stream splitters and makeup flow deliver appropriate concentrations of sample to the mass spectrometer that can be operated in either the APCI or ESI mode. Waters Corporation (Milford, Massachusetts) currently offers a commercially available instrument of this design.

Searle and Hochlowski [54] at Abbott Laboratories have modified a commercially available Agilent (Wilmington, Deleware) MW-triggered HPLC. This system is employed for the purification of parallel-format libraries generated by the HTOS group, sized from 10–50 mg, and is operated on a service basis. A chemiluminescent nitrogen-specific detector (CLND) was added to the system, which allows on-line determination of yield for each sample during library purification. Custom macros as well as a custom data browser were written for sample tracking and data viewing, and allow for the inclusion of CLND quantitation information in the final purification report that is returned to the synthetic chemist.

The newest member of the arsenal of MW-triggered, high throughput purification systems is a multiplexed ion-sourcing [55] format that was developed through collaborative efforts by researchers at Deltagen, Micromass (Manchester, UK), and Waters Corporation (Milford, Massachusetts). High efficiency is achieved on this system by directing the flow from each of four individual HPLCs into a multiplexed (MUX) ESI ion source. This MUX ion source is composed of four separate electrospray needles that are set around a rotating disc. This arrangement allows for independent sampling from each of the individual sprays by a single mass spectrometer. Fraction collection is MW-triggered and eluant from each of the four individual HPLCs is directed to a dedicated fraction collector, respectively. Expansion of up to eight individual channels is possible and allows for the throughput to be doubled. This system is set up with the capability to purify sample sizes that range from 1 mg to 10 mg. The authors describe applications for the Deltagen chemistry libraries, wherein recovery values of over 80% and typical sample purities of 90% are achieved.

SFC is a technique similar to HPLC, where the mobile phase is a compressed gas with a gradient of organic modifier. Carbon dioxide is the most commonly used gas in SFC. Due to the lower viscosity of supercritical fluids, the chromatography can be run at much higher flow rates compared to HPLC. As a consequence, SFC can typically achieve higher chromatographic resolution or faster separation speeds than HPLC. SFC and HPLC can be viewed as a continuum, both part of a larger unified chromatography approach [56]. In practice, a wide variety of different compositions of carbon dioxide and the organic modifier have been employed to achieve separation of components of interest. SFC has traditionally been applied to chiral separations to provide the increased resolution relative to HPLC. SFC has only recently been applied to the purification of high throughput synthesis libraries as SFC instrumentation begins to catch up to HPLC instrumentation in terms of operational facility and robustness. For the reader unfamiliar with this technology, SFC and HPLC techniques are similar in that each achieves separation by the adsorption of crude mixtures onto a stationary-phase column followed by the preferential elution of components by a mobile phase. SFC differs from HPLC in the nature of the mobile phase employed. SFC is composed of high-pressure carbon dioxide and miscible co-solvents such as methanol. The advantage of a carbon dioxide mobile phase is that after a pressure drop postchromatography, the carbon dioxide reverts to the gas phase, and thus, the sample dry-down (a significant bottleneck step in the purification process) is reduced to the evaporation of small amounts of volatile co-solvent.

Berger Instruments and Alanax co-developed [57,58] a semipreparative-scale supercritical-fluid chromatography instrument that has been employed for the purification of combinatorial chemistry libraries. This UV-triggered commercially available instrument is made up of several developed and modified components to achieve chromatography and sample recovery, wherein the mobile phase is carbon dioxide—a compressible fluid with high expansion upon conversion to the gas phase. Back-pressure regulation controls the column outlet pressure requisite for phase maintenance, and a "separator" prevents aerosol formation as the fluid-phase carbon dioxide and methanol mixture expands to gas-phase carbon dioxide and liquid-phase methanol. Fractions are collected in this system at elevated pressure into a "cassette" system composed of four individual compartments, each with gas injection tube inserts, that allow for high recovery collection of up to four components from a reaction mixture.

The purification of combinatorial libraries on this system is described by Farrell et al. [59] to accommodate the purification of solution-phase combinatorial libraries synthesized at Pfizer. Their overall protocol employs as well an analytical SFC used in combination with mass spectrometry and CLND off-line to the semipreparative SFC purification system. Prepurification analytical LC-MS/UV/CLND allows the triage of samples for purification. An in-house–developed software package analyzes data for predicted quality based upon and evaluation of the UV and MS data for the possibility of co-eluting peaks during purification. This same software package selects a collection window for purification that is necessitated by the ability to collect only a total of four UV-active fractions from each sample processed. Postpurification analytical SFC-MS/CLND is used to validate and assess the quality of purified samples.

Engineers at Abbott Laboratories have modified a Berger Instruments (Newark, Deleware) semipreparative SFC for use in support of the purification of libraries synthesized by the HTOS group [60,61]. A manual version of the commercially available instrument was integrated with a Gilson (Middleton, Wisconsin) autosampler and a Cavero (Sunnyville, California) two-arm pipetting instrument customized in-house to serve as a fraction collector. A custom-designed fluid/gas outlet or "shoe" on the fraction collector arms enables the collection of samples at atmospheric pressure. A methanol system is incorporated into the fraction collection line to assure high recovery and eliminate cross-contamination between fractions. Fraction collection is UV-triggered, and postpurification validation is subsequently accomplished by loop-injection APCI mass spectrometry in the positive- or negative-ion mode. An in-house software package written for the UV-triggered, MW-validated HPLC system already in place was expanded to accommodate SFC data, sample tracking, labeling, and report generation. Samples processed on this instrument range from 10 mg to 50 mg. Chromatography methods have been developed for SFC [62] for each of standardized reactions routinely carried out by the HTOS group at Abbott Laboratories. Method selection is performed by trained technicians based solely on structural information. Each time the HTOS group deploys new standardized chemistries, pilot reactions are used to explore the relative amenability of products to SFC vs. HPLC such that subsequent libraries that are generated by his protocol are triaged to the most appropriate chromatographic technique. Typical yields vary from 72% to 98%, and are comparable to that achieved by HPLC. Purity is comparable when the more appropriate of these two HPLC techniques is selected.

Ontogen scientists custom designed and patented [63,64] an MS-triggered semipreparative SFC system employed for the purification of the companies' parallel synthesis libraries. This instrument is a parallel four-channel system capable of simultaneously processing samples from four microplates. Each of the four SFC column eluants is monitored by a dedicated UV detector, but a unique protocol is used wherein UV detection is employed only to identify the eluting peaks from each column. The identified peaks are then diverted to a single TOF mass spectrometer for MW determination. Fraction collection is initiated only when the MS detects the desired MW, which then initiates collection into a "target" plate. In addition to collection of the desired component, reaction by-products are diverted to an ancillary plate when a MW other than the target is detected. In-house software tracks all components collected with respect to plate and location. The collection-plate format is a 2-mL-deep well microtiter that is custom designed with expansion chambers to accommodate the evaporation of carbon dioxide as it reverts to the gas phase.

DuPont scientists have reported [65] the development of an MS-directed preparative-scale SFC system capable of the purification of crude samples up to 50 mg. This system is used for chiral separations as well as general library purifications. The system is comprises a Gilson (Middletown, Wisconsin) preparative SFC and a Gilson autosampler, which is used for both injection and collection. The standard valves are replaced with high-pressure-rated valves. A PE Sciex (Foster City, California) mass spectrometer with ESI source is used with this system and is operated in the positive-ion mode. To improve recovery into the fraction-collection system, a simple foil seal is placed over the collection tubes to provide a loose seal to encourage retention of the methanol portion of the eluant and departure of the carbon dioxide. Two solvent makeup pumps were added to the system; one pump was used to provide additional flow to the fraction collector as carbon dioxide departs, and a second pump was used to provide auxiliary flow of methanol, which contained formic acid additive to the mass spectrometer stream to dilute sample concentration as well as to improve ion signal and peak shape. Fraction collection is triggered when signal for a component of the desired molecular weight from each sample goes above a preset threshold value and is controlled through the mass spectrometer software package. Kassel from the DuPont group has also reported preliminary efforts toward a MS-triggered preparative SFC system on the Berger platform [66].

MASS SPECTROMETRY IN LIBRARY ANALYSIS

The large numbers of compounds synthesized by HTOS and purified by HTP require high throughput analysis and quality assessment. Mass spectrometry is commonly used for this purpose. Research and development in the analysis of libraries have been focused primarily on the improvement of throughput and automation in the analysis, interpretation, and reporting of data. Morand et al. have recently reviewed development of high throughput mass spectrometry in pharmaceutical research [20] for the analysis of chemical library compounds. Also, the use of one or more auxiliary detection techniques, such as UV or ELSD in addition to mass spectrometry, allows for better assessment of the purity of library compounds.

Flow-Injection Mass Spectrometry Analysis

FIA is the simplest form of sample introduction into the mass spectrometer, and this injection format has been widely used in the analysis of combinatorial library samples. This technique offers the highest throughput combined with ease of use and facile automation. Richmond et al. [67–69] reported methods to minimize sample carryover for the FIA-MS analysis of combinatorial libraries. Samples were sorted before the analysis to maximize the molecular-weight difference between samples in the analysis queue and to minimize the conditions where consecutively measured wells contain samples similar to building blocks. Cycle times of less than a minute were reported with a carryover of 0.01%. A software application was developed to automatically report the sample purity and calculate sample carryover by an automatic spectrum comparison method [70,71]. A quasi-molecular ion discovery feature was also implemented [72] in the automated data-processing program. Automated FIA-MS analysis and reporting were also used in the analysis of fractions from the purification of combinatorial libraries [73]. Whalen et al. developed software to allow automated FIA-MS analysis from 96-well plates [74]. The system optimizes the interface for mass spectrometry and MS/MS conditions, and reports the results in an unattended fashion.

High Throughput FIA/MS

One important advantage of FIA-MS analysis is the high throughput capabilities. Samples are analyzed routinely with less than a minute per injection with a single injector. Higher throughput is typically achieved by parallel injection from multiple injectors when injector washing becomes the rate-limiting step. Wang et al. [75] used a parallel eight-probe injector system to achieve an effective throughput of 7.5 s/sample. The method was used for the analysis of combinatorial libraries synthesized in microtiter plates and compounds purified in deep-well plates. Morand et al. [76] reported an improved method of parallel sampling and serial injection of samples with an eight-probe injector system from microtiter plates with a throughput of 4 s/sample. Improved analysis rates were achieved without a sacrifice in the integrity of the flow-injection peak profile, and baseline resolution was maintained for all samples.

A different approach to achieve increased sample throughput into a mass spectrometer is via the use of a multichannel device. A device with an array of electrospray tips was developed by Liu et al. [77] for FIA-MS analysis. The device uses an independent electrospray exit port for each sample to minimize sample cross-contamination and increase throughput. Their results demonstrated a throughput of 5 s/sample. An alternative multichannel system was developed based upon a subatmospheric ESI interface that couples 96-well plates directly with an ESI interface. This system enabled the analysis of samples at a speed of 10 s/sample with a 120-nL sample consumption [78]. This infusion system was used in the analysis of preparative HPLC fractions from a library synthesis.

Liquid Chromatography Mass Spectrometry

Despite its speed and ease of use, FIA-MS results can be affected by ionization suppression due to unresolved components and impurities in the sample. The use of short columns, high flow rates, and generic elution-gradient HPLC methods provide purification and separation of library compounds before detection with mass spectrometry while a reasonable throughput is maintained. Such fast LC-MS techniques have become a crucial enabling technology in the analysis of combinatorial libraries that are generated from parallel synthesis.

Optimization of Fast HPLC and LC-MS

Weller et al. [43] developed generic, high-flow, reverse-phase gradient HPLC methods for application in the analysis and purification of parallel synthesis libraries. The methods enable the analysis of over 300 compounds/day or the purification of up to 200 compounds/day on a single system. Hardware and software modifications allowed for the continuous unattended use for maximum efficiency and throughput. Mutton [79] described fast HPLC methods with short (20–100 mm) columns swept by fast, yet shallow gradients. The results were compared with those obtained with 150-mm columns and slow gradients. The resolution losses incurred with shorter columns were minimized by the use of fast flow rates. High-quality performance was obtained with turnaround times of 5–10 min. An overall fivefold enhancement in the rate of information generation was obtained. Goetzinger et al. [80] investigated different packing materials, gradient methods, and sample solvents for the development of ultrafast gradient HPLC methods. Commercially available equipment and short columns (<30 mm) packed with small particles (<4 μm) were used to achieve a one minute total analysis time with a peak capacity of 49. These methods were used for quality control of spatially addressable combinatorial libraries. Pereira et al. [81] investigated different buffer types and column geometry effects in high-throughput LC-MS. All the tested buffers, phosphate, acetate, and acetic acid, exhibited good resolution, while the separation time varied from 4 minutes to 12 minutes. The use of acetic acid resulted in the shortest separation time. As the column length was shortened and the flow rate was increased, the separation time was reduced with no change in selectivity.

Fast LC-MS methods have been used to assess library quantity and purity, as well as to triage purification of compounds. Zeng et al. [51] developed one of the first fully automated analytical/preparative LC-MS systems for the characterization and purification of compound libraries derived by parallel synthesis. The system incorporated fast, reverse-phase LC/ESI-MS analysis (5–10 minutes). Post-data-acquisition purity assessment of compound libraries was performed automatically with software control. Compounds that were below a threshold level of purity were automatically purified with HPLC. The real-time purity assessment eliminated the need for postpurification analysis or pooling of fractions collected.

With the widespread application of combinatorial chemistry in drug discovery, there are an increased number of compounds that are tested for pharmaceutical profiling (solubility and ADME; see Chapter 15). The need for high throughput analysis of these compounds stimulated active research in the improvement of LC-MS–based quantitation techniques. Wu et al. [82] investigated monolithic columns for high throughput bioanalysis application. Due to the lower pressure-drop on a monolithic column than on a particulate column, a high flow rate (6 mL/min) was used for a 4.6 $\times$ 50-mm monolithic column. The separation efficiency and signal-to-noise ratios (S/N) for this separation remained almost constant at flow rates of 1, 3, and 6 mL/min. The chromatographic retention time, separation quality, peak response, and sensitivity were highly reproducible throughout a run of 600 plasma extracts. Romanyshyn et al. [83] examined the effects of column length and gradient time on ultrafast chromatographic resolution. By judicious adjustment of column length and gradient slope, the chromatographic integrity of chemically diverse analytes was maintained at a much faster elution speed. This optimized method development strategy enabled separations on 2 $\times$ 20-mm HPLC columns at flow rates of 1.5 mL/min to 2 mL/min with full linear gradients that could be achieved in one minute. Cheng et al. [84] described a simple and comprehensive LC/MS/MS strategy for the rapid analysis of a wide range of pharmaceutical compounds. The authors started with a column that provided a good peak capacity at short gradient run times; then employed high flow rates to achieve a good gradient peak capacity. This fast LC-MS/MS method was used to separate and identify a wide range of analytes with one-minute gradient analyses. Zhang et al. [85] described an isocratic LC-ESI/TOF-MS method for quantitation and accurate mass measurement of five tricyclic amine drugs fortified in human plasma with a per-sample run time of 18 seconds, with a short C-18 column (15 × 2.1-mm Id). The authors used a highly aqueous mobile phase at a flow rate of 1.4 mL/min. Samples were prepared by off-line liquid–liquid extraction. An acquisition speed of 0.2 s/spectrum accommodates these fast separation conditions. Accurate masses were determined by two-point internal mass calibration with postcolumn addition of standard. Results showed a mass error not greater than 9 ppm for all the target compounds. Zweigenbaum and Henion [86] demonstrated more than 2000 samples in 24 hours with LC-MS/MS separation and quantitation for compounds in control human plasma. The method includes sample preparation with liquid–liquid extraction in the 96-well format, an LC separation of the five compounds in less than 30 seconds. Hsieh et al. [87] developed a direct injection bioanalytical method based on a single column LC-MS/MS for pharmacokinetic analysis. Each plasma sample was mixed with a solution that contained an internal standard. The sample was directly injected into a polymer-coated mixed-function column for sample cleanup, enrichment, and chromatographic separation. The stationary phase incorporates both hydrophilic and hydrophobic groups that allow proteins and macromolecules to pass through the column due to restricted access to the surface of the packing, while the drug molecules are retained on the bonded hydrophobic phase. The analytes retained in the column were eluted with a strong organic mobile phase. The total analysis time was 5 min/sample. Yu et al. [88] developed a fast LC-MS method and compared the approach with an FIA-MS method for effective quantitation. With the application of fundamental concepts of fast LC such as the use of a small column (30-mm × 2.1-mm, 2.6-μm particle) at an elevated temperature (40°C) and the use of a high flow rate (1.0 mL/min), the authors were able to reduce the LC cycle time from more than 20 minutes to 2.7 minutes. The authors compared the limits of detection and quantitation, linearity, precision, and accuracy for each analyte. The results indicate that fast LC-MS is generally better than FIA-MS analysis. Romanyshyn et al. [89] developed an LC-MS/MS method for quantitation with rapid ("ballistic") gradients on narrow bore, short HPLC columns and compared the fast-gradient approach with the more traditional high-organic isocratic LC-MS/MS methods. Fast isocratic methods frequently elute the analytes of interest at the solvent front, the region of unretained salts.

The fast-gradient method, in contrast, retains analytes on-column until well after the solvent front has eluted. Overall sample throughput is increased with fast-gradient methods due to reduced analytical run time, decreased method development time, and fewer repeat analyses. Onorato et al. [90] used a multiprobe autosampler for parallel sample injection, short, small-bore columns, high flow rates, and elevated HPLC column temperatures to perform LC separations of idoxifene and its metabolite at 10 s/sample. Sample preparation employed liquid–liquid extraction in the 96-well format. An average run time of 23 s/sample was achieved for human clinical plasma samples.

LC-MS and Hyphenated Techniques for the Library Characterization

Frequently one or more auxiliary detection methods such as UV, nuclear magnetic resonance (NMR), ELSD, and more recently, CLND, are combined with LC-MS to provide a better overall assessment of library purity. These different detection methods often provide complementary capabilities for detection selectivity, range of sensitivity, and linearity of response.

Kibbey [91] compared HPLC quantitation of combinatorial libraries with ELSD and UV detection methods. The ELSD detector response is independent of chemical structural and requires no chromophore, which makes it well suited to HPLC analyses of mixtures of dissimilar compounds. Furthermore, the ELSD exhibits a nearly equivalent response to compounds within a structural class. Hence, rapid quantitation of compound libraries may be carried out with the use of a single external standard. Hsu et al. [92] reviewed the theory of ELSD and the design of commercial instruments. The application of ELSD to library analysis was illustrated with examples from the authors' library synthesis program. Complemented by UV detection for purity assessment and mass spectrometry for product identification, ELSD was the only technique that afforded sufficient accuracy and sensitivity for high throughput library analysis. Fang el al. [93] examined 42 compounds from seven different combinatorial libraries with a high throughput LC-MS/UV/ELSD method and 33 commercial and standard compounds with both high throughput and standard quantitation methods. It was demonstrated that compounds with low molecular weight (<300 Da) are generally less responsive to ELSD and can result in a purity measurement by ELSD that appears higher than when measured by UV. Quantitation with a general UV calibration curve generally gives higher precision than with a general ELSD calibration curve. A calibration curve from structurally related compounds is needed for better quantitation. Fitch et al. [94] made one of the earliest descriptions on the use of LC/CLND for the assessment of solid-phase synthesis of combinatorial libraries that contain nitrogen in the structure. Taylor et al. [95] evaluated a CLND detector as a nearly universal quantitation tool for nitrogen-containing compounds. CLND produced a linear response from 25 pmol to 6400 pmol of nitrogen in the molecule for a set of chemically and structurally diverse compounds with FIA and gradient HPLC. In addition, the response was independent of mobile-phase composition. These results demonstrate that the CLND can be used with FIA or on-line, with HPLC for rapid and accurate quantitation down to low-picomole levels with only a single external standard. The authors also demonstrated the combination of LC/UV/CLND/MS as a generic method for rapid identification, quantitation, and purity assessment of small organic compounds. Shah et al. [96] developed a method for high throughput quality control of combinatorial library compounds from parallel synthesis with a combination of FIA-MS and flow-injection CLND. Compounds were characterized by mass spectrometry and concentration was determined by CLND with a throughput of 60 compounds/h. Dulery et al. [97] developed a strategy of a generic fast HPLC method with diode-array detection (DAD) and MS to provide structure and purity information. In addition, complementary NMR analyses were performed on selected compounds to provide a better structural characterization of the expected compounds and their potential side products. In a recent report, Yurek et al. [98] described the development and use of a new system for the simultaneous determination of identity, purity, and concentration of library components produced by parallel synthesis. The system makes use of HPLC with DAD, ELSD, CLND, and TOF-MS detectors (HPLC/DAD/ELSD/CLND/TOF-MS). The use of exact mass capability of TOF-MS along with CLND provides a synergistic combination that enables identification of target and side-product structures and determination of concentrations and purities in a single analysis.

Parallel LC-MS

The MS instrumentation is the most expensive part of the LC-MS system, hence efforts to improve the throughput of the LC-MS analysis often involve the use of parallel multiple columns that feed into a single mass spectrometer. Zeng and Kassel [99] developed an automated parallel analytical/preparative LC-MS workstation to increase the throughput for the characterization and purification of combinatorial libraries. The system incorporates two columns operated in parallel for both LC-MS analytical and preparative LC-MS purifications. A multiple-sprayer ESI interface was designed to support flows from multiple columns. The system is under complete software control and delivers the crude samples to the two HPLC columns from a single autosampler. The authors demonstrated characterization of more than 200 compounds per instrument per day, and purification of more than 200 compounds per instrument per night. De Biasi et al. [100] described a four-channel multiplexed electrospray LC interface coupled with an orthogonal TOF-MS capable of rapid data acquisition. The flow from a single HPLC pump was split to feed four columns simultaneously and the flows from each of the four columns were coupled to the four-way multiplexed electrospray interface. The use of a high-resolution TOF-MS instrument provided unambiguous molecular-weight assignment to both major components and synthetic by-products. Wang et al. [101] described a similar parallel four-column four-way multiplexed electrospray interface. This setup enabled effluent flow streams from an array of HPLC columns to be sampled independently and sequentially into a single mass spectrometer. Effluent flow streams from an array of four HPLC columns are connected to a parallel arrangement of electrospray needles co-axial to the mass spectrometer entrance aperture. Sage et al. [102] reported a similar four-channel, as well as an eight-channel, source with the orthogonal acceleration TOF-MS system. A 96-well microtiter plate could be analyzed in one hour with a 5 minute HPLC gradient with the eight-channel parallel LC-MS multiplexed electrospray interface system. Tolson et al. [103] reported a parallel LC system for a commercial multichannel electropray source. The performance of the Jasco PAR-1500 pumping system, which delivers an equal-gradient flow to eight individual flow channels, was compared to that of a conventional system, where the flow is split postpump. Both systems were coupled to a LC/ESI/TOF-MS system equipped with a multichannel inlet system and used for the analysis of combinatorial libraries. In addition to flow stability, the Jasco system has resulted in a dramatic improvement in chromatographic performance. Fang et al. [104] reported an eight-channel parallel LC/UV/MS system for analysis of combinatorial library compounds with a throughput of 3200 compounds/day. The system employs eight parallel HPLC columns and UV detectors with eluant from a UV detector that feeds into an eight-channel multiplexed ESI source of a MUX-LCT mass spectrometer.

Parallel LC-MS methods are also widely used in high throughput quantitation for in vitro and in vivo pharmaceutical profiling of compounds. Bayliss et al. [105] described a parallel ultrahigh flow rate LC system with four columns in parallel and a four-way multiple-sprayer interface to the mass spectrometer. This method was applied on both the narrow-bore and capillary scale, and enabled sensitive quantification of drugs from four plasma samples simultaneously without sample preparation and with a throughput of up to 120 samples per hour. Jemal et al. [106] investigated parallel-column liquid LC in conjunction with a conventional single-source ESI-MS. Within a single chromatographic run time, sample injections were made alternately onto each of two analytical columns in parallel at specified intervals. The sample throughput was increased by a factor of 2 when compared with a conventional single-column approach. The dual-column and single-column methods were found to be equivalent in terms of accuracy and precision. Wu [107] developed a similar method of dual-column single mass spectrometer strategy and coupled the method with on-line extraction. This allowed for the direct injection of biofluids onto the system. The performance and capability of this system were shown to be comparable to those obtained on a conventional single-column system. Yang et al. [108] evaluated a four-channel multiplexed electrospray interface for the simultaneous validation of LC-MS/MS methods for the quantitation of the same compounds in four different biological matrixes. Performances in limit of quantitation, precision, accuracy, and intersprayer cross talk for the multiplexed ESI source were all satisfactory. Van Pelt et al. [109] developed a method incorporating four LC columns into a conventional system composed of one binary LC pumping system, one autosampler, and one mass spectrometer. Increased sample throughput was achieved by staggered injections onto the four columns and allowed the mass spectrometer to continuously analyze the chromatographic window of interest. Precision and accuracy were demonstrated to be well within the acceptance criteria. The parallel chromatography system decreased the overall run time from 4.5 minutes to 1.65 minutes in comparison to a conventional LC-MS-MS analytical method. Recently Xu et al. [110] reported an eight-channel parallel LC/MS system in combination with custom automated data-processing applications for high throughput early ADME study. The parallel LC-MS system was configured with one set of gradient LC pumps and an eight-channel multiple-probe autosampler. The flow was split equivalently into eight streams before the multiple probe autosampler and recombined after the eight columns and just prior to the mass spectrometer ion source. The parallel LC-MS system is capable of analyzing up to 240 samples per hour. Deng et al. also reported a four-channel parallel LC-MS system in application of drug analysis from plasma [111].

Step Gradient LC-MS

Further improvement of throughput in LC-MS analysis may be achieved by step-gradient elution. This elution format is essentially an on-line solid-phase extraction (SPE) process, where the samples are loaded onto the column, washed with aqueous mobile phase to remove water soluble impurities, and compounds are eluted with a mobile phase of high organic content. The technique combines the simplicity of FIA with the benefit of the removal of impurities and buffer components before mass spectrometry detection. In this case, selectivity is achieved by mass spectrometry alone without chromatographic separation. The technique has been used for compound purity assessment and quantitation. An on-line back-flush SPE-MS technique has been used by Marshall for quality assessment of the combinatorial libraries [112]. This back-flush elution procedure provides a very effective in-line removal method for the aqueous components in the matrix that are often responsible for ion suppression. Bu et al. [113] used a high throughput analytical method for permeability screening of drug candidates. This simple and sensitive method was based on direct injection coupled with on-line guard-cartridge extraction with a very steep gradient. The method relied on MS/MS to achieve selectivity without chromatographic separation. The authors used this method in a comparison study of apparent permeability across Caco-2 cells measured by sample pooling or cassette dosing strategy with results from single-drug dosing and discrete sample analysis. Janiszewski et al. [114] used a step-gradient strategy in their dual-column LC-MS application on automated high throughput quantitation. The strategy of collecting all data for a compound into a single file greatly reduced the number of data files collected, increased the speed of data collection, allowed rugged and complete review of all data, and provided facile data management. The described systems have analyzed over 40,000 samples per month for 2 years and have the capacity for over 2000 samples per instrument per day.

Two developments in the on-line SPE-MS methodology improved the throughput of the method in the application to analysis of compounds in biological matrix. The first is the use of the turbulent flow chromatography (TFC) for extraction of compounds from in vitro or in vivo samples to remove the biological matrix and elute the compounds for mass spectrometry analysis. The use of columns packed with large particles (30–60 μm) and fast flow (turbulent flow condition) facilitated removal of biological-matrix components, such as proteins and buffers, while the compounds of interests were retained and subsequently eluted out for mass spectrometry analysis [115,116]. The second method is the use of SPE card made in the format of a disposable 96-elution zone compatible with 96-well sampling. The compounds are cleaned up in an off-line device through SPE and the retained compounds are eluted directly into mass spectrometry one zone at a time with high throughput [117,118]. These methods have the common feature that biological matrices are removed by on-line or off-line SPE and compounds are rapidly eluted and analyzed without further separation. These SPE-based methods are useful in the analysis of a large number of samples in the early stage of drug discovery where high throughput is more desirable, while a moderate level of sensitivity is acceptable. When higher sensitivity is desirable, an analytical column can be coupled to these on-line SPE elution devices through a column-switching strategy [119,120].

Analytical SFC-MS

Recently, there has been a growth in the application of SFC and SFC-MS to the analysis of compound libraries. The combination of mass spectrometry with

SFC offers the potential benefit of rapid, high-resolution analysis, separation, and purification with sensitive detection and identification. In addition, the SFC mobile phase is considerably more volatile than the aqueous-based mobile phases that are typically used with reverse-phase LC-MS. This condition allows the entire effluent to be directed into the MS interface and simplifies the coupling of the SFC with mass spectrometry with ESI and APCI.

One of the earliest reports of SFC interfaced with APCI was by Huang et al. [121]. The authors used a pin-hole restrictor to maintain supercritical fluid conditions in a packed-column (pcSFC) system. Results for a mixture of five corticosteoids were described with an injection of 25 ng of each of the components. The system was also amenable for capillary SFC/MS applications with minimum modification. Sadoun and Virelizier [122] reported an SFC interface with ESI in which a two-pump SFC and a packed column were used with the outlet directly interfaced to an ESI source of a quadrupole mass spectrometer. Also, 1–30% (v/v) of polar organic modifier (MeOH-H_2O 95:5) was added to CO_2 mobile phase to help elute polar organic compounds. The setup was shown to allow analysis of polar organic compounds that were difficult to analyze with earlier implementations of SFC-MS with a chemical ionization interface. A recent review article is available on pcSFC-MS [123].

Baker and Pinkston [124] modified an LC/ESI-MS interface for use with pcSFC. The use of a concentric sheath-flow liquid provided ESI modifiers to assist the ionization of neutral, pcSFC-separated components. Postcolumn chromatographic fidelity was preserved with a pressure-regulating fluid, supplied under pressure control to the effluent and positioned just ahead of the sprayer. This modified interface has been used to characterize a variety of mixtures of compounds. Spectra produced with the pcSFC/MS interface are similar to LC/ESI-MS spectra. Hoke et al. [125] compared a pcSFC-MS/MS method to an LC-MS/MS method for quantitation of enantiomers in human plasma. Samples were prepared with automated solid-phase extraction in the 96-well format. Generally, most analytical attributes, such as specificity, linearity, sensitivity, accuracy, precision, and ruggedness, were comparable for both of these methods, with the exception that the pcSFC separation provided a roughly threefold reduction in analysis time. A 2.3-minute pcSFC separation and a 6.5-min LC separation provided equivalent, near-baseline-resolved peaks, and demonstrated significant time savings for the analysis of a large batch of samples with pcSFC. Hoke et al. [126] demonstrated the use of pcSFC for high throughput bioanalytical quantitation with dextromethorphan as a model compound. Plasma samples were prepared by automated liquid–liquid extraction in the 96-well format prior to pcSFC-MS/MS analysis. A throughput of ~10 min/plate was achieved with acceptable relative standard deviation (RSD). Pinkston et al. [127] described a comparative study of 2266 diverse organic compounds with generic pcSFC (CO_2 with 5–60%

MeOH gradient) and HPLC (3–95% ACN gradient) methods with ESI-MS detection, and concluded that the range of coverage is comparable for the two techniques.

The Markides group at Uppsala University in Sweden was also among the earliest in the development of SFC-MS, and demonstrated the application of the technique in the analysis of small organic molecules. Tyrefors et al. [128] described an APCI-MS interface for an open tubular SFC system. The interface was designed to permit transport of the supercritical mobile phase into the ionization region of the mass spectrometer while the temperature is maintained to within one degree of the chromatographic oven temperature. Temperature control of the interface–transfer line was achieved with preheated gas streams from the chromatographic oven and an active electrical insulation. An average retention-time reproducibility of 0.24% was demonstrated with an average 2.6% precision in relative peak height. Sjoberg and Markides [129] described an SFC interface probe for API-MS (including ESI and APCI). A sheath–liquid flow of 20 μL/min in ESI provides optimal conditions for both separation and ionization. The new probe also allows for an easy ionization-mode conversion between ESI and APCI. An improvement to the previous setup was made [130] to obtain stable ion signals and better sensitivity. Factors that influence the ion signal intensity and stability have been studied, and include corona needle position; nebulizer gas flow; gas additives; spray capillary assembly dimension and position; liquid flow-rate; and composition. The achievable detection limits were in the 50–0.1-pg (i.e., 290 fmol–140 amol) range. The detection limit in APCI mode was improved by a factor of about 20–25 compared to an earlier design [129].

Ventura et al. [131] have interfaced an SFC system to a mass spectrometer and evaluated the system for applications requiring high sample throughput. The authors demonstrate the high-speed separation and accurate quantitation capability of SFC-MS. The LC-MS analysis cycle time was reduced threefold with a general SFC-MS high throughput method. Unknown mixture characterization was improved due to the increased selectivity of SFC-MS compared to LC-MS, and was demonstrated with the analysis of combinatorial library mixtures with negative-ion APCI-MS analysis. Rapid elution of SFC was also shown to reduce both sample carryover and cycle time. In an extension of their early work [132], positive-ion APCI-MS was used for the analysis of compounds with a wide range of polarities. The use of SFC-MS instead of LC-MS resulted in substantial time savings, increased chromatographic efficiency, and more precise quantitation of sample mixtures. The instrumental setup also allows for facile conversion between LC-MS and SFC-MS modes of operation.

Morgan et al. [133] described an optimized interface for coupling SFC with APCI-MS. Data presented demonstrate that the internal diameter and length of the transfer line between the SFC unit and the APCI source are not critical to maintain peak shape or retention time under the set of conditions tested. A comparison of responses from an in-line UV detector, two quadrupole mass spectrometers, and an ion trap was presented to demonstrate limits of detection and linear ranges for the SFC separation of a six-compound test mix. Villeneuve and Anderegg [134] developed an automated analytical SFC method to separate enantiomers based on a commercial instrument and column-selection valves. Similar racemic compounds, even those from the same molecular class, were separated with different column and modifier combinations. The optimal chiral separation of several compounds can be obtained within 24 hours with the fully automated system. Garzotti et al. [135] described a simple and economical method to couple a commercial SFC system with a high-resolution hybrid mass spectrometer (Q-TOF-MS). The setup provided on-line accurate mass SFC-MS measurements, and the fast spectral acquisition rate of TOF-MS facilitated data acquisition from rapid SFC separations.

Capillary Electrophoresis–Mass Spectrometry

Capillary electrophoresis (CE) is a powerful separation technique. It is especially useful for separation of ionic compounds and chiral mixtures. Mass spectrometry has been coupled with CE to provide a powerful platform for separation and detection of complex mixtures such as combinatorial libraries. However, the full potential of CE in the application of routine analysis of samples has yet to be realized. This is in part due to perceived difficulty in the use of the CE technique compared to the more mature techniques of HPLC and even SFC. Dunayevskiy et al. [136] analyzed a library of 171 theoretically disubstituted xanthene derivatives with a CE/ESI-MS system. The method allowed the purity and makeup of the library to be determined: 160 of the expected compounds were found to be present, and 12 side products were also detected in the mixture. Due to the ability of CE to separate analytes on the basis of charge, most of the xanthene derivatives could be resolved by simple CE-MS procedures even though 124 of the 171 theoretical compounds were isobaric with at least one other molecule in the mixture. Any remaining unresolved peaks were resolved by MS/MS experiments. The method shows promise for the analysis of small combinatorial libraries with fewer than 1000 components. Boutin et al. [137] used CE-MS along with NMR and MS/MS to characterize combinatorial peptide libraries that contain 3–4 variable positions. The CE-MS method was used to provide a rapid and routine method for initial assessment of the construction of the library. Simms et al. [138] developed a micellar electrokinetic chromatography method for the analysis of combinatorial libraries with an open-tube capillary and UV detection. The quick analysis time of the method made it suitable for the analysis of combinatorial library samples. CE-MS was also used in the analysis of combinatorial libraries in affinity screening through the use of affinity capillary electrophoresis (ACE) [139–142].

COMPOUND-REPOSITORY MANAGEMENT

Collections of large numbers of organic compounds are commonly used in pharmaceutical research efforts such as in HTS. These collections are stored in a variety of formats and under a broad range of environmental conditions. The compounds are synthesized internally, purchased from commercial sources, or obtained as natural products. Currently, more and more of the compounds are made by combinatorial or parallel syntheses. The size and the diversity of the repository compound collection are generally deemed important for the success of the initial lead discovery efforts. Recently, there is an increased interest in the quality and stability of these compounds. The prolonged storage of compounds can cause sample degradation of the screening collection that result from the exposure of the compounds to a variety of environmental factors. Therefore, the up-front analytical and synthetic efforts to generate high-quality samples for the compound supply can be readily lost due to the poor selection of storage conditions. Compound-stability studies have been carried out to find appropriate conditions and formats for the storage and handling of compounds to minimize the degradation. At the same time, there is also a need to assess the quality of the compounds before and during the storage. These quality-control and validation studies are important to assure that good-quality compounds are used for drug-discovery activities. Because the number of compounds to be analyzed can be very large in these compound quality-assessment and stability studies, mass spectrometry–based methods have been used extensively.

Assessment of Library Purity in Compound Repository

The validation of materials upon entering the compound repository is assured by various analytical and separation techniques. Mass spectrometry is often featured for this application. A recent example of this by Pharmacia researchers [98] details an integrated HPLC system with detectors, including UV, ELSD, and CLND in addition to TOF-MS. More references on the use of hyphenated methods for library-quality assessment can be found in [91]–[97] in an earlier section, where library characterization is described.

Compound-Stability Studies

The integrity of repository compounds can degenerate due to environmental effects. Compounds can also be lost due to absorption and precipitation when stored in dimethyl sulfoxide (DMSO). There are a number of factors that can affect repository-compound stability. These include physiochemical properties, humidity, atmosphere and oxygen content, storage container materials, salt form, or the presence of impurities. Information about the relative importance and actual magnitude of these factors is desirable to select the optimal format and conditions for repository compound storage and handling. Numerous studies [143–155] have been performed to monitor compound stability and recovery either under controlled conditions or in practical storage and handling. Mass spectrometry–based methods have been employed in many instances of these studies. FIA-MS is used when qualitative and semiquantitative results are required for high throughput applications. On the other hand, LC-MS in combination with other detection methods such as UV and ELSD are used when more accurate quantitation is needed.

Darvas et al. [143,144] used kinetic modeling approaches to develop a systematic method to assess the stability of compounds that are generated with combinatorial chemistry. With their approach, Stabex™, compound stability is measured experimentally at several elevated temperatures for a selected set of compounds from a chemical library to determine an average range for the Arrhenius parameters. The experimental Arrhenius parameters were subsequently used to predict the expiry rate of the library at selected storage temperatures. The prediction model was verified with experimental high throughput LC-MS measurements at various intermediate temperatures. The authors found that the observed shelf life for the majority of the compounds investigated at an intermediate temperature agreed well with the predictions based on kinetic parameters obtained at elevated temperatures. Milgram and Greig [145] used LC/UV/MS to characterize 17 combinatorial libraries that were composed of more than 5000 compounds. All compounds were identified by MS, but only the total-ion-chromatogram (TIC) ratio was used to determine the compound purity. On average, 15% compounds showed a >20% decrease after six months storage in frozen DMSO. Kozikowski et al. used several different analytical methods in the study of compound stability in room-temperature storage and compound loss in repeated freeze/thaw experiments [146,147]. The large sample number included in the room-temperature stability study dictated that a high throughput analytical method such as FIA-MS be used instead of LC-MS–based quantitative methods [146]. In the study that examined the effect of freeze/thaw cycles on the stability for compounds stored as 20 mM DMSO solutions, approximately 320 compounds were selected from a number of commercial vendors. The number of compounds selected was much less than the room-temperature stability study. Therefore, LC-MS and LC/ELSD were used to quantify the compounds [147]. The set of compounds had an average initial purity of 96%. Compounds were stored at 4°C under argon in pressurized canisters to simulate a low-humidity environment and subjected to 25 freeze/thaw cycles while exposed to ambient atmospheric conditions after each thaw cycle to simulate the time and manner by which compound plates are exposed to the atmosphere during typical liquid-handling processes. Control plates were stored either at room temperature under argon or at 4°C under argon without freeze/thaw cycles, and were evaluated only at the midpoint and end of the study. The percentage of compound that remained was found to decrease for all three storage methods with the freeze/thaw samples, and decreased more than the room-temperature samples and frozen control samples. Additional peaks were not observed in any of the HPLC chromatograms and indicated the absence of soluble degradation products. However, solid precipitate was observed in many of the solutions at the end of the study.

Cheng et al. have carried out a comprehensive study on repository compound stability in DMSO under various conditions [148]. The chemical diversity of the entire repository compound collection was represented by the selection of 644 compounds. An accelerated study was conducted and the results demonstrate that most compounds are stable for 15 weeks at 40°C. The presence of water was determined to be more important than the presence of oxygen for the instigation of compound loss. A freeze/thaw cycle study was done with freezing at −15°C and thawing under nitrogen atmosphere at 25°C. The results indicate no significant compound loss after 11 freeze/thaw cycles. Compound recovery was also measured from glass and polypropylene containers for a period of five months at room temperature, and no significant difference was found for these two types of containers. LC/UV/MS technique was used for the quantitation of compounds, and an optimum wavelength was selected for each compound to balance UV absorption between the internal standard and the target compound. The UV and MS data were processed with Waters Openlynx software with predetermined parameters (mass-to-storage ratio (m/z) values for protonated and sodiated species and UV wavelength) for each compound. UV data (retention time and peak area) were converted to ASCII files and were transferred to an Oracle database. Target and internal standard peaks were assigned based on a lookup table, and the target-to-internal standard ratio was calculated automatically for each compound. Other applications of mass spectrometry and LC-MS in compound-stability studies were also reported recently [149–154]. These applications were also discussed in a recent review on the repository-compound stability studies [155].

Compound-library compounds are most often subjected to purification prior to assay or storage. The most widely employed method is RP-HPLC in solvent system gradients of aqueous acetonitrile. Trifluoroacetic acid (TFA) is the most commonly used tertiary modifier for these systems due to good chromatographic behavior for the widest variety of compounds. The implication of residual TFA that results from this purification method must be considered for potential affects of compounds in DMSO for storage and handling. Recently, Yan [151] reported an example of compound degradation through a hydrolysis mechanism for which the presence of TFA salt accelerated the reaction. Hochlowski et al. have undertaken a study [149] wherein the relative stability of TFA and non-TFA analogues of a diverse set of compounds was investigated. Preliminary results indicated that for a certain subset of structural types, the potential for acid-catalyzed decomposition existed with long-term storage conditions in DMSO for TFA salts or adducts. Newer compound-purification methods, such as supercritical-fluid chromatography, that do not use TFA would likely eliminate these issues, and we look for the continued development of these techniques in the future.

CONCLUSIONS AND PROSPECTIVE

Compound-library creation, analysis, profiling, and management are integral components of modern drug-discovery efforts. With well-validated targets and assays, the success of drug lead discovery and optimization has largely depended upon the diversity and quality of the compound library. As an analytical technique, mass spectrometry has played an important role in these drug-discovery activities. The widespread application of mass spectrometry for chemical-library analysis is due to the unique features such as high speed, sensitivity, selectivity, general applicability to diverse compounds, as well as the ease of coupling of modern ionization techniques with liquid handling and separation techniques. Thus, mass spectrometry–based methods have become an important enabling technology for compound-library analysis, purification, and profiling. Parallel processing, open access, and automation (i.e., analysis, data processing, interpretation, and reporting of results) have been the major trends in recent years. Continued development of mass spectrometry–based techniques in the analysis of compound libraries is likely in the foreseeable future.

Improvement in LC-MS and LC-MS/MS analysis throughput has been reported by the use of monolithic silica columns to increase the speed of chromatography separation [156–158]. The substitution of an SFC front end to MS in lieu of HPLC has been a growing trend in compound library analysis. It is possible that the use of SFC-MS will be extended to in vitro and in vivo evaluation of library compounds such as ADME and DMPK. CE-MS has not been widely used in the analysis of combinatorial libraries. To date, the application of CE-MS has been frequently in the analysis of mixture-component libraries derived from split synthesis and in the affinity screening of libraries through ACE. However, with the miniaturization of biological screening in the "lab-on-a-chip" format, CE-MS may find renewed interest due to the ease of coupling CE with miniaturized biochip separation platforms [159–162]. Thus, miniaturization and chip-based technology may be an important new development for the analysis of compound libraries in conjunction with the biological testing of these compounds. Several new developments addressed chip-based analysis with different strategies. Researchers from Advion Biosceinces developed the NanoMate device for chip-based nano-ESI mass spectrometry analysis. Applications in proteomics and pharmacokinetcs have been reported [163]. An alternative nano-ESI device with polymer chips has been reported with Fourier transform ion cyclotron resonance mass spectrometry (FT-ICR-MS) [164]. In addition, a chip-based parallel LC device has been developed. Researchers from Nanostream have reported a 24-channel parallel HPLC chip device with UV detection [165]. The device will need to be coupled with mass spectrometry detection to make full use of the power of the microscale separation and analysis. Trends in library purification seem to be directed more toward mass-triggered collection technologies in place of UV-, ELSD-, or CLND-triggered collection. Further, SFC has seen a growth in acceptance for the preparative purification of combinatorial libraries. The format has already been formatted to provide a MS-triggered format. In compound-library management, the routine quality control of compounds will be facilitated by development of high throughput LC-MS technique in combination with good quantitation modules such as UV, ELSD, or CLND. It is also possible that these will be implemented in a parallel separation format with miniaturized chip devices [165].

REFERENCES

1. Furka, A. "Combinatorial Chemistry: 20 Years On . . ." *Drug Disc. Today* **7,** 1–4 (2002).
2. Lebl, M. "Parallel Personal Comments on 'Classical' Papers in Combinatorial Chemistry," *J. Comb. Chem.* **1,** 3–24 (1999).
3. Thompson, L.A.; Ellman, J.A. "Synthesis and Applications of Small Molecule Libraries," *Chem. Rev.* **96,** 555–600 (1996).
4. Gordon, E.M.; Gallop, M.A.; Patel, D.V. "Strategy and Tactics in Combinatorial Organic Synthesis. Applications to Drug Discovery," *Acc. Chem. Res.* **29,** 144–154 (1996).
5. Yan, B., ed. *Analytical Methods in Combinatorial Chemistry*, Technomic, Lancaster, PA (2000).
6. Swartz, M.E., ed. *Analytical Techniques in Combinatorial Chemistry*, Dekker, New York (2000).
7. Jung, G., ed. *Combinatorial Chemistry. Synthesis, Analysis, Screening*, Wiley-VC, Weinheim, Germany (1999).

8. Gordon, E.M.; Kerwin Jr., J.F., eds. *Combinatorial Chemical and Molecular Diversity in Drug Discovery*, Wiley-Liss, New York (1998).
9. Kassel, D.B. "Combinatorial Chemistry and Mass Spectrometry in the 21st Century Drug Discovery Laboratory," *Chem. Rev.* **101,** 255–268 (2001).
10. Triolo, A.; Altamura, M.; Cardinali, F.; Sisto, A.; Maggi, C.A. "Mass Spectrometry and Combinatorial Chemistry: A Short Outline," *J. Mass Spectrom.* **36,** 1249–1259 (2001).
11. Shin, Y.G.; van Breemen, R.B. "Analysis and Screening of Combinatorial Libraries Using Mass Spectrometry," *Biopharm. Drug Disposition* **22,** 353–372 (2001).
12. Hoke II, S.H.; Morand, K.L.; Greis, K.D.; Baker, T.R.; Harbol, K.L.; Dobson, L.M. "Transformations in Pharmaceutical Research and Development, Driven by Innovations in Multidimensional Mass Spectrometry-based Technologies," *Int. J. Mass Spectrom.* **212,** 135–196 (2001).
13. Enjalbal, C.; Martinez, J.; Aubagnac, J.L. "Mass Spectrometry in Combinatorial Chemistry," *Mass. Spectrom. Rev.* **19,** 139–161 (2000).
14. Sussmuth, R.D.; Jung, G. "Impact of Mass Spectrometry on Combinatory Chemistry," *J. Chromatogr. B* **725,** 49–65 (1999).
15. Swali, V.; Langley, G.J.; Bradley, M. "Mass Spectrometric Analysis in Combinatorial Chemistry," *Curr. Opinion Chem. Biol.* **3,** 337–341 (1999).
16. Lee, M.; Kerns, E.H. "LC/MS Applications in Drug Development," *Mass Spectrom. Rev.* **18,** 187–295 (1999).
17. Kyranos, J.N.; Hogan, J.C. Jr. "High-Throughput Characterization of Combinatorial Libraries Generated by Parallel Synthesis," *Anal. Chem.* **70,** 389A–395A (1998).
18. Loo, J.A. "Mass Spectrometry in the Combinatorial Chemistry Revolution," *Eur. Mass Spectrom.* **3,** 93–104 (1997).
19. Hughes, I.; Hunter, D. "Techniques for Analysis and Purification in High-throughput Chemistry," *Curr. Opin. Chem. Biol.* **5,** 243–247 (2001).
20. Morand, K.L.; Burt, T.M.; Regg, B.T.; Tirey, D.A. "Advances in High-throughput Mass Spectrometry," *Curr. Opin. Drug Disc. Devel.* **4,** 729–735 (2001).
21. LaCourse, W. "Column Liquid Chromatography: Equipment and Instrumentation," *Anal. Chem.* **72,** 37R–51R (2000).
22. LaCourse, W. "Column Liquid Chromatography: Equipment and Instrumentation," *Anal. Chem.* **74,** 2813–2832 (2002).
23. Reetz, M.T.; Becker, M.H.; Klein, H.-W.; Stöckigt, D. "A Method for High-Throughput Screening of Enantioselective Catalysts," *Angew. Chem. Int. Ed.* **38,** 1758–1761 (1999).
24. Guo, J.; Wu, J.; Siuzdak, G.; Finn, M.G. "Measurement of Enantiomeric Excess by Kinetic Resolution and Mass Spectrometry," *Angew. Chem. Int. Ed.* **38,** 1755–1758 (1999).

25. Szewczyk, J.W.; Zuckerman, R.L.; Bergman, R.G.; Ellman, J.A. "A Mass Spectrometric Labeling Strategy for High-Throughput Reaction Evaluation and Optimization: Exploring C-H Activation," *Angew. Chem. Int. Ed.* **40,** 216–219 (2001).
26. McKeown, S.C.; Watson, S.P.; Carr, R.A.E.; Marshall, P.A. "Photolabile Carbamate Based Dual Linker Analytical Construct for Facile Monitoring of Solid Phase Chemistry: 'TLC' for Solid Phase?" *Tetrahedron Lett.* **40,** 2407–2410 (1999).
27. Lorthioir, O.; Carr, R.A.E.; Congreve, M.S.; Geysen, M.H.; Kay, C.; Marshall, P.; McKeown, S.C.; Parr, N.J.; Scicinski, J.J.; Watson, S.P. "Single Bead Characterization Using Analytical Constructs: Application to Quality Control of Libraries," *Anal. Chem.* **73,** 963–970 (2001).
28. Congreve, M.S.; Jamieson, C. "High-throughput Analytical Techniques for Reaction Optimization," *Drug Discovery Today* **7,** 139–142 (2002).
29. Enjalbal, C.; Maux, D.; Combarieu, R.; Martinez, J.; Aubagnac, J.-L. "Imaging Combinatorial Libraries by Mass Spectrometry: from Peptide to Organic-Supported Syntheses," *J. Comb. Chem.* **5,** 102–109 (2003).
30. Pullen, F.S.; Perkins, G.L.; Burton, K.J.; Ware, R.S.; Teague, M. S.; Kiplinger, J.P. "Putting Mass Spectrometry in the Hands of the End User," *J. Am. Soc. Mass Spectrom.* **6,** 394–399 (1995).
31. Taylor, L.C.E.; Johnson, R.L.; Raso, L. "Open-access Atmospheric Pressure Chemical Ionization Mass Spectrometry for Routine Sample Analysis," *J. Am. Soc. Mass Spectrom.* **6,** 387–393 (1995).
32. Spreen, R.C.; Schaffter, L.M. "Open-access MS: A Walk-up MS Service," *Anal. Chem.* **68,** 414A–419A (1996).
33. Mallis, L.M.; Sarkahian, A.B.; Kulishoff Jr, J.M.; Watts Jr., W.L. "Open-access Liquid Chromatography/Mass Spectrometry in a Drug Discovery Environment," *J. Mass Spectrom.* **37,** 889–896 (2002).
34. Greaves, J. "Operation of an Academic Open-access Mass Spectrometry Facility with Particular Reference to the Analysis of Synthetic Compounds," *J. Mass Spectrom.* **37,** 777–785 (2002).
35. Daley, D.J.; Scammell, R.D.; James, D.; Monks, I.; Raso, R.; Ashcroft, A.E.; Hudson, A.J. "High-throughput LC-MS Verification of Drug Candidates from 96-Well Plates," *Am. Biotechnol. Lab.* **15,** 24–28 (1997).
36. Greig, M. "Use of Automated HPLC-MS Analysis for Monitoring and Improving the Purity of Combinatorial Libraries," *Am. Lab.* **31,** 28–32 (1999).
37. Tong, H.; Bell, D.; Tabei, K.; Siegel, M.M. "Automated Data Massaging, Interpretation, and E-mailing Modules for High Throughput Open-access Mass Spectrometry," *J. Am. Soc. Mass Spectrom.* **10,** 1174–1187 (1999).
38. Choi, B.K.; Hercules, D.M.; Michelotti, E.; Martinez, B.; Eisenschmied, M.; Zhang, T.; Gusev, A.I. "Application of Visual Basic for Automation of LC-MS Analysis," *JALA* **5,** 102–105 (2000).

39. Choi, B.K.; Hercules, D.M.; Sepetov, N.; Issakova, O.; Gusev, A.L. "Intelligent Automation of LC–MS Analysis for the Characterization of Compound Libraries,"*LC GC* **20,** 152–162 (2002).
40. Williams, J.D.; Weiner, B.E.; Ormand, J.R.; Brunner, J.; Thornquest Jr., A.D.; Burinsky, D.J. "Automated Molecular Weight Assignment of Electrospray Ionization Mass Spectra," *Rapid Commun. Mass Spectrom.* **15,** 2446–2455 (2001).
41. Yates, N.; Wislocki, D.; Roberts, A.; Berk, S.; Klatt, T.; Shen, D.-M.; Willoughby, C.; Rosauer, K.; Chapman, K.; Griffin, P. "Mass Spectrometry Screening of Combinatorial Mixtures, Correlation of Measured and Predicted Electrospray Ionization Spectra," *Anal. Chem.* **73,** 2941–2951 (2001).
42. Klagkou, K.; Pullen, F.; Harrison, M.; Organ, A.; Firth, A.; Langley, G.J. "Approaches Towards the Automated Interpretation and Prediction of Electrospray Tandem Mass Spectra of Non-peptidic Combinatorial Compounds," *Rapid Commun. Mass Spectrom.* **17,** 1163–1168 (2003).
43. Weller, H.N.; Young, M.G.; Michalczyk, J.D.; Reitnauer, G.H.; Cooley, R.S.; Rahn, P.C.; Loyd, D.J.; Fiore, D.; Fischman, J.S. "High Throughput Analysis and Purification in Support of Automated Parallel Synthesis," *Mol. Divers.* **3,** 61–70 (1997).
44. Kibby, C.E. "An Automated System for Purification of Combinatorial Libraries by Preparative LC/MS," *Lab. Robotics Automation* **9,** 309–321 (1997).
45. Routburg, M.; Swenson, R.; Schmitt, B.; Washington, A.; Mueller, S.; Hochlowski, J.; Maslana, G.; Minn, B.; Matuszak, K.; Searle, P.; Pan, J. "Implementation of an Automated Purification/Verification System," presented at The International Symposium on Laboratory Automation and Robotics, Boston, MA, October, 1996.
46. Searle, P. "An Automated Preparative HPLC-MS System for the Rapid Purification of Combinatorial Libraries," presented at The Strategic Institute Conference: High-Throughput Compound Characterization, Dallas, TX, March, 1998.
47. Coffey, P. "Parallel Purification for Combinatorial Chemistry," *Lab. Automation News* **2,** 7–13 (1997).
48. Schultz, L.; Garr, C.D.; Cameron, L.M.; Bukowski, J. "High Throughput Purification of Combinatorial Libraries," *Bioorg. Med. Chem. Lett.* **8,** 2409–2414 (1998).
49. Edwards, C.; Hunter, D. "High-throughput Purification of Combinatorial Arrays," *J. Comb. Chem.* **5,** 61–66 (2003).
50. Zeng, L.; Burton, L.; Yung, K.; Shushan, B.; Kassel, D.B. "Automated Analytical/Preparative High-performance Liquid Chromatography-Mass Spectrometry System for the Rapid Characterization and Purification of Compound Libraries," *J. Chromatog. A* **794,** 3–13 (1998).
51. Zeng, L.; Kassel, D.B. "Development of a Fully Automated Parallel HPLC/Mass Spectrometry System for the Analytical Characterization and Preparative Purification of Combinatorial Libraries," *Anal. Chem.* **70,** 4380–4388 (1998).

52. Isbell, J.; Xu, R.; Cai, Z.; Kassel, D.B. "Realities of High-Throughput Liquid Chromatography/Mass Spectrometry Purification of Large Combinatorial Libraries: A Report on Overall Sample Throughput Using Parallel Purification," *J. Comb. Chem.* **4,** 600–611 (2002).

53. Kiplinger, J.P.; Cole, R.O.; Robinson, S.; Roskamp, E.J.; Ware, R.S.; O'Connell, H.J.; Brailsford, A.; Batt, J. "Structure-controlled Automated Purification of Parallel Synthesis Products in Drug Discovery," *Rapid Commun. Mass Spectrom.* **12,** 658–664 (1998).

54. Searle, P.; Hochlowski, J. "High Throughput Purification of Parallel Synthesis Samples: An Integrated Preparative-LC/MS System with Quantitation by Chemiluminescent Nitrogen Detection," presented at The HPLC 2002, Montreal, Canada, June, 2002.

55. Xu, R.; Wang, T.; Isbell, J.; Cai, Z.; Sykes, C.; Brailsford, A.; Kassel, D. "High-Throughput Mass-Directed Parallel Purification Incorporating a Multiplexed Single Quadrupole Mass Spectrometer," *Anal. Chem.* **74,** 3055–3062 (2002).

56. Chester, T.L.; Pinkston, J.D. "Supercritical Fluid and Unified Chromatography," *Anal. Chem.* **72,** 129R–135R (2000).

57. Berger, T.; Fogelman, K.; Staats, L.; Nickerson, M.; Bente, P. "Apparatus and Method for Preparative Supercritical Fluid Chromatography," US patent, July, 2002, US 6,413,428 B1.

58. Berger, T.A.; Fogleman, K.; Staats, T.; Bente, P.; Crocket, I.; Farrell, W.; Osonubi, M. "The Development of a Semi-preparatory Scale Supercritical-fluid Chromatograph for High-throughput Purification of 'Combi-Chem' Libraries," *J. Biochem. Biophys. Methods* **43,** 87–111 (2000).

59. Farrell, W.; Ventura, M.; Aurigemma, C.; Tran, P.; Fiori, K.; Xiong, X.; Lopez, R.; Osbonubi, M. "Analytical and Semi-Preparative SFC for Combinatorial Chemistry," presented at The International Symposium on Supercritical Fluid Chromatography, Extraction, and Processing, Myrtle Beach, SC, August, 2001.

60. Olson, J.; Pan, J.; Hochlowski, J.; Searle, P.; Blanchard, D. "Customization of a Commercially Available Prep Scale SFC System to Provide Enhanced Capabilities," *JALA* **7,** 69–74 (2002).

61. Hochlowski, J.; Searle, P.; Gunawardana, G.; Sowin, T.; Pan, J.; Olson, J.; Trumbull, J. "Development and Application of a Preparative Scale Supercritical Fluid Chromatography System for High Throughput Purification," presented at Prep 2001, Washington, DC, May, 2001.

62. Hochlowski, J.; Olson, J.; Pan, J.; Sauer, D.; Searle, P.; Sowin, T. "Purification of HTOS Libraries by Supercritical Fluid Chromatography," *J. Liquid Chromatog.* **26,** 333–354 (2003).

63. Maiefski, R.; Wendell, D.; Ripka, W.D.; Krakover, J.D. "Apparatus and method for multiple channel high throughput purification," PCT Int. Appl., May, 2000, WO 0026662.

64. Ripka, W.C.; Barker, G.; Krakover, J. "High Throughput Purification of Compound Libraries," *Drug Disc. Today* **6,** 471–477 (2001).

65. Wang, T.; Barber, M.; Hardt, I.; Kassel, D.B. "Mass-directed Fractionation and Isolation of Pharmaceutical Compounds by Packed-column Supercritical Fluid Chromatography/Mass Spectrometry," *Rapid Commun. Mass Spectrom.* **15,** 2067–2075 (2001).
66. Kassel, D. "Possibilities and Potential Pitfalls of SFC/MS as a Complementary Tool to HPLC/MS for High throughput Drug Discovery," presented at The International Symposium on Supercritical Fluid Chromatography, Extraction, and Processing, Myrtle Beach, SC, August, 2001.
67. Richmond, R.; Gorlach, E. "Sorting Measurement Queues to Speed Up the Flow Injection Analysis Mass Spectrometry of Combinatorial Chemistry Syntheses," *Anal. Chim. Acta* **394,** 33–42 (1999).
68. Richmond, R.; Gorlach, E. "The Automatic Visualization of Carry-over in High-throughput Flow Injection Analysis Mass Spectrometry," *Anal. Chim. Acta* **390,** 175–183 (1999).
69. Richmond, R. "The Analytical Characterisation of Sub-minute Measurement Duty Cycles in Flow Injection Analysis Mass Spectrometry, by Their Carry-over," *Anal. Chim. Acta* **403,** 287–294 (2000).
70. Hegy, G.; Gorlach, E.; Richmond, R.; Bitsch, F. "High Throughput Electrospray Mass Spectrometry of Combinatorial Chemistry Racks with Automated Contamination Surveillance and Results Reporting," *Rapid Commun. Mass Spectrom.* **10,** 1894–1900 (1996).
71. Gorlach, E.; Richmond, R.; Lewis, I. "High-Throughput Flow Injection Analysis Mass Spectroscopy with Networked Delivery of Color-Rendered Results. 2. Three-Dimensional Spectral Mapping of 96-Well Combinatorial Chemistry Racks," *Anal. Chem.* **70,** 3227–3234 (1998).
72. Gorlach, E.; Richmond, R. "Discovery of Quasi-Molecular Ions in Electrospray Spectra by Automated Searching for Simultaneous Adduct Mass Differences," *Anal. Chem.* **71,** 5557–5562 (1999).
73. Richmond, R.; Gorlach, E.; Seifert, J.-M. "High-throughput Flow Injection Analysis-Mass Spectrometry with Networked Delivery of Color Rendered Results: The Characterization of Liquid Chromatography Fractions," *J. Chromatogr. A* **835,** 29–39 (1999).
74. Whalen, K.M.; Rogers, K.J.; Cole, M.J.; Janiszewski, J.S. "AutoScan: An Automated Workstation for Rapid Determination of Mass and Tandem Mass Spectrometry Conditions for Quantitative Bioanalytical Mass Spectrometry," *Rapid Commun. Mass Spectrom.* **14,** 2074–2079 (2000).
75. Wang, T.; Zeng, L.; Strader, T.; Burton, L.; Kassel, D.B. "A New Ultra-high Throughput Method for Characterizing Combinatorial Libraries Incorporating a Multiple Probe Autosampler Coupled with Flow Injection Mass Spectrometry Analysis," *Rapid Commun. Mass Spectrom.* **12,** 1123–1129 (1998).
76. Morand, K.L.; Burt, T.M.; Regg, B.T.; Chester, T.L. "Techniques for Increasing the Throughput of Flow Injection Mass Spectrometry," *Anal. Chem.* **73,** 247–252 (2001).

77. Liu, H.; Felten, C.; Xue, Q.; Zhang, B.; Jedrzejewski, P.; Karger, B.L.; Foret, F. "Development of Multichannel Devices with an Array of Electrospray Tips for High-Throughput Mass Spectrometry," *Anal. Chem.* **72,** 3303–3310 (2000).
78. Felten, C.; Foret, F.; Minarik, M.; Goetzinger, W.; Karger, B.L. "Automated High-Throughput Infusion ESI-MS with Direct Coupling to a Microtiter Plate," *Anal. Chem.* **73,** 1449–1454 (2001).
79. Mutton, I.M. "Use of Short Columns and High Flow Rates for Rapid Gradient Reversed-Phase Chromatography," *Chromatographia* **47,** 291–298 (1998).
80. Goetzinger, W.K.; Kyranos, J.N. "Fast Gradient RP-HPLC for High-throughput Quality Control Analysis of Spatially Addressable Combinatorial Libraries," *Am. Lab.* **30,** 27–37 (1998).
81. Pereira, L.; Ross, P.; Woodruff, M. "Chromatographic Aspects in High Throughput Liquid Chromatography/Mass Spectrometry," *Rapid Commun. Mass Spectrom.* **14,** 357–360 (2000).
82. Wu, J.T.; Zeng, H.; Deng, Y.; Unger, S.E. "Automated Analytical/Preparative High-performance Liquid Chromatography–Mass Spectrometry System for the Rapid Characterization and Purification of Compound Libraries," *Rapid Commun. Mass Spectrom.* **15,** 1113–1119 (2001).
83. Romanyshyn, L.A.; Tiller, P.R. "Ultra-short Columns and Ballistic Gradients: Considerations for Ultra-fast Chromatographic Liquid Chromatographic–Tandem Mass Spectrometric Analysis," *J. Chromatog. A* **928,** 41–51 (2001).
84. Cheng, Y.-F.; Lu, Z.; Neue, U. "Ultrafast Liquid Chromatography/Ultraviolet and Liquid Chromatography/Tandem Mass Spectrometric Analysis," *Rapid Commun. Mass Spectrom.* **15,** 141–151 (2001).
85. Zhang, H.; Heinig, K.; Henion, J. "Atmospheric Pressure Ionization Time-of-flight Mass Spectrometry Coupled with Fast Liquid Chromatography for Quantitation and Accurate Mass Measurement of Five Pharmaceutical Drugs in Human Plasma," *J. Mass Spectrom.* **35,** 423–431 (2000).
86. Zweigenbaum, J.; Henion, J. "Bioanalytical High-Throughput Selected Reaction Monitoring-LC/MS Determination of Selected Estrogen Receptor Modulators in Human Plasma: 2000 Samples/Day," *Anal. Chem.* **72,** 2446–2454 (2000).
87. Hsieh, Y.; Bryant, M.S.; Gruela, G.; Brisson, J.-M.; Korfmacher, W.A. "Direct Analysis of Plasma Samples for Drug Discovery Compounds Using Mixed-function Column Liquid Chromatography Tandem Mass Spectrometry," *Rapid Commun. Mass Spectrom.* **14,** 1384–1390 (2000).
88. Yu, K.; Balogh, M.A. "Protocol for High-Throughput Drug Mixture Quantitation: Fast LC-MS or Flow Injection Analysis-MS?" *LC GC* **19,** 60–72 (2001).
89. Romanyshyn, L.; Tiller, P.R.; Alvaro, P.; Pereira, A.; Cornelis, E.C.A. "Ultra-fast Gradient vs. Fast Isocratic Chromatography in Bioanalytical Quantification by Liquid Chromatography/Tandem Mass Spectrometry," *Rapid Commun. Mass Spectrom.* **15,** 313–319 (2001).

90. Onorato, J.M.; Henion, J.D.; Lefebvre, P.M.; Kiplinger, J.P. "Selected Reaction Monitoring LC-MS Determination of Idoxifene and Its Pyrrolidinone Metabolite in Human Plasma Using Robotic High-Throughput, Sequential Sample Injection," *Anal. Chem.* **73,** 119–125 (2001).
91. Kibbey, C.E. "Quantitation of Combinatorial Libraries of Small Organic Molecules by Normal-phase HPLC with Evaporative Light-scattering Detection," *Mol. Divers.* **1,** 247–258 (1996).
92. Hsu, B.H.; Orton, E.; Tang, S.Y.; Carlton, R.A. "Application of Evaporative Light Scattering Detection to the Characterization of Combinatorial and Parallel Synthesis Libraries for Pharmaceutical Drug Discovery," *J. Chromatogr. B* **725,** 103–112 (1999).
93. Fang, L.; Pan, J.; Yan, B. "High-throughput Determination of Identity, Purity, and Quantity of Combinatorial Library Members Using LC/MS/UV/ELSD," *Biotech. Bioeng.* **71,** 162–171 (2001).
94. Fitch, W.L.; Szardenings, A.K.; Fujinari, E.M. "Chemiluminescent Nitrogen Detection for HPLC: An Important New Tool in Organic Analytical Chemistry," *Tetrahedron Lett.* **38,** 1689–1692 (1997).
95. Taylor, E.W.; Qian, M.G.; Dollinger, G.D. "Simultaneous Online Characterization of Small Organic Molecules Derived from Combinatorial Libraries for Identity, Quantity, and Purity by Reversed-Phase HPLC with Chemiluminescent Nitrogen, UV, and Mass Spectrometric Detection," *Anal. Chem.* **70,** 3339–3347 (1998).
96. Shah, N.; Gao, M.; Tsutsui, K.; Lu, A.; Davis, J.; Scheuerman, R.; Fitch, W.L. "A Novel Approach to High-Throughput Quality Control of Parallel Synthesis Libraries," *J. Comb. Chem.* **2,** 453–460 (2000).
97. Dulery, B.D.; Verne-Mismer, J.; Wolf, E.; Kugel, C.; Hijfte, L.V. "Analyses of Compound Libraries Obtained by High-throughput Parallel Synthesis: Strategy of Quality Control by High-performance Liquid Chromatography, Mass Spectrometry and Nuclear Magnetic Resonance Techniques," *J. Chromatogr. B* **725,** 39–47 (1999).
98. Yurek, D.A.; Branch, D.L.; Kuo, M.-S. "Development of a System to Evaluate Compound Identity, Purity, and Concentration in a Single Experiment and Its Application in Quality Assessment of Combinatorial Libraries and Screening Hits," *J. Comb. Chem.* **4,** 138–148 (2002).
99. Zeng, L.; Kassel, D.B. "Developments of a Fully Automated Parallel HPLC/Mass Spectrometry System for the Analytical Characterization and Preparative Purification of Combinatorial Libraries," *Anal. Chem.* **70,** 4380–4388 (1998).
100. De Biasi, V.; Haskins, N.; Organ, A.; Bateman, R.; Giles, K.; Jarvis, S. "High Throughput Liquid Chromatography/Mass Spectrometric Analyses Using a Novel Multiplexed Electrospray Interface, *Rapid Commun. Mass Spectrom.* **13,** 1165–1168 (1999).

101. Wang, T.; Zeng, L.; Cohen, J.; Kassel, D.B. "A Multiple Electrospray Interface for Parallel Mass Spectrometric Analyses of Compound Libraries," *Comb. Chem. High Throughput Screening* **2,** 327–334 (1999).
102. Sage, A.B.; Little, D.; Giles, K. "Using Parallel LC-MS and LC-MS-MS to increase sample throughput," *LC GC* **18,** S20–S29 (2000).
103. Tolson, D.; Organ, A.; Shah, A. "Development of a High-pressure Gradient Pumping System for Parallel Liquid Chromatography/Mass Spectrometry for the Analysis of Combinatorial Libraries," *Rapid Commun. Mass Spectrom.* **15,** 1244–1249 (2001).
104. Fang, L.; Cournoyer, J.; Demee, M.; Zhao, J.; Tokushige, D.; Yan, B. "High-throughput Liquid Chromatography Ultraviolet/Mass Spectrometric Analysis of Combinatorial Libraries Using an Eight-channel Multiplexed Electrospray Time-of-flight Mass Spectrometer," *Rapid Commun. Mass Spectrom.* **16,** 1440–1447 (2002).
105. Bayliss, M.K.; Little, D.; Mallett, D.N.; Plumb W.S. "Parallel Ultra-high Flow Rate Liquid Chromatography with Mass Spectrometric Detection Using a Multiplex Electrospray Source for Direct, Sensitive Determination of Pharmaceuticals in Plasma at Extremely High Throughput," *Rapid Commun. Mass Spectrom.* **14,** 2039–2045 (2000).
106. Jemal, M.; Huang, M.; Mao, Y.; Whigan, D.; Powell, M.L. "Increased Throughput in Quantitative Bioanalysis Using Parallel-column Liquid Chromatography with Mass Spectrometric Detection," *Rapid Commun. Mass Spectrom.* **15,** 994–999 (2001).
107. Wu, J.-T. "The Development of a Staggered Parallel Separation Liquid Chromatography/Tandem Mass Spectrometry System with On-line Extraction for High-throughout Screening of Drug Candidates in Biological Fluids," *Rapid Commun. Mass Spectrom.* **15,** 73–81 (2001).
108. Yang, L.; Mann, T.D.; Little, D.; Wu, N.; Clement, R.P.; Rudewicz, P.J. "Evaluation of a Four-Channel Multiplexed Electrospray Triple Quadrupole Mass Spectrometer for the Simultaneous Validation of LC/MS/MS Methods in Four Different Preclinical Matrixes," *Anal. Chem.* **73,** 1740–1747 (2001).
109. Van Pelt, C.K.; Corso, T.N.; Schultz, G.A.; Lowes, S.; Henion, J. "A Four-Column Parallel Chromatography System for Isocratic or Gradient LC/MS Analyses," *Anal. Chem.* **73,** 582–588 (2001).
110. Xu, R.; Nemes, C.; Jenkins, K.M.; Rourick, R.A.; Kassel, D.B.; Liu, C.Z.C. "Application of Parallel Liquid Chromatography/Mass Spectrometry for High Throughput Microsomal Stability Screening of Compound Libraries," *J. Am. Soc. Mass Spectrom.* **13,** 155–165 (2002).
111. Deng, Y.; Wu, J.-T., Lloyd, T.L.; Chi, C.L.; Olah, T.V.; Unger, S.E. "High-speed Gradient Parallel Liquid Chromatography/Tandem Mass Spectrometry with Fully Automated Sample Preparation for Bioanalysis: 30 Seconds per Sample from Plasma," *Rapid Commun. Mass Spectrom.* **16,** 1116–1123 (2002).

112. Marshall, P.S. "Development and Applications of a Rapid Back-flush Microseparation System Coupled to a Mass Spectrometer for the Quality Control of Combinatorial Libraries," *Rapid Commun. Mass Spectrom.* **13,** 778–781 (1999).

113. Bu, H.-Z.; Poglod, M.; Micetich, R.G.; Khan, J.K. "High-throughput Caco-2 Cell Permeability Screening by Cassette Dosing and Sample Pooling Approaches Using Direct Injection/On-line Guard Cartridge Extraction/Tandem Mass Spectrometry," *Rapid Commun. Mass Spectrom.* **14,** 523–528 (2000).

114. Janiszewski, J.S.; Rogers, K.J.; Whalen, K.M.; Cole, M.J.; Liston, T.E.; Duchoslav, E.; Fouda, H.G. "A High-Capacity LC/MS System for the Bioanalysis of Samples Generated from Plate-Based Metabolic Screening," *Anal. Chem.* **73,** 1495–1501 (2001).

115. Grant, R.P.; Cameron, C.; Mackenzie-McMurter S. "Generic Serial and Parallel On-line Direct-injection Using Turbulent Flow Liquid Chromatography/Tandem Mass Spectrometry," *Rapid Commun. Mass Spectrom.* **16,** 1785–1792 (2002).

116. Herman, J.L. "Generic Method for On-Line Extraction of Drug Substances in the Presence of Biological Matrices Using Turbulent Flow Chromatography," *Rapid Commun. Mass Spectrom.* **16,** 421–426 (2002).

117. Wachs, T.; Henion, J.A. "Device for Automated Direct Sampling and Quantitation from Solid-Phase Sorbent Extraction Cards by Electrospray Tandem Mass Spectrometry," *Anal. Chem.* **75,** 1769–1775 (2003).

118. Olech, R.M.; Pranis, R.A.; Cole, M.J.; Janiszewski, J.S. "Elutri-Zone Chromatography: Applications in a Drug Discovery," in *Proceedings of the 50th ASMS Conference on Mass Spectrometry and Allied Topics*, Orlando, FL, June, 2002, Abstract No. MPF 205.

119. Mallet, C.R.; Mazzeo, J.R.; Neue, U. "Evaluation of Several Solid Phase Extraction Liquid Chromatography/Tandem Mass Spectrometry On-line Configurations for High-throughput Analysis of Acidic and Basic Drugs in Rat Plasma," *Rapid Commun. Mass Spectrom.* **16,** 421–426 (2002).

120. Jemal, M. "High-throughput Quantitative Bioanalysis by LC/MS/MS," *Biomed. Chromatogr.* **14,** 422–429 (2000).

121. Huang, E.C.; Wachs, T.; Conboy, J.J.; Henion, J.D. "Atmospheric Pressure Ionization Mass Spectrometry Detection for the Separation Science," *Anal. Chem.* **62,** 713A–725A (1990).

122. Sadoun, F.; Virelizier, H. "Packed-column Supercritical Fluid Chromatography Coupled with Electrospray Ionization Mass Spectrometry," *J. Chromatog.* **647,** 351–359 (1993).

123. Combs, M.T.; Ashraf-Khorassani, M.; Taylor, L.T. "Packed Column Supercritical Fluid Chromatography-Mass Spectrometry: A Review," *J. Chromatogr. A* **785,** 85–100 (1997).

124. Baker, T.R.; Pinkston, J.D. "Development and Application of Packed-Column Supercritical Fluid Chromatography/Pneumatically Assisted Electrospray Mass Spectrometry," *J. Am. Soc. Mass Spectrom.* **9,** 498–509 (1998).

125. Hoke II, S.H.; Pinkston, J.D.; Bailey, R.E., Tanguay, S.L.; Eichhold, T.H. "Comparison of Packed-Column Supercritical Fluid Chromatography-Tandem Mass Spectrometry with Liquid Chromatography-Tandem Mass Spectrometry for Bioanalytical Determination of (*R*)- and (*S*)-Ketoprofen in Human Plasma Following Automated 96-Well Solid-Phase Extraction," *Anal. Chem.* **72,** 4235–4241 (2000).
126. Hoke II, S.H.; Tomlinson, J.A.; Bolden, R.D.; Morand, K.L.; Pinkston, J.D.; Wehmeyer, K.R. "Increasing Bioanalytical Throughput Using pcSFC-MS/MS: 10 Minutes per 96-Well Plate," *Anal. Chem.* **73,** 3083–3088 (2001).
127. Pinkston, J.D.; Wen, D.; Morand, K.L.; Tirey, D.A.; Stanton, D.T. "Screening a Large and Diverse Library of Pharmaceutically Relevant Compounds Using SFC MS and a Novel Mobile Phase Additive," presented at The International Symposium on Supercritical Fluid Chromatography, Extraction, and Processing, Myrtle Beach, SC, August, 2001.
128. Tyrefors, L.N.; Moulder, R.X.; Markides, K.E. "Interface for Open Tubular Column Supercritical Fluid Chromatography/Atmospheric Pressure Chemical Ionization Mass Spectrometry," *Anal. Chem.* **65,** 2835–2840 (1993).
129. Sjoberg, J.R.; Markides, K.E. "New Supercritical Fluid Chromatography Interface Probe for Electrospray and Atmospheric Pressure Chemical Ionization Mass Spectrometry," *J. Chromatogr. A* **785,** 101–110 (1997).
130. Sjoberg, J.R.; Markides, K.E. "Capillary Column Supercritical Fluid Chromatography–Atmospheric Pressure Ionisation Mass Spectrometry; Interface Performance of Atmospheric Pressure Chemical Ionisation and Electrospray Ionisation," *J. Chromatogr. A* **855,** 317–327 (1999).
131. Ventura, M.C.; Farrell, W.P.; Aurigemma, C.M.; Greig, M.J. "Packed Column Supercritical Fluid Chromatography/Mass Spectrometry for High-Throughput Analysis," *Anal. Chem.* **71,** 2410–2416 (1999).
132. Ventura, M.C.; Farrell, W.P.; Aurigemma, C. M.; Greig, M.J. "Packed Column Supercritical Fluid Chromatography/Mass Spectrometry for High-Throughput Analysis. Part 2," *Anal. Chem.* **71,** 4223–4231 (1999).
133. Morgan, D.G.; Harbol, K.L.; Kitrinos, P.N. "Optimization of a Supercritical Fluid Chromatograph–Atmospheric Pressure Chemical Ionization Mass Spectrometer Interface Using an Ion Trap and Two Quadrupole Mass Spectrometers," *J. Chromatogr. A* **800,** 39–49 (1998).
134. Villeneuve, M.S.; Anderegg, R.J. "Analytical Supercritical Fluid Chromatography Using Fully Automated Column and Modifier Selection Valves for the Rapid Development of Chiral Separations," *J. Chromatogr. A* **826,** 217–225 (1998).
135. Garzotti, M.; Rovatti, L.; Hamdan, M. "Coupling of a Supercritical Fluid Chromatography System to a Hybrid (Q-TOF 2) Mass Spectrometer: On-line Accurate Mass Measurements," *Rapid Commun. Mass Spectrom.* **15,** 1187–1190 (2001).

136. Dunayevskiy, Y.M.; Vouros, P.; Wintner, E.A.; Shipps, G.W.; Carell, T.; Rebek Jr., J. "Application of Calipoary Electrophoresis-Electrospray Ionization Mass Spectrometry in the Determination of Molecular Diversity," *Proc. Nat. Acad. Sci. USA* **93,** 6152–6157 (1996).

137. Boutin, J.A.; Hennig, P.; Lambert, P.H.; Bertin, S.; Petit, L.; Mahieu, J.P.; Serkiz, B.; Volland, J.P.; Fauchère, J.L. "Combinatorial Peptide Libraries: Robotic Synthesis and Analysis by Nuclear Magnetic Resonance, Mass Spectrometry, Tandem Mass Spectrometry, and High-performance Capillary Electrophoresis Techniques," *Anal. Biochem.* **234,** 126–141 (1996).

138. Simms, P.J.; Jeffries, C.T.; Huang, Y.; Zhang, L.; Arrhenius, T.; Nadzan, A.M. "Analysis of Combinatorial Chemistry Samples by Micellar Electrokinetic Chromatography," *J. Comb. Chem.* **3,** 427–433 (2001).

139. Chu, Y.-H.; Dunayevskiy, Y.M.; Kirby, D.P.; Vouros, P.; Karger, B.L. "Affinity Capillary Electrophoresis-Mass Spectrometry for Screening Combinatorial Libraries," *J. Am. Chem. Soc.* **118,** 7827–7835 (1996).

140. Lyubarskaya, Y.V.; Carr, S.A.; Dunnington, D.; Prichett, W.P.; Fisher, S.M.; Appelbaum, E.R.; Jones, C.S.; Karger, B.L. "Screening for High-Affinity Ligands to the *Src* SH2 Domain Using Capillary Isoelectric Focusing-Electrospray Ionization Ion Trap Mass Spectrometry," *Anal. Chem.* **70,** 4761–4770 (1998).

141. Little, J.N.; Hughes, D.E.; Karger, B.L. "A Powerful Screening Technology Utilizing Capillary Electrophoresis," *Am. Biotech. Lab.* **17,** 36 (1999).

142. Guzman, N.A.; Stubbs, R.J. "The Use of Selective Adsorbents in Capillary Electrophoresis-Mass Spectrometry for Analyte Preconcentration and Microreactions: A Powerful Three-dimensional Tool for Multiple Chemical and Biological Applications," *Electrophoresis* **22,** 3602–3628 (2001).

143. Darvas, F.; Karancsi, T.; Slegel, P.; Dorman, G. "Estimating Stability for HTS Library Compounds," *Genetic Eng. News* **7,** 30–31 (2000).

144. Darvas, F.; Dorman, G.; Karansci, T.; Nagy, T.; Bagyi, I. "Estimation of Stability and Shelf Life for Compounds, Libraries and Repositories in Combination with Systematic Discovery of New Rearrrangement Pathways," pp. 806–828 in Nicolaou, K.; Hanko, R.; Hartwig, W., eds., *Handbook of Combinatorial Chemistry*, Wiley-VCH, New York, Wheinheim (2002).

145. Milgram, K.E.; Greig, M.J. "Assessing the Stability of Combinatorial Libraries Stored in Frozen DMSO," in *Proceedings of 48th ASMS Conference on Mass Spectrometry and Allied Topics*, Long Beach, CA, June, 2000, Abstract No. MPF 218.

146. Kozikowski, B.A.; Burt, T.M.; Tirey, D.A.; Williams, L.E.; Kuzmak, B.R.; Stanton, D.T.; Morand, K.L.; Nelson, S.L. "The Effect of Room-temperature Storage on the Stability of Compounds in DMSO," *J. Biomol. Screening* **8,** 205–209 (2003).

147. Kozikowski, B.A.; Burt, T.M.; Tirey, D.A.; Williams, L.E.; Kuzmak, B.R.; Stanton, D.T.; Morand, K.L.; Nelson, S.L. "The Effect of Freeze/Thaw Cycles

on the Stability of Compounds in DMSO," *J. Biomol. Screening* **8,** 210–215 (2003).

148. Cheng, X.; Hochlowski, J.; Tang, H.; Hepp, D.; Beckner, C.; Kantor, S.; Schmitt, R. "Studies on Repository Compound Stability in DMSO under Various Conditions," *J. Biomol. Screening* **8,** 292–304 (2003).
149. Hochlowski, J.; Cheng, X.; Sauer, D.; Djuric, S. "Studies of the Relative Stability of TFA Adducts vs. non-TFA Analogues for Combinatorial Chemistry Library Members in DMSO in a Repository Compound Collection," *J. Comb. Chem.* **5,** 345–349 (2003).
150. Heaton, Z. "A Summary of Stability Experiments Carried Out at GlaxoSmithKlein," presented at the Drug Discovery Technology World Congress post-conference session on Repository Compound Stability Studies, Boston, MA, August, 2002.
151. Yan, B. "An Evaluation of the Stability of Purified Combinatorial Libraries," presented at the Drug Discovery Technology World Congress post-conference session on Repository Compound Stability Studies, Boston, MA, August, 2002.
152. Yaskanin, D. "Quality Perspectives for Compound Storage," presented at the Drug Discovery Technology World Congress post-conference session on Repository Compound Stability Studies, Boston, MA, August, 2002.
153. Turmel, M. "Compound Stability under Various Different Storage Conditions," presented at the Drug Discovery Technology World Congress post-conference session on Repository Compound Stability Studies, Boston, MA, August, 2002.
154. Savchuk, N. "Comparative Stability of Small Molecules under a Variety of Storage and Handling Conditions," presented at the Drug Discovery Technology World Congress post-conference session on Repository Compound Stability Studies, Boston, MA, August, 2002.
155. Morand, K.; Cheng, X. "Organic Compound Stability in Large, Diverse Pharmaceutical Screening Collections," in Yan, B., ed., *Analysis and Purification Methods in Combinatorial Chemistry*, Wiley, New York, 2003.
156. Zeng, H.; Deng, Y.; Wu, J.-T. "Fast Analysis Using Monolithic Columns Coupled with High-flow On-line Extraction and Electrospray Mass Spectrometric Detection for the Direct and Simultaneous Quantitation of Multiple Components in Plasma," *J. Chromatogr., B* **788,** 331–337 (2003).
157. Hsieh, Y.; Wang, G.; Wang, Y.; Chackalamannil, S.; Korfmacher, W.A. "Direct Plasma Analysis of Drug Compounds Using Monolithic Column Liquid Chromatography and Tandem Mass Spectrometry," *Anal. Chem.* **75,** 1812–1818 (2003).
158. Peng, S.X.; Barbone, A.G.; Ritchie, D.M. "High-throughput Cytochrome P450 Inhibition Assays by Ultrafast Gradient Liquid Chromatography with Tandem Mass Spectrometry Using Monolithic Columns," *Rapid Commun. Mass Spectrom.* **17,** 509–518 (2003).

159. Wachs, T.; Henion, J. "Electrospray Device for Coupling Microscale Separations and Other Miniaturized Devices with Electrospray Mass Spectrometry," *Anal. Chem.* **73,** 632–638 (2001).

160. Deng, Y.; Henion, J. "Chip-Based Capillary Electrophoresis/Mass Spectrometry Determination of Carnitines in Human Urine," *Anal. Chem.* **73,** 639–646 (2001).

161. Schultz, G.A.; Corso, T.N.; Prosser, S.J.; Zhang, S. "A Fully Integrated Monolithic Microchip Electrospray Device for Mass Spectrometry," *Anal. Chem.* **72,** 4058–4063 (2000).

162. Oleschuk, R.D.; Harrison, D.J. "Analytical Microdevices for Mass Spectrometry," *TrAC Trends in Anal. Chem.* **19,** 379–388 (2000).

163. Dethy, J.M.; Ackermann, B.L.; Delatour, C.; Henion, J.D.; Schultz, G. "A. Demonstration of Direct Bioanalysis of Drugs in Plasma Using Nanoelectrospray Infusion from a Silicon Chip Coupled with Tandem Mass Spectrometry," *Anal. Chem.* **75,** 805–811 (2003).

164. Rossier, J.S.; Youhnovski, N.; Lion, N.; Damoc, E.; Becker, S.; Reymond, F.; Girault, H.H.; Przybylski, M. "Thin-Chip Microspray System for High-Performance Fourier-Transform Ion-Cyclotron Resonance Mass Spectrometry of Biopolymers," *Angew. Chem. Int. Ed.* **42,** 53–58 (2003).

165. Patel, P.; Anderson, K.; Hobbs, S.A. "Microfluidic Device for High Throughput Separations," presented at the Drug Discovery Technology World Congress, Boston, MA, August, 2002.

8

A COMBINATORIAL PROCESS FOR DRUG DISCOVERY

DAVID S. WAGNER, RICHARD W. WAGNER, FRANK SCHOENEN, AND H. MARIO GEYSEN

INTRODUCTION

Combinatorial chemistry is widely used to describe high throughput synthesis of compounds, both organic and inorganic. In reality, high throughput synthesis activities represent the industrialization of chemistry in much the same way that commodity items, from appliances to motor cars, are made today within a highly industrialized society. Computers and automation have replaced the labor-intensive aspects of chemical synthesis. As is often the case with faster and cheaper technology, this new approach to chemistry has necessitated some concessions in terms of the traditional requirements of characterization and purity. Initially, the emphasis was on the numbers of compounds made and the use of solid-phase synthesis methods. Chemical libraries with essentially no characterization were assayed for biological activity.

In summary, the early experience with chemical libraries was less than stellar. The general belief that smaller rounds of more controlled synthesis would resolve any ambiguity about active "hits" found in the initial screen proved to be overoptimistic, and accounts of failure were common. It quickly became apparent that the use of solid phase, necessary with an encoding strategy, was in itself a challenge for organic chemistry procedures and very different from a chemist's experience of solution phase. The numerically smaller libraries of discreetly synthesized compounds (parallel synthesis) that followed

Integrated Strategies for Drug Discovery Using Mass Spectrometry, Edited by Mike S. Lee

by purification became the norm, and very few companies that use a viable "real" combinatorial approach to drug discovery remain. Pharmacopia (still encoding)[1], Affymax (hard-tag encoding)[2], and GlaxoWellcome (mass encoding)[3–5] were among the few with a technology that allowed the use of a "mix and split" procedure that generated bead-based, numerically large compound libraries.

It is worthwhile to note that three consecutive reports in *Tetrahedran* [6–8] summarized both the progress made in combinatorial chemistry and the challenges that require a technological solution to more fully realize its full potential. In each of these reports, quantitation and characterization of chemistry carried out on the solid phase during the synthesis and at the completion before any biological testing were stated to be urgent priorities. In the absence of the foregoing and due to the widespread belief that solid-phase chemistry was limited with respect to the diversity of chemistries that could be carried out, the industry has invested significantly in equipment-intensive facilities geared to high throughput solution-phase synthesis followed by purification.

The inherent power of a bead-based solid-phase approach to compound libraries for hit identification is in the ability to use a mix- and-split process for numerically large libraries at an economically feasible scale, typically about 1–2 nmole per bead. This chapter outlines the culmination of several years of research by a dedicated team comprising a wide range of expertise for the development of an inherently simple technology based on mass spectrometry. The described technology overcomes many of the perceived disadvantages of a bead-based approach to compound generation, as well as provides the opportunity to rapidly discover and validate new chemistries on solid phase. When coupled with "statistical decoding" [9], this bead-based technology virtually eliminates any possibility of obtaining an ambiguous outcome from either a failed or incomplete synthesis or from a false-positive that occurs during the screening procedure.

ANALYTICAL CONSTRUCTS

Two of the difficulties encountered with solid-phase synthesis is quantitation and identification of all the products that result from resin-based solid-phase methods. "Analytical constructs" were developed to facilitate rapid quantitative and qualitative analytical measurements during compound synthesis and library production [10,11]. The analytical construct designed by Diversity Sciences possesses multiple functions. The construct serves as the physical link between the solid support and the site of synthesis, promotes the analytical analysis of a library synthesis at each step of the procedure, and also provides a facile method of bead encoding.

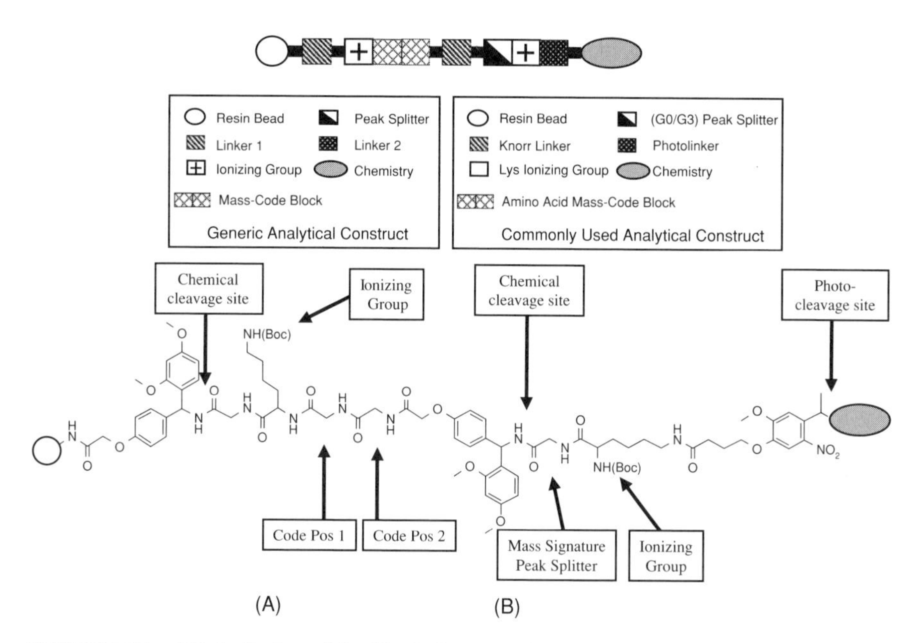

FIGURE 8.1 (A) A display of the block diagram of the "analytical construct" designed by Diversity Sciences. The keys represent the nonspecific version and the general version used for library production. (B) Displays the molecular structure of the commonly used construct during library production and reaction optimization.

A generic representation of the analytical construct designed by GlaxoSmithKline is shown in Figure 8.1. The diagram shows the construct has a linear format that consists of three linker groups, two of which have the same cleavage mechanism, and a third linker that has an orthogonal cleavage condition. In addition, there are two groups (mass-spectrometric levelers) that ionize readily for mass-spectral analysis, a signature peak element (peak splitter), and a mass-coding region. Both the mass-code block and the peak-splitting elements are incorporated into the construct with isotopically stable enriched reagents. The ionizable group or mass-spectrometric leveler can be positive or negative in nature, but is usually positive, because it has a higher sensitivity by mass-spectral analysis. Production of the analytical construct uses amide-bond formation chemistries with nearly 100% yields.

The standard analytical construct used for library production in Diversity Sciences at GlaxoSmithKline is shown in Figure 8.1. The analytical construct uses the common Knorr acid cleavable linkers at the first and second linker positions. The terminal linker is photocleavable and is released upon exposure to 350-nM light [12–15]. The construct contains the mass-code block and uses isotopically labeled Gly as a peak splitting element to facilitate compound identification by mass spectrometry. Also, the construct has two lysine amino acid groups to aid the ionization process of both the code block and the ligand block during mass spectrometry. The side chain of the lysine in the ligand block is used to provide a chemical spacer between the signature peak and the photocleavable linker. The N-terminal amine is butoxycarbonyl (Boc) protected during synthetic chemistry and serves as the ionization site upon acid cleavage. This spacer ensured that the desired library-specific organic chemistry is not hindered by the construct elements. At the completion of the analytical construct, the library-specific chemistry is carried out on the photocleavable link, as illustrated in Figure 8.1.

The initial concept and practical use of analytical constructs was designed by Diversity Sciences to facilitate mass-spectral analysis during combinatorial processes [10]. Several internal and external groups adopted the original concept and published similar results [16–20]. Each element of the analytical construct developed by Diversity Science is outlined below in more depth in the following sections.

Mass-Encoding Technology

The term *combinatorial chemistry* refers to a collection of techniques for the synthesis of large numbers of compounds, which rely on encoding and decoding strategies. The encoding strategies range from spatial orientation for parallel synthesis to complex chemical tags, optical tags, and radio-frequency tags when mixing solid supports [21]. Synthetic methods that use mixing and splitting of solid supports (normally resin beads) provide the advantage of synthesizing large numbers of compounds, while the number of synthetic reactions is minimized, as shown in Figure 8.2. The resin-bead solid support is divided into equal portions; each portion of beads undergoes a reaction with a single but unique monomer (i.e., one bead, one compound). After the synthetic step, all the bead portions are combined into a single batch, thoroughly mixed, and divided into equal portions for the next synthetic step Each portion is identical and is comprises all possible combinations. Repetition of the divide, couple, and recombine processes creates a combinatorial library that consists of resin beads, each containing a single library member. The example in Figure 8.2 produces 27 unique compounds that use only nine synthetic reactions (the same number to traditionally make three compounds).

A major complication in biological screening of libraries synthesized by mixing the solid supports is the identification of the compound on a given bead because of the mixing that occurs between synthetic steps. Encoding strategies address this problem and track the synthetic history or the monomer added at each step. However, most of the encoding/decoding strategies have disadvantages associated with lengthy analytical procedures; number of possible codes; restrictions on compatible organic chemistry; and the requirement of orthogonal chemistry during synthesis for the encoding tag. Mass-encoding

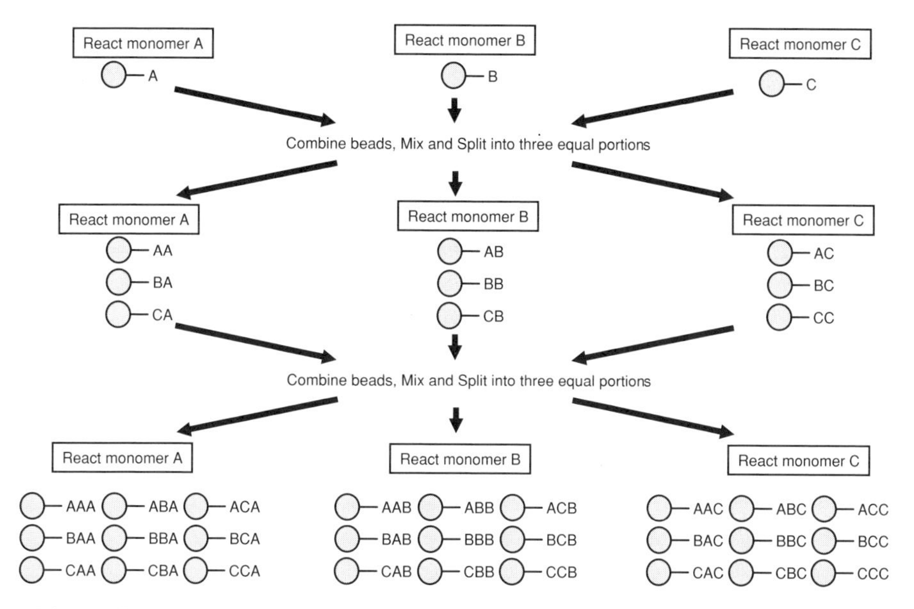

FIGURE 8.2 The generic representation of "split–mix–recombine" synthetic strategy with three unique monomers (A, B, and C) at each synthetic step. The final outcome for this example is the synthesis of all possible combinations, 27 unique compounds after three synthetic rounds.

technology overcomes most of the difficulties encountered by these other methods.

The mass-encoding technology developed by Diversity Sciences uses mass spectrometry to resolve isotopic contributions in a molecule and precisely determine their abundance [3–5,22–27]. Stable isotopes are incorporated into molecular tags to produce distinct isotopic patterns when analyzed by mass spectrometry. These patterns serve to record the chemical history of the bead. The mass coding system employs amino acids as the encoding monomers. Isotopic variants (^{13}C and ^{15}N) of glycine and alanine are used to carry the synthetic information (Figure 8.3). The code block is a peptide that consists of Gly–Lys–**(*Gly and/or *Ala)–(*Gly and/or *Ala)**. The first Gly serves as a chemical spacer between the chemical linker and the rest of the code block, and promotes the chemical cleavage during the decoding process. The lysine provides a basic site to ensure a reasonable mass-spectral response during decoding. The side chain of the Lys is Boc is protected until the decoding step, when the side chain is converted to a free amine. The last two code positions (isotopic variants of *Gly or *Ala) are used to track the synthetic history of the beads. Although a code may contain the same amino acids, they are made up of isotopic variants that generate a unique pattern.

The prominent encoding strategy for library synthesis has been "binary coding" [3–4]. A mixture of amino acid monomers is reacted in equivalent

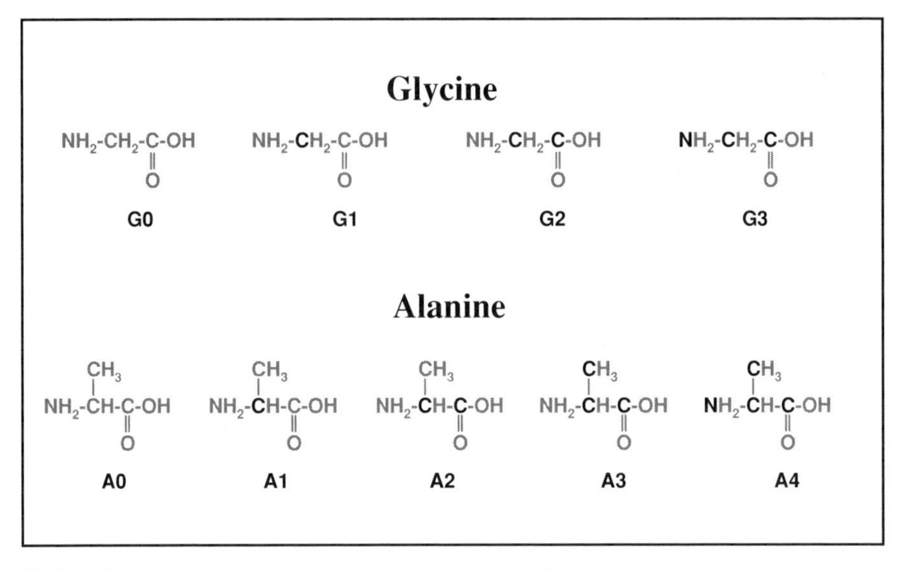

FIGURE 8.3 The amino acid monomers (isotopic variants of glycine and alanine) used to synthesize the mass code. The boldface letters indicate the position of C^{13} and N^{15}in each code monomer.

concentrations in order that each is incorporated uniformly. The resulting isotopic pattern by mass spectral analysis consists of peaks with equal intensities. A typical binary code is shown in Figure 8.4. The code set encompasses a small, specified mass range, 15 unique masses in a 23-amu window in which code peaks can be located. The coding system is binary (on or off) by the presence or absence of a peak at each of the 15 possible sites in the code window. The line divides the naturally occurring isotopes for the code monomers and the intensities and the intensity of the labeled amino acids necessary to be a peak in the mass code. The intensity of the peak must be above the red line (for a peak in the code region to be "on" (1)), while peaks that appear below the red line are designated "off" (0).

The binary code can range from the presence of a single peak to all 15 peaks. However, a unique prerequisite was incorporated to provide a built-in checking advantage: each code is made up of four peaks. The presence of four peaks in the code region ensures that a product peak or impurity has not been superimposed on one of the code peaks, and that only a single code is analyzed. Should a reaction product interfere in the code region, it is still possible to read the code by inference or implement an analytical separation method. Any combination of the labeled amino acids that produces four distinct masses in the code region can be used in this coding system. Figure 8.5 displays several four-peak codes that demonstrate the ease of visual decoding. Individual codes are analyzed by flow-injection analysis into a quadrupole mass spectrometer with

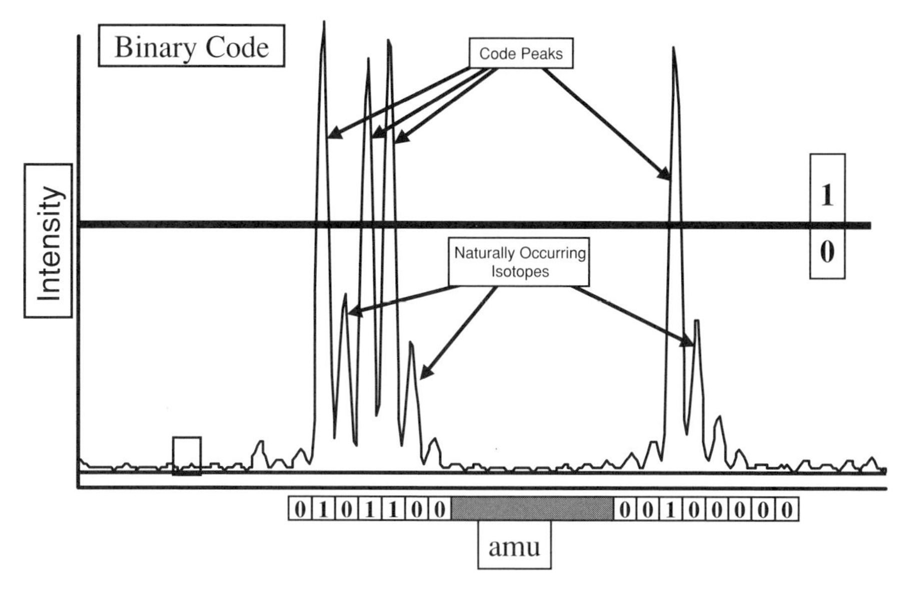

FIGURE 8.4 A display of the mass-spectral region of a typical mass code. The boxes at the bottom of the spectrum outline the 15 amu mass-code region in which a peak can be present, each box is labeled with a 1 or 0 to represent the presence or absence of a peak at that particular mass. The line across the spectrum demonstrates the difference in intensities between the naturally occurring isotopic distributions from the intentional code peaks.

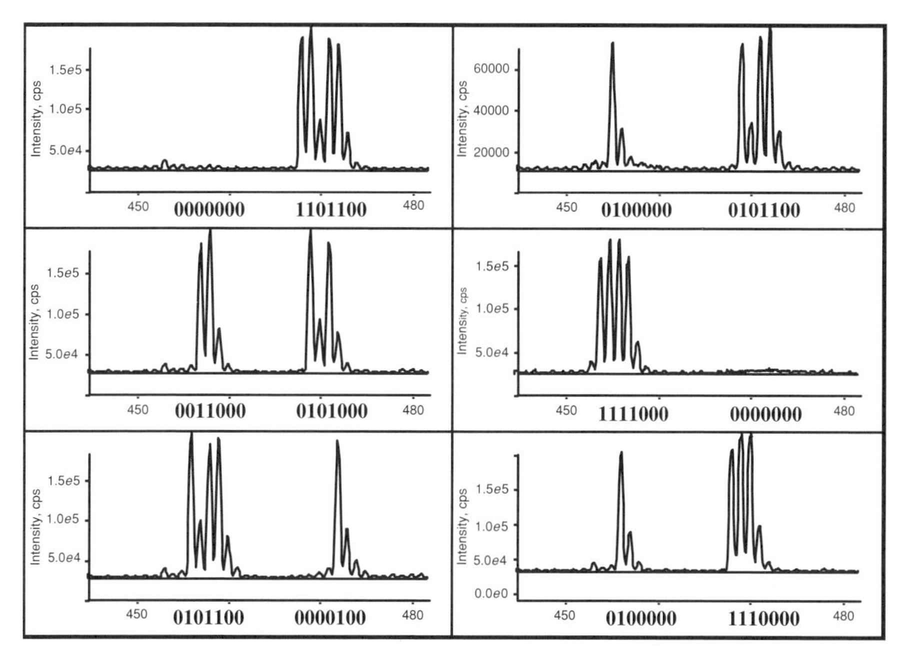

FIGURE 8.5 The mass-spectral region from the mass spectra obtained from several four-peak codes.

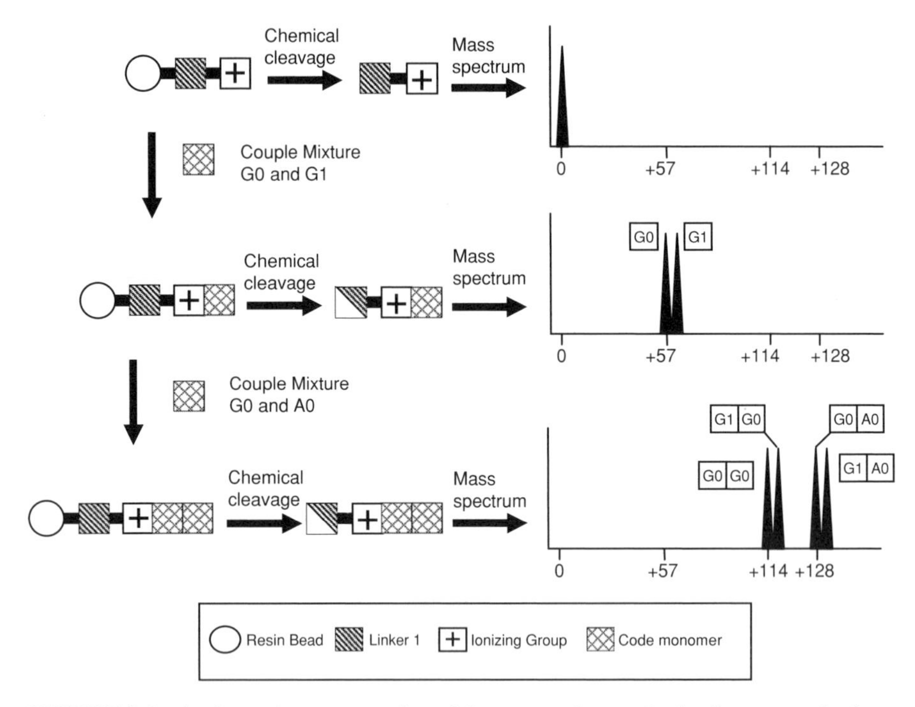

FIGURE 8.6 A schematic representation of the progressive synthesis of a mass code along with the corresponding mass spectrum. Initially the ionizing group is added to the solid phase; here the mass spectrum scale is normalized to zero for illustrative purposes. The isotopic variants of Gly and Ala are added in the next two synthetic stages to produce the four-peak mass code.

electrospray ionization, typically with a 25–35-second data-acquisition cycle. Approximately 2800 codes can be processed per mass spectrometer in a 24-hour period. Code identification and a dual verification system have been automated by in-house software where 2800 codes can be completed in a few minutes.

The assembly of the code block is performed in a progressive fashion. Figure 8.6 represents the code production and the corresponding mass spectrum where the mass scale characterizes the change in mass after each synthetic step of the code block. The first step is the addition of the first-position Gly and Lys amino acids. The mass of the code block at this stage has been normalized to zero for illustrative purposes. The first code position is the equal-molar coupling of G0 and G1. The mass spectrum displays two peaks that are separated by 1 amu and 57 amu (residue weight of Gly) higher than the previous step. The second code position couples an equal-molar mixture of G0 and A0 and yields four peaks in the corresponding mass spectrum that relate to all possible combinations of the binary mixtures. Variation of the amino acid mixtures during code production generates all possible codes.

The advantages of mass spectrometry as a decoding platform include sensitivity, accuracy, and speed of decoding. The rapid rate of decoding allows for the identification of all potentially active compounds from a biological assay and outperforms most other decoding technologies, which are time-consuming and only permit the decoding of a small subset of the identified active compounds. The capacity to decode all the active compounds from a biological screen leads to a more comprehensive structure–activity relationship (SAR). In addition, the speed and accuracy of the mass-decoding platform allows for a statistical approach to biological data analysis and will be discussed later in detail.

Mass-Spectrometric Signature Peak

It is often difficult to determine the degree to which the chemistry proceeded on the entire library population and whether peaks in a mass spectrum are due to the product, side reactions, reagents, solvents, or impurities. Diversity Sciences developed mass-spectral methods to distinguish all components that are cleaved from a solid support and implemented the method into the analytical construct. While early studies demonstrated promising results for fragmentation methods with tandem mass spectrometry (MS/MS), stable isotopes were routinely implemented as signature peaks for the identification of compounds that are produced from solid-phase reactions [27].

Mass-spectrometric "peak-splitter" signature methodology incorporates stable isotopes into compounds to produce a distinctive isotopic pattern similar to that produced by the presence of a bromine atom. The ionization of an organic compound that contains one atom of bromine will display two molecular-ion peaks of approximately equal intensities that are separated by two mass units because the ions contain bromine 79 and bromine 81 in approximately equal proportions. This feature gives rise to a set of easily identified goalpost-like peak intensities separated by two mass units. Similarly, the *N*-acetylation of peptides with a 1:1 ratio of monoisotopic and D3 labeled forms of reagent has facilitated the identification of N-terminal fragment ions during MS/MS studies [28,29]. Peaks that have equal intensity and are separated by three mass units identify the N-terminal fragments. Diversity Sciences extended this concept to facilitate the identification of synthetic products during solid-phase library production.

The addition of a mass-signature element that contains a 1:1 ratio of one or more stable isotopes (^{13}C, ^{15}N, or ^{2}H) splits the molecular ion into two equal peaks (similar to bromine). All compounds that contain the peak-splitting component are easily recognized in the spectrum as goalpost-like peaks separated by the mass difference between the natural and isotopically labeled components. The peak-splitting element provides a valuable means for recognition of the compounds synthesized on the solid phase.

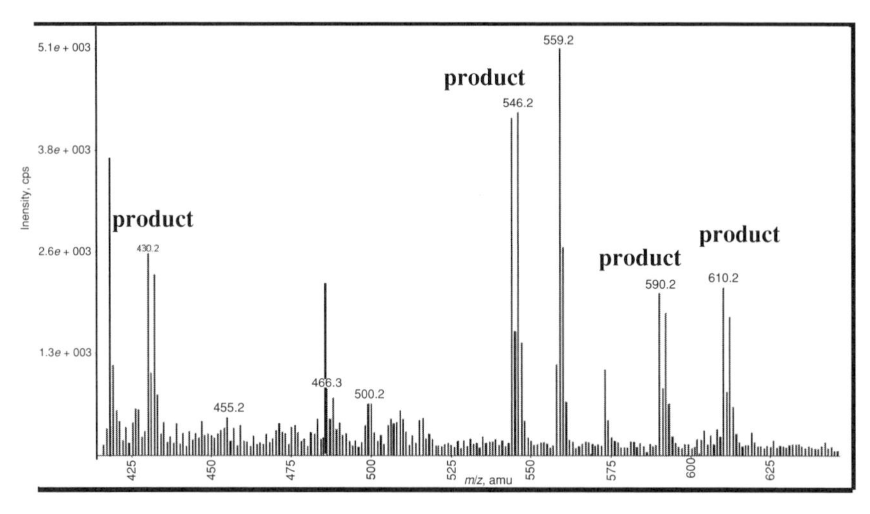

FIGURE 8.7 A mass spectrum that displays the use of the peak-split signature. All products that originate from the resin bead demonstrate the goalpost-like signature peaks, while the normal single peaks are not related to the synthetic chemistry on the solid phase.

One simple method used to incorporate the peak-splitting component for solid-phase synthesis is with the use of a linker that transfers at least one atom to the cleavage product. However, it was experimentally determined that the most facile method for library production was to use a separate component in the analytical construct to produce the bromine effect. Investigations determined that a 3 amu peak split was optimal for multiply charged ions and the subsequent automation of the analysis with software. An equal-molar mixture of G0 and G3 is used to generate the mass signature of the synthetic compounds. Figure 8.7 shows the mass spectrum from a reaction with peak splitting. All compounds that are related to the chemistry display the goalpost-like signature-split peaks, while the single-peak components are due to solvent or impurities in the sample.

Mass-Spectrometric Ionization Leveler

Mass spectrometry is sensitive, rugged, and has the ability to provide extensive qualitative and quantitative information. Mass spectrometry also provides a large dynamic range, can be automated, and has a fast analysis time. These characteristics have made mass spectrometry a very powerful and attractive analysis tool for medicinal chemistry and combinatorial chemistry. However, not all synthetic organic compounds are readily detected by mass spectrometry, and the sensitivity can vary with the mode of ionization. As a result, sensitivity is dependent on the physical characteristics of the compound.

To assist with the ionization of synthetic compounds, the concept of the attachment of an easily charged moiety to the library compound was introduced into the analytical construct [27]. This concept also enabled the consistent use of electrospray-ionization mass spectrometry as the ionization mode for library quality control (QC). The standard analytical construct incorporates the amino acid lysine that has a free amine on the amino acid's side chain. The lysine's side chain amine is protected with Boc throughout the synthesis. During acid–chemical cleavage, the protecting group on the side chain of the lysine is removed and a free amine remains. This process provides a basic moiety that is covalently attached to the molecules subjected to analysis by mass spectrometry and ensures a strong mass-spectral signal for all synthetic compounds cleaved from the resin. It is important to note that the basic moiety is for QC purposes only and does not proceed to biological assay. The intent of the ionization leveler is to normalize the mass-spectral response across all the compounds in the library. The ionization leveler does not ensure that all compounds have the same ionization efficiencies. However, this assumption is a good approximation for compounds within a given library that have an identical core structure.

Orthogonal Linker System

The general design features of the analytical construct for library production and reaction screening include three cleavable linkers: a code region, ionization leveler; mass-signature element; and the ligand. During library QC, the first and second cleavable linkers from the resin are released under identical reaction conditions while the terminal linker is released by an orthogonal condition and used for the preparation of library compounds for biological assay.

Two orthogonal cleavage protocols can be performed on the analytical construct, as illustrated in Figure 8.8. Chemical cleavage, used for QC, is performed by treatment of a single bead with the appropriate acidic conditions to promote cleavage of the first and second Knorr linkers that generate the code block and the ligand block in solution. The code block consists of the mass code that serves to record the chemical history and also serves as an internal standard for ligand yield measurement. The ligand block consists of the mass signature, the mass spectrometry ionization leveler, and the ligand region. The second possible cleavage is photocleavage. Upon exposure to 350-nM light, only the ligand is released. The ligand is suitable for biological assay while the code remains on the bead, where it can be used after obtaining positive assay results are obtained.

LIBRARY QUALITY CONTROL

A significant problem with the use of biological data obtained on compounds with minimal analytical characterization is the interpretation of assay results.

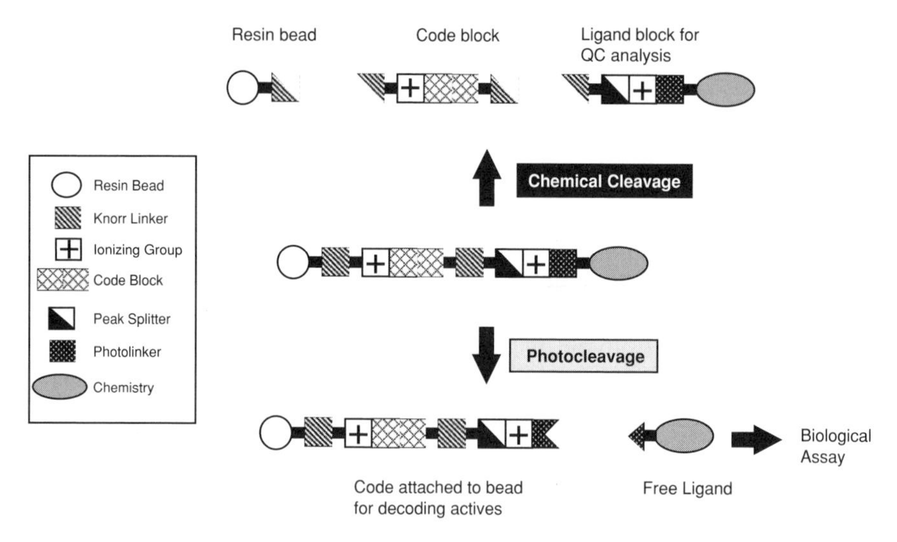

FIGURE 8.8 The two possible cleavage mechanisms for the orthogonal linker system. The top mechanism illustrates the acid–chemical cleavage process used for quality control of library production. The bottom mechanism illustrates the photocleavage process used to release the synthetic compound for biological assay. (QC = quality control.)

Without analytical proof of the presence of the designed compound, it is dangerous to assume that the compound is inactive, as it may not have been present in the assay at all. Also, it is risky to assume that the designed compound is active without verification, that it, was efficiently synthesized, as side products have been known to cause biological activity. Uncertainty in the compound synthesis leads to the inability to fully exploit biological screening data. This is especially true in the case of the negative assay results. Conventionally active compounds identified by bead-based libraries require resynthesis, purification, and rescreening efforts to confirm the activity of the positive results. In many cases, the purified compounds are not active in the rescreening process and may be indicative of either a false-positive or that a low-level impurity is the active compound.

To obtain a valuable SAR and use the biological data to its full potential without additional synthesis or screening effort, each library must be sufficiently characterized. Rapid quantitation and identification of all the products that result from chemistry via resin-based solid-phase methodologies can be achieved with analytical constructs. This process uses flow-injection mass-spectral analysis (FIA-MS) with a single quadrupole mass spectrometer to generate a throughput of 20–30 samples per hour for quality assessment/quality control (QA/QC) of library production. The result is the preparation of high-quality, analytically characterized bead-based libraries, from which the synthetic failures have been removed prior to biological assay.

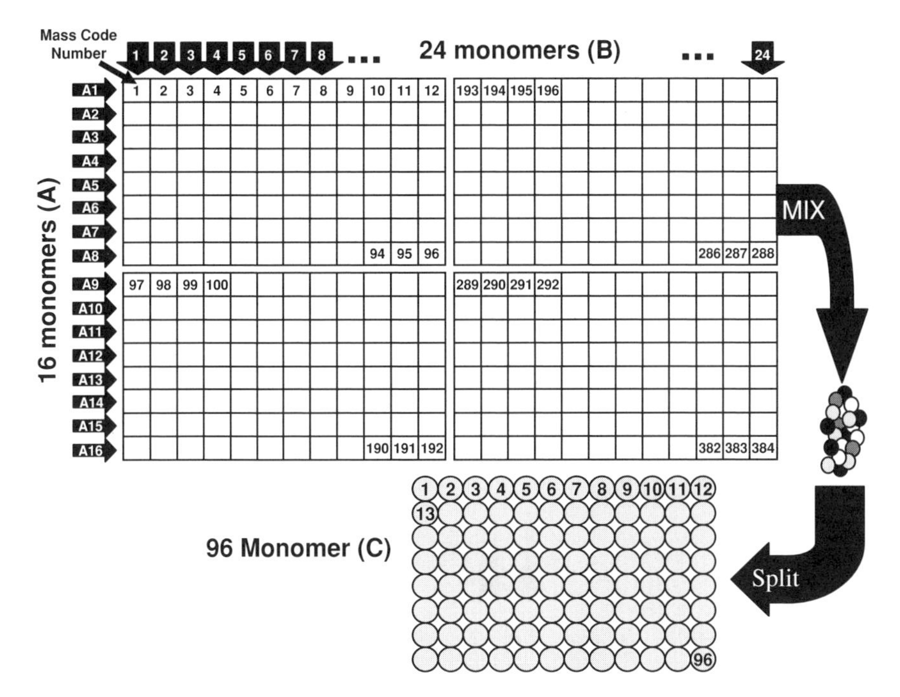

FIGURE 8.9 An outline of the standard library production scheme with four 96-well reaction blocks. Each of the 384 wells is preloaded with beads with a unique mass-encoded analytical constructed resin. Normally 16 monomer As are added across the rows and 24 monomer Bs down the columns. After a mix and split procedure, 96 monomer Cs are reacted in individual wells to generate 36,864 compounds.

Diversity Sciences developed a library synthesis strategy that combines the simplicity of parallel synthesis and the power of resin-mixing techniques. The general format is four 96-well plates that give rise to 384 synthetic wells, as shown in Figure 8.9. The layout of the synthesis blocks enables 16 unique monomers in monomer position A (across rows) and 24 unique monomers in monomer position B (down the columns). All of the 384 wells are preloaded with off-the-shelf resin where each well has a unique binary code embedded in the analytical construct. The first two points of diversity (monomer A and monomer B) is added in all possible combinations by parallel synthesis. Each spatial location has a unique binary-mass code that encodes for a particular combination of monomer A and monomer B. For example, binary code number 8 represents monomer A1 and monomer B8. After the addition of monomer B, the resin from all 384 wells is mixed together and split into 96 identical pools, to which monomer C is added. The third monomer, monomer C, is spatially encoded, since the 96 pools are not mixed after the last step and screened as pools. Upon decoding, the identification of the binary code reveals the combination of monomer A and monomer B on each bead.

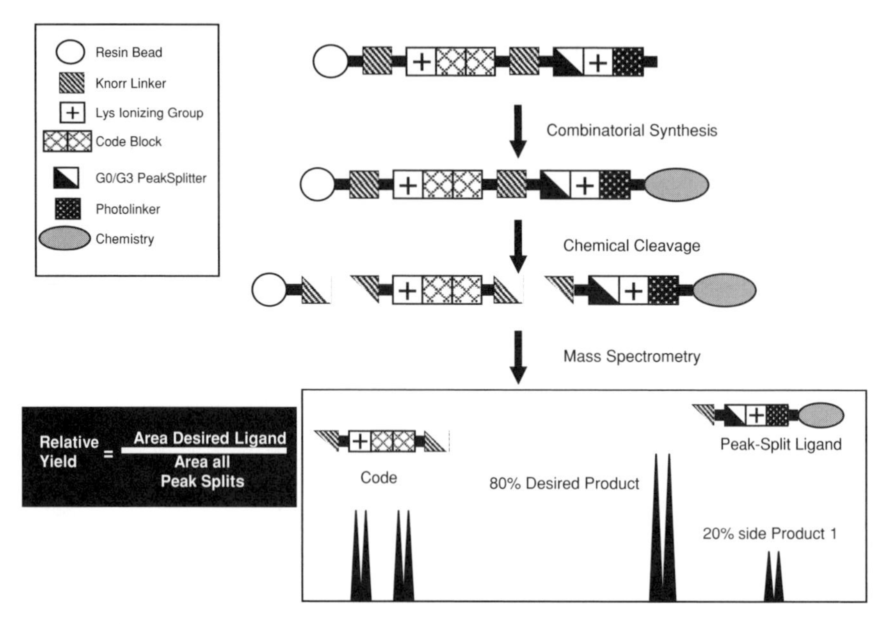

FIGURE 8.10 An illustration of QC process during combinatorial synthesis. An individual bead undergoes acid–chemical cleavage and the ensuing compounds are analyzed by mass spectrometry. The mass spectrum displays the presence of the mass code, the desired product, and side products. The desired product yield is calculated by the ratio of the desired product peak-split peak area with the total peak-split area from all identified signature peaks. The mass-code peaks are used to give a quality assessment of the of the acid–chemical cleavage process.

Mass-spectral data are used to ascertain the synthetic yields after each synthetic step of the library production. The total area of the peaks in a spectrum related to the desired peak-split product is compared to the total area of the peaks from all mass-signature–identified side products (peak-split peaks) in the spectrum, as illustrated in Figure 8.10. This ratio of the peak area of the desired product to the peak area of all peak-split products is defined as the relative yield. This information is used to make critical synthetic decisions and identify specific wells to take forward in the library synthesis.

Library production starts with the addition of the first position of diversity (monomer A). Four beads from each of the 384 wells are used for the QC assessment to ensure homogeneity in the coupling reaction across all beads in a given well and identify any automated synthesizer errors in the delivery of reagents. In cases where specific monomers do not couple properly to the resin or fail the QC analysis, these monomers are eliminated from the rest of the library. After completion of the second step (monomer B), there is maximally 384 unique compounds spatially and mass encoded in the reaction plates. Again, four beads from each well are used in the QC process.

At this point, all 384 wells (i.e., 384 compounds) have been analytically characterized by mass spectrometry. A narrow yield distribution is one of the criteria in the decision as to which wells to move forward in the synthesis to produce a tight concentration distribution of the ligands for biological assay. A tight-ligand concentration range will help facilitate the interpretation of biological data. Typically 25% of the library is removed at this point in the library synthesis based on relative yield calculations.

This procedure marks the end of the spatial component of the library synthesis. The resin beads in the wells identified to proceed in the library synthesis are mixed and then split out into a 96-well reaction block. All 96 reaction wells are indistinguishable and consist of compounds that have all possible A–B combinations. Monomer set C typically consists of 96 unique monomers, where one unique monomer C is coupled in each well for the third point of diversity. QC is conducted after completion of the final synthesis step by the selection of a minimum of 12 beads from each well. The sampling rate does not permit the calculation of relative synthetic yields for all compounds in the library; however, a global assessment on synthesis for each monomer C is produced. A narrow bandwidth of mass spectral–relative yields of the final product is selected for assay and assures a tight-ligand concentration band. Those monomers that fail after the last synthetic step are not forwarded to biological assays. This synthetic scheme can produce a 36,864 compound library that is characterized by approximately 4200 mass-spectral data points. Normally, the final library size is between 20 and 30K after removal of the identified synthetic failures during the QC process.

This QC procedure allows the first two steps of the synthesis to be completely analyzed for QC. The third step is statistically analyzed to give a good estimation of quality. Only quality compounds with similar concentrations are sent to biological assay with this library synthesis and QC method. This procedure eliminates the problem of interpretation of the results without analytical evidence of the compound, and therefore, provides the ability to generate a more reliable SAR.

AUTOMATED DATA MINING SOFTWARE

A complete library QC or reaction optimization requires approximately 4000 single-bead mass spectra. The time and effort required to manually interpret data from a given library synthesis or reaction screen would be enormous. To make efficient use of the collected data, the interpretation must be automated as well. In addition, several calculations are required to prepare the library for biological screening. Diversity Sciences wrote several software programs to aid in the process. A brief description of a few of these programs follows.

CAPTURE Analysis Program

An analysis program (CAPTURE) was developed to be a mass-spectral browser and to automate mass-spectral characterization for a large number of samples [30–33]. This analytical application routinely calculates the yield and purity of products from a given reaction or combinatorial synthesis with mass-spectral data. The CAPTURE program is versatile and can use FIA-MS or liquid chromatography mass spectrometry (LC-MS) data for qualitative analysis and FIA-MS, LC-MS, ultraviolet (UV), or chemiluminescent nitrogen-specific detector (CLND) data for quantitative analysis. In addition, the program identifies all side products (known and unknown) from signature peaks in the mass spectrum and calculates synthetic yields. The use of CAPTURE has reduced the time for data interpretation from days (manual inspection) to less than a minute.

Daily Standardization

Since the peak areas in the mass spectra are used for quantitation, the mass spectrometers must produce reliable and reproducible data to calculate meaningful product yields and allow a comparison of data collected on multiple mass spectrometers and collected at different times. Therefore, each mass spectrometer undergoes a comprehensive daily standardization and calibration, which has been automated by in-house software.

Determining Side Transformations-Delta

During the library QC process, it is desirable to ascertain all synthetic compounds cleaved from the resin bead. Identification of side products facilitates the understanding of chemistry, helps determine product yields for a particular library, and lends the possibility of discovering novel chemistries for subsequent libraries. A computer program, DELTA, was written to facilitate the interpretation of side-product characterization and determine their yields using mass-spectral data. The DELTA program uses the output from CAPTURE and associates unexpected and unknown chemistries with the desired products and the monomers used in the library synthesis. This program facilitates the chemist's understanding and ability to identify the side transformations. Ordinarily, a chemist would spend one month identifying side reactions and finishing the QC analysis, and in most cases, the analysis would not be thorough. This process provides a complete synthetic pathway related to specific monomers with the ability to interpret the results in a matter of one day.

Dilution for Biological Assay

The amount of compound cleaved from a 160-μm single bead (~1.5 nmol) is sufficient for a number of different high throughput screens. The purpose of this program is to aid the biologist in the determination of the library dilution for their particular assay. While beads are purchased with highly uniform bead size for synthesis, there remains a distribution of load on any given batch of resin beads. There is also a distribution related to the yield of the library compounds, because organic chemistry rarely produces 100% yields. Lastly, there is a distribution related to the photocleavage of the products from the resin beads, as this step is rarely done with 100% fidelity. An understanding of these three distributions (bead load, synthetic yield, and photocleavage yield) permits a calculation for the amount of the cleaved material available for the biological screen. A fraction of the solution that is dependent on the concentration requirements for the biological assay is used for a given screen. A software program was written to automate this calculation for each combinatorial library and or reaction screen.

BIOLOGICAL SCREENING OF COMBINATORIAL LIBRARIES

Solution-Phase Assay

Encoded bead-based libraries represent a unique problem for assay when the separation of the ligand from the corresponding resin bead and code is required, as is the case for a solution-phase assay. The overall procedure can be broken down into three discrete steps. First is the distribution of the beads, cleavage of the compounds from the beads, and separation of the solution that contains the ligand from the bead but retains the association (should decoding be required). Second is the distribution of the ligand solution into the requisite number of subplates and completion of the assay. The final step involves thorough analysis of the biological assay results and the identification of the active compound(s).

Screening the library as discrete compounds would provide unambiguous activity for each compound; however, the time and cost of this exercise is too enormous to be feasible. Pooling strategies have been implemented to reduce assay load and make the screening of numerically large libraries and large compound sets more practical. Pooling requires multiple compounds, in our case multiple beads, to be placed in a single well and assayed as a mixture or pool. This process is used by Diversity Sciences and is highlighted in Figure 8.11. The Diversity Sciences process typically pooled five beads from the reaction block in each well for assay. The synthetic compounds are photocleaved from the bead and subsequently separated from the beads. At this point there are two sets of plates that correspond to the beads and the cleaved synthetic compound in solution where the association between the beads and solution must be maintained for decoding. A fraction of the ligand solution will be directly used for screening in biological assays. The second set of plates

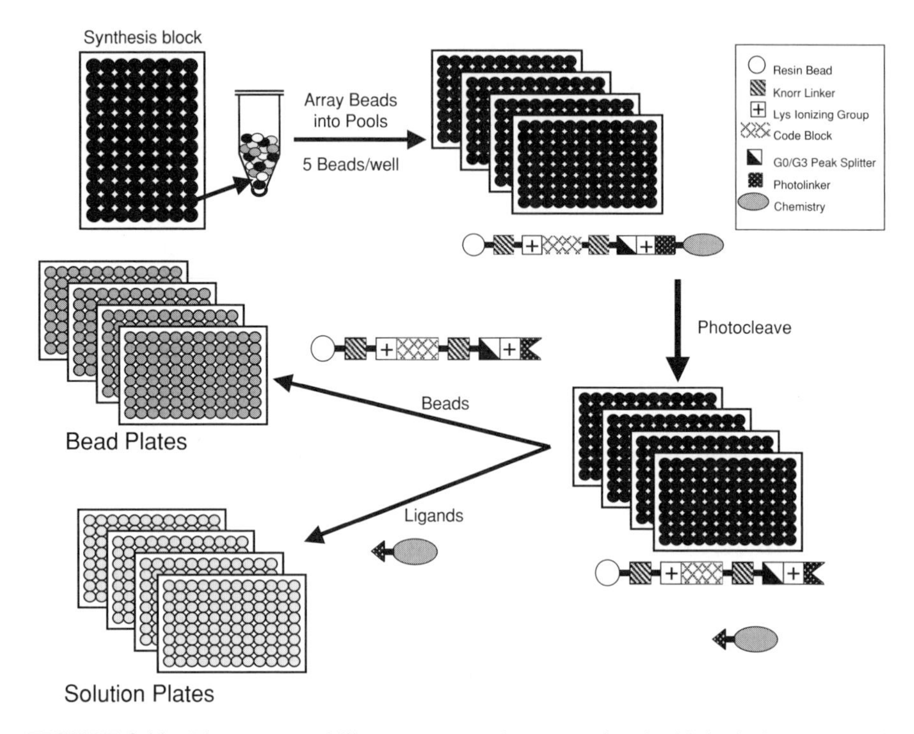

FIGURE 8.11 The process of library compounds preparation for biological assay. Beads from the synthetic reaction block are arrayed into cleavage plates (5 beads/well). The library compounds are photocleaved from the resin beads, and then compound solution is separated from the beads, where the relationship between the resin beads and the cleaved compound is preserved. The compound's solution can then be taken on to biological assay.

that hold the beads is used to determine the specific compounds associated with the high assay values, as shown in Figure 8.12. Each well of the biological assay represents a specific set of five beads. After analysis of the assay results, the beads that correspond to active wells are identified and redistributed as single beads. The individual beads are chemically cleaved and the code is determined by mass spectrometry to reveal the identity of the compounds.

Melanaphore Cell Lawn Assay

Xenopus laevis (African Clawed frogs) melanophores have a dark-brown pigment, melanin, which is contained in intracellular-membrane–bound organelles called *melanosomes*. The pigment can either be isolated in a single location near the nucleus or dispersed uniformly throughout an individual cell. Stimulating receptors that increase levels of cyclic adenylic acid (cAMP) generate a dispersion of melanosome or darkens the frog, while stimulating receptors that decreases cAMP levels leads to melanosome aggregation

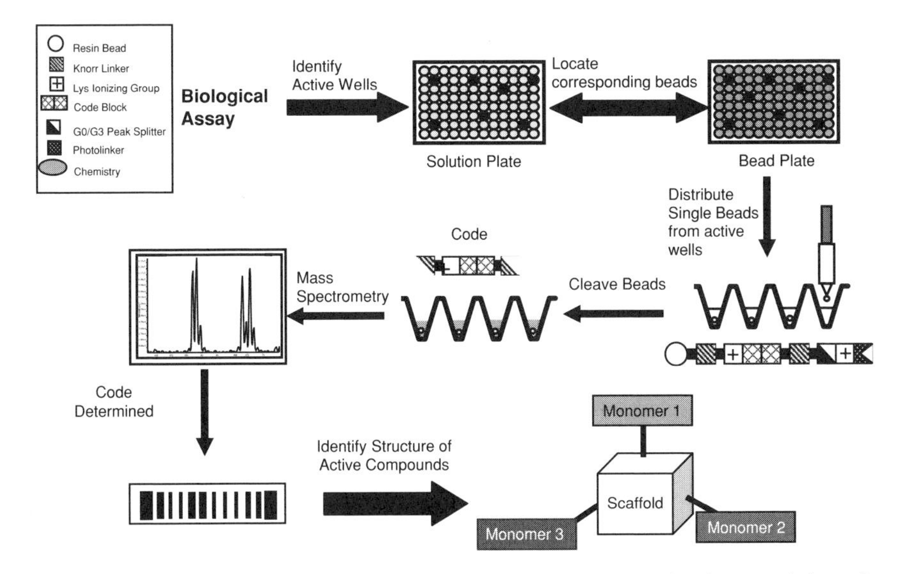

FIGURE 8.12 The decoding process from the biological assay results that reveal the active wells from the solution plates. The beads that correspond to these wells are identified in the bead plates. The five beads in each active well are distributed as single beads and the code on each bead is acid–chemically cleaved and analyzed by mass spectrometry. Each code is determined and then identified with the appropriate compound structure.

or lightens the frog. Recombinant melanophore receptor assay relies on the melanosomes' ability to migrate upon external stimuli, where pigment translocation in melanophores can be detected within a few minutes following the activation [34].

Typically the melanophore technology is performed in a lawn format, which is attractive for the determination of responses that arise from individual beads [35–37]. Melanophores are plated as a lawn on a petri dish, as shown in Figure 8.13. The library beads from individual pools are spread across the lawn of cells transfected with various G-proteins [38]. Release of the synthetic compound is accomplished by irradiation of a light source for a few minutes. A partial cleavage of the photolinker releases a small amount of the synthetic compound on each bead. The compound migrates to the cell lawn for biological analysis. Partial cleavage is acceptable for this assay because the ligand is concentrated in a small area and is in direct contact with the receptors. A change in color of the cell lawn around individual beads indicates activity. The beads that indicate active compounds are isolated as single beads and undergo acid cleavage for decoding. The code identified by mass spectrometry reveals the synthetic history of the bead, and, thus, the active compounds. This approach is similar to the screening of discrete compounds, since the beads are isolated on the cell lawn and each bead produces its own assay result.

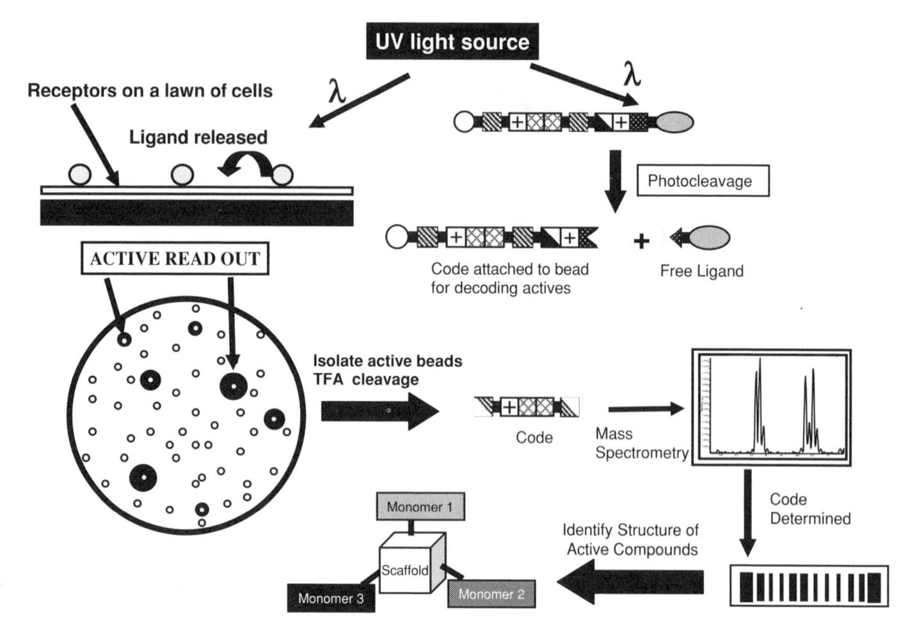

FIGURE 8.13 A representation of the melanophore lawn assay. A lawn of receptors is grown on a petri dish. Library beads are spread across the surface of the assay plate. The compounds interact with receptors on the cell lawn after photocleavage. Active compounds generate a color change on the cell lawn. The active beads are identified, isolated as single beads, and the code on each bead is chemically cleaved and analyzed by mass spectrometry. Each code is determined and associated with the compound's structure.

STATISTICAL DECODING METHODOLOGY AND INTERPRETATION OF BIOLOGICAL ASSAY RESULTS

A statistical approach essentially eliminates the false-positive problem that occurs in all biological assays. A strategy that applies statistics to the decoding process is used to establish the active compounds from the pooled beads [5,10]. The beads are arrayed from the reaction block to the assay format in a random fashion. The total number of beads selected from each final pool was sufficient to have an average of six copies of each compound to be present in a solution assay. Lawn assays used an average of three copies of each compound. Statistical decoding relies on repetition of the mass codes to indicate activity. The results from biological assay are compared to that of random selection of wells from a pool that employs a software program. The frequency of repeating codes from the assay will differ from the random case when biological activity is due to a compound in the library. Since there is an average of six copies of each compound in the solution assay, then an active compound is expected to be present in six of the higher biologically active wells. The decoding of all the beads associated with a positive assay result

(~five beads for each well) generates a table of codes whereby the common code equates to the active compound and occurs with an average frequency of six (number of times a compound is in the assay). This approach permits the observed frequency of individual compounds in the various wells to identify the biologically active component.

Because the beads are random in the reaction well when the beads are dispensed for biological assay, there is a distribution of the number of copies of each compound that is present in the assay. To obtain a code with a frequency of greater than three is a clear indication that the corresponding compound is active in the assay, as this frequency is significantly higher than could be accounted for by chance alone. In this way, false-positives, either associated with the assay itself or from some unknown factor that occurs during the synthesis or the cleavage procedure, have been eliminated.

Since the beads are dispensed in the plates in a random fashion, the activity in the assay plate should display a random pattern. The active wells from the assay are verified to be randomly distributed with in-house software. This should always be true, except for the case where the activity is associated with monomer C. In this case, every compound in a pool contains the active monomer and the entire plate will be active.

The data and results obtained from the sequential processing routine are illustrated in Figure 8.14. The solution plates of library compounds are screened by biologists. The data are processed through a series of algorithms to determine the wells that contain a probable lead or a "structure-activity-relationship" upon decoding. The library pools were rank ordered with a scoring function, (Figure 8.14A) that takes into account several factors from the biological screen data. The higher the scoring function, the more likely the pool contains an active compound. One pool stood out from the rest, and the individual assay values from that pool were again ranked to show 11 high-activity values significantly above the remaining wells (Figure 8.14B). A total of 58 beads from the corresponding wells were arrayed as singles beads, chemically cleaved, and the codes read by mass spectrometry. The data for these eleven wells were tabulated and compared to that obtained at random by Monte Carlo prediction of an average number of repeating codes (MC (avg)) and a standard deviation (MC (std)) by the random selection of 11 wells from the assay. The assay results show that code 45 is present in all eleven wells and that the probability of 11 codes repeating purely by chance is zero (Figure 8.14B). This compound was resynthesized in a 10-mg scale, high-performance liquid-chromatographic (HPLC) purified, and biologically rescreened to verify the activity. This strategy leads to the identification of an active compound without the task of resynthesis of all compounds in the active wells or rescreening discrete beads of the active pool. The active pool can be identified without the requirement of resynthesis and rescreening efforts, because the statistical decoding methodology ensures activity.

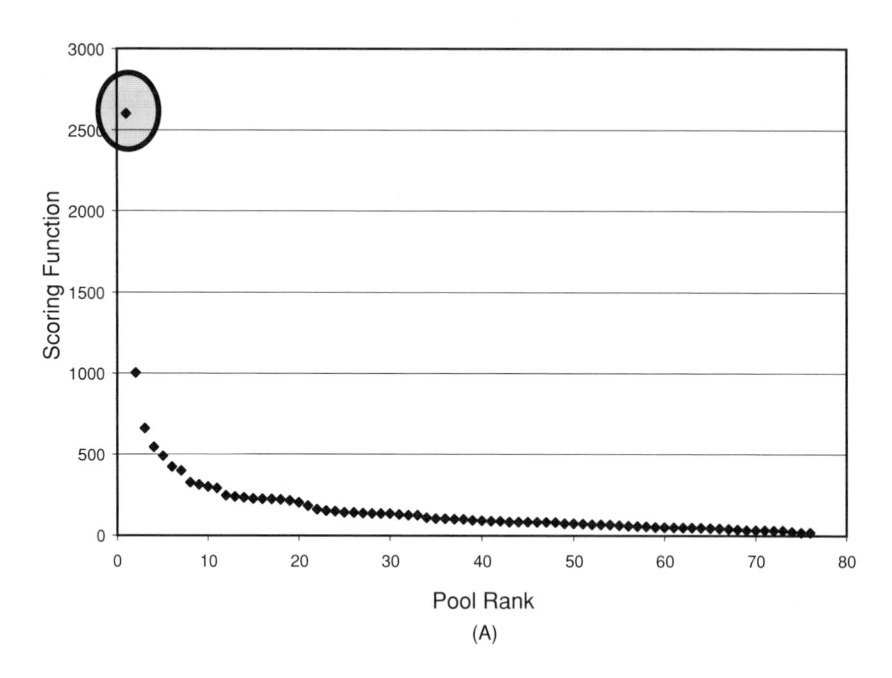

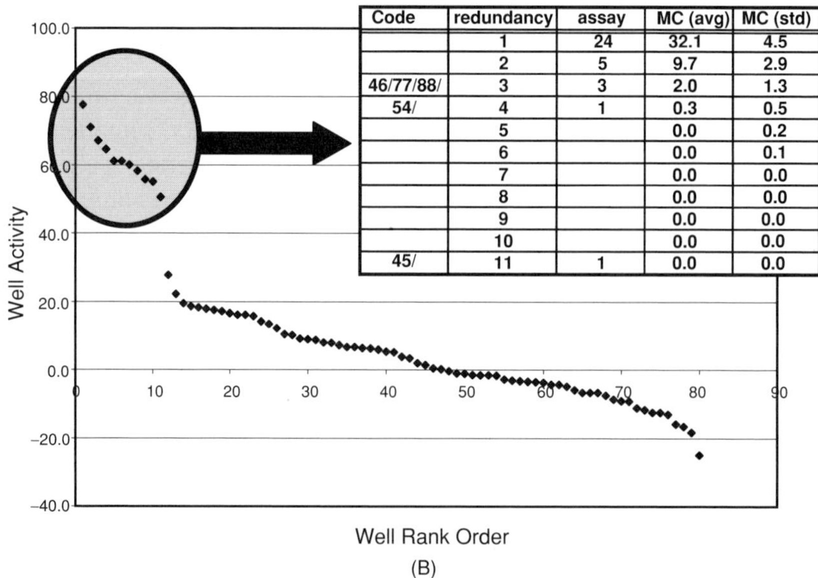

Code	redundancy	assay	MC (avg)	MC (std)
	1	24	32.1	4.5
	2	5	9.7	2.9
46/77/88/	3	3	2.0	1.3
54/	4	1	0.3	0.5
	5		0.0	0.2
	6		0.0	0.1
	7		0.0	0.0
	8		0.0	0.0
	9		0.0	0.0
	10		0.0	0.0
45/	11	1	0.0	0.0

FIGURE 8.14 (A) A demonstration of the statistical decoding process for solution-phase assays. Each library pool is rank ordered by an in-house scoring function that predicts the probability of discovering an active compound by decoding the active wells in the pool. In this example, the score of one pool stands well above the rest of the library pools. (B) The biological activities from the wells of the highly scored pool are rank ordered. The presence of two activity distributions would indicate a good probability of finding an active compound. Here the higher active distribution consists of eleven wells. Upon decoding of the beads from the eleven highly active wells reveals that all eleven wells have one compound in common. This observation is well outside the possibility of happening by pure chance as calculated by a Monte Carlo program. MC avg. refers to the average number of occurrences by a Monte Carlo program. MC Std. refers to the standard deviation of each MC avg.

Structure Activity Relationship

To obtain SAR for monomer A and monomer B, the decoding information is organized to depict the number of times a particular monomer was present in the active wells. The decode information is organized to resemble the synthesis format. A record is made of each observation of a particular monomer A and monomer B combination. An example of the data obtained from a lawn assay is shown in Figure 8.15. The synthesis of the library shown in Figure 8.15 used an 8 × 48 × 96 matrix, where 8 unique monomers were used at the first step, 48 monomers at the second step, and 96 unique monomers used on the third and final synthetic coupling. The two-dimensional representation depicted is 8 × 48, and is associated with the first two points of diversity. The 8 monomer As are depicted across the rows, and the 48 monomer Bs are represented down the columns. The frequency of monomer combinations determined by decoding are shown in the boxes. For example, the combination of monomer A3 and monomer B2 occurred nine times during the decoding process, whereas the combination of monomer A3 and monomer B39 comprises 21 decodes that occurred from all the library pools that contained this particular monomer A and monomer B.

SAR for a particular library screen is illustrated across the columns and down the rows that incorporate the high-frequency occurrences. In this example, there are three columns that have a high frequency of a particular

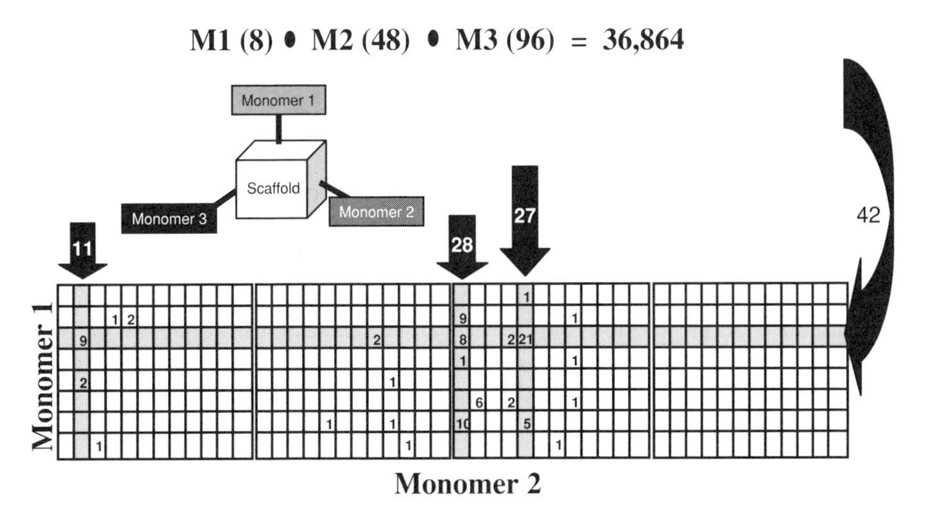

FIGURE 8.15 A graphic representation of the process used to determine SAR from biological results. This example uses the results from a melanophore assay of a library with 8 monomer As, 48 monomer Bs, and 96 monomer Cs. A high number of compounds that are decoded with a particular monomer A (across the rows) or monomer B (down the columns) indicates SAR for those monomers.

monomer B; these are monomer B2 with 11, monomer B25 with 28, and monomer B29 with 27. On the other axis only one row is associated with a high frequency, thus only one monomer A (monomer A3) displays any SAR information against this biological target. This example shows some evidence for SAR with these two points of diversity. SAR data for C has not been shown; however, similar graphs are used to obtain this information.

Reaction Screening/ Reaction Optimization

It is generally accepted that the translation of solution-phase reactions to solid-phase procedures for the production of compound libraries is usually time-consuming and not always successful. The most robust solid-phase synthesis instrumentation has operational temperature and solvent compatibility limitations that impinge on the direct translation of solution-phase chemistries to the solid phase. These instrument restrictions necessitate a considerable investment in experimentation to redefine the solution-phase reaction protocol for the solid phase. In spite of this apparent obstacle, a substantial number of reactions for solid-phase library production has been published. Many of these reactions cannot be exploited because the overall product yields are too low (which could lead to ambiguity in the interpretation of the eventual assay results). Also, some reactions have a restricted range of potential monomers that lead to a limited diversity of the compounds in the resulting library.

Analogous to a combinatorial library that probes all possible combinations of the monomers in the chemical protocol, the rapid optimization or discovery of the chemical protocol itself can be carried out in a combinatorial fashion. The dimensions of the combinatorial scheme are reaction variables such as solvent, temperature, time, catalyst, reagent, and the physical properties of possible monomers. To employ the analytical-construct methodology, a combinatorial exploration of a large number of the possible reaction variables becomes a relatively simple exercise to perform and rapidly leads to an optimized chemistry that is suitable for the generation of a compound library. A typical approach for the rapid exploration of the many permutations of the parameters that affect the outcome of a candidate reaction sequence is illustrated in Figure 8.16. This example illustrates the optimization of the reaction of monomer type A with monomer type B to give a product. The process evaluates six different solvents, five different reagents, five different physical characteristics for monomer A, and five different physical characteristics for monomer B, for a total of 750 unique reactions ($5 \times 5 \times 5 \times 6 = 750$). The coding prerequisites for this process are appreciably more modest than those required for encoding a chemical library. In this case, only five different codes are necessary. At the conclusion of the reaction, resin beads are sampled from each well and analyzed by mass spectrometry after chemical cleavage of the

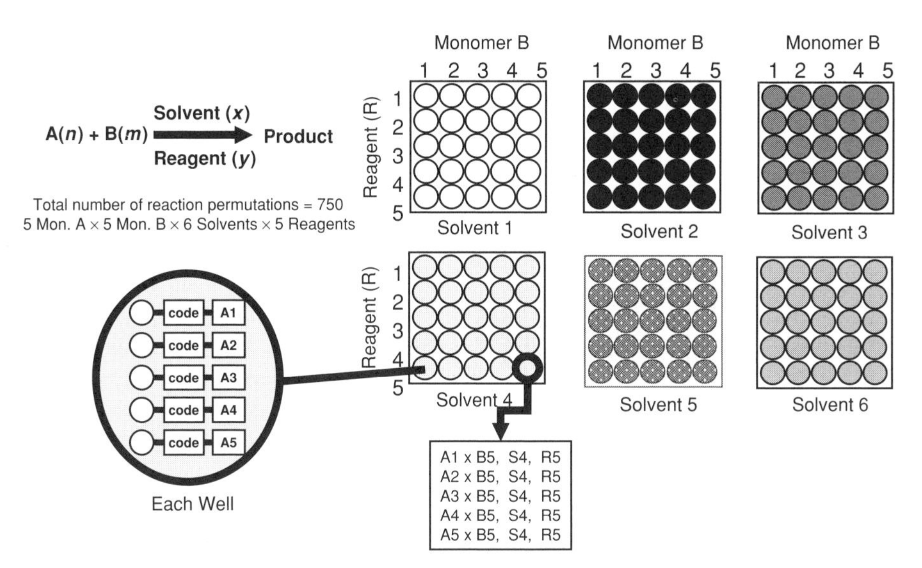

FIGURE 8.16 An illustration of a reaction optimization scheme as a function of monomer A, monomer B, solvent, and reagent.

construct. Sampling the beads from the well highlighted would give the reaction outcome for all variations of monomer A with monomer B5 in solvent 4, and with reagent R5, after a specified time and temperature.

CONCLUSION

Combinatorial chemistry offers the ability to generate large numbers of diverse compounds. Solid-phase chemistry has the advantage to drive reactions to completion without a sacrifice in purification. Most, if not all, of the negative aspects associated with solid-phase synthesis have been overcome. Technological advances for the synthesis of compounds now permit numerically large, quality libraries, on a functional scale. Encoding strategies have been developed that eliminate the uncertainty obtained in the biological assay results. A novel method for rapid optimization of chemical synthesis, as well as discovery of novel synthetic routes to chemical classes, has been demonstrated. The sheer size of small-molecule space is so enormous it requires a systematic search of this diverse space for future drug molecules as opposed to restricting synthesis around a current hit or lead or compound in a company's store. The notion that a few hundred compounds are adequate to determine whether a compound class is active against a target is risky and has not been supported by experiment. The general translation of classic solution-phase organic chemistry to the solid-phase generates unique challenges;

however, the fundamental advantages obtained by numerically large–diverse compound libraries makes solid-phase library production necessary for drug discovery.

ACKNOWLEDGMENT

The authors thank the members of Diversity Sciences for their hard work and dedication during the development of the technologies.

REFERENCES

1. Ohlmeyer, M.H.J.; Swanson, R.N.; Dillard, L.W.; Reader, J.C.; Asouline, G.; Kobayashi, R.; Wigler, M.; Still, W.C. "Complex Synthetic Chemical Libraries Indexed with Molecular Tags," *Proc. Natl. Acad. Sci. USA* **90,** 10922–10926 (1993).
2. Ni, Z-J; Maclean, D.; Holmes, C.P.; Murphy, M.M.; Ruhland, B.; Jacobs, J.W.; Gordon, E.M.; Gallop, M.A. "Versatile Approach to Encoding Combinatorial Organic Synthesis Using Chemically Robust Secondary Amine Tags," *J. Med. Chem.* **39,** 1601–1608 (1996).
3. Geysen, H.M.; Wagner, C.D.; Bodnar, W.M.; Markworth, C.J.; Parke, G.J.; Schoenen, F.J.; Wagner, D.S.; Kinder, D.S. "Isotope or Mass Encoding of Combinatorial Libraries," *Chem. Biol.* **3,** 679–688 (1996).
4. Geysen, H.M.; Kinder, D.S.; Wagner, C.D. "Mass-based Encoding and Qualitative Analysis of Combinatorial Libraries," PCT Int. Appl. WO 9737953 (1997).
5. Geysen, H.M.; Schoenen F.; Wagner, D.; Wagner, R. "Combinatorial Compound Libraries for Drug Discovery: An Ongoing Challenge," *Nature Rev. Drug Disc.* **2,** 222–230 (2003).
6. Hermkens, P.H.H.; Ottenheijm, H.C.J.; Rees, D. "Solid-phase Organic Reactions: A Rreview of the Recent Literature," *Tetrahedron* **52,** 4527–4554 (1996).
7. Hermkens, P.H.H.; Ottenheijm, H.C.J.; Rees, D. "Solid-phase Organic Reactions II: A Review of the Literature November 95–November 96," *Tetrahedron* **53,** 5643–5678 (1997).
8. Booth, S.; Hermkens, P.H.H.; Ottenheijm, H.C.J.; Rees, D. "Solid-phase Organic Reactions III: A Review of the Literature November 96," *Tetrahedron* **54,** 15385–15443 (1998).
9. Wagner, D.S.; Schoenen, F.J.; Geysen, H.M. "Mass Encoding and Statistical Decoding of Combinatorial Libraries," *Amer. Pharm. Rev.* **2,** 60–67 (1999).
10. Wagner, D.S.; Schoenen, F.J.; Wagner, C.D.; Wild, J.J.; Geysen, H.M. "Mass Encoding and Quality Control of Combinatorial Libraries," the 50th Annual Pittsburgh Conference Abstracts, pp. 1024, Orlando, FL, (1999).

11. Wagner, D.S.; Schoenen, F.J.; Shampine, L.J.; Wagner, C.D.; Geysen, H.M. "Decoding of Combinatorial Libraries," in *Proceeding of the 47th Conference on Mass Spectrometry and Allied Topics*, Dallas, TX (1999).
12. Holmes, C.P. "Model Studies for New o-Nitrobenzyl Photolabile Linkers: Substituent Effects on the Rates of Photochemical Cleavage," *J. Org. Chem.* **62,** 2370–2380 (1997).
13. Holmes, C.P.; Jones, D.G. "Reagents for Combinatorial Organic Synthesis: Development of a New o-Nitrobenzyl Photolabile Linker for Solid Phase Synthesis," *J. Org. Chem.* **60,** 2318–2319 (1995).
14. Brown, B.B.; Wagner, D.S.; Geysen, H.M. "A Single-bead Decode Strategy Using Electrospray Ionization Mass Spectrometry and a New Photolabile Linker: 3-Amino-3-(2-Nitrophenyl)Propionic Acid," *Mol. Divers.* **1,** 4–12 (1995).
15. Wagner, D.S.; Brown, B.B.; Geysen, H.M. "A Single Bead Decode Strategy Using Electrospray Ionization Mass Spectrometry and a New Photolabile Linker: 3-Amino-3-(2-Nitrophenyl)Propionic Acid," American Society for Mass Spectrometry Annual meeting at Atlanta, pp. 488 (1995).
16. Murray, P.J.; Kay, C.; Scicinski, J.J.; McKeown, S.C.; Watson, S.P.; Carr, R.A.E. "Rapid Reaction Scanning of Solid Phase Chemistry Using Resins Incorporating Analytical Constructs," *Tetrahedron Lett.* **40,** 5609–5612 (1999).
17. Williams, G.M.; Carr, R.A.E.; Congreve, M.S.; Kay, C.; McKeown, S.C.; Murray, P.J.; Scicinski, J.J.; Watson, S.P. "Analysis of Solid-phase Reactions: Product Identification and Quantification by Use of UV-Chromophore-Containing Dual-Linker Analytical Constructs," *Angew. Chem., Intern. Ed. Eng.* **39,** 3293–3296 (2000).
18. Lorthioir, O.; Carr, R.A.E.; Congreve, M.S.; Geysen, M.H.; Kay, C.; Marshall, P.; McKeown, S.C.; Parr, N.J.; Scicinski, J.J.; Watson, S.P. "Single Bead Characterization Using Analytical Constructs: Application to Quality Control of Libraries," *Anal. Chem.* **73,** 963–970 (2001).
19. Kiser, P.F.; Collupy, G.C.; Eichenbaum, G.M.; Rush, C.; Baumann, S.; Kust, M.P.; Greaves, M.D.; Frederick, E.D.; Soltmann, B.; Vasina, H.; Orji, C.; Bienfiat, B.; Brown, J.D.; Davies, J.; Atkins, M.A; Lee, J.R.; Labow, E.; Harp, M.; Goodrich, A.F.; Conyers, H.; Geysen, H.M. "Engineering an experimental platform for high throughput reactivity Screening," *JALA* **6,** 99–106 (2001).
20. Congreve, M.S.; Ley, S.V.; Scicinski, J.J. "Analytical Construct Resins for Analysis of Solid-phase Chemistry," *Chem. Europ. J.* **8,** 1768–1776 (2002).
21. Czarnik, A.W. "Encoding Strategies for Combinatorial Chemistry," *Curr. Opin. Chem. Biol.* **1,** 60–66 (1997).
22. Wagner, D.S.; Bodnar, W.; Schoenen, F.; Geysen, H.M."Automated Mass Spectral Data Analysis for Combinatorial Chemistry," in *Proceedings of the 44th Conference on Massx Spectrometry and Allied Topics*, Portland, Oregon, 1996, pp. 1034.
23. Wagner, D.S.; Geysen, H.M.; Wagner, C.D.; Bodnar, W.M.; Markworth, C.J.; Parke, G.J.; Kinder, D.S.; Schoenen, F.J. "Isotope Encoding of Combinatorial Libraries," in *Proceedings of the 45th Conference on Mass Spectrometry and Allied Topics*, Palm Springs, CA, 1997, pp. 408.

24. Wagner, D.S.; Geysen, H.M.; Markworth, C.J.; Schoenen, F.J. "Mass Encoding of Combinatorial Libraries," Asilmar Conference in Mass Spectrometry, 1997.
25. Wagner, D.S.; Markworth, C.J.; Wagner, C.D.; Schoenen, F.J.; Rewerts, C.E.; Kay, B.K.; Geysen, H.M. "Ratio Encoding Combinatorial Libraries with Stable Isotopes and Their Utility in Pharmaceutical Research," *Comb. Chem. High Throughput Screening* **1,** 143–153 (1998).
26. Schoenen, F.J.; Geysen, H.M.; Bodnar, W.M.; Craddock, L.; Jeganathan, I.; Kinder, D.; Kinsey, K.E.; Markworth, J.A.; Nelson, J.C.; Parke, G.; Patel, H.; Salovich, J.M.; Shampine, L.; Wagner, C.D.; Wagner, D.S.; Wild, J.J.; Wilgus, R.W. "Mass Encoding and the Drug Discovery Process," in *Proceedings of the 46th ASMS Conference on Mass Spectrometry and Allied Topics*, Orlando, FL, 1998, pp. 1139.
27. Wagner, D.S.; Bodnar, W.M.; Shampine, L.J.; Patel, H.; Wagner, C.D.; Wild, J.J.; Geysen, H.M. "Novel Methods for QA/QC of Combinatorial Libraries," in *Proceedings of the 46th Conference on Mass Spectrometry and Allied Topics*, Orlando, Florida, 1998, pp. 1029.
28. Morris, H.R.; Panicao, M.; Barber, M.; Bordoli, R.S.; Garner, G.V.; Gordon, D.B.; Sedgewick, R.D.; Tyler, A.N. "Fast Atom bombardment: a new mass spectrometric method for peptide sequence analysis," *Biochem. Biophys. Res. Com.* **101,** 623–631 (1981).
29. Stultz, J.T. "Peptide Sequencing by Mass Spectrometry, Biomedical Applications of Mass Spectrometry," **34,** 145–194 (1990).
30. Wilgns, R; Geysen, M; Wagner, D; Schoenen, F; Wagner, C; Bodner, W. "Automated Data Interpretation and Analysis: the CAPTUE Program," in *Proceeding of the 45th Conference on Mass Spectrometry and Allied Topics*, Palm Springs, CA, 1997, pp. 1255.
31. Wagner, D.S. "High Throughput Analysis of Combinatorial Libraries," International ISSX Meeting, Hilton Head, SC (1997).
32. Shah, N.; Gao, M.; Tsutsui, K.; Lu, A.; Davis, J.; Scheuerman, R.; Fitch, W.L.; Wilgus, R.L. *J. Comb. Chem.* **2,** 453–460 (2000).
33. Fitch, W.L.; Zhang, J.J.; Shah, N.; Ouchi, G.; Wilgus, R.L.; Muskal, S. "Software for Automating Analysis of Encoded Combinatorial Libraries," *Comb. Chem. High Through. Screen.* **5,** 531–543 (2002).
34. McCLintock, T.S.; Graminski, G.F.; Ptenza, M.N.; Jayawickreme, C.K.; Roby-Shemkovitz, A.; Lerner, M.R. "Functional Expression of Recombinant G Protein Coupled Receptors Monitored by Video Imaging of Pigment Movement in Melanophores," *Anal. Biochem.* **209,** 298–305 (1993).
35. Carrithers, M.D.; Marotti, L.A.; Yoshimura, A.; Lerner, M.R. "A Melanophore-based Screening Assay for Erythropoietin Receptors," *J. Biom. Screen.* **4,** 9–14 (1999).
36. Jayawickreme, C.K.; Sauls, H.; Bolio, N.; Ruan, J.; Moyer, M.; Burkhart, W.; Marron, B.; Rimele, T.; Shaffer, J. "Use of a Cell-based, Lawn Format Assay to Rapidly Screen a 4,442,368 Bead-based Peptide Library," *J. Pharm. Tox. Meth.* **42,** 189–197 (1999).

37. Haizlip, J.E.; Iganr, D.M.; Jayawickreme, C.K.; King, H.K.; Liacos, J.A.; Mills, K.; Ruan, J.J.; Sauls, H.R.; Shaffer, J.E. "High Throughput Method for Screening Candidate Compounds for Biological Activity," PCT Int. Appl. WO 200107 (2001).

38. Chen, G.; Way, J.; Armour, S.; Watson, C.; Queen, K.; Jayawickremem, C.K.; Chen, W.J.; Kenakin, T. "Use of Constitutive G Protein-coupled Receptor Activity for Drug Discovery," *Mol. Pharmacol.* **57,** 125–134 (2000).

9

APPLICATION OF TECHNOLOGICAL ADVANCES IN BIOTRANSFORMATION STUDIES

CARMEN L. FERNÁNDEZ-METZLER, RAJU SUBRAMANIAN, AND RICHARD C. KING

INTRODUCTION

The traditional role of preclinical drug metabolism in pharmaceutical research has been to define absorption, distribution, metabolism, and excretion (ADME) of potential drug compounds for regulatory filings. This role has expanded over the past 5 years to include support of early drug discovery. Consequently, the number of compounds whose ADME characteristics need to be defined has expanded; the work now encompasses thousands of compounds annually rather than the traditional tens of compounds. At the same time, the focus has evolved from merely defining a compound's own fate in the body to include the identification of potential liabilities, such as parameters that limit in vivo pharmacokinetics (PK).

The landscape of pharmaceutical research has changed. There is increased pressure to discover medicines and to get those medicines into patient populations as quickly and inexpensively as possible. This pressure drives efforts to predict ADME properties in humans with human in vitro metabolism data, and correlations between in vitro and in vivo animal data. These same factors, speed and cost, also fuel research to use in vitro data alone to predict in vivo ADME properties in animals. While the exclusive use of in vitro data would

Integrated Strategies for Drug Discovery Using Mass Spectrometry, Edited by Mike S. Lee

greatly reduce compound and animal requirements, the ultimate goal is to use in silico predictions to eliminate very poor compounds before any experiment has been conducted. However, the development of useful in silico screening models first requires the generation of large amounts of high-quality in vitro and ADME data over a broad region of drug-molecule structural space. The consequence for the preclinical drug-metabolism scientists is the need to provide more data, at a faster pace, with only modest growth in resources. To meet this challenge, the development of new methods and techniques for drug-metabolism experiments is required to keep up with the demand for information.

An important component of understanding the in vivo PK behavior of potential drug candidates is knowledge of their biotransformations. Biotransformation can be generally defined as a chemical change in a compound's primary structure mediated by a biological system. A key process in drug metabolism, biotransformation influences many PK parameters. Biotransformation can govern the clearance of a compound, and thereby directly impact its half-life. The bioavailability of a drug can be controlled by biotransformation at the gut wall or in the first pass through the liver.

In addition to affecting the PK profile of a drug candidate, the products of biotransformation pathways may be responsible for undesirable drug interactions independent of those of the parent compound. The metabolite(s) produced can be inhibitors of either metabolizing enzymes or transport proteins involved in the distribution and elimination of the parent or co-administered drugs. Alternatively, biotransformation can produce reactive intermediate species capable of covalent modification and inactivation of important metabolizing enzymes or other proteins. Reactive intermediate formation and subsequent covalent modification of protein have been proposed to be an important component in idiosyncratic drug toxicity and drug sensitization [1]. The identification and the structural characterization of metabolites provides a first step in overcoming the undesirable effects that they may cause.

This chapter highlights the technological advances in both instrumentation and experimental systems that have affected the important role of biotransformation studies to provide early drug discovery ADME information. Emphasis is placed on the emergence and application of mass spectrometry, automation, and other new methodologies to provide key drug-metabolism data at the speed required to support early drug discovery. This chapter then concludes with a view toward future methodologies for the study of biotransformation.

CURRENT PRACTICES FOR STUDYING BIOTRANSFORMATION

Drug-metabolism scientists have traditionally focused on detecting and understanding the effects of biotransformation only as compounds proceed into clinical development. More recently, these studies have impacted the *drug-discovery* process as well. The end result has been an increase in the "quality" of the drug as it proceeds through the pipeline, along with a simultaneous reduction in time to clinical trials. The structural identification of the metabolites has proved to be extremely valuable in understanding how the primary structure of a drug candidate might be synthetically modified to alter the biotransformation that is the cause of undesirable effects. Researchers are able to evaluate simple structural modifications, based on in vitro biotransformation results, and iteratively improve a compound's biotransformation profile and metabolic stability.

In this manner, metabolism-guided studies provide direction for the synthetic effort in structural optimization of a lead compound [2]. Biotransformation studies have also impacted the design of in vivo preclinical animal studies. Metabolite profiles that show the number and relative amount of metabolites formed have value in interspecies comparisons. In some cases, significant differences can be observed in the in vitro metabolism of drug candidates. These studies enable scientists to choose a suitable animal species for preclinical safety studies [2]. Similarly, the ability to test in preclinical animal species the exposure to metabolites formed by in vitro human systems is a very important part of deciding whether a compound can be successfully taken through safety assessment. After identification and characterization of metabolites using in vitro systems, investigators can then monitor these metabolites in the safety species and determine plasma exposure. Knowledge of the structure of metabolites also aids in the definition and extension of the intellectual property associated with a compound. The structural identification of metabolites is required to protect novel chemistries during compound registration and the patent application process. The latter is particularly important, since metabolites are often pharmacologically active and, in some cases, show more desirable properties than the parent compound.

Until recently, the detailed characterization of biotransformation pathways and metabolite structures was not attempted on the wide array of discovery-stage compounds. Some of the earlier difficulties associated with the assessment of the extent and significance of biotransformations stemmed from the lack of suitable biotransformation model systems, and the absence of analytical instrumentation capable of the detection of the compounds under study in very complex matrices. Traditionally, biotransformation products and pathways were studied in drug development as part of investigational new drug (IND) submissions. Most of this work was done with radiolabeled compound; few metabolite structures were actually elucidated in detail, mostly because metabolite profiling and identification were difficult, time-consuming, and expensive. Structure elucidation required isolation of the potential metabolite from in vivo samples, often by many steps of high-performance liquid-chromatography (HPLC) separation and fraction collection. Purified fractions were usually analyzed by mass spectrometry to determine molecular weight and by one-dimensional nuclear magnetic resonance spectroscopy (NMR) to provide more complete structural identification. The final proof of a structure was obtained from the synthesis of the proposed metabolite followed by comparison of its spectroscopic properties with those of the isolated metabolite [3]. The time and effort required to obtain this information prevented the routine use of biotransformation studies in drug discovery. However, with the recent advances in experimental systems and in both software and instrumentation, structural identification of even minor metabolites has become nearly routine.

EXPERIMENTAL APPROACHES

Biotransformation studies have benefited from advances in both molecular biology and biochemistry. Molecular biology provided new reagents for mechanistic studies and screening experiments in the form of expressed drug-metabolizing enzymes [4]. Some of these expression systems include cytochrome P450, glucuronide transferases, and sulfotransferases [5,6]. More recently, drug-transport proteins, such as *p*-glycoprotein, have been cloned and expressed. Transport proteins are now recognized as a major factor that contributes to the ADME of drugs [7]. Drug interactions traditionally thought to occur most frequently with P450 metabolizing enzymes are beginning to be recognized with transport systems as well [7]. Improved cell/tissue culture and handling techniques allow for the harvest and cryopreservation of primary hepatocytes that are an important tool for the approximation of in vivo biotransformation [8]. Additionally, the development of human intestinal epithelial cancer cell lines ($CaCO_2$ cells) allowed the generation of in vitro absorption models such that the role of transport and biotransformation in absorption may be studied now on a routine basis [7].

Biotransformation studies and metabolite identification are now feasible at the screening stages of drug discovery. Related screening paradigms have been established to characterize compounds with respect to important metabolic properties, such as metabolic stability, P450 inhibition, P450 induction, P450 phenotyping, and membrane permeability [9–13]. Increasingly, liquid-handling robots are used to conduct these in vitro assays and to perform sample preparation for subsequent rapid liquid chromatography–mass spectrometry (LC-MS) analyses. The implementation of automation and parallel processing [14] have increased the throughput of metabolic screening and compound characterization to the point where drug-metabolism scientists can start to analyze a significant fraction of the compounds that are synthesized by combinatorial chemistry and related parallel synthetic approaches.

Taken as a whole, the characterization of metabolic pathways in early drug discovery has become both an important and realistic objective. Recognition of the value of these experiments has been only one half of the equation; the availability of greatly improved hardware and software for the characterization of biotransformations has made it possible to ask more questions.

HARDWARE INNOVATION

Separation Advances

HPLC is the most commonly used separation technique for metabolite identification and other biotransformation studies, regardless of the detection method. Certainly, the use of HPLC separation prior to introduction to the mass spectrometer is vital for successful analysis. Chromatographic separations have changed over the past years to take advantage of the selectivity, sensitivity, and more universal detection of mass spectrometers. The high sensitivity of mass spectrometers has enabled the reduction of HPLC column size from the 4.6-mm × 25-cm columns used with ultraviolet (UV) detection, to the 2-mm × 5-cm columns used with most triple quadrupole and ion-trap instruments. Similarly, 0.1-mm packed-capillary columns are now used with low-flow electrospray ionization (ESI) sources and the time-of-flight (TOF) instruments. Along with the reduction in column size has also come an increase in flow rate to over fivefold the normal flow rate. The increased flow rate allows for more rapid column equilibration, and therefore, faster separations. Chromatographic separations that traditionally take 40 to 60 minutes are now done in less than 10 minutes [15]. Recent developments that hold promise for faster separations without a sacrifice in chromatographic resolution include ultrahigh-pressure chromatography [16] and the use of monolithic stationary phases [17,18].

Less selective stationary-phase materials are now available that allow generic gradient conditions for early screening-stage biotransformation work that involve many different chemical structures and their metabolites. These generic stationary-phase materials also have shown applicability in off-line sample cleanup procedures [19], and in on-line sample extraction systems [20,21]. The generic nature of the interactions makes it possible to retain compounds with the wide polarity range that might be created through biotransformation, while a large fraction of the unwanted matrix components is removed.

While the majority of separations are carried out with chromatographic processes, recent advances in the design of ion-mobility instruments provide a crude means to fractionate samples that enter the mass spectrometer based on molecular size and three-dimensional (3D) shape, rather than mass to charge or polarity. The use of high-field asymmetric-waveform ion-mobility spectrometers (FAIMS) coupled with mass spectrometry and HPLC provides an added dimension of selectivity. The FAIMS device acts as a discriminator that allows only a small subpopulation of the ions produced in the ion source of the mass spectrometer to be transmitted into the vacuum chamber of the mass spectrometer. The result is selectivity based on gas-phase ion-mobility separation, in addition to the separation achieved by chromatography and mass spectrometry [22,23]. To date, the systems have been used primarily for quantitative analysis of small molecules as a means of reducing or eliminating the need for chromatographic separation [24–27]. In the biotransformation arena, the advantage of the FAIMS is in the ability to produce simplified full-scan mass spectra. At any given point along the chromatographic separation, the ions that enter the mass spectrometer can be further fractionated with the FAIMS device to transmit only ions of similar size and shape. The result is another dimension to the separation and simplified full-scan spectra. With these simplified spectra, the identification of important ions is a much simpler task. Ultimately, the advantage may be to provide simplified data sets for automated interrogation.

Mass Spectrometry

The single most important advance for the study of biotransformation has been the development of atmospheric pressure ionization (API) that enabled HPLC separations to be coupled to mass spectrometric detection (LC-MS). Prior to this advancement, HPLC-UV was the primary technique used to obtain information on the number of metabolites formed and their relative intensities [2]. The metabolite profiles generated by HPLC-UV usually provide minimal structural information. HPLC retention time (t_R) and perhaps a wavelength shift in the UV absorption spectrum are the only indications that some structural change has occurred on the parent-drug molecule. In contrast, LC-MS provides the molecular-weight information on potential metabolites, as well as the molecular weight of fragment ions. This information can be used to elucidate structural features of the analyte.

Equally important to the development of API as an interface for mass spectrometry was the recognition that triple-quadrupole tandem mass spectrometers were ideally suited to biotransformation studies [28–30]. Triple-quadrupole mass spectrometers offer the ability to collect three major types of tandem mass spectrometry (MS/MS) spectra that can be used to help assign structures to metabolites in addition to full-scan mass spectra that allow assignment of metabolite molecular weights. In a product-ion MS/MS scan, a single molecular ion is selected in the first quadrupole, fragmented in the second quadrupole, and the resulting fragment ions are mass analyzed in the third quadrupole. In precursor and constant neutral-loss MS/MS scan modes, a fragment ion or a constant neutral-loss generated in the collision cell and observed in the third quadrupole is related back to the ions in the first quadrupole that gave rise to those fragment ions or neutral losses, respectively. All of these MS/MS scan modes are very specific for the ions they monitor, thus signal from most of the endogenous material present in the samples is eliminated, and the signal-to-noise (S/N) of the target species is increased. MS/MS scan modes address one of the most difficult aspects of metabolite identification: identification of drug-related material in complex matrices [31]. Precursor and neutral-loss MS/MS scans have been shown to be invaluable for identification of conjugate metabolites in complex in vivo samples [29,32]. One of the conjugate metabolic pathways most often examined is glutathione (GSH) adduction, which often serves as an indicator of reactive or toxic metabolite formation. One strategy has been to identify the GSH conjugates, typically in bile, with constant neutral-loss MS/MS scanning. The sample is reanalyzed to acquire product-ion MS/MS spectra of the conjugate species [33]. With the advent of faster scanning instruments and information-dependent acquisition, it is now possible to perform both analyses in a single chromatographic run.

Another significant advance that has greatly influenced the role of biotransformations in drug discovery was the introduction of the benchtop ion-trap LC-MS instruments [34,35]. The ease of use, low cost, and small size of the ion trap brought liquid chromatography–tandem mass spectrometry (LC-MS/MS) capability out of the specialized mass spectrometry laboratory and into the general drug-metabolism laboratory, to give the drug-metabolism scientist easy access to MS/MS analysis for biotransformation studies. The main advantage of the ion-trap mass spectrometer over the triple-quadrupole mass spectrometers is the ability to collect sequential stages of ion fragmentation data. Such fragmentation pathway data often simplifies or clarifies structure elucidation [36].

The commercialization of TOF and quadrupole time-of-flight (Q-TOF) instruments has added accurate mass capabilities to help with rapid metabolite identification by LC-MS. The TOF mass analyzers routinely provide resolution (50% valley) of 5000 to 15,000 with a mass accuracy of better than 5 ppm with internal calibration. The resolution and mass accuracy are sufficient to determine a molecular formula for small drug molecules. For larger compounds, the mass-to-charge ratio (m/z) > 200 Da, the possible combinations of atoms that comprise drug molecules within the mass-accuracy window of the TOF analyzers generally becomes too large to identify a compound molecular formula based on isotope ratio and mass alone. However, the latter is rarely a problem in biotransformation studies of drugs, since the molecular formula and structure of the starting compound is typically well known. Information obtained from the parent drug is used to set boundary conditions for the possible molecular formula and extends the usefulness of accurate mass measurements for metabolism.

The quantitative abilities of the LC-TOF, although limited in the linear dynamic range, often allow measurement of metabolic profiles and PK profiles from the same samples. Zhang and co-workers [37] provide such an example. Quantitative results for a co-administration of five compounds to rats, with assay limits of quantitation (LOQ) between 1 and 5 ng/mL, and an upper limit of quantitation (ULQ) at 100 ng/mL were described. In this study, precision and accuracy were better than 20% for all five analytes, and comparable to triple-quadrupole quantitative data. More importantly, in addition to following the levels of dosed compound, the authors were able to identify several metabolites from the same full-scan data using the accurate mass capabilities of the LC/TOF-MS.

The tandem Q-TOF instruments add the ability to obtain accurate mass product-ion spectra from LC-MS/MS analysis [38]. The improved accuracy of the measured fragment masses allows confirmation of molecular formula for proposed fragment ion structures. Pilard and co-workers [39] used accurate mass fragmentation data to unambiguously identify an unknown metabolite of a preclinical drug candidate as the hydroxylamide sulfoconjugate of the parent molecule. Accurate mass data may also simplify product-ion MS/MS spectral interpretation for metabolite identification. A recent example published by Hop and co-workers [40] demonstrated the use of accurate mass product-ion MS/MS spectra to elucidate a fragmentation mechanism involving a three-hydrogen atom transfer. Elucidation of this mechanism led to the structural assignment of metabolites from rat urine, which was not possible with other LC-MS techniques.

One limitation to the current LC/Q-TOF tandem instruments is the inability to select the precursor ion with high resolution. The commercially available instruments use a low-resolution quadrupole mass filter for precursor ion selection. The low-resolution precursor ion selection may allow more than one species to be transferred to the collision cell where fragmentation occurs. The resulting product-ion MS/MS spectrum may then be a combination of products from more than a single nominal mass precursor. Recent commercialization of a high-resolution triple-quadrupole mass analyzer with good ion transmission characteristics (Quantum™, ThermoFinnigan, San Jose, California) paves the way for future instruments capable of high-resolution precursor ion selection and accurate mass fragment ion spectra.

Perhaps the biggest liability with the TOF analyzers is the very large volume of data collected during LC-MS analysis. The large file size is not practical in rapid applications. Processing speed and archival of the large amounts of collected data must be addressed before the LC-TOF instruments will be of practical use in routine-discovery metabolism applications.

More recently, a linear ion trap (LIT) on a triple-quadrupole platform was introduced, which provides the best of both the ion trap and the triple quadrupole [41]. Unique to the AB/MDS Sciex QTrap hybrid instrument, is the ability to function as a normal resolving quadrupole in a triple-quadrupole configuration and as a LIT with mass-analysis capability. The very fast scanning speed of the LIT permits very rapid switching between triple quadrupole and ion-trap scan modes such that the analyst is able to conduct traditional selected reaction monitoring (SRM) analyses with the true triple-quadrupole capabilities of the instrument, while collecting full-scan data (enhanced mass spectrometry, EMS) with the instrument used as a LIT. King and co-workers demonstrated the ability of the hybrid triple–quadrupole-linear ion trap to alternate scans between SRM and full scan on a time scale that allowed suitable definition of chromatographic peaks for quantification of target compounds while collecting full-scan data [42]. The authors compared the quantitative results from SRM analysis, and from combined full-scan and SRM analysis (SRM/EMS). The quantitative data in both scan modes was acceptable in terms of sensitivity, accuracy, and precision. Additionally, they presented examples of the additional information that can be obtained from plasma samples analyzed primarily for target-compound concentrations; these include detection and characterization of circulating metabolites, dosing vehicle, interfering matrix components, and potential interfering drug conjugates. Their examples highlight the potential problems with co-eluting glucuronide conjugates and polyethylene glycol (PEG) dosing vehicle and demonstrate how full-scan data can be used to improve the quality of quantitative results. An example from an in vivo PK study shows the possibility of gaining information about circulating plasma metabolites from early PK studies on a routine basis. Such information might lead to the identification of compounds with better PK properties, or provide insight into pharmacokinetic–pharmacodynamic (PK-PD) relationships not gained from a targeted analysis of parent drug alone. The time profiles of circulating metabolites might also provide a better understanding of the PK results themselves. Such information might help explain PK differences between compounds at the discovery and lead optimization stages. The data presented show the potential for collecting more information from each sample without an increase in mass spectrometer run times.

The fast scanning speed (4000 amu/s), rapid cycle time, and high sensitivity afforded by the AB/MDS Sciex QTrap allow the use of combined scan functions, such as precursor or neutral-loss MS/MS scans followed by product-ion MS/MS scans. Therefore, the hybrid triple-quadrupole/LIT could be the ideal platform for acquiring more useful information from every drug-metabolism sample currently analyzed by LC-MS/MS.

A new and potentially very powerful hybrid mass spectrometer was recently introduced. The LTQ-FTMS from Thermo Electron Corporation combines the speed and sensitivity of the LIT with the high resolution and accurate mass capabilities of a Fourier transform ion cyclotron resonance (FTICR) mass spectrometer. Mass accuracies better than 1 ppm are obtained with external mass calibration at resolutions as high as 100,000 for an acquisition rate of one scan per second. Coupling of the LIT with the FTICR allows the instrument to use rapidly acquired full-scan data from the LIT as a survey scan for more detailed analysis in the FTICR cell. The LTQ-FTMS is the first FTICR system to show the ability to practically keep up with the chromatographic separations normally used for biotransformation studies [43]. The instrument has the potential to make metabolite identification as simple as a mathematical calculation based on accurate mass measurements and theoretical atomic weights of the elements known to compose the parent drug under study. The practical limitation is the ability to rapidly store, retrieve, and process the large volumes of data the system can generate.

Electrochemistry On-Line with Mass Spectrometry

Electrochemistry has been used with mass spectrometry for a number of years to mimic oxidative metabolism [44]. The combined limitations of the early ionization methods and electrochemical cells of the day made the experiments challenging to conduct on a routine basis. The advent of flowthrough carbon frits as electrodes in electrochemical cells allows simple and reliable set up and conduct of electrochemistry on-line with mass spectrometry. Jurva and co-workers recently conducted a systematic study of the types of oxidation reactions that can be observed for both cytochrome P450 metabolizing enzymes and the flowthrough electrochemical cell [45]. The studies indicate that reactions initiated by a one-electron process such as N-oxidation, S-oxidation, P-oxidation, alcohol oxidation, dehydrogenation, and N-dealkylation are likely to be observed with the electrochemical cell. Reactions thought to involve direct hydrogen atom abstraction such as O-dealkylation and hydroxylation of unsubstituted aromatic rings were not mimicked [44].

Another use for the on-line electrochemical cell was described by King and co-workers [46] for the generation and characterization of GSH–drug adducts. The experimental design allowed the oxidation of the drug in the electrochemical cell to be performed in the presence of excess GSH. Reactive intermediates were shown to be effectively trapped by the GSH. The experimental system provides a simple and effective way to generate GSH–drug adducts for characterization by mass spectrometry. Additionally, an experimental design was described that gave information about the potential pathways involved in the formation of GSH–drug adducts. The phase I metabolites from a microsomal incubation of drug were separated by HPLC and then passed through the on-line electrochemical flow cell along with an infusion of GSH. The reactive species formed by electrochemical oxidation of the metabolites were trapped as GSH conjugates and characterized on-line by mass spectrometry at the chromatographic retention time of the metabolite itself. The formation of acetaminophen by O-dealkylation of phenacetin in microsomes was used as a model system to demonstrate the potential of the approach [46].

While it is clear that electrochemical reactions represent only a portion of the reactions observed with the cytochrome P450 drug-metabolizing enzymes, electrochemical generation of both phase I and phase II metabolites can be a valuable means for the study of metabolic events, some of which may occur through mechanisms other than the action of cytochrome P450 enzymes.

NMR and LC-NMR

Although mass spectra can provide structural information, the determination of the exact location of biotransformation on a drug molecule is best done by NMR. NMR is considerably more information-rich than MS, but less sensitive. This means more material and longer acquisition times are needed to obtain interpretable data. To determine a metabolite structure, a number of NMR experiments are available. An interpretable one-dimensional (1D) ^{1}H spectrum for a metabolite can be generated in a few hours from about 1 μg of metabolite with conventional ambient-temperature 3-mm probes. However, approximately 50 μg of the same metabolite would be required to acquire a two-dimensional (2D) ^{1}H–^{13}C multibond correlation data set in a few days. These sample requirements oftentimes exceed the levels generated in most typical metabolism studies.

In recent years, two different technologies, cryoprobes and low-volume flow cells, have yielded improvements in the sensitivity of an NMR experiment. In the case of cryoprobes or cold-probes, the entire probehead housing the copper radio frequency (RF) coil and the preamplifier(s) is cooled to a cryogenic temperature of 20 to 25 K [47]. The cryogenic temperature significantly reduces the noise factor of the detector, thereby improving the S/N ratio of an NMR spectrum by 3.3-fold over ambient temperature probes. The improved S/N translates to a 10-fold reduction in time for an NMR experiment on equivalent amounts of metabolites.

An example of the low-volume flow cells is the CapNMR™ probe (Protasis/MRM Corp., Savoy, Illinois). The CapNMR probe is operated at ambient temperature and has a flow-cell volume of 5 μL, where the solenoidal RF coil encloses an observed volume of about 1.5 μL. Typically, the mass-limited sample is dissolved in a few microliters of an appropriate NMR solvent and is transported to the active probe volume via a microsyringe pump or a capillary HPLC system. The mass sensitivity [48] of the CapNMR probe is reported to be similar to that of a cold-probe [49].

The availability of liquid chromatography–nuclear magnetic resonance (LC/NMR) flow probes [50–52], such as microcoil/CapNMR probes [48,49,

53,54] and the cryocooled flow probes [55,56] is changing the use of NMR in biotransformation studies. LC/NMR facilitates the structural determination of biotransformation products that previously were very difficult to handle. Rather than undergo the task of metabolite purification prior to analysis, a combination of mass-directed fraction collection for analyte peak purification [57] and flow-injection sample delivery vastly improves the traditional low throughput of NMR. The ability to obtain NMR spectra on minimally processed samples on-line has been useful in cases where it has been difficult to isolate the species of interest. Some of these include unstable oxidative products [58], unstable conjugates [59,60], and small volatile [61], and polar fragments [62–64] of a parent drug.

An additional step of concentrating the HPLC analyte peaks with an on-line solid-phase extraction (SPE) cartridge also has been recently introduced [56]. In this approach, the analytes of interest that elute from an HPLC column are each directed to individual SPE columns. There are three advantages to this approach. The first advantage is the reduction in the amount of endogenous components of the biological matrix that can overload a normal LC column. The second advantage is that HPLC separation can be performed off-line with protonated solvents. The third and perhaps most valuable advantage is that the metabolite of interest can be collected and concentrated on an SPE column from multiple HPLC runs and then eluted with a small amount of deuterated solvent with the flow directed to the NMR probe. In such a setup, the chromatography step is capable of concentrating the analytes of interest into a much smaller volume that is matched to the volume of the flow cell in an NMR probe.

The selectivity of capillary HPLC coupled with the sensitivity and improved S/N of microcoil cryoprobes may allow rapid analysis of biotransformation products with minimal purification or preconcentration. The increase in sensitivity and speed offered by these approaches may bring NMR metabolite structure elucidation into a timescale useful for discovery screening.

SOFTWARE ADVANCES

Instrument advances in recent years have led to an everincreasing quantity of data that require examination. The automated identification of metabolites and their structures will be crucial to the success of biotransformation studies in drug discovery. Several software packages have been developed that accelerate various stages of this process. One such program is *Metabolite ID 1.4 for Analyst* from Applied Biosystems/MDS Sciex. Other instrument vendors offer similar software packages to help in metabolite structure elucidation. *Metabolite ID 1.4 for Analyst* will acquire full-scan chromatographic data of an experimental and a control sample, compare the two samples and determine which peaks in the experimental sample are potential metabolites, generate a list of potential metabolites and their corresponding retention time and ion *m/z*, and program the acquisition of product-ion MS/MS spectra for each of the potential metabolites. A key feature of *Metabolite ID 1.4* is the use of correlation analysis of the product-ion MS/MS spectra to determine whether the ion identified by the software from the full-scan data is drug related rather than a background component of the sample. In this case, correlation analysis is used as a pattern-recognition routine that compares the product-ion spectrum of the parent compound to that of the potential metabolites, and scores the similarity between the spectra compared. The analyst is able to very quickly distinguish metabolite masses from endogenous material based on the correlation values. To increase throughput, samples for metabolite identification can be pooled and analyzed simultaneously. Each metabolite can subsequently be assigned to its corresponding substrate with correlation analysis [65].

Metabolite ID 1.4 operates in both interactive and batch mode. In the interactive mode, the user reviews the full-scan data prior to MS/MS generation. In batch mode, the user submits a list of samples to be analyzed and starts automated acquisition. With such automated approaches, the metabolic profile of a single compound can be evaluated in approximately 1.5 hours, provided that adequate separation can be achieved with short, narrow-bore columns and fast-gradient chromatography.

An alternate approach for automated data collection is data-dependent acquisition, frequently referred to as a data-dependent experiment (DDE) [66]. DDE is particularly useful for the automated collection of biotransformation data from in vitro samples. DDE typically uses information from a preliminary survey scan mode to trigger other scan modes within the elution time of a single LC peak. Using DDE, several types of MS data can be acquired on each LC peak detected. For example, full-scan data can be acquired to establish a component's molecular weight, which in turn triggers the collection of product-ion MS/MS spectra for that component from a single injection. Building on this idea, López and co-workers [67,68] have developed an integrated metabolite identification (ID)/DDE algorithm for data acquisition that includes user-defined masses for triggering the product-ion MS/MS scanning mode. The advantage of DDE is speed. Most of the desired data can be acquired in a single LC run. One disadvantage of DDE is the possibility of missing important components in a mixture while the mass spectrometer gathers product-ion MS/MS data rather than monitor the full-scan data for incoming peaks. Unfortunately, the technique also has the potential to generate very large amounts of data on mixture components that are unrelated to the drug.

In response to the latter problem, a new software approach to remove background from LC-MS full-scan data has been developed and implemented under the name of dynamic background subtraction (DBS) [69]. The procedure sums previous scans to generate a composite background spectrum to be subtracted from the last scan acquired. The procedure works on-the-fly as data are acquired, and results in collected data composed only of ions that change on the chromatographic timescale. Since the subtraction of background takes place scan by scan as the data are acquired, the background- subtracted data can be used as a survey scan for a subsequent data-dependent scan in a dynamic fashion. Most often, this type of automation is used to acquire full-scan spectra and product-ion MS/MS spectra in a single LC-MS run. The dynamic background-subtraction improves metabolite identification and increases the likelihood that drug and metabolites will be identified in the survey scan as candidates for MS/MS acquisition. DBS was shown to increase coverage of even low-level isobaric metabolites with little chromatographic separation in a single injection.

The automated approaches described previously focus on the identification of potential metabolites and methods to establish drug-relatedness. However, the spectra must be interpreted to arrive at a possible structure for the metabolite. This generally requires detailed interpretation of the substrate product-ion MS/MS spectrum. Expert systems for the interpretation of fragment ion spectra are being developed for use in conjunction with automated metabolite identification.

One such approach uses a combination of molecular modeling and a simple rule set to predict initial sites of fragmentation and the subsequent decay products that give rise to the fragmentation pathways observed in low-energy collision-induced dissociation (CID) [70]. Bond energies are calculated for each bonded atom pair in the molecule using molecular-mechanics force fields developed specifically for the task. The bond energies are ranked to predict the initial sites of fragmentation. To rank atom pairs according to bond energy, one assumes that the quasi-equilibrium theory applies to the fragmentation reaction [71]. If the assumption holds, then the weakest bonds will be preferentially cleaved to produce the most frequently observed initial fragments. The energy that remains in the fragments is often observed to be released by chemical reactions driven by formation of neutral leaving groups. A second modeling algorithm allows a list of common neutral losses to be searched as substructures in the parent drug. Elimination of substructures as neutral leaving groups provides possible structures for fragment ions not generated by the initial fragmentation event. The predicted fragments can then be compared to the experimentally observed fragments to provide an interpretation of the product-ion MS/MS spectrum.

Other expert systems under development include metabolite prediction software that can be used with LC-MS data. The structure of the starting compound and some knowledge of the metabolic processes active in the experimental system are used to predict the more common metabolic transformations. The use of full-scan data with high resolution and accurate-mass measurements, allows the presence or absence of the predicted metabolites to be rapidly confirmed without operator intervention. Application of expert systems to metabolite identification has the potential not only to dramatically increase the speed of metabolite identification, but also to move routine structural identification out of the specialized laboratories and into general drug-metabolism areas.

FUTURE DIRECTIONS

Technology advances not only increase the speed and reduce the cost of analysis, but they also enable one to contemplate experiments for problems previously believed to be intractable. Such new experimental challenges in turn drive the development of additional technologies that will continue to enhance processes for exploring the mechanisms of drug action. On the horizon, advances in analytical sensitivity, compound-ionization procedures, and instrument design will dramatically impact the study of the ADME properties of drugs. Technological advances will also enhance analytical processes through miniaturization and parallel processing. New biological reagents made available through molecular biology and cell/tissue culture should provide a new level of understanding of the mechanisms that control the fate of compounds in vivo. With this understanding, better prediction of the fate of drugs before testing in humans has the potential to improve the speed of drug discovery, and assure that safer drugs are available to the patients who need them. We now discuss some of the advances that are expected to have the largest impact on drug-metabolism research.

Ionization Methods

A significant challenge that still faces LC-MS and LC-MS/MS analysis of biologically important molecules is compound ionization and transfer into the gas phase. API, the predominant method of choice for biotransformation work, has room for improvement. To improve the transfer of liquid from the LC to the gas phase, instrument manufacturers have redesigned API sources to increase the efficiency of solvent evaporation. In the ESI source, attention paid to gas-flow dynamics in the interface has led to increased ionization efficiency and improved transfer of the ions from atmospheric regions of the ion source to the vacuum system of the mass analyzer. The atmospheric-pressure chemical ionization (APCI) source incorporates ceramic surfaces to improve thermal evaporation without subjecting the samples to extreme heat. An alternate method of ionization is atmospheric-pressure photoionization (APPI) [72]. Because the photoionization mechanism is fundamentally different from the other API techniques, APPI offers the potential to detect species not easily observed with APCI and ESI.

More recently, porous carbon electrode technology is used to enhance the range of chemical structures that can be ionized by ESI and to reduce background often observed with electrospray. By control of the redox chemistry at the ESI emitter, Van Berkle and co-workers have demonstrated the ability to change the species observed in the full-scan mass spectrum of Reserpine [73,74]. Further understanding of the redox chemistry that occurs at the electrospray emitter is critical to both its use as an analytical tool and the interpretation of ESI-MS data for biotransformation studies.

While ESI and APCI have been the most commonly used means for the separation and mass analysis of biotransformation products, matrix-assisted laser desorption/ionization (MALDI) and related techniques, such as surface-enhanced laser desorption/ionization (SELDI) [75] and direct ionization off silicon (DIOS) [76], hold promise for rapid screening of biotransformation samples, and may one day replace much of the work now done by LC-MS for qualitative studies. The advantage of these techniques over LC-MS is rapid sample analysis, and high-density sample formats that can be prepared robotically.

Accelerator Mass Spectrometry

The improved sensitivity afforded by accelerator mass spectrometry (AMS) [77] has the potential to revolutionize ADME in humans. At present, ADME studies in humans have been limited because of the need to incorporate radioisotopes into a drug to enable detection. However, AMS can detect radioisotopes at extremely low levels—below those considered radioactive, yet above the natural abundance of the isotope. The sensitivity of AMS will enable very low-level radiotracers to be incorporated in all drug syntheses. Because these radiotracers are not, by definition, radioactive, they can provide markers for the detection of drug-related material not only in human-metabolism studies but also in drug-discovery studies. Radiolabeled compound dosed to humans at such low levels would make distribution and metabolism studies in man a much less expensive and time-consuming task. Likewise, drug candidates labeled at these low levels could be employed in all routine drug-discovery studies. The challenges to be overcome before AMS can be used routinely are significant, however. Advances are needed to facilitate sample preparation, which at present is performed off-line in a lengthy and labor-intensive manner. In addition, the large size of the AMS instruments (warehouse-size) precludes their general use and location. This latter challenge is not much different than that which occurred with the deployment of magnetic resonance imaging (MRI) spectrometers, and it is expected that this challenge should be solved with time.

Microfabricated Devices

Microfabricated devices are becoming increasingly important in biochemical analysis, and their use has been particularly valuable for the miniaturization of enzymatic assays. While detection has most often been accomplished with fluorescence, electrochemical detection, or ion-selective electrode detection, several groups have successfully coupled microfabricated devices to mass spectrometry, and demonstrated both qualitative and quantitative results for many different classes of molecules, including drug molecules [78–83]. Application of microfabricated devices to in vitro metabolism may provide the solution to the increased demand for more in vitro data in drug screens, and for the investigation of relationships between parameters that control the ADME of potential drugs. Chemical analyses, such as sample processing, mixture separation, and analyte detection, can be accomplished in parallel on a single chip the size of a credit card or compact disk.

Microfabrication techniques originally developed for electronic devices, and applied to lab-on-a-chip, or miniaturized total chemical analysis systems (μTAS) rely on microfluidics to move and mix solutions in very small volumes on etched or embossed substrates. Although the manufacture of silicon chips is well developed, the generation of reproducible flows and separations on the microscale (in mass-produced μTAS) remains a significant challenge. Various approaches have been developed to address this challenge. The most widely used means of moving liquids is via electrodynamically driven flows [84]. Flows have also been created by application of centrifugal force, generated by spinning etched compact disks at varying speeds [85]. Reverse-phase chromatographic separations typical for analysis of drug molecules are now incorporated into chip systems with in situ polymerization of stationary phase [86,87]. Temperature-driven pumping is being investigated as a means to supply regulated, pressure-driven flows and solvent gradients on microfabricated devices [88].

Miniaturization has already enabled significant inroads to be made toward the development of parallel analytical systems. For example, microfabrication techniques have been used to produce ESI emitters in an array format [89,90]. Miniaturized arrays of ion-trap mass spectrometers are being tested and developed with some success, and promise at least mass selective detection in a miniature parallel device in the near future [91].

While miniaturization is a revolution in biochemical analysis, much remains to be resolved. Some of the greatest challenges to be overcome are rapid evaporation of nanoliter volumes, analyte interaction with surfaces, and universal detection of very small samples [84]. Nevertheless, the combination of parallel formats and miniature components is expected to reduce the time and cost required per experiment by orders of magnitude. More significantly, miniaturization will allow experiments not currently possible with large-scale instruments, and should facilitate understanding of fundamental biochemistry on the scale of single cells and perhaps organelles.

Artificial Organs

To increase the efficiency of drug discovery, drug-metabolism scientists would like to assess the extent and significance of biotransformation in humans, before a compound gets to a human. Biotransformations are carried out for the most part in the liver. Since it is very difficult to perform early biotransformation studies in humans, drug-metabolism scientists have used preparations of hepatic microsomes, liver tissue slices, or freshly isolated hepatocytes to carry out these studies. However, microsomal incubations do not contain appropriate co-factors for metabolism, and tissue slices and isolated hepatocytes are either not readily available or not stable for more than a few hours. In addition, none of these in vitro systems represents the dynamic nature of the liver. Artificial organs, on the other hand, are being developed for research and transplant, and might one day be the ultimate in vitro tool for drug-metabolism studies [92]. Bioengineering researchers have demonstrated the seeding of biodegradable polymers with human cells to generate functional blood vessels and organs [93]. These researchers envision biodegradable polymers that can be molded into networks of capillary-like grooves. The inside of the grooves would be seeded with lining cells and the outside with liver cells. As blood passes through the capillaries, biotransformation would take place in the liver cells. Wastes (i.e., biotransformation products) would be removed across the polymer into lining cells designed as vessels to collect bile. A small array of the liver cells and bile-collecting vessels would be enough to perform repeated biotransformation studies in a fashion similar to ultrafiltration MS [94], where time courses can be generated by discrete on-line sampling. This approach would combine the benefits of the sensitivity afforded by the mass spectrometer with human "in vivo" biotransformation studies at the discovery stage.

CONCLUSION

The immediate value for the collection of so much biotransformation data is to help make decisions at the drug-discovery stage on a daily basis. Compounds that are most likely to have the desired properties might be chosen based in part on data from biotransformation and metabolism studies. The long-term goal is to gain understanding that enables the prediction of the metabolic properties of a molecule based on the chemical structure. Metabolic characteristics such as cytochrome P450 inhibition and induction, and reactive metabolite formation might be predicted in many cases before the molecule is synthesized [95]. Such in silico approaches are very attractive as pharmaceutical companies search for ways to reduce the time and cost associated with the discovery and development of new medicines.

Today, researchers continue the quest to establish correlations between in vitro assays and in vivo observations [96,97]. Work is also underway to use animal data collected from in vivo experiments along with data generated by human in vitro systems to predict the metabolic behavior of compounds in humans [98]. The amount and diversity of data collected will be used to develop computational models for ADME properties. In the future, computational screens may replace many of the experimental screens run today.

The trend to bring more detailed information into earlier stages of drug discovery will continue to drive improvements in technology, and in experimental and analytical procedures for the study of biotransformation of drugs. The challenges are significant, but so is the promise of the contributions that can be made by biotransformation studies.

ACKNOWLEDGMENTS

The authors acknowledge J.A. Yergey for his scientific review and helpful discussions.

REFERENCES

1. Uetrecht, J.P. "Is It Possible to More Accurately Predict Which Drug Candidates Will Cause Idiosyncratic Drug Reactions?" *Curr. Drug Metab.* **1**, 133–141 (2000).
2. Li, C.; Chauret, N.; Ducharme, Y.; Trimble, L.; Nicoll-Griffith, D.A.; Yergey, J.A. "Integrated Application of Capillary HPLC/Continuous-flow Liquid Secondary Ion Mass Spectrometry to Discovery Stage Metabolism Studies," *Anal. Chem.* **57,** 2931–2936 (1995).
3. Balani, S.K.; Pitzenberger, S.M.; Schwartz, M.S.; Ramjit, H.G.; Thompson, W.J. "Metabolism of L-689,502 by Rat Liver Slices to Potent HIV-1 Protease Inhibitors," *Drug Metab. Dispo.* **23**(2), 185–189 (1995).
4. McGinnity, D.F.; Griffin, S.J.; Moody, G.C.; Voice, M.; Hanlon, S.; Friedberg, T.; Riley, R.J. "Rapid Characterization of the Major Drug-metabolizing Human Hepatic Cytochrome P-450 Enzymes Expressed in *Escherichia coli*," *Drug Metab. Dispo.* **27**(9), 1017–1023 (1999).

5. Crespi, C.L.; Penman, B.W. "Use of cDNA-expressed Human Cytochrome P450 Enzymes to Study Potential Drug–drug Interactions," *Adv. Pharmacol.* **43,** 171–188 (1997).
6. Ethell, B.T.; Beaumont, K.; Rance, D.J.; Burchell, B. "Use of Cloned and Expressed Human UDP-glucuronosyltransferases for the Assessment of Human Drug Conjugation and Identification of Potential Drug Interactions," *Drug Metab. Dispo.* **29**(1), 48–53 (2000).
7. Hochman, J.H.; Chiba, M.; Nishime, J.; Yamazaki, M.; Lin, J.H. "Influence of P-glycoprotein on the Transport and Metabolism of Indinavir in Caco-2 Cells Expressing Cytochrome P-450 3A4," *J. Pharmacol. Exp. Ther.* **292**(1), 310–318 (2000).
8. Rodrigues, A.D. and Lin, J.L. "Screening of Drug Candidates for Their Drug–Drug Interaction," *Curr. Opin. Chem. Biol.* **5,** 396–401 (2001).
9. Masimirembwa, C.M.; Thompson, R.; Andersson, T.B. "In Vitro High Throughput Screening of Compounds for Favorable Metabolic Properties in Drug Metabolism," *Comb. Chem. High Throughput Screening* **4,** 245–263 (2001).
10. Silva, J.M.; Morin, P.E.; Day, S.H.; Kennedy, B.P.; Payette, P.; Rushmore, T.; Yergey, J.A.; Nicoll-Griffith, D.A. "Refinement of an *In Vitro* Cell Model for Cytochrome P40 Induction," *Drug Metab. Dispo.* **26**(5), 490–496 (1998).
11. Bertrand, M.; Jackson, P.; Walther, B. "Rapid Assessment of Drug Metabolism in the Drug Discovery Process," *Eur. J. Pharm. Sci.* **11,** suppl 2, s61–s72 (2000).
12. Rodrigues, A.D. "Preclinical Drug Metabolism in the Age of High-throughput Screening: An Industrial Perspective," *Pharml. Res.* **14**(7), 1504–1510 (1997).
13. Eddershaw, P.J.; Dickens, M. "Advances in *In Vitro* Drug Metabolism Screening," *Pharm. Sci. Technol. Today* **2**(1), 13–19 (1999).
14. Korfmacher, W.A.; Palmer, C.A.; Nardo, C.; Dunn-Meynell, K.; Grotz, D.; Cox, K.; Lin, C-C.; Elicone, C.; Liu, C.; Duchoslav, E. "Development of an Automated Mass Spectrometry System for the Quantitative Analysis of Liver Microsomal Incubation Samples: A Tool for Rapid Metabolite Screening of New Compounds for Metabolic Stability," *Rapid Commun. Mass Spectrom.* **13,** 901–907 (1999).
15. Lim, H.K.; Stellingweif, S.; Sisenwine, S.; Chan, K.W. "Rapid Drug Metabolite Profiling Using Fast Liquid Chromatography, Automated Multiple-stage Mass Spectrometry and Receptor Binding," *J. Chromatogr. A* **831,** 227–241 (1999).
16. MacNair, J.E.; Patel, K.D.; Jorgenson, J.W. "Ultrahigh-Pressure Reversed-Phase Capillary Liquid Chromatography: Isocratic and Gradient Elution Using Columns Packed with 1.0-μm Particles," *Anal. Chem.* **71,** 700–708 (1999).
17. Tanaka, N.; Kobayashi, H.; Nakanishi, K.; Minakuchi, H.; Ishizuka, N. "Monolithic LC Columns. A New Type of Chromatographic Support Could Lead to Higher Separation Efficiencies," *Anal. Chem.* **73** (15), 421A–429A (2001).
18. Dear, G.; Plumb, R.; Mallett, D. "Use of Monolithic Silica Columns to Increase Analytical Throughput for Metabolite Identification by Liquid Chromatography/Tandem Mass Spectrometry," *Rapid Comm. Mass Spectrom.* **15,** 152–158 (2001).

19. Kato, K.; Jingu, S.; Ogawa, N.; Higuchi, S. "Determination of Pibutidine Metabolites in Human Plasma by LC-MS/MS," *J. Pharm. Biomed. Anal.* **24,** 237–249 (2000).
20. Marches, A.; McHugh, C.; Kehler, J.; Bi, H. "Determination of Pranlukast and Its Metabolites in Human Plasma by LC/MS/MS with PROSPEKT TM On-line Solid-phase Extraction," *J. Mass Spectrom.* **33,** 1071–1079 (1998).
21. Lim, H.K.; Chan, K.W.; Sisenwine, S.; Scatina, J.A. "Simultaneous Screen for Microsomal Stability and Metabolite Profile by Direct Injection Turbulent-laminar Flow LC-LC and Automated Tandem Mass Spectrometry," *Anal. Chem.* **73**(9), 2140–2146 (2001).
22. Purves, R.W.; Guevremont, R. "Electrospray Ionization High-Field Asymmetric Waveform Ion Mobility Spectrometry-Mass Spectrometry," *Anal. Chem.* **71,** 2346–2357 (1999).
23. Krylov, E.V. "Comparison of the Planar and Coaxial Field Asymmetrical Waveform Ion Mobility Spectrometer (FAIMS)," *Int. J. Mass Spectrom.* **225,** 39–51 (2003).
24. Cui, M.; Ding, L.; Mester, Z. "Separation of Cisplatin and Its Hydrolysis Products Using Electrospray Ionization High-Field Asymmetric Waveform Ion Mobility Spectrometry Coupled with Ion Trap Mass Spectrometry," *Anal. Chem.* **75,** 5847–5853 (2003).
25. Gabryelski, W.; Froese, K.L. "Characterization of Naphthenic Acids by Electrospray Ionization High-Field Asymmetric Waveform Ion Mobility Spectrometry Mass Spectrometry," *Anal. Chem.* **75,** 4612–4623 (2003).
26. McCooeye, M.A.; Mester, Z.; Ells, B.; Barnett, D.A.; Purves, R.W.; Guevremont, R. "Quantitation of Amphetamine, Methamphetamine, and Their Methylenedioxy Derivatives in Urine by Solid-Phase Microextraction Coupled with Electrospray Ionization-High-Field Asymmetric Waveform Ion Mobility Spectrometry-Mass Spectrometry," *Anal. Chem.* **74,** 3071–3075 (2002).
27. McCooeye, M.; Ding, L.; Gardner, G.J.; Fraser, C.A.; Lam, J.; Sturgeon, R.E.; Mester, Z. "Separation and Quantitation of the Stereoisomers of Ephedra Alkaloids in Natural Health Products Using Flow Injection-Electrospray Ionization-High Field Asymmetric Waveform Ion Mobility Spectrometry-Mass Spectrometry," *Anal. Chem.* **75,** 2538–2542 (2003).
28. McLafferty, F.W. "Tandem Mass Spectrometry (MS/MS): A Promising New Analytical Technique for Specific Component Determination in Complex Mixtures," *Acc. Chem. Res.* **13,** 33–39 (1980).
29. Perchalski, A.D.; Yost, R.A.; Wilder, B.J. "Structural Elucidation of Drug Metabolites by Triple-quadrupole Mass Spectrometry," *Anal. Chem.* **54,** 1466–1471 (1982).
30. Lee, M.S.; Yost, R.A. "Rapid Identification of Drug Metabolites with Tandem Mass Spectrometry,". *Biomed. Environ. Mass Spectrom.* **15,** 193–204 (1988).

31. Jackson, P.J.; Brownsill, R.D.; Taylor, A.R.; Walther, B. "Use of Electrospray Ionization and Neutral Loss Liquid Chromatography/Tandem Mass Spectrometry in Drug Metabolism Studies," *J. Mass Spectrom.* **30,** 446–451 (1995).
32. Vrbanac, J.J.; O'Leary, I.A.; Bacynski L. "Utility of Parent-neutral Loss Scan Screening Technique: Partial Characterization of Urinary Metabolites of U-78875 in Monkey Urine," *Biol. Mass Spectrom.* **21,** 517–522 (1992).
33. Jin, L.; Davis, M.R.; Kharasch, E.D.; Doss, G.A.; Baillie, T.A. "Identification in Rat Bile of Glutathione Conjugates of Fluoromethyl 2,2-difluro-1-(trifluromethyl)vinyl Ether, a Nephrotoxic Degradate of the Anesthetic Agent Sevoflurane," *Chem. Res. Toxicol.* **9,** 555–561 (1996).
34. McLuckey, S.A.; Van Berkel, G.J.; Goeringer, D.E.; Glish, G.L. "Ion Trap Mass Spectrometry of Externally Generated Ions," *Anal. Chem.* **66**(3), 689A–696A (1994).
35. McLuckey, S.A.; Van Berkel, G.J.; Goeringer, D.E.; Glish, G.L. "Ion Trap Mass Spectrometry Using High-pressure Ionization," *Anal. Chem.* **66**(14), 737A–743A (1994).
36. Tiller, P.R.; Land, A.P.; Jardine, I.; Murphy, D.M.; Sozio, R.; Ayrton, A.; Schaefer, W.H. "Application of Liquid Chromatography–mass Spectrometry Analyses to the Characterization of Novel Glyburide Metabolites Formed In Vitro," *J. Chromatogr. A* **794,** 15–25 (1998).
37. Zhang, N.; Fountain, S.T.; Bi, H.; Rossi, D.T. "Quantification and Rapid Metabolite Identification in Drug Discovery Using API Time-of-flight LC-MS," *Anal. Chem.* **72,** 800–806 (2000).
38. Hopfgartner, G.; Chernushevich, I.V.; Covey, T.; Plomley, J.B.; Bonner, R. "Exact Mass Measurement of Product Ions for the Structural Elucidation of Drug Metabolites with a Tandem Quadrupole Orthogonal-acceleration Time-of-flight Mass Spectrometer," *J. Am. Soc. Mass Spectrom.* **10,** 1305–1314 (1999).
39. Pilard, S.; Caradec, F.; Jackson, P.; Luijten, W. "Identification of an N-(hydroxysulfonyl)oxy Metabolite Using In Vitro Microorganism Screening, High-resolution and Tandem Electrospray Ionization Mass Spectrometry," *Rapid Commun. Mass Spectrom.* **14,** 2362–2366 (2000).
40. Hop, C.E.C.A.; Yu, X.; Xu, X.; Singh, R.; Wong, B. "Elucidation of Fragmentation Mechanisms Involving Transfer of Three Hydrogen Atoms Using a Quadrupole Time-of-flight Mass Spectrometer," *J. Mass Spectrom.* **36,** 575–579 (2001).
41. Hager, J.W. "A New Linear Ion Trap Mass Spectrometer," *Rapid Commun. Mass Spectrom.* **16,** 512–526 (2002).
42. King, R.C.; Gundersdorf, R.; Fernández-Metzler, C.L. "Collection of Selected Reaction Monitoring and Full Scan Data on a Time Scale Suitable for Target Compound Quantitative Analysis by Liquid Chromatography–Tandem Mass Spectrometry," *Rapid Commun. Mass Spectrom.* **17,** 2413–2422 (2003).
43. Warrack, B.M.; Hnatyshyn, S.; Ott, K.; Ray, K.; Tymiak, A.; Zhang, H.; Sanders, M. "Application of LTQ-FTMS to Metabonomic Profiling," in *Proceedings of*

the 52nd ASMS Conference on Mass Spectrometry and Allied Topics, Nashville, TN, May 23–27, 2004.

44. Volk, K.J.; Yost, R.A.; Brajter-Toth, A. "Electrochemistry on Line with Mass Spectrometry. Insight into Biological Redox Reactions," *Anal. Chem.* **64,** 21–23 (1992).
45. Jurva, U.; Wikstrom, H.V.; Weidolf, L.; Bruins, A.P. "Comparison between Electrochemistry/Mass Spectrometry and Cytochrome P450 Catalyzed Oxidation Reactions," *Rapid Commun. Mass Spectrom.* **17,** 800–810 (2003).
46. King, R.C.; Dieckhaus, C.D.; Nitkowski, N. "On-line Electrochemical Oxidation Used with HPLC-MS for the Study of Reactive Drug Intermediates," in *Proceedings of the 52nd ASMS Conference on Mass Spectrometry and Allied Topics*, Nashville, TN, May 23–27, 2004.
47. The cryoprobe is called a CryoProbe™ or CryoFlowProbe™ by Bruker Biospin Corp. (Billerica, MA) and a Cold Probe by Varian Inc. (Palo Alto, CA).
48. Lacey, M.E.; Subramanian, R.; Olson, D.L.; Webb, A.G.; Sweedler, J.V. "High-resolution NMR Spectroscopy of Sample Volumes from 1 nL to 10 μL," *Chem. Rev.* **99**(10), 3133–3152 (1999).
49. Olson, D.L.; Norcross, J.A.; O'Neil-Johnson M.; Molitor, P.F.; Detlefson, D.J.; Wilson, A.G.; Peck, T.L. "Microflow NMR: Concepts and Capabilities," *Anal. Chem.* **76,** 2966–2974 (2004).
50. Silva Elipe, M.V. "Advantages and Disadvantages of Nuclear Magnetic Resonance Spectroscopy as a Hyphenated Technique," *Anal. Chim. Acta*, **497,** 1–25 (2003).
51. Burton, K.I.; Everett, J.R.; Newman, M.J.; Pullen, F.S.; Richards, D.S.; Swanson, A.G. "On-line Liquid Chromatography Coupled with High Field NMR and Mass Spectrometry (LC-NMR-MS): A New Technique for Drug Metabolite Structure Elucidation," *J. Pharm. Biomed. Anal.* **15**(12), 1903–1912 (1997).
52. Peng, S.X. "Hyphenated HPLC-NMR and Its Application in Drug Discovery," *Biomed. Chromatogr.* **14,** 430–441 (2000).
53. Subramanian, R.; Kelley, W.P.; Floyd P.D.; Tan Z.J.; Webb A.G.; Sweedler J.V. "A Microcoil NMR Probe for Coupling Microscale HPLC with On-line NMR Spectroscopy," *Anal. Chem.* **71**(23), 5335–5339 (1999).
54. Lacey M.E.; Tan Z.J.; Webb A.G.; Sweedler J.V. "Union of Capillary High-performance Liquid Chromatography and Microcoil Nuclear Magnetic Resonance Spectroscopy Applied to the Separation and Identification of Terpenoids," *J. Chromatogr. A* **922**(1–2), 139–149 (2001).
55. Spraul, M.; Freund, A.S.; Nast, N.E.; Withers, R.S.; Maas, W.E.; Corcoran, O. "Advancing NMR Sensitivity for LC-NMR-MS Using a Cryoflow Probe: Application to the Analysis of Acetaminophen Metabolites in Urine," *Anal. Chem.* **75,** 1546–1551 (2003).
56. Corcoran, O.; Spraul, M. "LC-NMR-MS in Drug Discovery," *Drug Disc. Today* **8,** 624–631 (2003).

57. Plumb, R.S.; Ayrton, J.; Dear, G.J.; Sweatman, B.C.; Ismail, I.M. "The Use of Preparative High Performance Liquid Chromatography with Tandem Mass Spectrometric Directed Fraction Collection for the Isolation and Characterisation of Drug Metabolites in Urine by Nuclear Magnetic Resonance Spectroscopy and Liquid Chromatography/Sequential Mass Spectrometry," *Rapid Commun. Mass Spectrom.* **13,** 845–854 (1999).
58. Cui, D.; Subramanian, R.; Shou, M.; Yu, X.; Wallace, M.A.; Braun, M.; Arison, B.; Yergey, J.A.; Prueksaritanont, T.P. "In vitro and In vivo Metabolism of a Potent and Selective Integrin Antagonist $\alpha_v\beta_3$ Antagonist in Rats, Dogs, and Monkeys," *Drug. Metab. Disp.* **32,** 848–861 (2004).
59. Prueksaritanont, T.; Subramanian, R.; Fang, X.; Ma, B.; Qiu, Y.; Lin, J.H.; Pearson, P.G.; Baillie, T.A. "Glucuronidation of Statins in Animals and Humans: A Novel Mechanism of Statin Lactonization," *Drug. Metab. Disp.* **30,** 505–512 (2002).
60. Corcoran, O.; Mortensen, R.W.; Hansen, S.H.; Troke, J.; Nicholson, J.K. "HPLC/^{1}H NMR Spectroscopic Studies of the Reactive α-1-O-acyl Isomer Formed During Acyl Migration of S-naproxen β-1-O-acyl Glucuronide," *Chem. Res. Toxicol.* **14,** 1363–1370 (2001).
61. Silva Elipe, M.V.; Huskey, S.E.; Zhu, B. "Application of LC-NMR for the Study of the Volatile Metabolite of MK-0869, a Substance P Receptor Antagonist," *J. Pharm. Biomed. Anal.* **30,** 1431–1440 (2003).
62. Dear, G.J.; Plumb, R.S.; Sweatman, B.C.; Parry, P.S.; Roberts, A.D.; Lindon, J.C.; Nicholson, J.K.; Ismail, I.M. "Use of Directly Coupled Ion-exchange Liquid Chromatography–Mass Spectrometry and Liquid Chromatography–Nuclear Magnetic Resonance Spectroscopy as a Strategy for Polar Metabolite Identification," *J. Chromatogr. B* **748**(1), 295–309 (2000).
63. Shockor, J.P.; Unger, S.E.; Savina, P.; Nicholson, J.K.; Lindon, J.C. "Application of Directly Coupled LC-NMR-MS to the Structural Elucidation of Metabolites of HIV-1 Reverse-transcriptase Inhibitor BW935U83," *J. Chromatogr. B* **748**(1), 269–279 (2000).
64. Singh, R.; Chen, I.W.; Jin, L.; Silva, M.V.; Arison, B.H.; Lin, J.H.; Wong, B.K. "Pharmacokinetics and Metabolism of a *ras* Farnesyl Transferase Inhibitor in Rats and Dogs. In vivo–In vitro Correlations," *Drug Metab. Dispos.* **29**(12), 1–10 (2001).
65. Fernández-Metzler, C.L.; Owens, K.G.; Baillie, T.A.; King, R.C. "Rapid Liquid Chromatography with Tandem Mass Spectrometry-based Screening Procedures for Studies on the Biotransformation of Drug Candidates," *Drug Metab. Dispos.* **27,** 32–40 (1999).
66. Decaestecker, T.N.; Clauwaert, K.M.; Van Bocxlaer, J.F.; Lambert, W.E.; Van den Eeckhout, E.G.; Van Peteghem, C.H.; De Leenheer, C.H. "Evaluation of Automated Single Mass Spectrometry to Tandem Mass Spectrometry Function Switching for Comprehensive Drug Profiling Analysis Using a Quadrupole Time-of-flight Mass Spectrometer," *Rapid Commun. Mass Spectrom.* **14,** 1787–1792 (2000).

67. López, L.L.; Yu, X.; Cui, D.; Davis, M. "Identification of Drug Metabolites in Biological Matrices by Intelligent Automated Liquid Chromatography/Tandem Mass Spectrometry," *Rapid Commun. Mass Spectrom.* **12,** 1756–1760 (1998).
68. Yu, X.; Cui, D.; Davis, M.R. "Identification of In Vitro Metabolites of Indinavir by 'Intelligent Automated LC-MS/MS' (INTAMS) Utilizing Triple Quadrupole Tandem Mass Spectrometry," *J. Am. Soc. Mass Spectrom.* **10,** 175–183 (1999).
69. Le Blanc, Y.C.Y.; Bloomfield, N. "Dynamic Background Subtraction to Improve Candidate Ion Selection for Information Dependant Acquisition LC-MS/MS Analysis," in *Proceedings of the 52nd ASMS Conference on Mass Spectrometry and Allied Topics*, Nashville, TN, May 23–27, 2004.
70. King, R.C. "Automated Product Ion Interpretation for Metabolite Identification," in *Proceedings of the 49th ASMS Conference on Mass Spectrometry and Allied Topics*, Chicago, IL, May 27–31.
71. Baer, T.; Mayer, P.M. "Statistical Rice-Ramsperger-Kassel-Marcus Quasiequilibrium Theory Calculations in Mass Spectrometry," *J. Am. Soc. Mass Spectrom.* **8,** 103–115 (1997).
72. Robb, D.B.; Covey, T.R.; Bruins, A.P. "Atmospheric Pressure Photoionization: An Ionization Method for Liquid Chromatography-Mass Spectrometry," *Anal. Chem.* **72**(15), 3653–3659 (2000).
73. Van Berkle, G.J.; Asano, K.G.; Granger, M.C. "Controlling Analyte Electrochemistry in an Electrospray Ion Source with a Three-Electrode Emitter Cell," *Anal. Chem.* **76,** 1493–1499 (2004).
74. Pretty, J.R.; Deng, H.; Goeringer, D.E.; Van Berkle, G.J. "Electrochemically Modulated Preconcentration and Matrix Elimination for Organic Analytes Coupled On-Line with Electrospray Mass Spectrometry," *Anal. Chem.* **72,** 2066–2074 (2000).
75. Merchant, M.; Weinberger, S.R. "Recent Advancements in Surface-enhanced Laser Desorption/Ionization-Time of Flight-Mass Spectrometry," *Electrophoresis* **21**(6), 1164–1177 (2000).
76. Shen, Z.X.; Thomas, J.J.; Averbuj, C.; Broo, K.M.; Engelhard, M.; Crowell, J.E.; Finn, M.G.; Siuzdak, G. "Porous Silicon as a Versatile Platform for Laser Desorption/Ionization Mass Spectrometry," *Anal. Chem.* **73**(3), 612–619 (2001).
77. Vogel, J.S.; Turteltaub, K.W.; Finkel, R.; Nelson, D.E. "Accelerator Mass spectrometry: Isoptope Quantification at Attomole Sensitivity," *Anal. Chem. June* **67**(1), 353A–359A (1995).
78. Oleschuk, R.D.; Harrison, J.D. "Analytical Microdevices for Mass Spectrometry," *Trends Anal. Chem.* **19**(6), 379–388 (2000).
79. Lazar, I.M.; Ramsey, R.S.; Jacobson, C.S.; Foote, R.S.; Ramsey, M.J. "Novel Microfabricated Device for Electrokinetically Induced Pressure Flow and Electrospray Ionization Mass Spectrometry," *J. Chromatogr. A* **892,** 195–201 (2000).
80. Deng, Y.; Zhang, H.; Henion, J. "Chip-based Quantitative Capillary Electrophoresis/Mass Spectrometry Determination of Drugs in Human Plasma," *Anal. Chem.* **73,** 1432–1439 (2001).

81. Deng, Y.; Henion, J.; Li, J.; Thibault, P.; Wang, C.; Harrison, D.J. "Chip-based Capillary Electrophoresis/Mass Spectrometry Determination of Carnitines in Human Urine," *Anal. Chem.* **73,** 639–646 (2001).

82. Wachs, T.; Henion, J. "Electrospray Device for Coupling Microscale Separations and Other Miniaturized Devices with Electrospray Mass Spectrometry," *Anal. Chem.* **73,** 632–638 (2001).

83. Chartogne, A.; Tjaden, U.R.; Van der Greef, J. "A Free-flow Electrophoresis Chip Device for Interfacing Capillary Isoelectric Focusing On-line with Electrospray Mass Spectrometry," *Rapid Commun. Mass Spectrom.* **14,** 1269–1274 (2000).

84. Bousse, L.; Cohen, C.; Nikiforov, T. Chow, A.; Kopf-sill, A.R.; Dubrow, R.; Parce, J.W. "Electrokinetically Controlled Microfluidic Analysis Systems," *Ann. Rev. Biophys. Biomol. Struct.* **29,** 155–181 (2000).

85. Duffy, D.; Gillis, H.L.; Lin, J.; Sheppard Jr., N.F.; Kellogg, G.J. "Microfabricated Centrifugal Microfluidic Systems: Characterization and Multiple Enzymatic Assays," *Anal. Chem.* **71,** 4669–4678 (1999).

86. Ngola, S.M.; Fintschenko, Y.; Choi, W-Y. Shepodd, T.J. "Conduct-as-cast Polymer Monoliths as Separation Media for Capillary Electrochromatography," *Anal. Chem.* **73,** 849–856 (2001).

87. Yu, C.; Svec, F.; Frechet, J.M.J. "Towards Stationary Phases for Chromatography on a Microchip: Molded Porous Polymer Monoliths Prepared in Capillaries by Photoinitiated In Situ Polymerization as Separation Media for Electrochromatography," *Electrophoresis* **21,** 120–127 (2000).

88. Handique, K.; Burke, D.T.; Mastrangelo, C.H.; Burns, M.A. "On-chip Thermopneumatic Pressure for Discrete Drop Pumping," *Anal. Chem.* **73,** 1831–1838 (2001).

89. Corso, T.N.; Prosser, S.J.; Zhang, S.; Taheri, A.; Schultz, G.A. "An Automated Microchip-based Electrospray Device for High Throughput MS Analysis," in *Proceedings of the 47th ASMS Conference on Mass Spectrometry and Allied Topics* Long Beach, CA, June 11–15, 2000.

90. Tang, K.; Lin, Y.; Matson, D.W.; Kim, T.; Smith, R.D. "Generation of Multiple Electrosprays Using Microfabricated Emitter Arrays for Improved Mass Spectrometric Sensitivity," *Anal. Chem.* **73,** 1658–1663 (2001).

91. Ouyang, Z.; Badman, E.R.; Cooks, G. "Characterization of a Serial Array of Miniature Cylindrical Ion Trap Mass Analyzers," *Rapid Commun. Mass Spectrom.* **13,** 2444–2449 (1999).

92. Langer, R.; Vacanti, J.P. "Tissue Engineering," *Science* **260**, 920–926 (1993).

93. Kim, S.S.; Utsunomiya, H.; Koski, J.A.; Wu, B.M.; Cima, M.J.; Sohn, J.; Mukai, K.; Griffith, L.G.; Vacanti, J.P. "Survival and Function of Hepatocytes on a Novel Three-dimensional Synthetic Biodegradable Polymer Scaffold with an Intrinsic Network of Channels," *Ann. Surgery* **228**(1), 8–13 (1998).

94. van Breemen, R.B.; Nikolic, D.; Bolton, J.L. "Metabolic Screening Using On-line Ultrafiltration Mass Spectrometry," *Drug Metab. Dispos.* **26**(2), 85–90 (1998).

95. Lewis, D.F.V. "On the Recognition of Mammalian Microsomal Cytochrome P450 Substrates and Their Characteristics: Towards the Prediction of Human P450 Substrate Specificity and Metabolism," *Biochem. Pharmacol.* **60,** 293–306 (2000).
96. Gedeck, P.; Willett, P. "Visual and Computational Analysis of Structure-activity Relationships in High-throughput Screening Data," *Curr. Opin. Chem. Biol.* **5,** 389–395 (2001).
97. von Moltke, Greenblat, D.J.; Duan, S.X.; Harmatz, J.S.; Shader, R.I. "*In vitro* Prediction of the Terfenadine–Ketoconazole Pharmacokinetic Interaction," *J. Clin. Pharmacol.* **34,** 1222–1227 (1994).
98. Lin, J.H. "Applications and Limitations of Interspecies Scaling and *In Vitro* Extrapolation in Pharmacokinetics," *Drug Metab. Disp.* **26**(12), 1202–1212 (1998).

10

APPLICATIONS OF MASS SPECTROMETRY FOR THE CHARACTERIZATION OF REACTIVE METABOLITES OF DRUGS IN DISCOVERY

JIM WANG, JACK WANG, MARGARET DAVIS, WILLIAM DEMAIO, JOANN SCATINA, AND RASMY TALAAT

INTRODUCTION

Extensive research carried out over a number of decades has shown that certain drugs undergo metabolic bioactivation to electrophilic intermediates. These intermediates can bind covalently to tissue macromolecules, such as lipids, nucleic acids, and proteins. It has long been believed that this covalent binding can lead to drug toxicity. With regard to the reactions of these intermediates with proteins in particular, sites of reaction are defined, but their biological significance has proved to be difficult to establish [1]. However, it has been proposed that cell toxicity may arise from covalent modification of proteins that are "critical" to specific cell functions [2–4]. In addition, drug–protein covalent adducts can elicit drug hypersensitivities, and there exists a body of knowledge that links reactive metabolite species to idiosyncratic drug reactions [5].

If toxicity is linked to the covalent binding of reactive drug metabolites to tissue macromolecules, then it is logical that the search for the end-products

Integrated Strategies for Drug Discovery Using Mass Spectrometry, Edited by Mike S. Lee

of metabolic activation and elucidation of the mechanisms responsible for the formation of the intermediates become active areas of research. Once a reactive metabolite has been identified, medicinal chemists can be made aware of which functional group or combination of functional groups should be avoided in the design of the next-generation of drug candidates.

A common approach for the identification of end-products of reactive metabolites is to trap them with a nucleophile of known structure that is much smaller and easier to work with than the target protein. These techniques mimic the conjugation of xenobiotic electrophiles with glutathione (2-amino-4-[1-(carboxymethyl-carbamoyl)-2-mercapto-ethylcarbamoyl]-butyric acid (GSH). GSH is present in mammalian cells at concentrations as high as 10 mM. With the aid of transferase enzymes (or without, in some cases), GSH forms conjugates with electrophilic metabolic intermediates via reaction with its "soft" nucleophilic sulfhydryl group. Unfortunately, GSH cannot scavenge all reactive intermediates, particularly those that possess "hard" electrophilicity. In this work, dipeptides that contain lysine or arginine were used to trap "hard" electrophiles such as aldehyde.

Prior to investigating the mechanism responsible for the formation of a reactive metabolite, elucidation of a structure of the trapped intermediate is essential. This process makes use of extensive expertise in certain technologies, in particular, high-performance liquid-chromatographic (HPLC)–mass spectrometry (LC-MS) and HPLC-tandem mass spectrometry (LC-MS/MS). Integral, monoisotopic molecular ions can be determined by LC-MS analysis, providing molecular-weight information for the trapped intermediate. Fragmentation of the molecular ion of the compound of interest with the product-ion MS/MS mode gives a mixture of fragment ions that are useful for elucidating the structure. This type of analysis is often performed on a triple-quadrupole mass spectrometer. If further fragmentation of product ions is desired, ion-trap mass spectrometers are used to perform not only MS/MS, but also MS^3 to MS^6 experiments. Analysis with quadrupole–time-of-flight (Q-TOF) tandem mass spectrometry is able to provide accurate mass determinations. This information is useful for the determination of possible elemental compositions of molecular and fragment ions. Straightforward techniques, such as chemical derivatization to confirm the presence or absence of certain functional groups and the use of deuterated water in place of water in the HPLC mobile phase to determine the number of exchangeable protons in a trapped product, can be used in conjunction with mass-spectrometric techniques.

The process for the identification of trapped reactive intermediates can be a slow, painstaking task. Strategies to "screen" for the occurrence of these reactive metabolites to increase the attrition of potentially toxic drugs can significantly reduce the cost of the development. Much of what follows in this chapter is focused on the development of such screening procedures.

Other groups have reported on their efforts to screen drug candidates for potential toxicity based upon their ability to generate nucleophilic adducts. Baughman et al. [6] screened a group of test compounds for GSH adduct production by LC-MS analysis of preparations of drugs (100 μM) that were incubated with human liver microsomes in the presence of GSH. Conjugates were detected by MS/MS experiments with a neutral loss of 129 Da for pyroglutamic acid. A group of 43 test compounds that were expected to generate GSH adducts tested positive. Another group of 16 compounds that were not expected to generate GSH adducts tested negative. Cai et al. [7] presented data with a quantitative method that used ^{3}H-labeled GSH to trap reactive intermediates of 18 compounds with known toxicity issues. Quantification of the radioactive GSH adducts was conducted by HPLC analysis with radiometric detection. Skonberg et al. [8] tested a series of carboxylic acid drugs that were derivatized to acyl CoA thioesters. Due to the nature of the thioester bond, the CoA thioesters of carboxylic acids were more reactive toward nucleophiles than the corresponding acyl-*O*-glucuronides.

In the first subsection of this chapter, we screened several marketed drugs, some of which contain structural moieties known to produce reactive species, for metabolic activation potential by LC/MS. Figure 10.1 shows the structures of the five test drugs, loxapine (LOX), diclofenac (DCL), quinidine (QIN), ticlopidine (TIC) and troglitazone (TGZ). Rat liver microsomes and horseradish peroxidase (HRP) were used as the test systems with substrate concentrations of 25 or 50 μM. The nuclephiles GSH and *N*-acetyl cysteine (NAC) (5 mM each) were added to the incubations to trap reactive

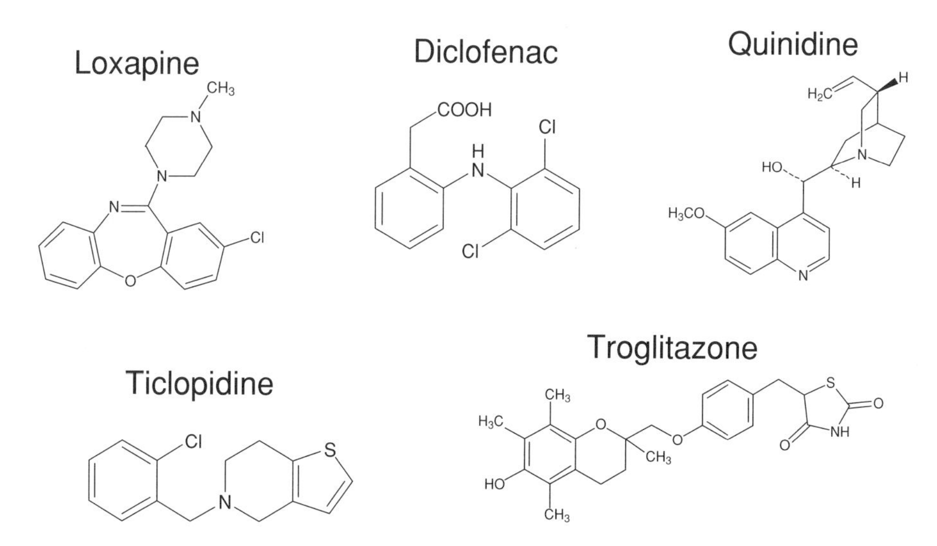

FIGURE 10.1 Structures of the drugs tested by LC-MS for metabolic activation potential.

intermediates generated by metabolic processes. The potential to form reactive metabolites was ranked based on estimates of the relative amount of GSH and NAC adducts generated.

In the second subsection the mechanism to form reactive metabolites and intermediates from a thiophene moiety was investigated. Recently, several drugs that contain the thiophene moiety have been removed from the market due to toxicity. Among these, tienilic acid, a uricosuric drug, was withdrawn due to immunoallergic hepatitis [9]. Tienilic acid is metabolized to 5-hydroxytienilic acid and a reactive sulfoxide intermediate that covalently binds to CYP2C9 appears to be responsible for the toxicity [10]. Suprofen, a nonsteroidal anti-inflammatory drug, was removed from the market because of serious renal toxicity [11]. O'Donnell et al. [12] demonstrated that suprofen was a mechanism-based inactivator of CYP2C9 and further suggested that the inactivation was due to the covalent binding of an electrophilic α,β-unsaturated aldehyde to the enzyme. Ticlopidine, an antiplatelet drug, is associated with a relatively high incidence of agranulocytosis [13,14] and aplastic anemia [15]. Lie and Uetrecht [16] have postulated that a thiophene-*S*-chloride reactive intermediate formed by activated neutrophils may be responsible for ticlopidine-induced agranulocytosis. In this study, rat liver and human microsomes were used as the test systems with substrate concentration of 50 μM. The nuclephiles glutathione (GSH, 5 mM), a dipeptide (leu-arg, 5 mM), dansylaziridine (1 μM), and iodoacetic acid (1 μM) were added to the incubations to trap reactive intermediates generated by metabolic processes. Three model compounds were selected, including methoxybenzyl thiophene and two other thiophene analogs bearing substitutions on the α carbon to the sulfur in the thiophene ring.

The third subsection focuses on acyl glucuronides (AG), which are the major metabolites for many drugs that contain a carboxylic acid moiety [18–25]. Observed clinical adverse effects of such drugs may arise from the formation of covalent binding of AG to proteins. The covalent binding may occur by two different mechanisms (Figure 12). The first is a transacylation mechanism, where a nucleophilic group ($-NH_2$, $-OH$, $-SH$) on a protein attacks the carbonyl group of the primary AG, leading to the formation of an acylated protein and free glucuronic acid. The second is a mechanism of Schiff base formation where condensation occurs between the aldehyde group of a rearranged AG and the primary amine group of the N-terminus or $-NH_2$ of a lysine residue of a protein. This mechanism results in the formation of a glycated protein [26].

In vitro assessment of the AG reactivity is likely to be of significant importance for the prediction of drug-candidate toxicity. Several procedures have been developed to assess the extent of AG covalent binding to proteins. For example, Benet et al. [26] have summarized a correlation between the extent of covalent binding of six drugs to protein and the "global degradation rate constant" (hydrolysis and intramolecular rearrangement) of the

1-O-β-glucuronide adduct. It was concluded that the nature of the -carbon substitution of the carboxylic acid group of the parent drug could affect the reactivity of the AG. Bolze et al. [27] established a correlation between the extent of covalent binding of eight drugs to protein by AG with the reappearance of the aglycone and weighted by the percentage of rearrangement of the glucuronic acid moiety. This correlation was done by complete hydrolysis of the protein adducts, followed by LC-MS/MS of released aglycone. It is relevant that the studies just mentioned used human serum albumin (HSA) as the binding target. In the procedures described, the protein adduct was precipitated and centrifuged to form protein pellets. The pellets required exhaustive wash procedures to remove reversible, noncovalently bound AG from the protein. The extent of covalent binding for the radiolabeled compounds studied was measured by scintillation counting. For studies with nonlabeled compounds, the protein pellets were subjected to hydrolysis in strong alkaline solution to release the aglycone prior to its quantitative analysis [27].

With respect to the study of AG reactivity in the drug-discovery process, the previously mentioned procedures have some limitations. The experimental procedures for the HSA studies described are indeed time-consuming. The majority of the procedures require the synthesis of the radiolabeled compounds of interest that in most cases are not available at the early stages of drug discovery. Also, the results obtained to date lacked the characterization information for the structures of the covalent adducts. To attempt to adapt such procedures to an automated method (a widely applied concept in drug discovery) would present a formidable if not impossible task with the techniques just described.

The objective of this work was to develop a novel, mechanistically well-defined, simple technique that could be easily adapted to automation. The technique uses LC-MS/MS to quantify a rearranged AG small peptide adduct instead of a protein adduct. The experiment was made up of two incubation steps followed by LC-MS analysis. Step one involved the biosynthesis of acyl glucuronides, where a drug candidate is incubated in human liver microsomes in the presence of UDPGA. The second step involved the formation of acyl glucuronide–peptide adduct, where acyl glucuronide from step one was incubated with a dipeptide (lys-phe). The reactivity of the acyl glucuronide was determined by the ratio of peptide adduct to total acyl glucuronide formed in the incubation.

Seven carboxylic acid-bearing drugs (Figure 13) that have been reported to bind covalently to protein, in vitro and in vivo, with different degrees [27–34] of reactivity, were selected to evaluate the technique. The degrees of reactivity of the AG metabolites of these seven drugs were determined based on a "reactivity index" developed for each drug. It was found that the relative reactivity rank of the AG metabolites of these seven compounds was consistent with that reported by other researchers who used HSA as the binding target [26,27]. Development of such a novel technique makes it possible to use automation to evaluate the reactivity of carboxylic acid–containing new chemical entities.

RESULTS AND DISCUSSION

Development of a Method to Screen Drug Candidates by LC-MS for Metabolic Activation Potential

Each drug was tested in four systems: (1) Rat Liver Microsome (RLM) with GSH; (2) RLM with NAC; (3) HRP with GSH; and (4) HRP with NAC. For illustrative purposes, the combined mass chromatograms obtained for each drug and their metabolites from the rat liver microsomal preparation in the presence of GSH are presented in Figure 10.2. Chromatograms that were obtained from the other three test systems are not shown. For this preliminary study, RLM and HRP were considered sufficient to demonstrate the use of surrogates for human CYP450 and peroxidase enzyme systems to generate reactive metabolites. In practice, a screen could be implemented with human liver micrsosomes, myeloperoxidase, other human enzyme systems, and possibly a single nucleophilic reagent mixture. An assumption was made that for each drug tested, all drug-related components would have the same LC-MS response. Product-ion MS/MS spectra were generated to obtain structural information for the drug adducts rather than perform constant neutral loss experiments to screen for drug adducts. This was done to take advantage of the ability of ion trap to perform data-dependent MS^n experiments. Figure 10.3 shows the product ion MS/MS spectrum (mass-to-charge ratio (m/z) 587) for hydroxy TIC GSH adducts (see Figure 10.2D, peaks 1, 2, and 3). The ion that corresponds to ($M–H^-$ pyroglutamic acid, loss of 129 Da)$^+$ was observed at m/z 458 for all three adducts, indicating conjugation with glutathione. The presence of the m/z 280 and 154 product ions in Figures 10.3A and 10.3C indicated a hydrated thiophene moiety. This suggested formation of an epoxide intermediate, which was followed by reaction with GSH to yield the observed GSH adducts. Metabolic activation of the thiophene ring was expected and is discussed in more detail later in this chapter. Figure 10.4 shows the product ions of m/z 747 spectrum and the proposed fragmentation scheme for the TGZ GSH adduct (see Figure 10.2E, peak 1). Again, neutral loss of 129 Da was observed (m/z 618), indicative of a glutathione conjugate. An unchanged chroman ring was indicated by the m/z 179 product ion.

Table 10.1 summarizes the estimated relative amounts of nucleophile adducts observed for each of the five test drugs. Calculations were based on LC-MS chromatographic peak areas. The nucleophile-adduct relative

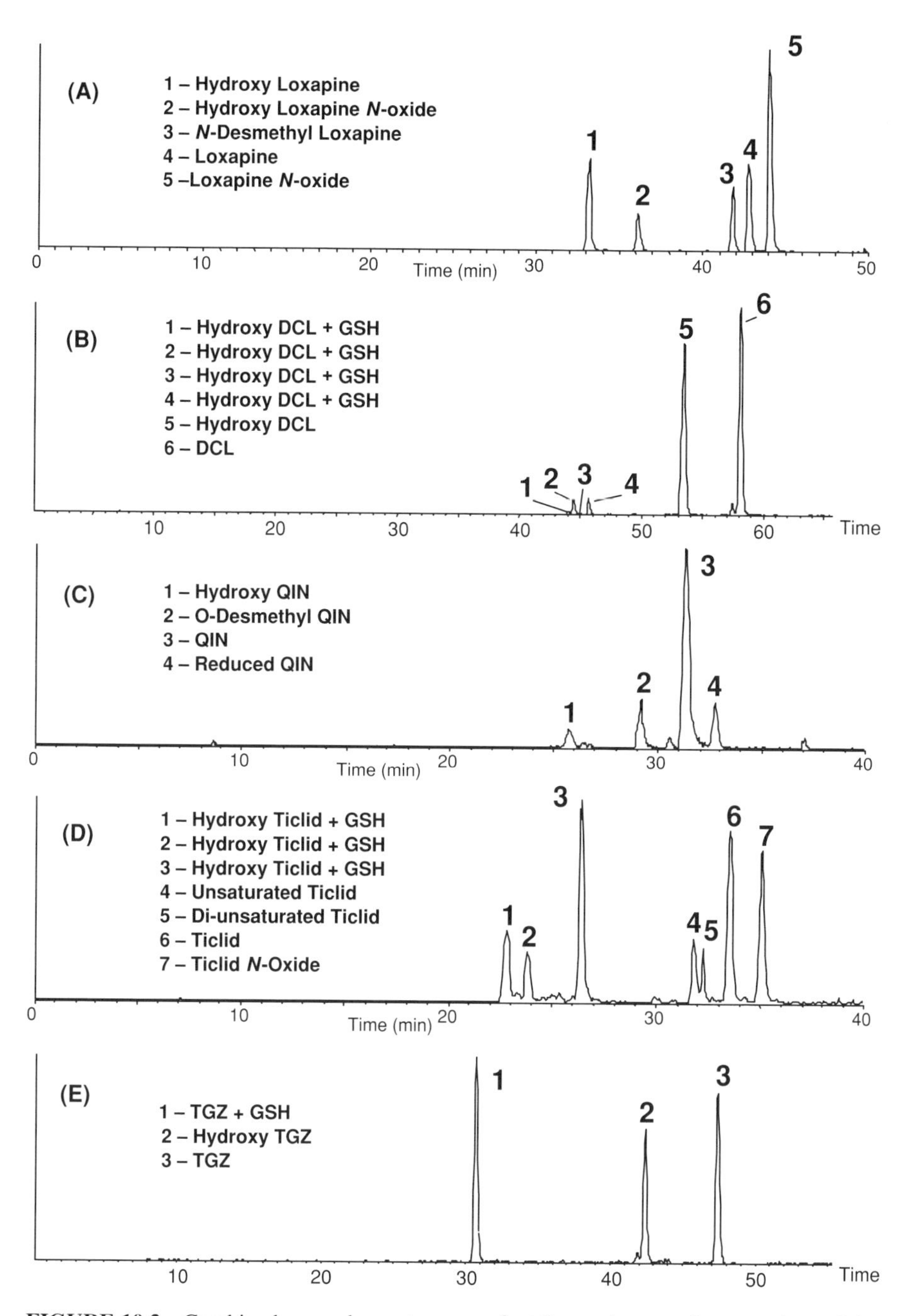

FIGURE 10.2 Combined mass chromatograms of rat liver microsomal preparations of the five tested drugs and their metabolites: (A) LOX, (B) DCF, (C) QIN, (D) TIC, and (E) TGZ.

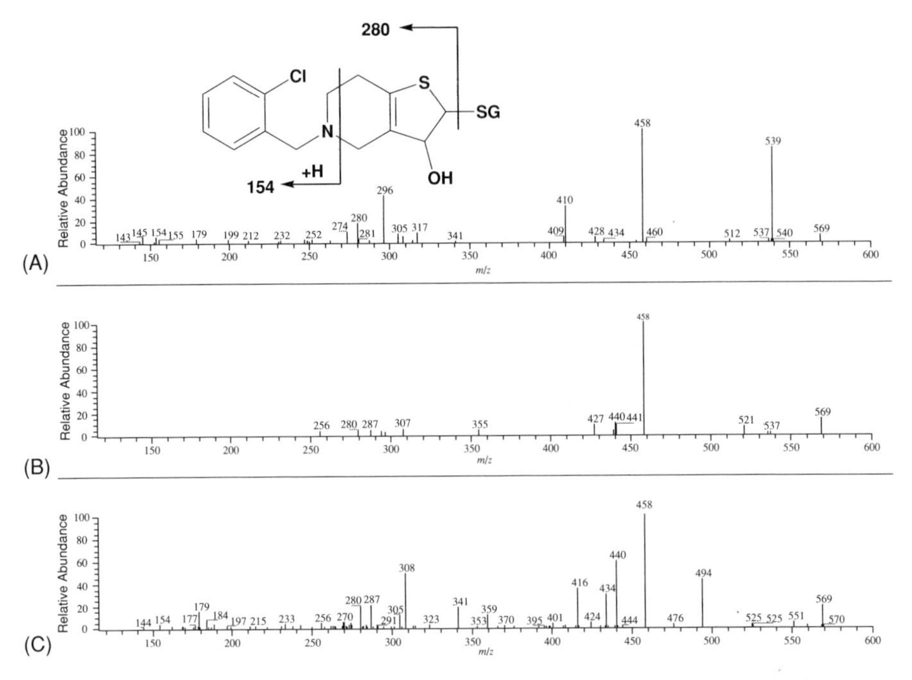

FIGURE 10.3 Product-ion MS/MS spectra (m/z 587) of hydroxy TIC GSH adducts.

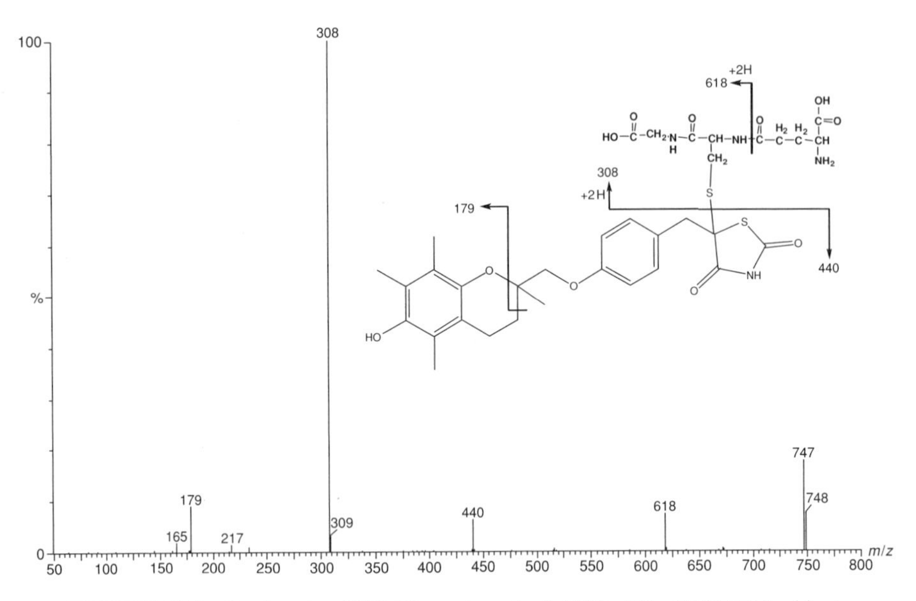

FIGURE 10.4 Product-ion MS/MS spectrum (m/z 747) of the TGZ GSH adduct.

TABLE 10.1 Summary of LC-MS Screening Results Obtained for Metabolic Activation of Five Drugs

Compound	Percent Nucleophile Adducts of Total Drug-Related Components					
	RLM +GSH	RLM +NAC	HRP +GSH	HRP +NAC	Total Nucleophile-Adduct %	Normalized Reactivity Index
Loxapine	0	0.7	0	0	0.7	0.56
Diclofenac	7.0	1.7	1.7	0.1	10.5	8
Quinidine	0	5.1	37.0	3.6	45.6	35
Ticlopidine	45.8	11.9	0	0	57.6	44
Troglitazone	42.9	4.3	74.5	9.6	131.0	100

percentages for each test system from each test drug were summed. The resulting total nucleophilic-adduct product percent for each test drug was then normalized to 100 to generate a reactive metabolite index (RMI).

As expected, LOX and TGZ had the lowest and highest RMI, respectively. The observation of a NAC adduct for LOX, a drug with no reported metabolic activation issues, demonstrates that the mere generation of nucleophilic adducts in and of itself does not indicate a potentially toxic drug. The apparent low RMI for DCL was probably due to the metabolic reactivity of DCL due to its acyl glucuronide metabolite (discussed later in this chapter), rather than from reactive oxygen species. TIC and TGZ, drugs with reported intrinsic toxicity issues, generated easily observable GSH adducts. NAC adducts for TIC and TGZ were also observed, but the LC-MS sensitivity for the GSH adducts was higher than for the NAC adducts. This would indicate that GSH has more utility than NAC for detecting reactive metabolite adducts by LC-MS. Evaluation of the actual yield of GSH versus NAC adducts could be determined with the use of radionuclide-labeled GSH and NAC. Development of a high throughput screen for metabolic activation should also include faster LC-MS analysis times, possibly based on monolithic HPLC columns.

Comparison of metabolites generated by RLM and HRP showed (1) HRP generated more reactive metabolites than RLM for QIN and TGZ, and (2) RLM generated more reactive metabolites than HRP for DCF and TIC. Therefore, peroxidases should be used in addition to, but not as a substitute for, liver subcellular fractions non-CYP450-mediated metabolic activation

Trapping Reactive Intermediates from Thiophene Moiety

Recently, several drugs that contain the thiophene moiety have been removed from the market due to toxicity. It was postulated that the toxicity was caused

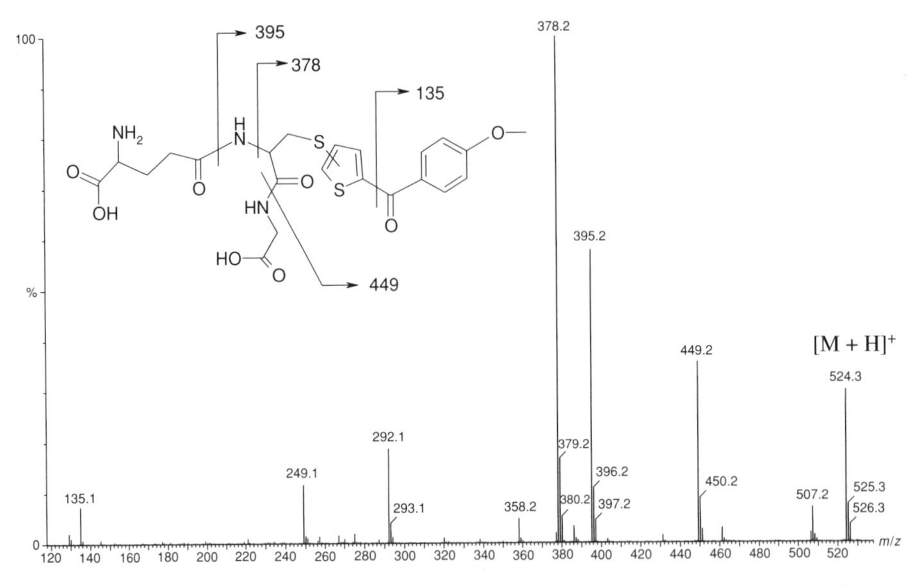

FIGURE 10.5 Product-ion MS/MS spectrum (m/z 524) and proposed fragmentation scheme of a GSH adduct of 2-(4-methoxybenzoyl)thiophene.

by the reactive intermediates formed from the thiophene moiety [9–12]. To understand the biactivation pathways of thiophene moiety, a model compound, 2-(4-methoxybenzoyl)thiophene, was incubated in rat or human liver microsomes. The major metabolites identified were an *O*-demethylated metabolite with $(M + H)^+$ at m/z 205, and a monohydroxylated metabolite with $(M + H)^+$ at m/z 235. Analysis of the incubation mixtures that contained 5-mM GSH revealed the formation of a GSH adduct at m/z 524 in both the human and rat liver microsomes. Figure 10.5 shows the product-ion MS/MS spectrum of the m/z 524 GSH adduct. Loss of glycine and pyroglutamic acid generated the product ions at m/z 449 and m/z 395, respectively. Further loss of NH_3 from m/z 395 generated the product ion at m/z 378. The product ion at m/z 135 corresponded to the 4-methoxybenzaldehyde portion of the molecule.

An additional GSH adduct at m/z 526 was observed in the rat liver microsomes, which could be formed from the addition of GSH to reduced thiophene moiety, or the addition of GSH to a ring-opened thiophene moiety. The fragmentation pattern for the second GSH adduct (m/z 526) observed in the rat liver microsomes was very different from the spectrum obtained on previous GSH adduct. The major product ions observed in the MS/MS spectrum (Figure 10.6A) were at m/z 219 and m/z 308, which corresponded to the molecular mass of 2-(4-methoxybenzoyl)thiophene and GSH, respectively. LC-MS, with D_2O (Figure 10.6B) substituted for H_2O in the mobile phase,

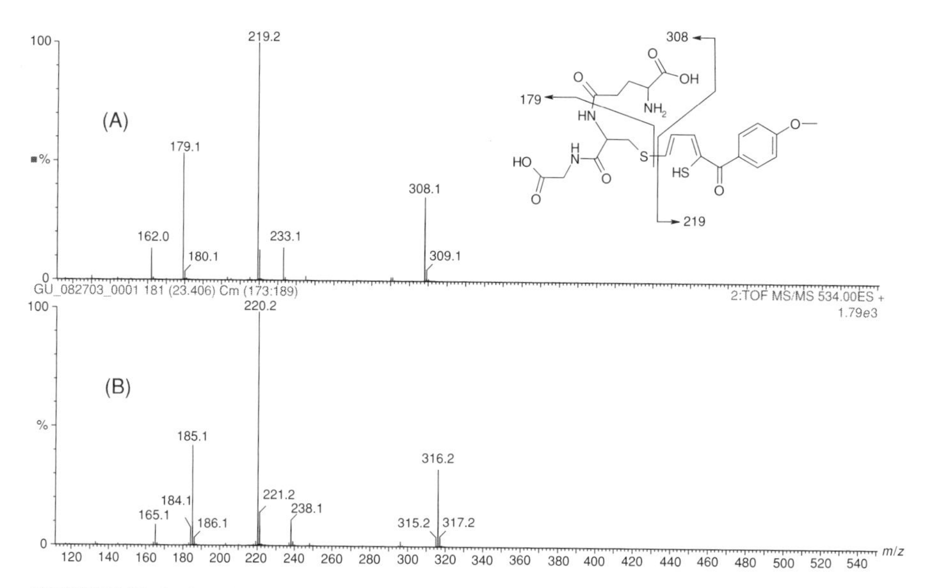

FIGURE 10.6 Product-ion MS/MS mass spectra of (A) m/z 526 (M + H), and (B) m/z 534 (M + D) of a GSH adduct of 2-(4-methoxybenzoyl)thiophene.

confirmed that the molecular weight increased to m/z 534, indicating that there were seven exchangeable protons in this GSH adduct, of which six were contributed by GSH. This result suggested that the metabolite was a GSH adduct of a free thiol formed through the opening of the thiophene ring. The free thiol provided the extra exchangeable proton and the opened ring could be envisioned to close during fragmentation to yield product ions that correspond to GSH and 2-(4-methoxybenzoyl)thiophene.

To further confirm the structure of this GSH adduct, dansylaziridine was used to derivatize the free-thiol group. Thiol groups react specifically with dansylaziridine to form dansyl-sulfonamide derivatives [17]. Figure 10.7 shows the product-ion MS/MS spectrum and the proposed fragmentation scheme of the dansylaziridine-derived ring-opened GSH adduct. Loss of glycine and GSH from the molecular ion ($[M + H]^+$ at m/z 802) generated the product ions at m/z 727 and m/z 495. The product ion at m/z 135 corresponds to the 4-methoxybenzaldehyde portion of the molecule. The fragmentation pattern was very similar for the iodoacetic acid–derived ring-opened GSH adduct shown in Figure 10.8. Loss of GSH from the molecular ion ($[M + H]^+$ at m/z 584) generated the product ion at m/z 277 and the product-ion at m/z 135, which corresponded to the 4-methoxybenzaldehyde portion of the molecule.

An electrophilic α,β-unsaturated aldehyde intermediate was also formed in rat and human liver microsomes in the presence of reduced form of nicotinamide adenosine dinucleotide phosphate (NADPH), and its structure

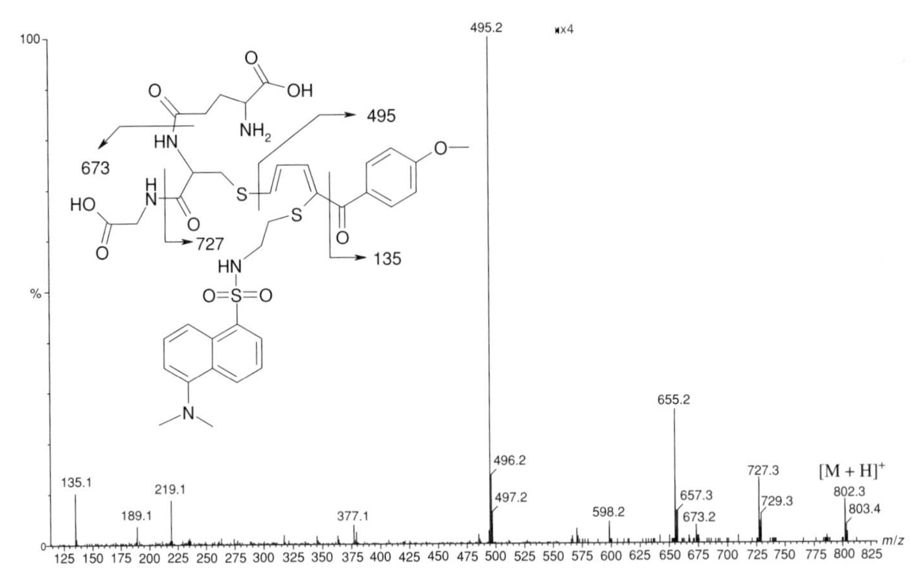

FIGURE 10.7 Product-ion MS/MS mass spectrum and the proposed fragmentation scheme of the dansylaziridine-derived ring-opened GSH adduct.

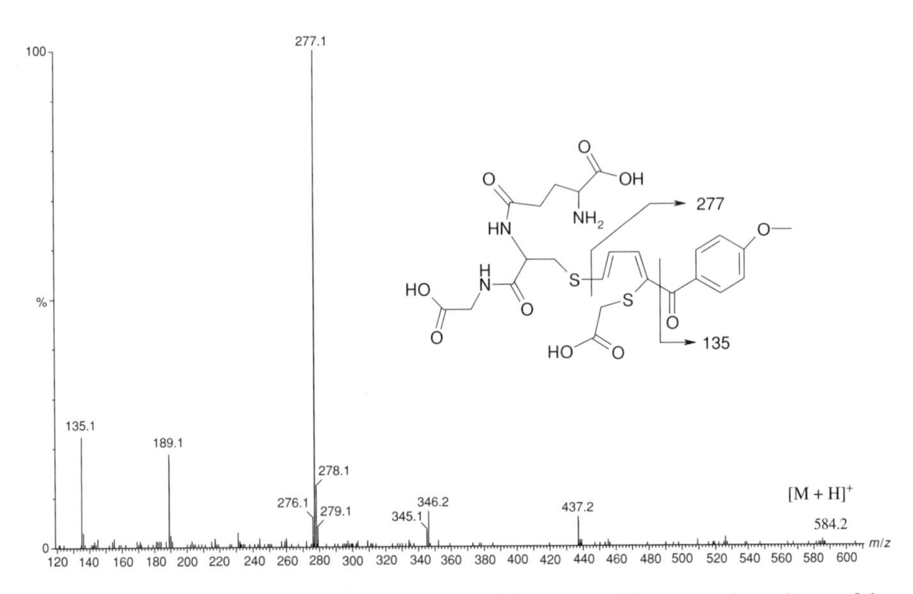

FIGURE 10.8 Product-ion MS/MS spectrum and the proposed fragmentation scheme of the iodoacetic acid-derived ring-opened GSH adduct.

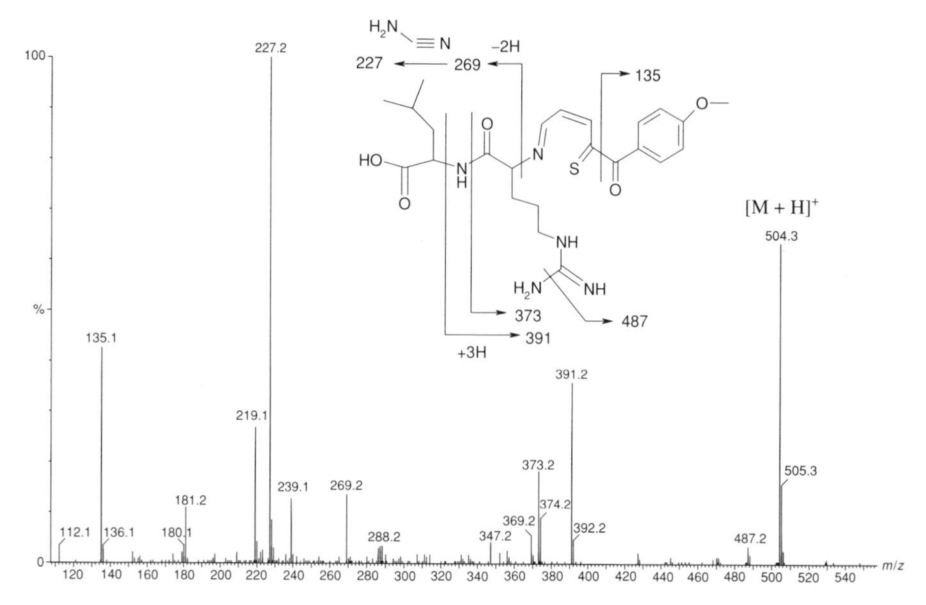

FIGURE 10.9 Product-ion MS/MS spectrum and the proposed fragmentation scheme of a leu-arg adduct of an α,β-unsaturated aldehyde intermediate, derived from a model compound.

was confirmed by the characterization of a stable adduct of leu-arg. Figure 10.9 shows the product-ion spectrum of this adduct. Loss of 4-methylpentanoic acid and leucine from the molecular ion (m/z 504) generated the product ions at m/z 391 and m/z 373, respectively. Loss of the N-terminal leu-arg generated the C-terminal ion at m/z 269, and further loss of aminonitrile (NH_2-CN) generated the product ion at m/z 227. The product ion at m/z 135 corresponded to the 4-methoxybenzaldehyde portion of the molecule.

The formation of an electrophilic α,β-unsaturated aldehyde intermediate was confirmed with the same methodology in the rat liver microsomal incubation of ticlopidine, an antiplatelet drug that causes severe hematological abnormalities. Figure 10.10 shows the product-ion spectrum and proposed fragmentation scheme of the trapped adduct. Loss of 4-methylpentanoic acid and leucine from the molecular ion (m/z 549) generated the product ions at m/z 436 and m/z 418 respectively. The product ion at m/z 288 corresponded to the molecular mass of leu-arg, and loss of leu-arg from the molecular ion generated the product ion at m/z 262. The product ion at m/z 125 corresponded to chloromethylbenzene.

Thiophene-2-3-epoxide intermediate may have been involved in the formation of the GSH adduct at m/z 524 (Figure 10.11). The GSH adduct at m/z 546

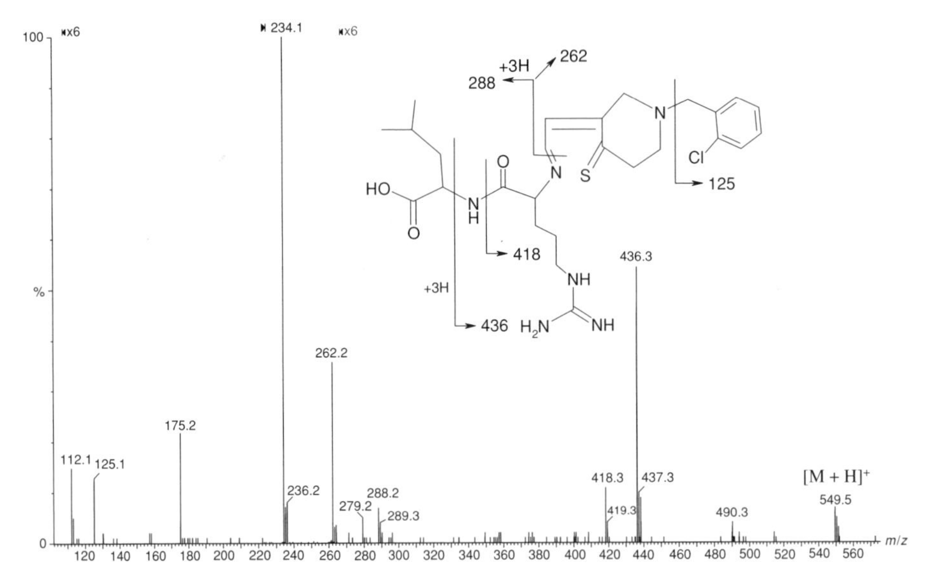

FIGURE 10.10 Product-ion MS/MS spectrum and the proposed fragmentation scheme of a leu-arg adduct of an α,β-unsaturated aldehyde intermediate from ticlopidine.

found in rat liver microsmes was formed by opening the thiophene ring following nucleophilic attack at the carbon α on the sulfur atom. The structure of this metabolite was confirmed by LC-MS and derivatization with dansylaziridine. The formation of the α,β-unsaturated carbonyl intermediate was confirmed by trapping with leu-arg. The formation of a GSH adduct through an epoxide intermediate and the trapping of an α,β-unsaturated carbonyl intermediate with a dipeptide could imply that the same type of binding might occur in vivo with proteins as targets. The formation of this adduct could contribute to the toxicity observed for this class of drugs. Two other thiophene analogs (5-bromo-2-thienyl-phenyl-methanone, and 2-[5-(4-methoxybenzyl)-2-thienyl] acetonitrile) with substitution on both α-carbons to sulfur in the thiophene ring were also evaluated. None of the reactive intermediates were observed. Blocking both α-carbons to sulfur in the thiophene ring prevented the formation of reactive metabolites. Medicinal chemists can use this information to block the active sites of thiophene moiety, therefore minimizing the formation of reactive metabolites.

Development of a Novel Technique to Predict Acyl Glucuronide Reactivity

AGs, are the major metabolites for many drugs that contain a carboxylic acid moiety [18–25]. Observed clinical adverse effects of such drugs may

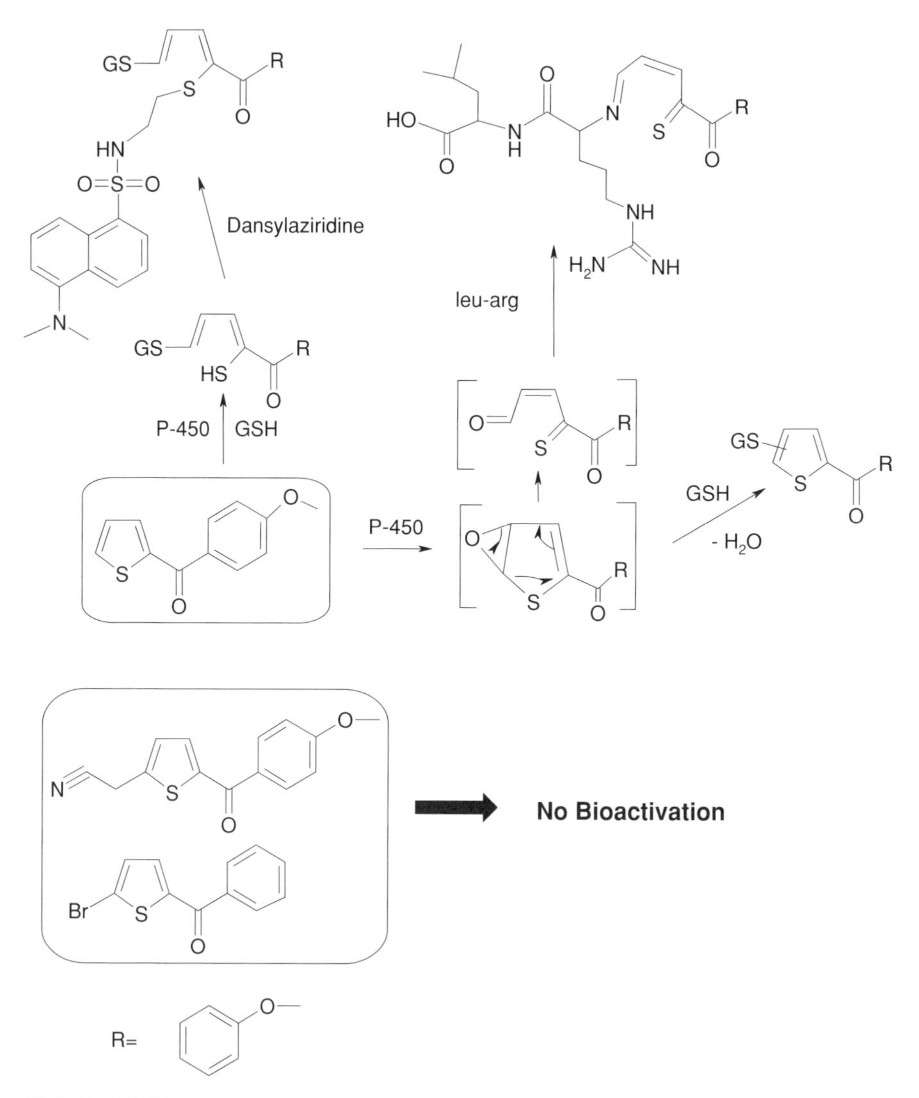

FIGURE 10.11 Proposed bioactivation pathways of thiophene-containing model compound and the schemes for trapping reactive intermediates.

arise from the formation of covalent binding of AG to proteins. The covalent binding may occur by a transacylation and Schiff base mechanisms (Figure 10.12) [26].

In vitro assessment of the AG reactivity is likely to be of significant importance for the prediction of drug-candidate toxicity. Several reported procedures [26,27] are not only time-consuming but require the synthesis of the radiolabeled compounds that are not available at early stages of drug discovery. The objective of this work was to develop a novel, mechanistically

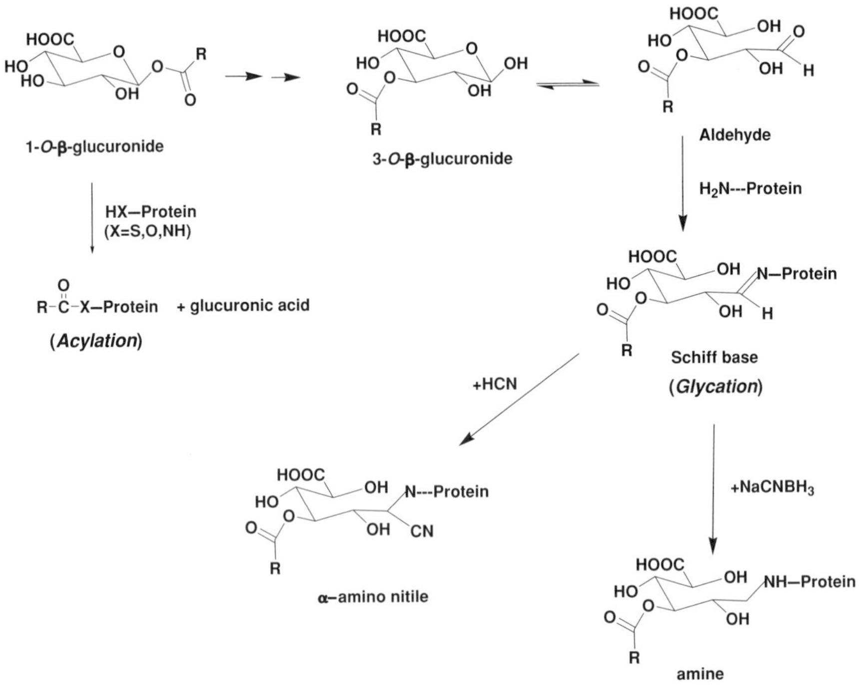
1-O-β-glucuronide
3-O-β-glucuronide
Aldehyde
H2N---Protein
HX—Protein
(X=S,O,NH)
R-C-X—Protein + glucuronic acid
(Acylation)
Schiff base
(Glycation)
+HCN
+NaCNBH3
α–amino nitile
amine

FIGURE 10.12 Schematic illustration of 1-O-β acyl glucuronide and its successive reactions: transacylation to form acylated protein; rearrangement by intramolecular acyl migration to form the 2-, 3-, and 4-C position isomers allows Schiff base formation between amine groups of proteins that lead to the formation of glycated proteins; nucleophilic addition of HCN at the imine bond in glycated proteins to form α-aminonitrile adducts; and reduction of the imine bond in glycated proteins by $NaCNBH_3$.

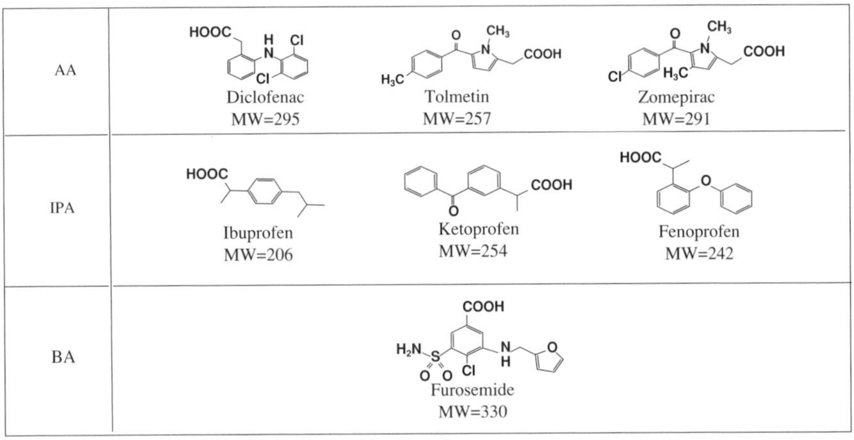
AA
Diclofenac
MW=295
Tolmetin
MW=257
Zomepirac
MW=291
IPA
Ibuprofen
MW=206
Ketoprofen
MW=254
Fenoprofen
MW=242
BA
Furosemide
MW=330

FIGURE 10.13 Structures of seven model compounds used to validate the new technique. (*Abbreviations*: AA: acetic acid derivative, IPA: isopropionic acid derivative, BA: benzoic acid derivative.)

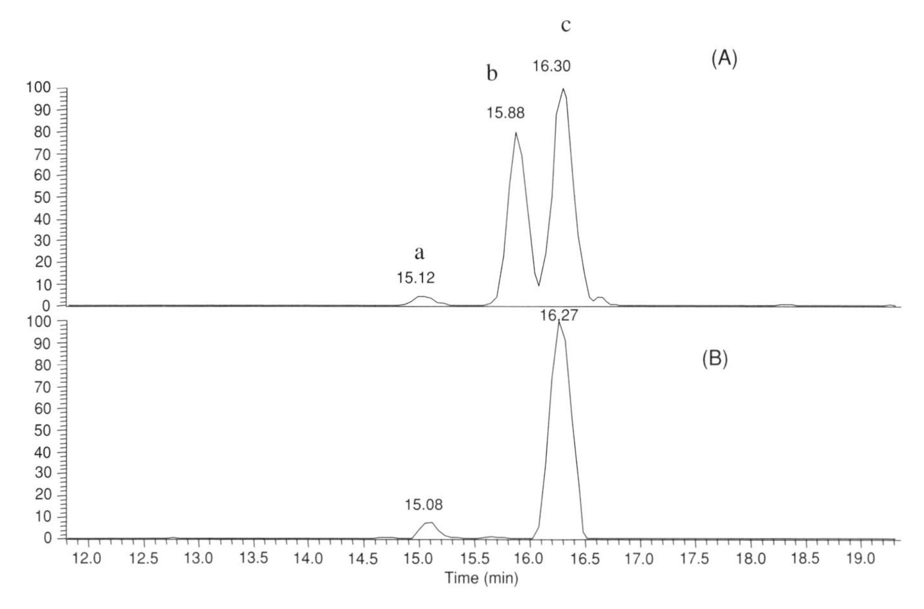

FIGURE 10.14 Molecular-ion chromatograms of deprotonated DCL-AG, m/z 470, in negative-ion ESI-MS mode. (A) One and a half (1.5) hours biosynthesis solution to generated primary AG and its isomers. (B) AG biosynthesis solution treated with β-glucuronidase, showing enzyme hydrolysis-resistant rearranged AG isomers.

well-defined, simple technique that could be easily adapted to automation. The technique uses LC-MS/MS to quantify a rearranged AG small peptide adduct instead of a protein adduct.

Figure 10.14A shows the extracted ion current chromatograms of all the AG: Acyl glucuronide isomers of DCL: Diclofenac. The two peaks shown in Figure 10.14B correspond to the rearrangement products of DCL-AGs, which were resistant to β-glucuronidase hydrolysis. Thus, the primary AG isomer was identified as peak "b." The AG rearrangement percentage ($AG_r\%$) of DCL-AG was determined by the sum of the ion-current peak areas of "a" and "c" divided by the sum of the peak areas of "a," "b," and "c" (Figure 10.14), multiplied by 100. The $AG_r\%$ values of the seven model compounds are listed in Table 10.2, and the calculation was based on the assumption that all AGs have similar ionization efficiency.

The structural characterization of the Schiff base AG adduct of DCL and the dipeptide (KF: lys-phe) B shown in Figure 10.15. This adduct displayed protonated molecular ions at m/z 747 and m/z 749 corresponding to two chlorine isotopes with an intensity ratio of 3:2 (Figure 10.15B). Data that corresponded to the peak at 18.87 minutes shows a similar spectrum (data not shown). The product ions at m/z 729 and m/z 711 correspond to the consecutive neutral losses of water molecules from the molecular ion. The

TABLE 10.2 Summary of the Results Obtained with Seven Drugs: Rearrangement Percentage and Reactivity Index of a Single Time Point (67 h)[1]

Compound Name	AGr %	C %
Tolmetin	80.0	1.5
Zomepirac	67.5	1.33
Diclofenac	58.4	0.88
Ketoprofen	36.4	0.47
Fenoprofen	33.8	0.46
Ibuprofen	15.8	0.33
Furosemide	2.0	0.03

ion at m/z 582 corresponds to a neutral loss of the phenylalanine moiety (165 Da). The ion at m/z 470 is derived from a neutral loss of dehydrated DCL (277 Da). Consecutive losses of water from m/z 470 generated ions at m/z 452, 434, and 416. The ion at m/z 294 is the protonated molecular ion of (KF: Lys-phe) (Figure 10.15C and 10.15D).

The amount of AG that is covalently bound to protein is dependent both on the concentration of AG and the degree of reactivity to form protein and peptide adducts. Historically, AG reactivity has been expressed as the percentage of total AG that is involved in the reaction with protein [26,27]. For this technique, the AG reactivity is calculated as reactivity index, "C%," which is the ratio of ion current peak areas of AG peptide adducts (peak area "a" + "b" in Figure 10.15A) to those that correspond to AGs from the same sample (peak area "a" +"b"+ "c" in Figure 10.14A) multiplied by 100. The C% for DCL-AG was determined to be 0.88%. The AG reactivities of the seven drugs were ranked based on their C% (Table 10.3).

As major metabolites of most carboxylic acid-bearing compounds, AGs have been shown to be labile electrophiles that covalently bind to the nucleophilic functional groups of tissue macromolecules. The chemical reactivity of these conjugates corresponds well with the toxicity observed for drugs that contain a carboxylic acid group [35–38]. It has been suggested that long-lived, drug-altered proteins may act as immunogens and produce cytotoxic T-cell-mediated or antibody-dependent, cell-mediated toxicity in susceptible patients (Boelsterli et al.) [39]. It has also been demonstrated with this method that Schiff base adducts of AGs showed a linear increase over the entire time range of 0–67 hours at physiological pH.

The formation of adducts through a "Schiff base mechanism" was the basis to assess the reactivity of the seven model compounds with this new technique. The reactivity index generated with our method was consistent with those reported by Benet et al. [26] and Bolze et al. [27] (Table 10.3), which validated this technique to evaluate AG reactivity. Schiff base adducts of AGs and proteins were obtained from the literature for TOL, ZOM, and DCL

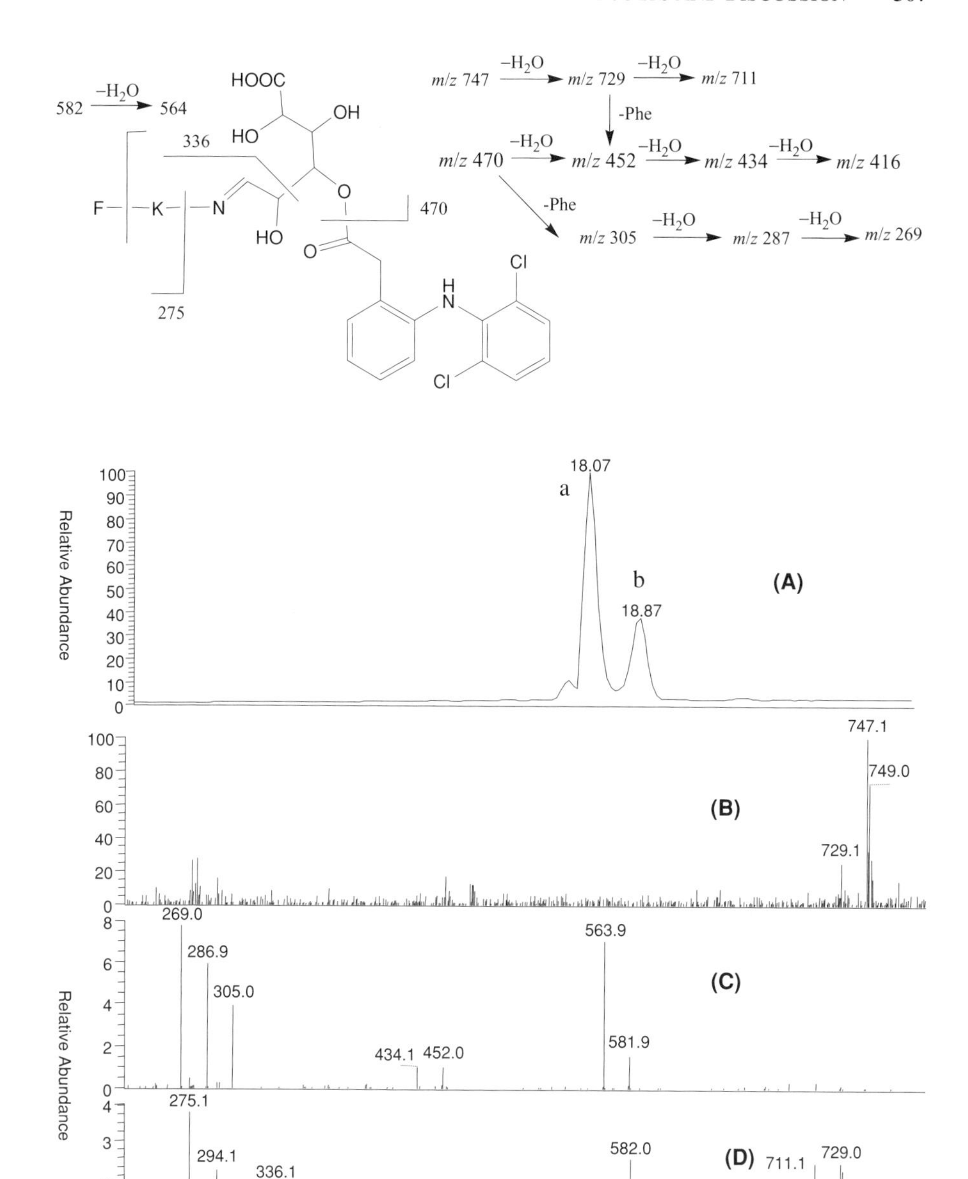

FIGURE 10.15 The LC-MS/MS analysis of DCL-AGP in the second incubation at a 67-h time point. (A) Ion chromatogram of the deprotonated molecular ion of DCL-AGP; m/z 745. (B) Positive-ion full-scan mass spectrum of the LC-MS peak at 18.07 min. (C) Product-ion MS/MS spectrum of m/z 747 at 18.07 min. (D) Product-ion spectrum of ions at m/z 747 at (RT) 18.9 min.

TABLE 10.3 Comparison of the Reactivity Ranking Obtained with This Method and Reported from the Literature (reactivity degree 7 > 6 > 5 > 4 > 3 > 2 > 1)

Compound Name	Our Result	Bolze et al.	Benet et al.
Tolmetin	7	7	7
Diclofenac	6	5	6
Zomepirac	5	6	—
Ibuprofen	4	3	—
Ketoprofen	3	2	3
Fenoprofen	2	4	—
Furosemide	1	1	1

[40–42]. Smith et al. [40] and Kretz-Rommel et al. [41] have confirmed the occurrence of Schiff base formation by the observation of increased covalent binding with the addition of NaCN to the reaction mixture of ZOM- and DCL-AGs in plasma.

The work described here demonstrated that the seven model compounds form AG peptide Schiff base adducts, and the yield is correlated with the rate of AG rearrangement. As shown in Figure 10.12, the rearrangement of AGs is a prerequisite for Schiff base–peptide adduct formation. The extent of the formation of Schiff base–peptide adducts from AGs should be proportional to the rearrangement rate of the primary AG. A correlation ($R^2 = 0.95$) between the reactivity index and percentage of rearrangement of the primary acyl glucuronide of the seven compounds is shown in Figure 10.16. Based on

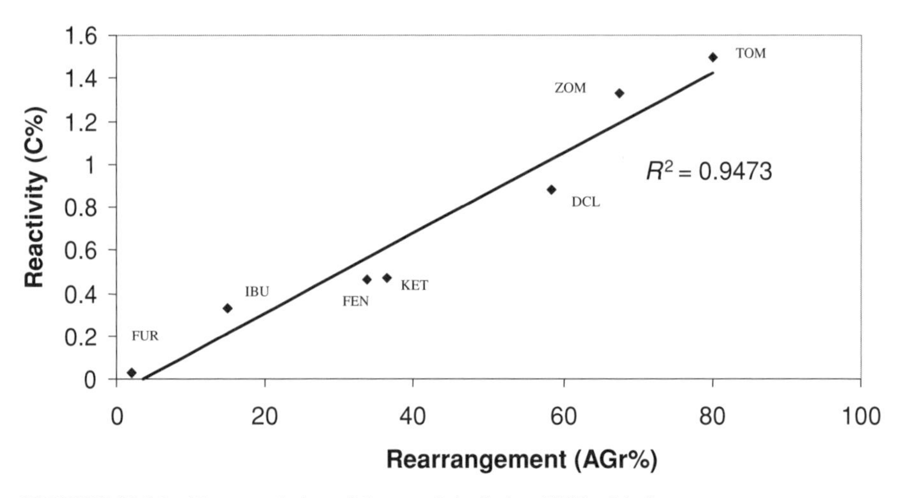

FIGURE 10.16 The correlation of the reactivity index (C%) with the rearrangement percentage (AGr%).

these results, it appears that the covalent binding reactivity of AGs to proteins can be predicted based on their rearrangement rate.

Further studies investigated the relationship between the structure of the aglycone and the rearrangement of AGs. In addition to well-known environmental factors, such as pH and temperature, the rate of AG rearrangement can be affected by the structure of the aglycone. Chemically, the rearrangement of an AG is an intramolecular transesterification process, where the acyl group migrates between the hydroxyl groups of the glucuronic acid moiety, driven by the nucleophilic attack from each adjacent –OH group. This migration process can be affected by the inherent electronic and steric properties of the aglycone. Although the number of compounds tested was limited to only seven, the order of rate of rearrangement observed was acetic acid > isopropionic acid > benzoic acid, which implies that inherent electronic and steric properties may play an important role in affecting the rate of primary AG rearrangement. It could be hypothesized that the drug containing the carboxylic acid group bound to an aromatic ring showed the lowest extent of rearrangement due to resonance stabilization provided by the aromatic moiety. Isopropionic acid derivatives display a slower rearrangement rate than those of acetic acid derivatives, possibly due to the higher steric hindrance capacity of the isopropyl group over the acetyl group.

Due to their low rate of covalent binding to protein, it might be suggested that compounds bearing benzoic acid or isopropionic acid substituents at the α-carbon of the carboxylic acid moiety should be considered when designing new chemical entities, provided these functional groups do not undermine desirable pharmacological activity.

CONCLUSION

We have demonstrated the feasibility of a method to trap reactive intermediates with the nucleophiles GSH and NAC as an approach to screening drug candidates for metabolic activation. LC-MS detection of drugs, their metabolites, and nucleophilic adducts appeared to provide a scale of the relative tendency to form reactive metabolites. We have shown the use of an integrated approach to trap and characterize the structures of reactive intermediates formed from the thiophene group. This approach encompasses a variety of techniques that include derivatization, deuterium exchange experiments, and LC-MS/MS. Finally, we have developed a simple technique to evaluate the propensity of carboxylate-type compounds to form protein adducts based on the investigation into the formation of an AG-peptide Schiff base adduct. These efforts make it possible to adapt automated methodologies to increase the throughput of reactive metabolites screening. This work will allow medicinal chemists the opportunity to minimize the formation of reactive metabolites through structural modification and develop better, safer drugs.

REFERENCES

1. Guengerich, F.P. "Common and Uncommon Cytochrome P450 Reactions Related to Metabolism and Chemical Toxicity," *Chem. Res. Toxicol.* **14**(6), 611–615 (2001).
2. Cohen, S.D.; Pumford, N.R.; Khairallah, E.A.; Boekelheide, K.; Pohl, L.R.; Amouzadeh, H.R.; Hinson, J.A. "Selective Protein Covalent Binding and Target Organ Toxicity," *Toxicol. Appl. Pharmacol.* **143,** 1–12 (1997).
3. Pumford, N.R.; Halmes, N.C. "Protein Targets of Xenobiotic Reactive Intermediates," *Ann. Rev. Pharmacol. Toxicol.* **37,** 91–117 (1997).
4. Koen, Y.M.; Hanzlik, R.P. "Identification of Seven Proteins in the Endoplasmic Reticulum as Targets for Reactive Metabolites of Bromobenzene," *Chem. Res. Toxicol.* **15,** 699–706 (2001).
5. Uetrecht, J. "Prediction of a New Drug's Potential to Cause Idiosyncratic Reactions," *Curr. Opin. Drug Disc. Devel.* **4**(1), 55–59 (2001).
6. Baughman, T.M.; Wells-Knecht, M.; Wells-Knecht, K.; Zhao, Z. "Method Validation for a Glutathione-trapping Reactive Metabolite Assay Using Drugs with Structural Moieties Known to Produce Reactive Species," *Drug Metab. Rev.* **35**(S2), 112 (2003).
7. Cai, C.; Stankovic, C.; Gifford, E.; Surendran, N.; Cai, H. "Radiolabeled Glutathione Assay: Method Development and Its Applications," *Drug Metab. Rev.* **35**(S2), 110 (2003).
8. Skonberg, C.; Sidenius, U.; Olsen, J.; Hansen, S.H. "In Vitro Reactivities of Carboxylic Acid-CoA Thioesters with Glutathione," *Drug Metab. Rev.* **35**(S2), 105 (2003).
9. Homberg, J.C.; Andre, C.; Abuaf, N. "A New and Anti-liver Kidney Microsome Antibody in Tienilic Acid Induced Hepatitis," *Clin. Exp. Immunol.* **55,** 561–570 (1984).
10. Valadon, P.; Dansette, P.M.; Girault, J.P.; Amar, C.; Mansuy, D. "Thophene Sulfoxides as Reactive Metabolites: Formation upon Microsomal Oxidation of a 3-Aroylthiophene and Fate in the Presence of Nucleophiles In Vitro and In Vivo," *Chem. Res. Toxicol.* **9,** 1403–1413 (1996).
11. Bakke, O.M.; Wardell, W.M.; de Abayo, F.; Kaitin, K.I.; Lasagna, L. "Drug Safety Discontinuations in the United Kingdom, the United States, and Spain from 1974 through 1993: A Regulatory Perspective," *Clin. Pharmacol. Ther.* **58,** 108–117 (1994).

12. O'Donnell, J.P.; Dalvie, D.K.; Kalgutkar, A.S.; Obach, R.S. "Mechanism-based Inactivation of Human Recombinant P450 2C9 by the Nonsteroidal Anti-inflammatory Drug Suprofen," *Drug Metab. Dispos.* **31,** 1369–1377 (2003).
13. Hass, W.K.; Easton, J.D.; Adams, H.P.; Pryse-Phillips, W.; Molony, B.A.; Anderson, S.; Kamm, B. "A Randomized Trial Comparing Ticlopidine Hydrochloride with Aspirin for the Prevention of Stroke in High Risk Patients," *N. Engl. J. Med.* **321,** 501–507 (1989).
14. Ono, K.; Kurohara, K.;Yoshihara, M.; Shimamoto, Y.; Yamaguchi, M. "Agranulocytosis Caused by Ticlopidine and Its Mechanism," *Am. J. Hematol.* **37,** 239–242 (1991).
15. Tsatala, C.; Chalkia, P.; Garyfallos, A.; Kakoulidis, I.; Xanthakis, I. " Ticlopidine Induced Aplastic Anemia: Case Report and Review of the Literature," *Clin. Drug Invest.* **9,** 127–130 (1995).
16. Liu, X.C.; Uetrecht, J.P. "Metabolism of Ticlopidine by Activated Neutrophils: Implications for Ticlopidine-induced Agranulocytosis," *Drug Metab. Dispos.* **28,** 726–730 (2000).
17. Scouten, W.H.; Lubcher, R.; Baughman, W. "N-Dansylaziridine: A New Fluorescent Modification for Cysteine Thiols" *Biochem. Biophys. Acta.* **336,** 421–426 (1974).
18. Bailey, M.J.; Dickinson, R.G. Acyl Glucuronide Reactivity in Perspective: Biological Consequences. [Review] [138 refs]. *Chemico-Biol. Interac.* **145**(2), 117–137 (2003).
19. Akira, K.; Uchijima, T.; Hashimoto, T. "Rapid Internal Acyl Migration and Protein Binding of Synthetic Probenecid Glucuronides," *Chem. Res. Toxicol.* **15**(6), 765–772 (2002).
20. Dansette, P.M.; Bonierbale, E.; Minoletti, C.; Beaune, P.H.; Pessayre, D.; Mansuy, D. "Drug-induced Immunotoxicity," *Eur. J. Drug Metab. Pharmacokinet.* **23,** 443–451 (1998).
21. Faed, E.M. "Properties of Acyl Glucuronides: Implications for Studies of the Pharmacokinetics and Metabolism of Acidic Drugs," *Drug Metab. Dispos.* **15,** 1213–1249 (1984).
22. Park, B.K.; Coleman, J.W.; Kitteringham, N.R. "Drug Disposition and Drug Hypersensitivity," *Biochem. Pharmacol.* **36,** 581–590 (1987).
23. Riley, R.J.; Leeder, J.S. " In Vitro Analysis of Metabolic Predisposition to Drug Hypersensitivity Reactions," *Clin. Exp. Immunol.* **99,** 1–6 (1995).
24. Spahn-Langguth, H.; Dahms, M.; Hermening, A. "Acyl Glucuronides, Covalent Binding and Its Potential Relevance," *Adv. Exp. Med. Biol.* **387,** 313–328 (1996).
25. Spahn-Langguth, H.; Benet, L.Z. "Acyl Glucuronides Revisited: Is the Glucuronidation Process a Toxification as Well as a Detoxification Mechanism?" *Drug Metab. Rev.* **24,** 5–48 (1992).

26. Benet, L.Z.; Spahn-Langguth, H.; Iwakawa, S.; Volland, C.; Mizuma, T.; Mayer, S.; Mutschler, E.; Lin, E.T. "Predictability of the Covalent Binding of Acidic Drugs in Man," *Life Sci.* **53,** 141–146 (1993).
27. Bolze, S.; Bromet, N.; Gay-Feutry, C.; Massiere, F.; Boulieu, R.; Hulot, T. "Development of an In Vitro Screening Model for the Biosynthesis of Acyl Glucuronide Metabolites and the Assessment of Their Reactivity Toward Human Serum Albumin," *Drug Metab. Dispos.* **30,** 404–413 (2002).
28. Castillo, M.; Smith, P.C. "Covalent Binding of Ibuprofen Acyl Glucuronide to Human Serum Albumin In Vitro," *Pharm Res.* **10**(S2), 42 (1991).
29. Dubois, N.; Lapicque, F.; Magdalou, J.; Abiteboul, M.; Netter, P. "Stereoselective Binding of the Glucuronide of Ketoprofen Enantiomers to Human Serum Albumin," *Biochem. Pharmacol.* **48,** 1693–1699 (1994).
30. Rachmel, A.; Hazelton, G.A.; Yergey, A.L.; Liberato, D.J. "Furosemide 1-*O*-acyl Glucuronide. In Vitro Biosynthesis and pH-dependent Isomerization to β-Glucuronidase-resistant Forms," *Drug Metab. and Dispos.* **13**(6), 705–710 (1985).
31. Ojingwa, J.C.; Spahn-Langguth, H.; Benet, L.Z. "Studies of Tolmetin Reversible and Irreversible Binding to Proteins via Its Glucuronide Conjugates," *Pharm. Res.* **5**(S2), 14 (1988).
32. Ojingwa, J.C.; Spahn-Langguth, H.; Benet, L.Z. "Irreversible Binding of Tolmetin to Macromolecules via Its Glucuronide: Binding to Blood Constituents, Tissue Homogenates and Subcellular Fractions In Vitro," *Xenobiotica* **24,** 495–506 (1994).
33. Tang, W. "The Metabolism of Diclofenac, Enzymology and Toxicology Perspectives," *Curr. Drug Metab.* **4**(4), 319–329 (2003).
34. Volland, C.; Sun, H., Dammeyer, J.; Benet, L.Z. "Stereoselective Degradation of the Fenoprofen Acyl Glucuronide Enantiomers and Irreversible Binding to Plasma Protein," *Drug Metab. Dispos.* **19**(6), 1080–1086 (1991).
35. Bailey, M.J.; Worrall, S.; de Jersey, J.; Dickinson, R.G. "Zomepirac Acyl Glucuronide Covalently Modifies Tubulin In Vitro and In Vivo and Inhibits Its Assembly in an In Vitro System," *Chemico-Biol. Interac.* **115**(2), 153–166 (1998).
36. Kretz-Rommel, A.; Boelsterli, U.A. "Diclofenac Covalent Protein Binding is Dependent on Acyl Glucuronide Formation and is Inversely Related to P450-mediated Acute Cell Injury in Cultured Rat Hepatocytes," *Toxicol. Appl. Pharmacol.* **120**(1), 155–161 (1993).
37. Dickinson, R.G.; King, A.R. "Studies on the Reactivity of Acyl Glucuronides. V. Glucuronide-derived Covalent Binding of Diflunisal to Bladder Tissue of Rats and Its Modulation by Urinary pH and Beta-glucuronidase," *Biochem. Pharmacol.* **46**(7), 1175–1182 (1993).
38. Seitz, S.; Boelsterli, U.A. "Diclofenac Acyl Glucuronide, a Major Biliary Metabolite, Is Directly Involved in Small Intestinal Injury in Rats," *Gastroenterol.* **115**(6), 1476–1482 (1998).

39. Boelsterli, U.A.; Zimmerman, H.J.; Kretz-Rommel, A. "Idiosyncratic Liver Toxicity of Nonsteroidal Antiinflammatory Drugs: Molecular Mechanisms and Pathology," *Crit. Rev. Toxicol.* **25**(3), 207–235 (1995).
40. Smith, P.C.; McDonagh, A.F.; Benet, L.Z. "Irreversible Binding of Zomepirac to Plasma Protein In Vitro and In Vivo," *J. Clin. Invest.* **77,** 934–939 (1986).
41. Kretz-Rommel, A.; Boelsterli, U.A. "Mechanism of Covalent Adduct Formation of Diclofenac to Rat Hepatic Microsomal Proteins. Retention of the Glucuronic Acid Moiety in the Adduct," *Drug Metab Dispos.* **22,** 956–961 (1994).
42. Van Breeman, R.B.L.; Fenselau, C. "Acylation of Albumin by 1-*O*-acyl Glucuronides," *Drug Metab. Dispos.* **13,** 318–320 (1985).

11

ADVANCES IN HIGH THROUGHPUT QUANTITATIVE DRUG DISCOVERY BIOANALYSIS

Bradley L. Ackermann, Michael J. Berna, and Anthony T. Murphy

INTRODUCTION

For pharmaceutical applications, the term *bioanalysis* refers to quantitative determination of a drug or its metabolites in a biological matrix. Although this term has traditionally been used to describe the analysis of in vivo samples (i.e., plasma or serum), current use of the term encompasses a broader range of applications that include the analysis of in vitro samples. Under this broader definition, possible bioanalytical sample types can range anywhere from transport media to tissue homogenate.

Despite the everincreasing number of applications associated with pharmaceutical bioanalysis, strong emphasis continues to be placed on the analysis of plasma samples from live-phase studies. The reason for this demand is that the concentration of a drug in plasma is the well-accepted surrogate marker for drug exposure and is essential to the study of pharmacokinetics (PK) and toxicokinetics (TK).

Through the use of liquid chromatography coupled with mass spectrometry (LC-MS), information on drug exposure is now routinely obtained at stages in drug discovery that were previously not thought possible. This capability has had a dramatic impact on the process of lead optimization, and ultimately,

Integrated Strategies for Drug Discovery Using Mass Spectrometry, Edited by Mike S. Lee

has resulted in more viable clinical candidates. The level of bioanalytical throughput associated with present-day LC-MS technology is not the result of a single event, but a steady stream of innovation. This chapter focuses on the evolution of this technology, with specific regard to plasma bioanalysis.

Historically, bioanalysis has been performed with chromatography coupled to one of several forms of detection that includes mass spectrometry (MS). Despite several advantages of liquid chromatography (LC) coupled to MS, growth in LC-MS applications did not occur until commercialization of the atmospheric pressure ionization (API) interface in the mid-1990s. This interface revolutionized bioanalysis and provided the first LC-MS format that was sufficiently robust to be considered for routine bioanalytical applications. The use of LC-MS with tandem MS detection (LC-MS/MS) rapidly emerged as the preferred method for bioanalysis of analytes in plasma and other biomatrices, due to the extraordinary selectivity offered by this approach against chemical background. This selectivity has in turn resulted in streamlined bioanalytical methods and has dramatically reduced analysis times and detection limits.

At the time of this writing, LC-MS/MS applications that involve API technology have been reported for more than a decade and have supplanted all other methods as the preferred approach to bioanalysis. It is interesting to note that from a historical perspective, early bioanalytical applications that involved LC-MS/MS were primarily reserved for drug development (i.e., toxicology or clinical analysis) [1,2], with comparatively little attention afforded to research. This trend quickly changed due to the well-chronicled proliferation of high throughput methods in drug discovery and the subsequent need to profile the absorption, distribution, metabolism, excretion properties (ADME) of an increased number of chemical leads. The demand for bioanalysis has firmly secured the positions of LC-MS and LC-MS/MS as essential tools for modern drug discovery. A survey of the innovations that have originated from this demand is the subject of this chapter. Interested readers are also directed toward a number of previous review articles on this subject [3–7].

In keeping with the theme of the book, this chapter only addresses bioanalytical support of drug-discovery research, as opposed to drug development. A clarification of the important distinctions between these two general areas of focus is briefly outlined in the section that follows. Also, the application of LC-MS/MS for in vitro analysis will not be covered, as this topic is addressed in other chapters.

Comparing Bioanalysis in Pharmaceutical Research and Development

It is generally agreed that drug development begins with the selection of a clinical candidate. Typically, the path that follows candidate selection is to examine the safety profile of the selected molecule in good laboratory practices

(GLP)–toxicology studies, after which viable candidates are advanced to clinical trials in humans. Bioanalysis in drug development is often subject to the regulatory rigor mandated by GLP or good clinical practices (GCP). In addition, the U.S. Food and Drug Administration (FDA) issued a guidance document that outlines scientific expectations for bioanalytical validation used to support studies that are intended for regulatory submission [8,9]. To meet these requirements, bioanalysis in drug development must focus on factors such as precision, accuracy, robustness, selectivity, and sensitivity.

While similar issues affect the success of bioanalysis during drug discovery, emphasis must be placed on fast turnaround and the ability to address a wide range of structural diversity. For example, it is not uncommon for bioanalysis to be requested for several new chemical entities (NCE) per week for any given project in the lead optimization stage. Not surprisingly, the time available for method development and validation is extremely finite compared to the drug-development setting. In addition, the conditions for analysis must accommodate the entire range in chemical diversity explored for a given structure–activity relationship (SAR) established around a particular target. Thus, extraction and chromatographic methods tend to be more generic.

Another fundamental difference, in addition to the number of NCEs encountered, is the number of samples per NCE. During the lead generation phase of drug discovery, exposure screening is often performed along with in vitro ADME screens to evaluate chemical scaffolds and to select molecules for lead optimization. It is not uncommon for most newly synthesized leads to be evaluated in a single live-phase study. This situation contrasts dramatically to drug development where several studies are performed for a single molecule.

A final difference between drug-discovery research and drug development is in the sensitivity required. As a general rule, the lower limit of quantitation (LLOQ) for research methods is in the range of 1–10 ng/mL. In contrast, most methods in drug development are performed in the pg/mL range, particularly for clinical bioanalysis. Each of the aforementioned differences has strong implications for issues that range from method development to lab automation. Interestingly, many previous discussions have not emphasized these differences and have resulted in a more generalized recommendation for bioanalysis. Although much of the material covered in this chapter has relevance for development-related bioanalysis, the intent is to provide the reader insight into the trends that have shaped current bioanalytical strategies for drug discovery.

OVERVIEW OF BIOANALYSIS

To gain insight into how LC-MS is used for bioanalysis for modern drug discovery, it is useful to break the overall process down into its component

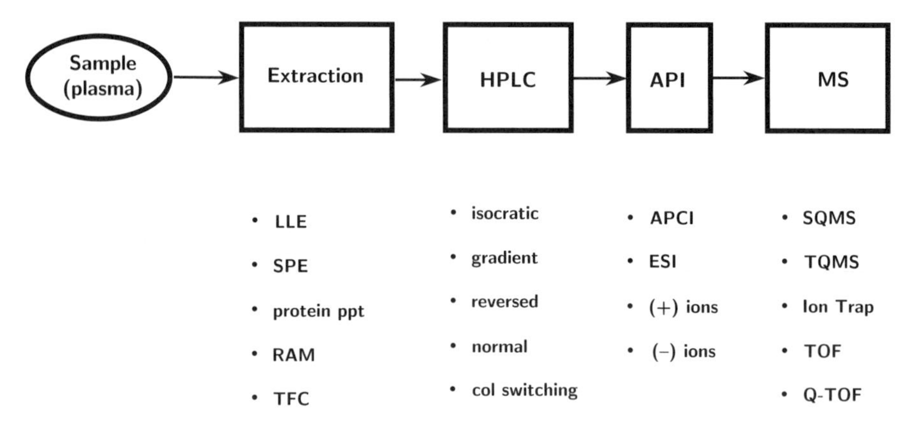

FIGURE 11.1 A generalized flow scheme that indicates the fundamental elements of LC-MS–based bioanalysis. *Abbreviations*: LLE = liquid–liquid extraction; SPE = solid-phase extraction; RAM = restricted-access media; TFC = turbulent flow liquid chromatography; API = atmospheric-pressure ionization; APCI = atmospheric-pressure chemical ionization; ESI = electrospray ionization; SQMS = single-quadrupole mass spectrometry; TQMS = triple-quadrupole mass spectrometry; TOF = time-of-flight; Q-TOF = quadrupole TOF. (Reprinted from Ackermann et al. [4], with permission from John Wiley & Sons, Inc.)

parts. The basic elements in the bioanalytical flow path are depicted in Figure 11.1. The process begins with samples that contain target analytes present at physiologically relevant concentrations in a given biomatrix. Although a number of biomatrices may be considered (e.g., tissue, urine, liver), plasma will be assumed as the sample matrix for the present discussion. Regardless of the path taken, the endpoint is the mass spectrometer. It is for this reason that method development commonly begins with procedures to establish conditions for ionization and detection prior to the investigation of conditions for chromatography and extraction. Each of the aforementioned components (Figure 11.1) is addressed in the following sections. The discussion begins with MS considerations and is followed by a review of advances in sample preparation strategies and LC.

Ionization

API interfaces have two available modes for operation, electrospray ionization (ESI) and atmospheric-pressure chemical ionization (APCI). The mechanistic aspects of ESI [10–12] and APCI [13] have been well covered in the literature. ESI, regarded as the "softer," more versatile of the two methods, is able to ionize extremely polar/nonvolatile molecules, sometimes difficult for APCI. APCI mass spectra often contain fragment ions from the analyte due to the harsher ionization conditions. In cases where metabolites (e.g., *N*-oxides) co-elute with the parent drug, such ions can give rise to mass-spectral interference, and artificially contribute to the parent signal [14]. On the other hand, APCI is generally regarded as the more robust ionization method and is less susceptible to matrix effects (e.g., ion suppression) from co-eluting endogenous components [15].

Both APCI and ESI allow for detection in either positive- or negative-ion mode. Although negative-ion detection is indicated for certain compound classes (carboxylic acids, hydoxramic acids, sulfates, phenols), most pharmaceutical molecules are relatively basic and are better suited to positive-ion mode. It is often possible to detect negative ions by ESI under the acidic conditions frequently employed with reverse-phase high-performance liquid chromatography (HPLC), although a reduction in sensitivity is to be expected. Consequently, ammonium acetate is commonly used as a mobile-phase additive for negative-ion applications to achieve a pH closer to neutral. Negative-ion APCI is probably the least frequently used of the four modes of operation, but from our experience this mode should not be overlooked. Schaefer and Dixon studied the influence of mobile-phase modifiers on negative-ion APCI and suggested the use of *N*-methylmorpholine as a modifier to enhance ion formation in this mode [16].

Positive-ion ESI with reverse-phase liquid chromatography (RP-LC) is the starting point for most applications. Aqueous mobile phases combined with either methanol (MeOH) or acetonitrile (ACN) are the most frequently used solvent systems. Mobile-phase additives, used to adjust pH or to enhance ionization, are also common. The most commonly used acids include formic, acetic, and trifluoroacetic acid (TFA). Although TFA can improve LC peak shape for basic compounds, it causes a dramatic reduction in signal intensity through an ion-pairing mechanism [17]. If it is necessary to use TFA, the concentration should not exceed 0.05% (v/v). Moreover, TFA should not be used with negative-ion ESI, as complete analyte signal suppression often occurs [18].

The relatively volatile ammonium buffers are commonly used in conjunction with ESI and APCI for separations that require elevated pH. Ammonium formate can be used in the approximate pH range of (3.5–5.0), while ammonium acetate can reach ranges that approximate neutral pH (4.5–6.0). For negative-ion applications, ammonium hydroxide can be used to adjust the pH above neutral if used with polymeric phases or newer base-tolerant silica-based columns. As a general rule, volatile buffers or additives should not exceed 20 mM, and nonvolatile buffers (K^+/Na^+ phosphate or acetate, tris (hydroxymethyl) amino methane (TRIS), etc.) should be avoided completely. In a recent study performed by King et al. that involved the study of ion suppression, it was shown that nonvolatile buffers inhibit ionization, in large degree by limited desolvation [19]. One would predict that this colligative effect would impair APCI response as well.

It is widely recognized that sensitivity in either ESI or APCI is improved as the percentage of organic modifier is increased, due to the improved facility for desolvation. Adequate reverse-phase retention ($k' = 1$ to 5) is therefore important when LC-MS is performed. Temesi and Law studied the effect of LC mobile-phase composition on ESI response for a series of 35 compounds [18]. In their investigation, the signal response for positive-ion ESI was on average 10–20% higher in MeOH than ACN.

Although mobile-phase composition can have a significant effect on ionization, full optimization cannot occur in the research setting due to time limitations. It is therefore desirable to have a few standard mobile-phase combinations that address most situations. Formic acid (20 mM), in combination with MeOH or ACN, is probably the best additive for positive-ion ESI (or APCI). Ammonium acetate (10–20 mM) is a useful alternative to formic acid, which may also be used for negative-ion applications of ESI or APCI.

Any comparison between APCI and ESI will ultimately conclude that both techniques are useful and that neither method can truly be considered a universal ionization source. Unfortunately, because of the time needed to switch between methods, most laboratories responsible for bioanalysis tend to rely on a single method, and develop new methods only when the current ionization method is not successful. To afford greater flexibility, instrument vendors now offer dual ESI/APCI sources that can switch between ionization modes on consecutive scans within a LC run [20,21]. The original concept can be traced to the work of Siegel et al., who demonstrated the first viable source of this type [22]. Although this concept is quite new, it could prove very useful for research laboratories required to produce expedient LC-MS conditions for a wide range of chemical structures.

Another recent variation in LC-MS source design is the introduction of a technique known as atmospheric-pressure photoionization (APPI) [23]. This technique closely resembles APCI, and the key modification is that the discharge needle needed for APCI is replaced by an ultraviolet (UV) lamp. In this mode, ionization may be enhanced by the co-introduction of a photoionizable liquid dopant, such as toluene. The highly abundant dopant ions are believed to participate in the ionization of the sample by charge exchange or proton transfer. Although this technique has yet to be fully tested for bioanalytical applications, Yang and Henion compared APPI with APCI for the analysis of idoxifene and its metabolites in human plasma and found that APPI was more sensitive for the analytes studied [24]. While early information suggests that APPI and APCI have considerable overlap in their capabilities, further evaluation is needed.

Mass Analysis

Several mass analyzers have been successfully applied for bioanalysis, including single quadrupole [25],triple quadrupole [2], quadrupole ion trap [26,27], time-of-flight (TOF) [28], and quadrupole time-of-flight (Q-TOF) [29,30]. A description of the operating principles for each mass analyzer is beyond the scope of this chapter. For bioanalytical applications, mass analyzers are rated on their ability to perform two functions: (1) discriminate against matrix interference, and (2) maximize the signal for the ion(s) of interest. The latter operation is a function of both transmission and duty cycle, where the duty cycle refers to the fraction of the scan function devoted to the transmission of the target ion of interest.

In a quadrupole mass analyzer, only a single mass-to-charge ratio (m/z) value is transmitted to the detector for any given combination of radio frequency (RF) and direct current (DC) potentials. Typically, the RF/DC ratio is held constant and scanned to provide a mass spectrum. If, for example, a quadrupole is scanned from m/z 1 to 1000 in 1 second, then any particular m/z is transmitted to the detector for only 1 millisecond, representing a duty cycle of 0.1%. Thus, a quadrupole mass analyzer has a *low* transmission duty cycle in the *full-scan* mode, which results in limited full-scan sensitivity. In contrast, ion-trap and TOF mass analyzers have the theoretical potential to transmit all ions that enter the mass analyzer and yield far better sensitivity across the entire mass spectrum. In reality, the pulse sequences associated with these analyzers devote significant time to functions such as ionization and detection. The actual duty cycles are generally between 10 and 25%, still far better than a scanning quadrupole mass spectrometer.

For quantitative applications, the sensitivity of a quadrupole mass analyzer can be greatly increased by only monitoring the target ion or ions of interest. This technique is known as *selected ion monitoring* (SIM). It is possible to achieve a duty cycle of 100% with SIM techniques (if only one m/z is monitored). Despite this gain in sensitivity with SIM, the signal-to-noise (S/N) ratio is ultimately limited by the chemical interference from the background ions that originate from the matrix. For this reason, mass analyzers that afford greater selectivity are generally sought for bioanalytical applications.

The most widely used tandem mass analyzer for bioanalysis is the triple-quadrupole mass spectrometer (TQMS). In the TQMS format, a characteristic fragmentation transition is monitored for an analyte with more than one mass filter in combination, hence the term MS/MS. A diagram of a TQMS is presented in Figure 11.2. A TQMS consists of two mass selecting quadrupoles, Q_1 and Q_3, and flanked by a center quadrupole (Q_2) that acts as a collision cell. Since Q_2 is operated in the RF-only mode, mass selection is not performed in this region. The Q_2 rods are important, however, since they refocus ions

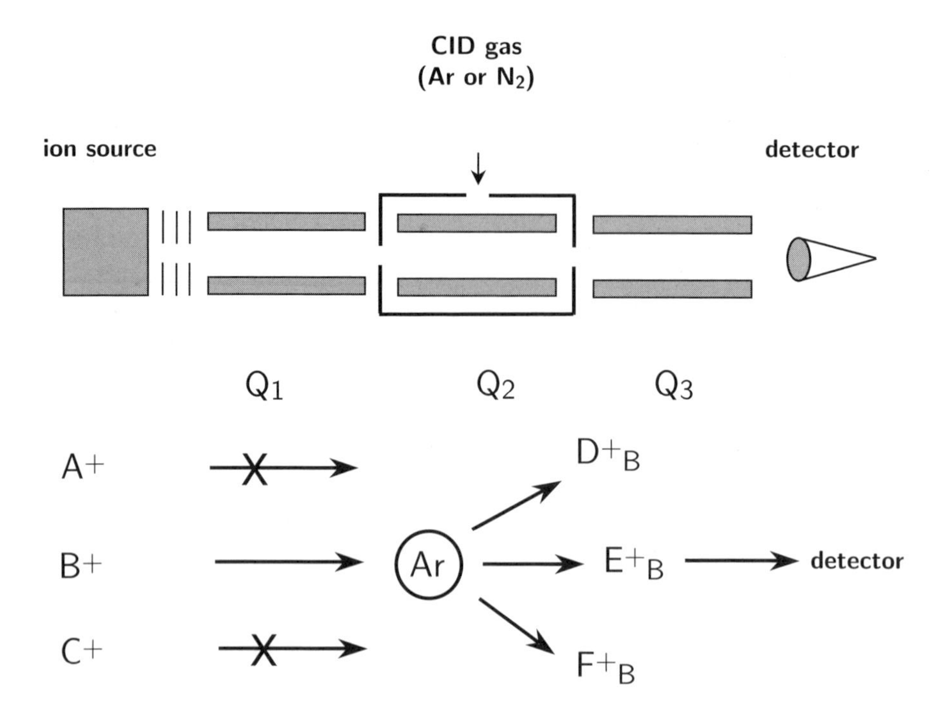

FIGURE 11.2 A schematic diagram of a TQMS that illustrates the process of selected reaction monitoring (SRM). In this process, quadrupole 1 (Q_1) is set to transmit only the precursor ion of the target analyte. This ion (B^+) is induced to fragment through collisions with gas molecules (argon or nitrogen) in a RF-only quadrupole collision cell (Q_2). Quadrupole Q_3 is set to transmit a single product ion m/z to the detector. The ability to monitor a fragment transition specific to the analyte confers sensitive detection in biological matrices. (Reprinted from Ackermann et al. [4], with permission from John Wiley & Sons, Inc.)

scattered by collisions with the target gas (e.g., argon or nitrogen). Thus, a molecular ion for the analyte is selected by Q_1 and induced to fragment in Q_2, after which a predominant fragment ion is monitored by Q_3. This process, termed *selected reaction monitoring* (SRM), offers increased S/N in cases where detection is limited by chemical background. Like SIM, SRM detection delivers a high duty cycle since the quadrupole filters are not scanned. It is important to note that the effective duty cycle is diluted proportionally as more channels are monitored to detect multiple analytes.

For most applications that involve quadrupoles, the resolution is set to resolve adjacent m/z values (i.e., unit resolution). The MS peak width in this case is generally about 0.7 atomic mass unit (u) full width half maximum (FWHM). Recently, a commercial TQMS was introduced that features enhanced mass resolution. Two citations have appeared in the literature on the application of this instrument for quantitative bioanalysis [31,32]. In these studies, the Q_3 resolution was set to give peaks widths as low as 0.1 u. From the data presented thus far, the S/N appears to remain constant as the mass resolution is changed. In addition, roughly equivalent bioanalytical results were obtained when low- and high-resolution modes were compared. Nevertheless, examples have been found where the added resolution has proved useful. An example reported by Yang et al. occurred when a major interfering peak had the same nominal m/z as the analyte of interest [32].

Another approach used to achieve discrimination against chemical background is to rely solely on enhanced mass resolution. Several reports have appeared on the potential use of ESI-TOF for bioanalysis with mixed success [28,33,34]. Needham et al. demonstrated the power of the improved mass resolution of TOF to achieve selectivity for the analysis of midazolam and its metabolites [33]. More recently, Zhang and Henion demonstrated the use of TOF for bioanalysis of idoxefine [34].

An inherent advantage of TOF relates to its ability to deliver a high duty cycle while ions are detected over a wide m/z range. This means that it is possible to detect multiple analyte ions without a concomitant loss in signal intensity. The advantage of TOF for simultaneous analysis of drugs and their metabolites was cited by Zhang et al., who used TOF for the bionanalysis of tricyclic antidepressants in rat plasma following cassette dosing [28]. However, in this study the LLOQ were on average fivefold higher than observed by TQMS, even though liquid-liquid extraction (LLE) was employed. The general consensus on TOF detection is that this mass spectrometry format does not achieve the same S/N levels as TQMS for analytes in biological matrices. Consequently, expanded sample cleanup may be necessary to achieve LLOQ values low enough to support lead optimization (1–10 ng/mL).

A number of comparisons have occurred between TQMS and other mass analyzers [27,28,30]. In all instances, TQMS gave superior performance in terms of sensitivity and linear dynamic range. It should be mentioned that, with the exception of Q-TOF technology, TQMS is more expensive than other options. Recently, a TQMS instrument was introduced and provides two-dimensional (2D) linear ion-trap capability in the Q_3 region [35,36]. Although this instrument has not been rigorously tested for bioanalysis, the expanded capabilities are primarily targeted for metabolism and proteomics, and it may be difficult to justify the added expense for laboratories that focus strictly on bioanalysis. Ultimately, the popularity of TQMS endures and can be attributed to its high selectivity against matrix background, which translates into faster method development and lower detection limits. Robustness and ease of use are other factors that contribute to this popularity.

Sample Preparation

Prior to the advent of routine LC-MS/MS, sample preparation was not the rate-limiting factor in bioanalysis. This was primarily due to the excessive run times associated with the chromatographic methods. As LC-MS/MS became more routinely used in the mid-1990s, attention was devoted to approaches to increase the sample preparation throughput. One of the most significant changes that occurred was the introduction of the 96-well format for sample preparation. In today's drug-discovery research environment, virtually all sample preparation techniques have migrated to a 96-well format and incorporate some degree of automation [37–39].

Although similar automation tools are used for both drug-discovery and drug-development applications, it is essential to understand that the key drivers for sample preparation differ. For instance, in drug-discovery research, it is more important for methods to be *generic* than *selective*, due to the limited time available for methods development, the high number of compounds investigated, and the fact that less rigorous guidelines govern drug-discovery bioanalysis. For these reasons, highly selective techniques, such as molecularly imprinted polymers (MIPS) [40] or immunoaffinity methods [41], are not widely applied to research bioanalysis. Expanded solid phase extraction (SPE) formats, such as 384-well [42,43], are also not used.

Another distinction about discovery bioanalysis is that *ease* of methods development is often more important than the net *speed* at which samples can be processed. This statement is again a manifestation of the high number of NCEs that are encountered in discovery. Consequently, enormous growth has occurred with on-line methods that combine sample preparation and analysis in a single injection format. Although several formats exist, the common link to all on-line methods is that they invoke column-switching techniques. The popularity of these methods can be traced, in part, to the ability to adjust extraction/analysis conditions on-the-fly and leads to extremely facile method development. In the section that follows, "off-line" and "on-line" methods are considered separately. Coverage of these subtopics can also be found in a number of review articles [4–7,44,45].

Off-Line Sample Preparation

The three main formats for sample preparation used in drug-discovery are protein precipitation (PPT), SPE, and LLE. Several examples of off-line sample preparation have been reported and involve SPE [37,38,46,47], LLE [38,48], and PPT [39,49]. In each of the examples cited, semi- or fully automated strategies for liquid handling were incorporated to enhance throughput. Even with the recent popularity of on-line methods, off-line techniques continue to be widely employed. The key advantage to off-line methods is that sample preparation may be independently optimized from the mass spectrometer and does not contribute overhead to the LC-MS injection duty cycle.

Of the techniques just listed, PPT is the most widely practiced method for plasma bioanalysis, due to its simplicity, low cost, and wide range of application. Despite these advantages, it must be acknowledged that PPT is not an extraction method and results in sample dilution. Due to these shortcomings, it is often necessary to use SPE or LLE to achieve sub-ng/mL quantitation.

Protein precipitation in 96-well format is commonplace and frequently involves 96-well pipetting stations. These techniques are considered semi-automated, since manual intervention is needed for protein removal. Protein removal usually occurs by centrifugation, although filtration has also been used [50]. Due to the sensitivity of the current-generation TQMS instruments, the amount of plasma taken for analysis is typically less than 100 μL. In our laboratory, protein precipitation with 50 μL of plasma is typically sufficient to achieve LLOQs in the range of 1 to10 ng/mL. Interested readers are referred to a recent paper by Polson and co-workers, who undertook a detailed investigation with regard to factors that affect optimization of PPT for LC-MS/MS [51].

SPE is a second widely used method for sample preparation. Unlike PPT, SPE allows for both sample cleanup and sample concentration. Other advantages to SPE are that it is relatively easy to optimize and offers several stationary phase options. Certainly, the single most significant change in SPE occurred with the introduction of a 96-well format [46,52]. However, since that time several noteworthy refinements have occurred, and an overall increase in the quality and characterization of the stationary phases used has been benchmarked [53].

One of the major advances in SPE technology has been the introduction of the particle-loaded membrane disk format [54]. In this design, SPE stationary-phase particles are impregnated within an inert polytetrafluoroethylene membrane (1 mm) and resulted in very finite bed volumes. The extraction can be accomplished with relatively little packing material (e.g., 20 mg), since research applications do not require large plasma sample volumes ($\leq$100 μL). The combined result of these factors is the ability to perform limited volume elution (e.g., 50, μL). The advantage to this approach is that it is often possible to avoid a dry-down step prior to injection [47].

A characteristic of SPE methods that supports research applications is that they tend to be universal. This would explain why C-18 continues to be so popular even though more selective extraction is often possible with other phases. In recent years, a polymeric phase, known as Oasis HLB,™ was introduced to accommodate the wide scope in analyte polarity encountered in drug discovery [55]. This particular phase is composed of both hydrophilic and hydrophobic functional groups and allows facile desalting for multiple compound types without method development. The ability to perform limited volume elution with this phase has also been shown [56].

Although SPE is a routine tool for the bioanalytical laboratory, it is generally not deployed as the front-line method for off-line sample preparation due to additional cost and labor relative to protein precipitation. The need for SPE is often realized when lower LLOQ concentrations are sought (i.e., sub-ng/mL). SPE is also useful in cases where additional sample cleanup is needed to address matrix-related issues often implicated in poor bioanalytical performance (i.e., interfering peaks or ion suppression).

The third sample preparation method used is LLE. LLE is known for providing clean extracts with high recovery, and is often the choice for assays in the low pg/mL range. Typical solvents used for LLE include methyl *tert*-butyl ether, cyclohexane, and ethyl acetate. Despite its long-standing use for bioanalysis, LLE is generally regarded as labor-intensive and is not as straightforward to optimize as SPE. Thus, LLE is typically reserved for more advanced leads to achieve greater sensitivity. An example would be to provide exposure data at the efficacy dose in definitive pharmacology studies.

Recently, a number of reports of semiautomated LLE have appeared in the literature [38,48], and LLE has experienced resurgence for research applications. A factor besides automation that makes LLE attractive is cost, since solvents are relatively inexpensive compared to SPE plates. Nevertheless, in most circumstances, LLE is not typically used as the front-line method for drug-discovery research applications due to the amount of method development that is required.

On-Line Sample Preparation

On-line methods refer to techniques where sample cleanup occurs on-line en route to the mass spectrometer. The common thread to all on-line techniques is column-switching (CS). CS refers to the use of HPLC valves to couple multiple LC columns in configurations that permit sample cleanup and analysis to occur on-line. Such techniques eliminate the manual wash and transfer steps associated with off-line sample preparation and are readily automated.

CS techniques are not new [57]. A resurgence of CS has occurred in response to the unique demands of the drug-discovery research environment where fast method development and high throughput are both needed. An inherent disadvantage to on-line methods is that the overall LC-MS cycle time must be expanded to include the time needed for sample preparation. It is generally accepted that, for drug-discovery research applications, this disadvantage is more than offset by the automation of sample preparation and the reduced time for methods development. A comprehensive overview will not be presented, as we recently published a review on the reemergence of CS methods [45].

CS bioanalysis methods generally involve two columns. The first column, referred to as a pre- or desalting column, is used for sample cleanup. Following on-line cleanup, the analytes of interest are eluted from the precolumn onto a second (analytical) column for chromatographic analysis. By far, the most common approach is to back-flush the retained material off of the

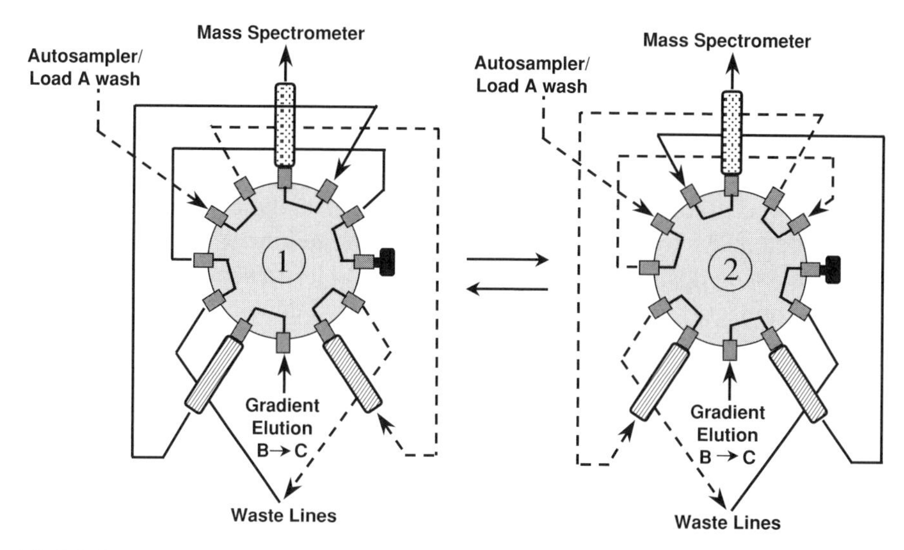

FIGURE 11.3 A diagram of a 12-port valve configured for column-switching with an alternate/regenerate back-flush configuration. This configuration incorporates dual desalting columns (bottom) that share a single analytical column (top). In the initial valve position (left), a single precolumn is loaded and washed with a weak mobile phase (dotted-line path). During this time, a sample loaded onto the alternate precolumn during the penultimate injection is back-flushed onto the analytical column with gradient elution for subsequent analysis and detection. (Reprinted from Ackermann et al. [4], with permission from John Wiley & Sons, Inc.)

precolumn. Figure 11.3 depicts a dual back-flush configuration often referred to as the alternate-regenerate mode. In this configuration, samples are alternatively loaded onto one of two precolumns that share a common analytical column. Several versions of this dual back-flush strategy have been used to increase throughput for each of the formats discussed later in this section [58–60]. On-line extraction methods can be divided into two categories: *on-line SPE* and *restricted-access methods*. On-line SPE methods may be further classified based on whether plasma is loaded directly onto the precolumn (direct) or whether a manual precipitation step precedes sample loading (indirect). Restricted-access methods emerged as a result of problems associated with direct plasma loading onto conventional chromatographic media. These techniques, which invoke size exclusion to affect protein removal, can be further subdivided into restricted-access media (RAM) and turbulent flow chromatography (TFC).

On-line SPE methods rely on the differential retention of small organic analytes relative to matrix materials (salts, proteins, lipids, etc.) on conventional reserve-phase media (d_p 3–10 μm). This strategy typically involves a back-flush configuration where the compounds of interest are eluted from the precolumn for subsequent chromatographic analysis on an analytical column.

A limitation to direct on-line SPE is that repeated injections of diluted plasma onto reverse-phase media often leads to column plugging. The Prospekt™ system is a commercially available instrument that was introduced to overcome this problem. This system automatically loads a new SPE cartridge into a back-flush-style CS apparatus prior to each injection. The Prospekt has been successfully applied to a number of bioanalytical applications, as highlighted by McLoughlin et al., who demonstrated the use of this technology for cassette dosing experiments ($N = 10$) [61]. In a separate application, Beaudry and co-workers successfully analyzed a mock cassette prepared from 64 compounds with a total run time of 4.5 minutes [62]. More recently, a dual Prospekt system that operated in an alternate-regenerate configuration was introduced for the analysis of paclitaxel in human serum to achieve a rate of 1.3 min/sample [59]. In all of the preceding examples, the LLOQs were typically <5 ng/mL, and in many cases, sub-ng/mL limits were achieved.

A closely related variation of on-line SPE involves off-line removal of proteins prior to sample injection. Usually this procedure is accomplished by the addition of an organic solvent (e.g., acetonitrile) that contains an internal standard [63]. Variations of this strategy have been applied in our own laboratory for plasma bioanalysis [49] as well as in vitro screens [60]. A number of variations on this theme have often appeared in the literature for the simultaneous analysis of multiple analytes. An example of this type of analysis is described by Zell and co-workers, who used perchloric acid pretreatment prior to on-line SPE for the bioanalysis of a platelet inhibitor, its ester prodrug, and an active metabolite [64]. Analytical ranges from 1 ng/mL to 1000 ng/mL are fairly typical for this technology, where most applications involve narrow-bore columns and begin with 50–100 μL plasma. Despite the need for a deliberate precipitation step, indirect SPE is generally more robust and less expensive than the other alternatives presented here.

Restricted-access methods provide a format for direct plasma injection, and thus, eliminate the need for sample preparation. The first format discussed, RAM, is named for the function of the precolumn that selectively retains small, organic analytes, while proteins and salts are allowed to flow through to waste. RAM chemistries have been classified in a review by Boos and Rudolphi [65]. The most widely adopted format for LC-MS/MS bioanalysis are internal surface reversed-phase (ISRP)–style columns. ISRP columns are prepared from porous silica (d_p 5–25 μm) that have 6–10-nm-diameter pores. The internal pores, which are not accessible to macromolecules, contain a bonded phase (e.g., C-18) to promote retention of small organic analyte molecules. The outer surface of the particle is covered with a hydrophilic, electroneutral material designed to be both nonadsorptive and nondenaturing to proteins. The first ISRP column was introduced by Hagestam and Pinkerton [66] in 1985, and a number of phases have applied for bioanalysis by LC-MS/MS. An example is the work by Boppana et al., who used an ISRP cleanup with subsequent analysis by a C-18 column to quantify the 5-HT_3 antagonist genisetron and its 7-hydroxy metabolite in dog plasma [67].

Two other ISRP phases have found extensive application in the literature. The first is the alkyl-diol silica (ADS) format introduced by Boos and co-workers [68]. This phase (d_p 25 μm) is internally derivatized with either C-4, C-8 or C-18. The akyl-silica esters on the surface are then removed by lipase enzymatic hydrolysis to yield a hydrophilic diol moiety. This phase has found popularity in part due to its comparatively robust operation. According to Boos, it is possible for ADS media to accommodate more than 200 injections of plasma (500 μL each) on a single column [69].

Another widely adopted ISRP phase, called the BioTrap 500, was used by Needham and co-workers to investigate the utility of RAM media for rapid-discovery bioanalysis [58]. The exterior of this phase is coated with α1-acid glycoprotein (AGP), which prevents adsorption of plasma proteins. An alternate-regenerate system, similar to that depicted as Figure 11.3, was employed to yield an overall duty cycle of about 2.5 min/sample.

An important example of a non-ISRP phase is the shielded hydrophobic mixed function material known as Capcell Pak MF^{TM}. This phase consists of porous silica derivatized with a silicone polymer. The polymer is referred to as a mixed phase, since it contains hydrophobic moieties dispersed at intervals in the polymer chain. Unlike small molecules, matrix proteins are unable to penetrate the hydrophilic polymer and are eluted to waste. A recent application by Hsieh et al. demonstrated the utility of this phase for direct plasma analysis without the use of a secondary column [70].

TFC is a newer form of restricted access [71,72] that has become increasingly popular in recent years. TFC is characterized by the use of large irregularly shaped particles (d_p 30–50 μm), which readily accommodate the injection of plasma proteins. The large particles also permit the use of extremely high flow rates (linear velocity 8–10 cm/s), where it is believed that a change from laminar to turbulent flow conditions promotes eddy currents to allow more facile mass transfer within the phase. TFC is commonly performed with a mixed hydrophobic–hydrophilic polymer phase that accommodates direct injection of plasma volumes up to 100 μL. Injected proteins do not penetrate the pores of the phase and are eluted to waste, whereas small organic molecules migrate into the pores and are retained.

In typical TFC conditions, samples are loaded onto a large particle-size precolumn (1 $\times$ 50 mm, d_p 30–50 μm) at a flow rate of 4 mL/min followed by elution onto a conventional analytical column (2 $\times$ 30 mm, d_p 5–10 μm) at a flow rate of 1 mL/min. Although a single-column format is sometimes adopted for rapid bioanalysis, essentially no separation is achieved for the retained materials. Therefore, most reported applications involve a dual-column configuration, where overall run times of 3 min/sample are typical. Jemal and co-workers applied TFC for pravastatin analysis in rat and human plasma in both the single- and dual-column modes [73]. The same research group demonstrated the throughput of TFC by direct comparison with LLE for bioanalysis of a proprietary compound in human plasma. The authors reported that TFC achieved approximately a twofold advantage in overall sample throughput based largely on the reduced sample preparation time [74]. Wu and co-workers recently applied TFC for PK screening with cassette dosing and made similar claims [75].

Ayrton et al. have experimented with TFC with microbore (1 mm) and capillary (0.18 mm) formats for TFC. These columns reduce the excessive amount of mobile phase required for this technology and can provide greater sensitivity for sample-limited situations. At a flow rate of 130 μL/min, these authors demonstrated the ability to achieve sub-ng/mL quantitation for an assay based on only 2.5 μL of plasma [76]. More recently, this group performed TFC in a parallel four-column format. MS analysis was accomplished with a multiple sprayer–ESI interface and allowed for simultaneous LC-MS/MS detection [77]. In another example, King et al. used a commercial system based on four staggered TFC systems in conjunction with a modified autosampler to demonstrate increased throughput. These authors demonstrated the ability to pass a GLP-level validation with this system [78].

As a final note in this discussion of on-line methods, it is widely known that on-line methods are more prone to system carryover. The primary reason is that the valves used for CS present additional surfaces for contact with the analyte. In a paper recently published by Grant and co-workers, the issue of analyte carryover in TFC was addressed [79]. These authors incorporated a series of multiple-wash steps along with a switching valve made of polyarylethyl ketone (PAEK) to reduce carryover present in a generic method that provided a throughput of 1.5 min/sample.

Managing Reagent and Material Expense

As presented in the two previous subsections, several viable options exist for high throughput bioanalysis. It is interesting to note that an issue seldom discussed is the relative cost of the reagents and supplies associated with the various methods. For example, the issue of high mobile-phase consumption/waste-stream production in TFC may be a concern. A similar situation occurs with 4.6-mm inside diameter (ID) monolithic LC columns; hence, the push for smaller-bore monolithic columns.

Other costs are associated with the media used for sample preparation or extraction. As previously mentioned, there has been renewed interest in the use of LLE, in part, due to its low cost relative to 96-well SPE. Similarly, the costs of columns used in direct plasma-injection methods (i.e., RAM or TFC) also cannot be overlooked. Unfortunately, these columns can only sustain a finite number of injections and cannot easily be regenerated once fouled.

In our laboratory, we rely heavily on 96-well protein precipitation as the primary form of sample preparation. Not only have we found this approach to be the most universal, but it is also the most cost-effective method for discovery bioanalysis. To overcome limitations associated with protein precipitation, CS methods were developed (Figure 11.3) to permit on-line sample cleanup [49]. In addition, fast gradient elution has been used with this strategy to allow expedient method development and fast sample analysis.

Interestingly, in the past few years, we have found that straight "dilute and shoot" methods often succeed where in the past CS was necessary. We attribute this development to the increased sensitivity offered by present-generation mass spectrometers. In other words, it is possible to avoid on-line clean up by simply injecting a more dilute sample or even starting with smaller plasma aliquots. While this strategy is not permissible for all sample matrices, we have achieved consistent success with plasma.

Liquid Chromatography

The field of bioanalysis remains dominated by HPLC, despite the fact that several other chromatographic forms have been interfaced to MS, which include gas chromatography, supercritical fluid chromatography, and capillary electrophoresis. The popularity of RP-LC stems from its instrumental simplicity, wide scope of application, and relative ease of method development. This section primarily focuses on RP-LC, with attention also given to the recent resurgence in normal-phase methods (NP-LC).

Gradient versus Isocratic Elution

The most significant trend that affects the use of RP-LC for bioanalysis has been the migration from isocratic to gradient elution, and more recently toward high-flow gradient elution. Isocratic elution has historically been employed for bioanalytical applications of LC-MS/MS, because this elution format avoids the time associated with column re-equilibration. Although isocratic elution is still used to provide rapid bioanalysis [80], it is clear that gradient elution is far better suited for discovery applications. The trend toward gradient elution is consistent with the need to develop rapid methods and to accommodate a wide range of chemical diversity. As gradient elution became more widely practiced, methods were introduced to reduce the impact of column re-equilibration and gradient delay times. Examples include the implementation of various CS strategies along with the use of *n*-propanol in RP-LC to promote faster re-equilibration [81]. The more recent trend toward high-flow gradient methods (operation at linear velocities severalfold above the Van Deemter optimum) has dramatically reduced the significance of these issues.

The Concentration Dependence of ESI

With the arrival of pneumatically assisted ESI in the mid-1990s, a significant portion of bioanalysis has shifted away from APCI and toward ESI with a high reliance on narrow-bore LC columns (2.1 mm ID). Not only were narrow-bore flow rates compatible with most API interfaces (<0.3 mL/min), this format also afforded greater sensitivity than analytical bore chromatography (4.6 mm ID, 1 mL/min) due to the concentration dependence of ESI [82]. In other words, because ESI signal intensity depends on concentration, there is an inherent advantage with smaller-bore columns, since the concentration of a chromatographic band is inversely proportional to the square of the column internal diameter. The relative concentration enrichment scales as follows: 4.6 mm ID (1×); 2.1 mm ID (5×); 1.0 mm ID (20×); 0.32 mm ID (200×). It is interesting to note that despite this trend, smaller-bore formats, such as packed capillary (<0.5 mm ID), are seldom used for bioanalysis. There are several reasons for this observation, such as limited capacity (column overloading) and restrictions on injection volume. In addition, a number of logistical hurdles are inherent in low flow-rate methods including post-column band broadening and issues of gradient formation and delay. Fortunately, a number of packed-capillary LC systems are commercially available that readily address these issues. Plumb and co-workers published the use of packed-capillary LC to extend the sensitivity for certain applications [83]. The key point to remember is that small-bore LC-MS does not confer increased sensitivity for all applications, only those that are truly sample-limited.

Unlike ESI, APCI does not behave in a concentration-sensitive manner. Therefore, the aforementioned considerations that concern column diameter and flow rate do not directly apply. APCI methods are commonly run with either conventional-bore or narrow-bore columns and flow rates in the range of 0.5 to 2 mL/min. The signal strength with APCI depends largely on the amount of solute injected. As with ESI, the ability to achieve adequate desolvation is critical.

Fast-Gradient Elution

In an effort to speed up the rate of analysis, several researchers have experimented with gradient elution at flow rates that vastly exceed the optimum linear velocity predicted by the Van Deemter equation. Romanyshyn and co-workers have shown the ability to retain chromatographic integrity with ultrafast gradients (e.g., 1.5 mL/min on a 2 × 20-mm column) [84,85]. By direct comparison to isocratic methods, the authors further demonstrated that gradient elution provides superior resolution from co-extracted salts and metabolites and also reduces the overall level of ion suppression [85]. It is possible to have run times of less than one minute and peak widths of 4 s at baseline with this approach.

To many, an increased linear velocity without a sacrifice of chromatographic performance seems counterintuitive. Neue and co-workers described this phenomenon in terms of peak-capacity assessment (number of peaks that can be separated in a given gradient duration time). They demonstrated that vastly improved analysis times are possible provided that rapid gradients are coupled with short columns and high flow rates [86]. More recently, Hayward et al. examined the factors that affect the efficiency of fast-flow gradient elution [87]. Their analysis concluded that the contribution to plate height (H) imparted from operation at high linear velocities is compensated by a reduction in the influence of extra-column effects. They also cited the relative importance of secondary interactions. The bottom line is that operation at high flow rates does not unduly impair chromatographic performance and can yield great gains in sample throughput. The aforementioned reduction in the time associated with gradient delay and re-equilibration are additional benefits to this mode of operation.

To achieve the proper balance between speed and sensitivity requires a working knowledge of the various factors that influence fast-gradient-elution LC-MS/MS. Recently, our laboratory undertook a detailed investigation of narrow-bore gradient-elution LC-ESI/MS/MS for use with protein precipitation [88]. In this study, the effects of flow rate and column length were assessed for their impact on analyte response, cycle time, back pressure, and elution volume. Figure 11.4 displays a series of overlaid SRM traces for a series of four analytes (peaks 1–4) and an internal standard (peak 5). These compounds represent structural analogs of an undisclosed drug-discovery program. In this example, 10-ng/mL rat plasma standards were prepared by protein precipitation and chromatographed with an Intertsil ODS-3 column (2.1 × 30 mm). A gradient with methanol, water, and formic acid was used, and the following flow rates were examined: 0.25, 0.5, 1.0, and 1.5 mL/min. With the exception of 0.25 mL/min (Figure 4A), all of the other experiments incorporated a post-column effluent split to achieve a constant flow rate into the ESI interface of 0.25 mL/min.

Several conclusions were drawn from these experiments. First, the decrease in analyte signal experienced at higher flow rates was consistent with analyte band dilution in accordance with the known concentration dependence of ESI. Second, these data confirm that high-flow gradient elution can be used to significantly reduce LC cycle times without an unduly sacrifice of chromatographic fidelity. It is noted, however, that LC cycle times did not decrease proportionally with flow rate and that eventual deterioration occurred at

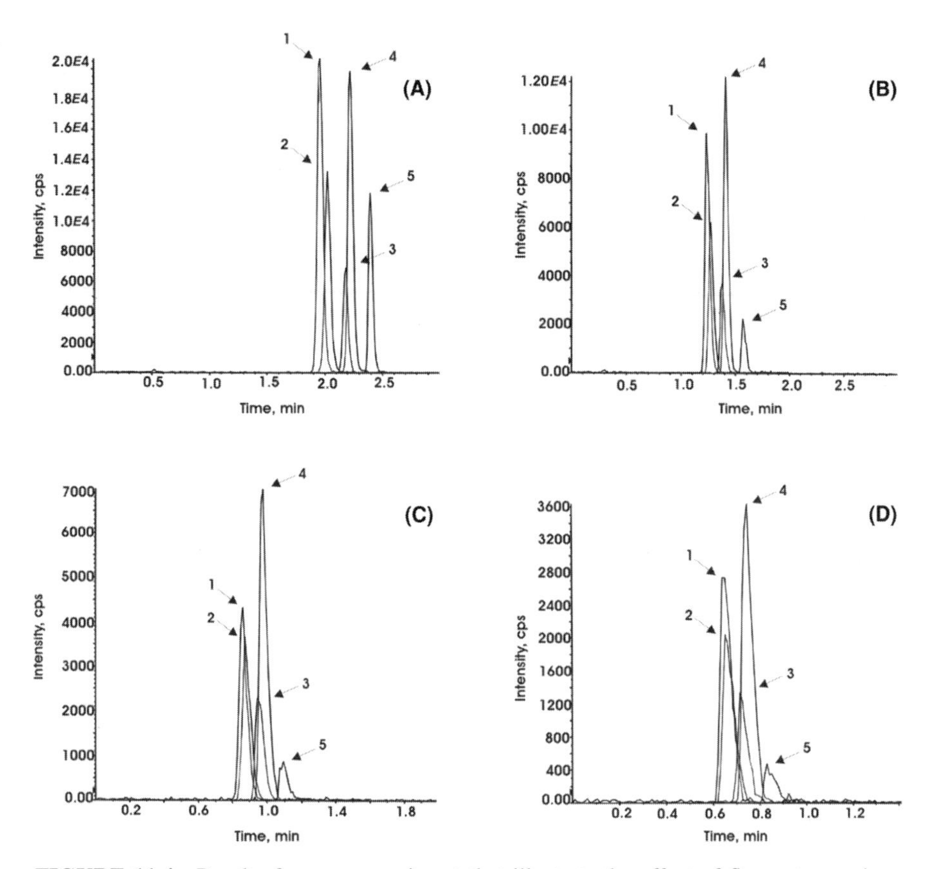

FIGURE 11.4 Results from an experiment that illustrate the effect of flow rate on chromatographic resolution and LC-MS/MS sensitivity. In this experiment, a series of five related compounds from an undisclosed drug-discovery project were analyzed on a 2.1 × 30-mm Interstil ODS-3 column at flow rates of 0.25 (panel A), 0.5 (panel B), 1.0 (panel C), and 1.5 (panel D) mL/minute. Postcolumn effluent splitting was incorporated for the data in panels B–D to achieve a constant flow rate of 0.25 mL/min to the ESI interface. The sample injected represents a 10-ng/mL rat plasma standard prepared with protein precipitation. These data, which are consistent with the concentration dependence of ESI, illustrate the balance between speed, LC resolution, and MS sensitivity during fast-gradient elution LC-MS/MS. (Reprinted from Murphy et al. [88], with permission from John Wiley & Sons, Inc.)

1.5 mL/min (Figure 4D). An important issue with high-flow gradient elution is the need to achieve adequate analyte retention. In the present example, all compounds had k' values greater than 6, which also served to minimize the influence of matrix-related ion suppression.

Although fast gradient elution may be used to provide shorter run times, the flow rates used often exceed the desolvation capacity of most API interfaces. Fortunately, because of the concentration dependence of ESI, a postcolumn effluent split can be employed without a proportional loss in signal. This strategy also extends instrument longevity, since a majority of the co-injected matrix material is diverted to waste. Postcolumn effluent splitting was used to acquire the data for Figure 11.4B–11.4D. In this example, the drop off in signal at higher flow rates was not due to the postcolumn effluent split. Rather, the decrease can be explained by the corresponding dilution in analyte peak concentration as the flow rate was increased.

The latest trend in API interface design is the introduction of sources with extremely high desolvation capacity [89]. It is now possible to alleviate the need for flow splitting with ESI even for flow rates considerably greater than 1 mL/min. Advances in source design also bring about gains in overall sensitivity. It can be argued that the extreme sensitivity offered by newer-generation interfaces is important for drug-discovery research as well as drug development. The use for drug-discovery research applications can be understood in terms of smaller plasma volumes to achieve low ng/mL bioanalysis, and thereby, afford greater instrumental robustness.

Monolithic LC

One of the problems inherent to high-flow operation is increased back pressure, particularly when methanol is used in the mobile phase. This difficulty has been circumvented by the introduction of monolithic stationary phases, which employ a unique contiguous biporous structure prepared from sol-gel chemistry [90,91]. Reduced flow resistance is accomplished by *throughpores* (2 μm), while smaller *mesopores* (13 nm) provide the surface-area capacity needed for adequate chromatographic separation. High flow rates can be employed with monolithic columns due to reduced back pressure and the ability for facile mass transfer.

Recently, Wu and co-workers demonstrated the use of monolithic high-flow LC-MS/MS for bioanalysis [92]. These investigators were able to achieve a 1-minute-gradient-elution cycle time with a 4.6 × 50-mm column at a flow rate of 6 mL/min (back pressure 61 bar), which permitted the analysis of over 600 samples in an overnight run (10 hours). Moreover, the resolving power obtained was commensurate with conventional particulate bonded phase columns ($d_p = 5$ μm). In one example, a glucuronide and an *O*-demethyl metabolite were resolved from the parent drug in approximately 30 seconds. Another example was published by Barbarin et al., where a 15 second separation of methylphenidate and its metabolite, ritalinic acid, was achieved and permitted the analysis of 768 rat plasma samples in under 4 hours [93]. The flow rate used in this example was 3.5 mL/min. Based on these and other examples, interest in monolithic LC has grown dramatically. Unfortunately, the number of commercially available column sizes (i.e., 4.6 mm ID) and phases (i.e., C-18) is extremely limited at the present time. As this technology matures, it is likely that monolithic LC will find expanded use as a routine tool for drug-discovery bioanalysis.

Achieving Adequate Retention

Perhaps the biggest limitation to RP-LC is the difficulty with adequate retention for very polar analytes. Three undesirable issues are associated with poor retention: (1) poor chromatographic separation, (2) greater ionization suppression, and (3) reduced sensitivity due to poor analyte desolvation in highly aqueous mobile phases. Column manufacturers have used several formats to improve retention of polar solutes that include extended alkyl phases, polar-endcapped alkyl phases, polar-embedded alkyl phases, nonendcapped short alkyl phases, and wide pore-diameter phases [94]. Interested readers are referred to a recent two-part review on this important subject [94,95].

The second general approach to overcome poor reverse-phase retention is to employ NP-LC. By comparison to RP-LC, the historic use of NP-LC for bioanalysis is negligible because of several limitations. The most notable limitation is the inability to perform reproducible gradient elution. Nevertheless, NP-LC can be a viable option for analytes too polar for RP-LC and produces far less back pressure. Perhaps, the most visible application of NP-LC is for the bioanalysis of chiral drugs [96]. This critical niche is largely the result of the frequent use of NP mobile phases with chiral stationary phases.

A development that has generated much interest and promise has been the introduction of aqueous-containing formats for NP-LC. This unique format is sometimes referred to as hydrophilic interaction liquid chromatography (HILIC), although a number of other descriptors have been applied. Regardless of the terminology, the use of aqueous-containing mobile phases with NP-LC has a number of significant advantages for LC-MS, which include the ability to perform both NP-LC and RP-LC with the same LC system (i.e, no exhaustive solvent changeover). Perhaps the most important advantage from a chromatographic perspective is the ability to perform gradient elution, a capability that is difficult to apply with conventional NP-LC. The importance of HILIC for drug-discovery applications was highlighted in an article by Strege et al. [97].

Several column types (silica, cyano, amino, diol) function in an NP mode when used with organic solvents, such as acetonitrile, that contain a small percentage of water (usually $<20\%$). In this mode of operation, gradient elution is performed by ramping the aqueous content of the mobile phase. Frequently, a small percentage of an ammonium buffer (<5 mM) is needed to disrupt secondary interactions that lead to peak tailing.

Naidong and co-workers have reported several bioanalytical applications of NP-LC that involve LC-MS/MS [98,99]. In addition to a positive effect on sensitivity (high organic content), these authors cite the ability to employ high flow rates due to low back pressure. An example of the power of this form of chromatography was demonstrated by the baseline resolution of morphine from its two isomeric glucuronides in less than one minute [98].

INSTRUMENTAL METHODS FOR INCREASED LC-MS/MS THROUGHPUT

One of the chief limitations to LC-MS/MS is that it is a serial technique. While typical gradients are on the order of a few minutes, the dwell time for an individual SRM channel is typically only 50 ms. Consequently, several reported methods have attempted to take advantage of this temporal disparity. In one example, Korfmacher et al. combined the postcolumn effluent from two LC systems to a single MS, which performed alternating detection of the analytes and internal standards injected in the two respective samples [100]. While stacked injections have been used for a number of years with isocratic systems, Van Pelt and co-workers published results from a parallel gradient system [101]. The method involves simultaneous injection of four samples with a multiprobe autosampler. Although injection occurs simultaneously, the arrival at the MS is staggered by plumbing a different gradient delay for each sample.

Another approach that has received great attention is the use of parallel chromatography coupled to a single MS equipped with multiple indexed ESI sprayers. A commercially available system, referred to as multiplexed ESI or MUX^{TM}, allows up to eight ESI sprayers to be consecutively sampled, each connected to an individual LC column. Sampling occurs by mechanical means, with a rotating device that occludes all but a single sprayer at any given instant.

Yang and co-workers demonstrated the viability of MUX for bioanalysis, and validation for loratadine and its metabolite descarboethoxyloratadine was performed in four preclinical species with a four-sprayer MUX interface [102]. Other examples that have been reported have been more closely linked to drug discovery. For instance, Bayliss et al. used a four-sprayer MUX in conjunction with TFC to achieve an overall throughput of 120 samples per hour [77]. With a somewhat similar design, Deng and co-workers used a parallel four-extraction column TFC system in combination with a four-sprayer MUX interface [103]. In this system, CS was used to couple an extraction column with an analytical column for each channel. Methotrexate was used as a test compound, and this technology achieved successful cross-validation with results obtained with a more conventional LC-MS/MS method. In previous work, this group demonstrated improved chromatography with a dual-column approach that was applied to cassette dosing [104].

The combination of MUX with monolithic LC has also been investigated as a means to increase bioanalytical throughput. Deng and co-workers used four monolithic C-18 columns (4.6 × 100 mm) with a four-sprayer MUX interface to analyze twelve 96-well plates in 10 hours [105]. In this example, the analyte, oxazepam, was chromatographically separated with a run time of 2 minutes to yield an overall throughput of 30 s/sample. Note that this rate did not include sample preparation, which was performed with automated SPE with human plasma as the matrix.

Despite the throughput provided by the MUX interface, several practical considerations should be noted. First, the MUX system cannot be used with flow rates that exceed 0.1 mL/min. In addition, a small amount of carryover exists between adjacent sprayers, typically on the order of 0.1%. While each of these issues can be tolerated, the overhead associated with sampling multiple effluent streams limits the number of points that can be acquired across a chromatographic peak. In a four-channel MUX system, each sprayer is accessed about every 1.2 seconds [77]. Thus, to permit the acquisition of 10 data points, the LC peak must be at least 12-seconds wide. Because typical drug-discovery runs include more than one analyte, as well as an internal standard, the MUX system is not truly compatible with fast LC methods that typically produce peak widths less than 6 seconds wide.

Because of the serial nature of LC-MS, much of the current discussion has centered on ways to reduce LC-MS/MS cycle time. Often the rate-limiting step for drug-discovery bioanalysis lies in the speed with which methods (sample preparation and chromatography) can be prepared for NCEs. One of the frequently overlooked steps is the need to tune and optimize MS/MS transitions for the various analytes studied. Fortunately, most MS vendors now offer semi- or fully automated procedures to perform this task. The origin for these procedures can be traced to the seminal work of Whalen et al., who published an automated procedure known as AUTOSCAN [106]. It is possible to establish experimental conditions with this procedure in the flow-injection analysis (FIA) mode for 96 analytes in less than one hour.

Direct Methods

In the interest of obtaining maximum bioanalytical throughput, some researchers have resorted to direct MS analysis without on-line chromatography. While there are obvious drawbacks to this approach, which include a greater likelihood for ion suppression and a higher occurrence of interfering peaks, the motivation behind FIA is basically twofold: (1) a reduction in LC injection duty-cycle time, and (2) the elimination of LC method development.

FIA methods have been used with a variety of sample cleanup methods. Chen and Carvey reported a validated assay for topiramate in human plasma with FIA/ESI-MS following LLE [107]. Recently, Zheng and co-workers demonstrated the viability of this approach, relative to LC-MS, for hepatic metabolic stability determination [108].

Another example of a direct method is the on-line SPE approach embodied by a novel commercial device known as the SPE Card™. This apparatus, originally developed by Olech and co-workers, couples 96-well disk SPE in a format that allows direct, serial elution into the mass spectrometer [109]. Note that this method should not be confused with the Prospekt system described earlier for on-line SPE. Sample preparation occurs off-line with the SPE-Card, and the card is placed in an on-line device that performs serial elution into the MS. This method is considered direct, since it avoids the use of LC columns and removes the various sample transfer steps that normally follow SPE.

A final example of direct bioanalysis was recently published by Dethy et al. and involves the application of infusion nanoelectrospray (nano-ESI) from a silicon chip [110]. In this example, supernatant obtained from protein precipitation was directly infused with an automated pipette-tip delivery system. Individual, conductive pipette tips that contain sample were sequentially introduced to the backplane of a silicon chip for analysis. The front plane of the chip that consisted of 100 individual nano-ESI nozzles, was positioned near the API orifice of a TQMS for direct serial analysis. Quantitation of verapamil and its metabolite norverapamil occurred in human plasma over a range of 5–1000 ng/mL. It is possible to achieve analysis times of less than 1 minute per sample with this technology. An important advantage, demonstrated by this work, is the unique ability to avoid system carryover with this device [110].

CONSIDERATIONS FOR EXPERIMENTAL DESIGN

Pooling Methods

In an effort to expand throughput for PK exposure screening, researchers have looked to improve LC-MS/MS efficiency. This trend has led to a number of novel formats for experimental design. Many of these approaches can be described as pooling strategies aimed at maximizing the data throughput per injection. The most common format is cassette or N-in-1 dosing. Since its introduction by Berman and Halm in 1997 [111], a number of applications and bioanalytical refinements have been published [112,113]. Typically, cassettes are limited to five or fewer compounds with one of the compounds that represent a previously examined lead. The advantages and limitations of this technique have been widely discussed and often debated. Readers interested in this topic are referred to a seminal paper by White and Manitpisitkul which examines the central issues involved with cassette dosing from a PK perspective

[114]. Clearly the primary risk associated with cassette dosing is the potential for drug–drug interactions that lead to questions that deal with the validity of the results. In addition, precautions must be exercised on the analytical end to avoid analytical interference from metabolites or between analytes. On the other hand, cassette dosing reduces the number of animals used for exposure screening and can dramatically increase the number of compounds that can be analyzed in live-phase experiments.

Several postdosing plasma-pooling strategies have been employed with single-compound administration. These strategies can be described as pooling by *time* or pooling by *compound*. The former involves pooling plasma for identical time points from different compounds. While this strategy reduces the overall demand for MS time, there is a trade-off in sensitivity due to sample dilution. In addition, this method requires extra pipetting and sample transfer. A second approach is to combine multiple time points for a given compound into a single sample that represents the "time-average" of exposure. This method, referred to by Hop as PK screening, allows comparisons of net exposure to be made from a single injection per compound, i.e. analogous to area under the curve (AUC) [115]. This approach dramatically limits the number of samples analyzed at the expense of temporal information (e.g., T_{max}, $T_{1/2}$). It should be noted that an advantage of cassette dosing over plasma pooling strategies is that the gain in efficiency translates beyond the MS by the reduction of animal use and handling.

Integrated Strategies for Bioanalysis

Many bioanalytical laboratories have looked beyond LC-MS/MS and have found gains in efficiency by the development of integrated strategies with animal-handling groups. One of the best examples was reported by Zhang et al., who described a comprehensive strategy that involves plasma collection in 96-well format followed by robotic plasma transfer and semiautomated sample preparation [116]. Similarly, we have found extensive benefit from implementation of a 96-well plasma collection strategy. A frequently overlooked problem in automated approaches that involve plasma transfer is difficulty in pipetting due to fibrin clots and lipid material that does not pellet upon centrifugation. Even though the degree of clot formation is improved with the use of ethylene diamine tetraacetic acid (EDTA) as the anticoagulant [117], this phenomenon leads to poor pipetting accuracy and the need to repipette the affected samples. To address this problem, we recently implemented a novel 96-well filter plate in our laboratory for plasma collection [118]. This plate, which has a low capacity for nonspecific binding, allows filtering to occur as the plasma is thawed, and ultimately allows for reliable automated plasma transfer. A manual pipetting step has been eliminated for all plasma samples collected in this fashion, with no restriction regarding the anticoagulant used.

Korfmacher and co-workers have increased efficiency through streamlined and standardized refinement of the bioanalytical process [119]. Their approach, known as cassette-accelerated rapid rat screen (CARRS), is based around a 96-well template that accommodates six compounds dosed via single-compound administration. Higher efficiency is achieved through the use of abbreviated standard curves (25, 250, and 2500 ng/mL), time points that are pooled from multiple animals, and the use of a standardized format.

Recognizing Potential Pitfalls

Strictly speaking, bioanalysis is about the *demonstration*, not the *presumption*, of quality. The pace of the drug-discovery research environment does not allow for excessive controls over quality assurance. In many settings, few if any guidelines may be warranted. In this subsection, the common pitfalls of LC-MS/MS–based bioanalysis are briefly discussed to provide insight into factors that affect precision and accuracy.

Logically, the level of characterization (e.g., method development, quality controls, etc.) is scaleable along the research continuum. This trend is illustrated in Table 11.1, which tracks typical experiments in lead optimization and the gradient toward increased assessment as molecules progress. For the purpose of this discussion, the transition from drug-discovery research to drug development is defined by whether the work is intended for regulatory submission. The remainder of the chapter highlights issues that affect the overall quality of bioanalysis. Although it is difficult to institute formal guidelines in the discovery setting, it is the responsibility of each scientist to be aware of pitfalls that can lead to the generation of spurious results.

TABLE 11.1 Quality Assessment Along the Research Continuum

Form of Bioanalysis	Setting	Level of Quality Assessment
Exposure Screening	Research	Low
PK of Lead Compounds (e.g., bioavailability, %F)	Research	Low to Moderate
Margin of Safety Assessment • Pilot toxicology • PK-PD (efficacy)	Research	Moderate
Work intended for regulatory submission (e.g., GLP tox)	Development	High[a]

[a] Subject to the *FDA Guidance on Bioanalytical Method Validation* [8,9].

Ionization Suppression

A unique property of LC/API/MS is the extent to which the analyte signal is affected by the sample matrix or the existence of co-eluting analytes. This property can have a profound influence on sensitivity and assay reproducibility. Because of matrix-ion suppression, it is not possible to estimate extraction recovery by comparison of the signal from a neat sample to an extracted sample. This is because the reduction in signal represents the combined effects of recovery and ion suppression. As first shown by Buhrman et al., quantitative assessment of extraction efficiency is made by spiking the neat sample into an extracted blank and comparison of the result to a similar sample spiked before extraction [120]. Conversely, the extent of ion suppression is obtained by the comparison of the signals for a neat unextracted sample to the same neat solution spiked into an extracted matrix blank.

Since this initial report, much attention has been given to this phenomenon, which is partially responsible for the extensive use of stable isotope internal standards for clinical bioanalysis. A general consensus has emerged that APCI is less prone to the effects of ion suppression than ESI [15]. In the case of APCI, ion suppression is in part related to the design of the APCI probe [121,122]. King et al. recently performed a detailed study on the origin and nature of ion suppression and showed that for ESI, ion suppression originates in solution (not the gas phase) [19]. In addition, the authors demonstrated that ion suppression for ESI not only occurs from competition for ionization at the surface of the droplets, but is also influenced by other factors. An example is the colligate effect of substances, such as salts, which inhibit the overall extent of desolvation in the spray. Previously, the same research group devised a now commonly used technique to monitor the extent of ion suppression over the course of a chromatogram [123]. A postcolumn tee is used to infuse an analyte, and the signal is detected while an extracted matrix blank is injected and chromatographically separated. A schematic of the associated instrumental configuration is shown in Figure 11.5. The output of this experiment, referred to as an *infusion chromatogram* [123], reveals regions in the chromatogram affected by ion suppression, as indicated by a negative deflection in the infused analyte signal. Figure 11.6 displays two infusion chromatograms that compare APCI (Figure 6A) and ESI (Figure 6B) following the injection of control dog-plasma supernatant obtained by protein precipitation [19]. The analyte in this case, urapidil (10 μM), was infused postcolumn at different rates for APCI and ESI to keep the concentration of urapidil delivered to each source constant for both experiments to compensate for the different flow rates used for APCI (0.5 mL/min) and ESI (0.25 mL/min). The greater deflection in the signal for ESI supports the widely accepted belief that ESI is more susceptible to the effects of ion suppression.

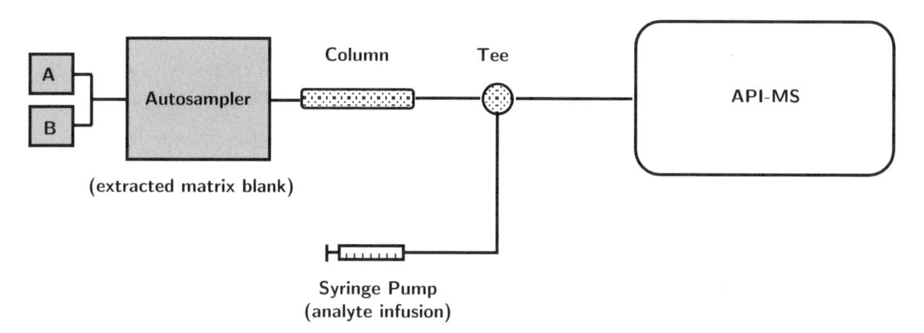

FIGURE 11.5 Schematic diagram that shows the instrumental configuration used to perform postcolumn infusion assessment of ionization suppression. In this experiment, a drug-free matrix blank is extracted and chromatographed with conditions developed for the analyte. The simultaneous infusion of the analyte postcolumn allows for an assessment of ionization suppression and is revealed by a negative deflection in the analyte signal.

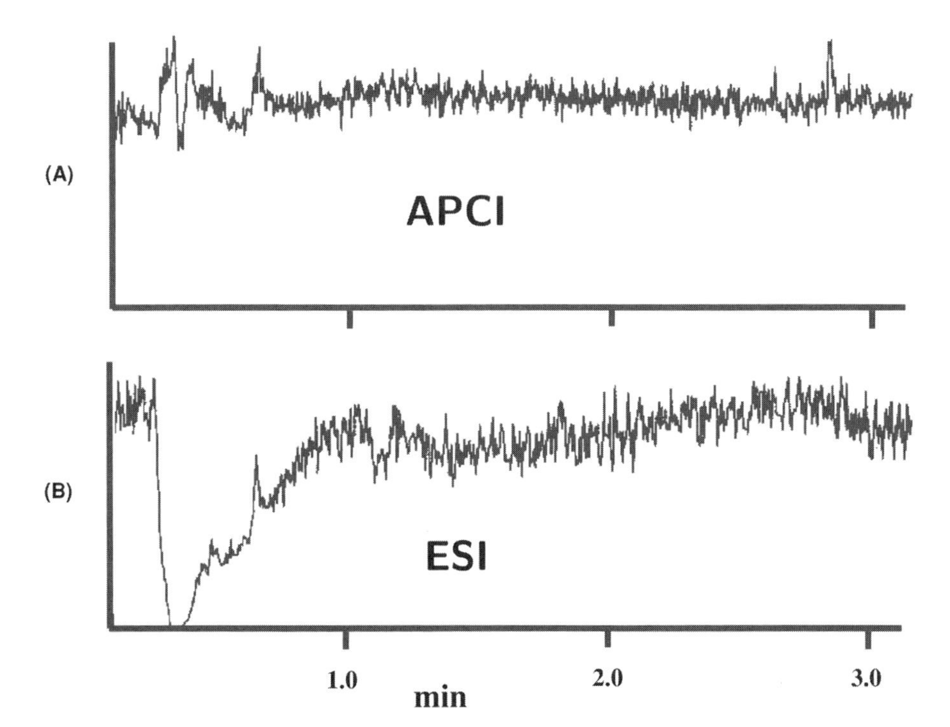

FIGURE 11.6 Comparison of infusion chromatograms generated with (A) APCI, and (B) ESI. The matrix in this case was dog plasma prepared by protein precipitation. Analysis was performed by isocratic LC-MS/MS with a 2.1 × 50-mm Zorbax SB-C18 column at a flow rate of 0.5 mL/min (APCI) or 0.25 mL/min (ESI). Postcolumn infusion of urapidil was adjusted to achieve similar analyte delivery to each source. Detection of urapidil occurred by SRM. (Reprinted from King et al. [19], with permission from Elsevier Science, Inc.)

It has been reported by Korfmacher et al. that significant contributions to ion suppression can come from sources other than endogenous co-extracted matrix components [124]. Their studies indicate that certain types of sample tubes, and even the choice of anticoagulant, can sometimes lead to ion suppression. A more recent investigation suggests than ion suppression can occur from the vehicle used in PK studies. Specifically, Tong and co-workers revealed that ion suppression attributed to excipients, such as PEG 400 and Tween 80, may result in spuriously low values for reported plasma concentrations (high clearance) [125]. It should be noted that these effects were only observed for compounds that have very little or no retention (i.e., $k' = 0.2$) and were ameliorated by adjustments with the chromatography to achieve k' values of at least 0.8. Similar phenomena were also reported by Shou and Naidong [126]. The authors concluded that polymeric vehicles are the most problematic and recommended spiking additional quality-control (QC) samples with the dosing vehicle as a way to evaluate the impact of the vehicle on analysis.

As illustrated by these and other examples, increased chromatographic retention is one way to reduce the impact of ion suppression. Switching from isocratic to gradient elution can also help and provide a wide difference in k' values for the analyte and weakly retained matrix materials. In cases where ion suppression is pronounced, more selective sample cleanup may be necessary. In addition, careful consideration should be given to internal standard selection. A procedure for the assessment of the ability of an internal standard to track the analyte was recently published by Avery [127]. In this procedure, fluctuation in the peak-area ratio of analyte to internal standard was examined for different lots of control plasma. As one would expect, this degree of fluctuation was not only related to the internal standard selection but also to the method used for extraction.

Metabolite Interference and Cross Talk

Cross talk refers to the situation where one analyte interferes with the detection of a second analyte due to specific MS or instrument-related phenomena. For TQMS, two types of cross talk exist. The first source is a blend of adjacent SRM transitions that can be attributed to the slower rate that product ions are scanned out Q_2 relative to the response rate of the detector. This effect is most pronounced when consecutive SRM transitions monitor the same product ion m/z and has been largely compensated for by newer generation TQMS instruments [89]. A second source of cross talk results from fragmentation within the ion source (in-source collision-induced dissociation (CID)) and gives rise to multiple sources for the precursor ion selected by Q_1. Conjugates, such as sulfates and glucuronides, are known to lose 80 u and 176 u, respectively, to yield ions identical to the parent ion for the drug of interest. *N*-Oxides may also produce cross talk due to a tendency to lose oxygen $(M + H - O)^+$. According to Ramanathan et al., the loss of 16 u is believed to be unique to APCI and was not observed in ESI for the compounds investigated [128]. This conclusion was supported by the work of Tong and co-workers, who suggest that the loss of oxygen occurs via thermal decomposition [129]. While this occurrence has been our general experience, we have also observed *N*-oxide cross talk in some instances that involve ESI (data not shown). We believe that the origin of this phenomenon is thermal and arises from heat employed during pneumatically assisted ESI. Nevertheless, because a number of examples of metabolite cross talk have been reported [14,130,131], the potential for in-source cross talk should be taken into account when an ionization mode is selected for LC-MS/MS bioanalysis.

Two points should be recognized about metabolite-induced cross talk. The first point is that cross talk requires co-extraction and co-elution with the drug of interest. The issue can be avoided with greater selectivity from sample preparation or improved chromatographic resolution. The second point is that the only way to assess selectivity is through the analysis of live-phase samples from dosed animals (incurred samples). Although such measures are not common in research bioanalysis, the analyst should become familiar with any known metabolism of the drug targeted for analysis. In cases where the metabolism is unknown, structural alerts prone to conjugation (i.e., alcohols, phenols, and carboxylic acids) always should be heeded. This topic was the subject of a paper by Jemal and co-workers who outlined a strategy for investigating metabolite interference [132]. Among other recommendations, they suggest the insertion of additional SRM transitions for likely metabolites (e.g., loss of 176 u) during the analysis of incurred samples.

Other Issues Related to Quality

In the drug-discovery research setting, often little regard is given to the stability of the test article. During drug development, several forms of stability are typically tested that include: specimen storage, freeze/thaw, stock solution, extract (autosampler) stability, and benchtop stability in the sample matrix. While it is not practical to institute a full stability-testing regimen in research, a few types of stability assessment, such as stock solution and benchtop stability, can be considered for compounds that progress toward candidate selection. An assessment of extract stability is easily obtained by bracketing study samples with duplicate standard curves.

The issue of stability is certainly more relevant in some cases than others. A common example would be the analysis of pro-drugs, where it is incumbent upon the analyst to demonstrate that the conditions of analysis do not contribute to release of the active drug. Ultimately, the best guide for stability is knowledge of the chemistry involved.

Care should also be taken to ensure that bioanalytical results are not unduly influenced by analyte carryover from the HPLC system. According to the FDA guidance on bioanalytical method validation, interference is defined as any signal from an injected blank that exceeds 20% of the analyte response at the LLOQ [8,9]. This criterion, which often limits the dynamic range of GLP methods, is hard to meet in the drug-discovery research setting. Nevertheless, an estimate of carryover should be made during bioanalysis by the injection of a single blank (blank + internal standard) following the injection of the most concentrated standard. This practice allows the contribution from carryover to be estimated for individual samples in the run.

CONCLUSION AND FUTURE DIRECTION

The supply of timely information on drug exposure is critical to modern drug discovery. The advances described in this chapter are part of a revolution in the way that bioanalysis is conducted in support of drug discovery. As a result of these efforts, ADME issues are considered far earlier in the drug-discovery process. This revolution, which is closely aligned with the growth and maturation of LC-MS technology, has influenced both the efficiency of pharmaceutical research as well as the quality of drug candidates.

With all of the tools and capabilities described in this chapter, one might conclude that LC-MS is a mature science and that we should turn our attention to bottlenecks elsewhere in drug-discovery. We would argue that only the second half of this statement is true. For all of the attention directed toward LC-MS, the technique suffers from two fundamental drawbacks. The first is that LC-MS is a serial technique. Novel parallel formats are needed, particularly for screening where MS is at an inherent disadvantage to other forms of detection that support parallel detection (e.g., plate readers). The second issue relates to ionization. Despite the wide scope of API techniques, LC-MS cannot be considered a universal technique. Moreover, the everpresent issue of ion suppression serves to undermine the ultimate quality and robustness that may be achieved in LC-MS applications. Contrary to popular belief, the need still exists for improved ionization methods.

Finally, it is anticipated that the trend toward miniaturization will continue and that applications that truly benefit from nanotechnology and microfluidics will become apparent over time. While much attention has recently been given to Lab-on-a-Chip technology, MS has lagged behind other forms of detection. Current advances toward the interface between ESI and microchip technology suggest that MS applications will soon be more prevalent [133].

Advantages of microfluidics for bioanalysis could potentially include lower costs for laboratory reagents and equipment and the avoidance of system carryover. Perhaps the biggest advantage will be in the introduction of novel formats for innovation that are not readily envisioned at the present time. For instance, the solution to improved ionization or parallel detection may lie in nanotechnology. As nanotechnology is reduced to practice, we may witness a dramatic shift in how future bioanalysis is conducted in drug discovery.

REFERENCES

1. Gilbert, J.D.; Hand, E.L.; Yuan, A.S.; Olah, T.V.; Covey, T.R. "Determination of L-365,260, a New Cholecystokinin Receptor (CCK-B) Antagonist, in Plasma by Liquid Chromatography/Atmospheric Pressure Chemical Ionization Mass Spectrometry," *Biol. Mass Spectrom.* **21,** 63–68 (1992).
2. Murphy, A.T.; Kasper, S.C.; Gillespie, T.A.; DeLong, A.F. "Determination of Xanomeline and Active Metabolite, N-Desmethylxanomeline, in Human Plasma by Liquid Chromatography-Atmospheric Pressure Chemical Ionization Mass Spectrometry," *J. Chromatogr B. Biomed Appl.* **668,** 273–280 (1995).
3. Lee, M.S.; Kearns, E.H. "LC/MS Applications in Drug Development," *Mass Spectrom. Rev.* **18,** 187–279 (1999).
4. Ackermann, B.L.; Berna, M.J.; Murphy, A.T. "Recent Advances in Use of LC/MS/MS for Quantitative High-Throughput Bioanalytical Support of Drug Discovery," *Curr. Top. Med. Chem.* **2,** 56–66 (2002).
5. Brewer, E.; Henion, J. "Atmospheric Pressure Ionization LC/MS/MS Techniques for Drug Disposition Studies," *J. Pharm. Sci.* **87,** 395–402 (1998).
6. Plumb, R.S.; Dear, G.J.; Mallett, D.N.; Higton, D.M.; Pleasance, S.; Biddlecombe, R.A. "Quantitative Analysis of Pharmaceuticals in Biological Fluids Using High-Performance Liquid Chromatography Coupled to Mass Spectrometry: A Review," *Xenobiotica* **31,** 599–617 (2001).
7. Jemal, M. "High-Throughput Quantitative Bioanalysis by LC/MS/MS," *Biomed. Chromatogr.* **14,** 422–429 (2000).
8. FDA Guidance for Industry. Center for Drug Evaluation and Research. Biopharmaceutics, Bioanalytical Method Validation. Issued 5/2001. http://www.fda.gov/cder/guidance/index.htm
9. Shah, V.P.; Midha, K.K.; Findlay, J.W.; Hill, H.M.; Hulse, J.D.; McGilveray, I.J.; McKay, G.; Miller, K.J.; Patnaik, R.N.; Powell, M.L.; Tonelli, A.; Viswanathan, C.T.; Yacobi, A. "Bioanalytical Method Validation—A Revisit with a Decade of Progress," *Pharm. Res.* **17,** 1551–1557 (2000).
10. Kebarle, P.; Tang, L. "From Ions in Solution to Ions in the Gas Phase: The Mechanism of Electrospray Mass Spectrometry," *Anal. Chem.* **65,** 972A–986A (1993).

11. Enke, C.G. "A Predictive Model for Matrix and Analyte Effects in Electrospray Ionization of Singly-Charged Ionic Analytes," *Anal. Chem.* **69,** 4885–4893 (1997).
12. Bruins, A.P. "Mechanistic Aspects of Electrospray Ionization," *J. Chromatogr. A* **794,** 345–357 (1998).
13. Sunner, J.; Nicol, G.; Kebarle, P. "Factors Determining Relative Sensitivity of Analytes in Positive Mode Atmospheric Pressure Ionization Mass Spectrometry," *Anal. Chem.* **60,** 1300–1307 (1988).
14. Berna, M.; Shugert, R.; Mullen, J. "Quantitative Determination of Olanzapine in Human Plasma and Serum by Liquid Chromatography/Tandem Mass Spectrometry," *J. Mass Spectrom.* **33,** 1003–1008 (1998).
15. Matuszewski, B.K.; Constanzer, M.L.; Chavez-Eng, C.M. "Matrix Effect in Quantitative LC/MS/MS Analyses of Biological Fluids: A Method for Determination of Finasteride in Human Plasma at Picogram Per Milliliter Concentrations," *Anal. Chem.* **70,** 882–889 (1998).
16. Schaefer, W.H.; Dixon, F. Jr. "Effect of High-Performance Liquid Chromatography Mobile Phase Components on Sensitivity in Negative Atmospheric Pressure Chemical Ionization Liquid Chromatography-Mass Spectrometry," *J. Am. Soc. Mass Spectrom.* **7,** 1059–1069 (1996).
17. Kuhlmann, F.E.; Apffel, A.; Fischer, S.M.; Goldberg, G.; Goodley, P.C. "Signal Enhancement for Gradient Reverse-Phase High-Performance Liquid Chromatography—Electrospray Ionization Mass Spectrometry Analysis with Trifluoroacetic and Other Strong Acid Modifiers by Post-Column Addition of Proprionic Acid with Isopropanol," *J. Am. Soc. Mass Spectrom.* **6,** 1221–1225 (1995).
18. Temesi, D.; Law, B. "The Effect of LC Eluent Composition on MS Responses Using Electrospray Ionization," *LC-GC* **17,** 626–632 (1999).
19. King, R.; Bonfiglio, R.; Fernandez-Metzler, C.; Miller-Stein, C.; Olah, T. "Mechanistic Investigation of Ionization Suppression in Electrospray Ionization," *J. Am. Soc. Mass Spectrom.* **11,** 942–950 (2000).
20. Gallagher, R.T.; Balogh, M.P.; Davey, P.; Jackson, M.R.; Sinclair, I.; Southern, L.J. "Combined Electrospray Ionization-Atmospheric Pressure Chemical Ionization Source for Use in High-Throughput LC-MS Applications," *Anal. Chem.* **75,** 973–977 (2003).
21. Liu, C.C.; Kovarik, P.; LeBlanc, Y.; Sakuma, T.; Garofolo, F.; Marland, A.; Pang, H.; McIntosh, M.; Wong, E.; Kennedy, M.; Covey, T. "Use of Fast Settling De-Solvation Heaters and Rapid Ionization Process Selection for Enhanced LC-MS/MS Performance," in *Proceedings of the 50th ASMS Conference on Mass Spectrometry and Allied Topics*, Orlando, FL, June 2–6, 2002.
22. Siegel, M.M.; Tabei, K.; Tong, H.; Lamber, F.; Candela, L.; Zoltan, B. "Evaluation of a Dual Electrospray Ionization/ Atmospheric Pressure Ionization Source at Low Flow Rates (~50 μL/min) for the Analysis of Both Highly and Weakly Polar Compounds," *J. Am. Soc. Mass Spectrom.* **9,** 1196–1203 (1998).

23. Robb, D.B.; Covey, T.R.; Bruins, A.P. "Atmospheric Pressure Photoionization: An Ionization Method for Liquid Chromatography-Mass Spectrometry," *Anal. Chem.* **72,** 3653–3659 (2000).

24. Yang, C.; Henion, J. "Atmospheric Pressure Photoionization Liquid Chromatographic-Mass Spectrometric Determination of Idoxifene and Its Metabolites in Human Plasma," *J. Chromatogr. A* **13,** 155–165 (2002).

25. Doerge, D.R.; Fogle, C.M.; Paule, M.G.; McCullagh, M.; Bajic, S. "Analysis of Methylphenidate and its Metabolite Ritalinic Acid in Monkey Plasma by Liquid Chromatography/Electrospray Ionization Mass Spectrometry," *Rapid Commun. Mass Spectrom.* **14,** 610–623 (2000).

26. Tiller, P.R.; Cunniff, J.; Land, A.P.; Schwartz, J.; Jardine, I.; Wakefield, M.; Lopez, L.; Newton, J.F.; Burton, R.D.; Folk, B.M.; Buhrman, D.L.; Price, P.; Wu, D. "Drug Quantitation on a Benchtop Liquid Chromatography–Tandem Mass Spectrometry System," *J. Chromatogr. A.* **771,** 119–125 (1997).

27. Weiboldt, R.; Campbell, D.A.; Henion, J. "Quantitative Liquid Chromatographic-Tandem Mass Spectrometric Determination of Orlistat in Plasma with a Quadrupole Ion Trap," *J. Chromatogr. B.* **708,** 121–129 (1998).

28. Zhang, N.; Fountain, S.T.; Bi, H.; Rossi, D.T. "Quantification and Rapid Metabolite Identification in Drug Discovery Using API Time-of-Flight LC/MS," *Anal. Chem.* **72,** 800–806 (2000).

29. Clauwaert, K.M.; Van Bocxlaer, J.F.; Major, H.J.; Claereboudt, J.A.; Lambert, W.E.; Van den Eeckhout, E.M.; Van Peteghem, C.H.; De Leenheer, A.P. "Investigation of the Quantitative Properties of the Quadrupole Orthogonal Acceleration Time-of-Flight Mass Spectrometer with Electrospray Ionisation Using 3,4-Methylenedioxymethamphetamine," *Rapid Commun. Mass Spectrom.* **13,** 1540–1545 (1999).

30. Yang, L.; Wu, N.; Rudewicz, P.J. "Applications of New Liquid Chromatography-Tandem Mass Spectrometry Technologies for Drug Development Support," *J. Chromatogr. A* **926,** 43–55 (2001).

31. Jemal M.; Ouyang, Z. "Enhanced Resolution Triple-Quadrupole Mass Spectrometry for Fast Quantitative Bioanalysis Using Liquid Chromatography/Tandem Mass Spectrometry: Investigations of Parameters That Affect Ruggedness," *Rapid Commun. Mass Spectrom.* **17,** 24–38 (2003).

32. Yang, L.; Amad, M.; Winnik, W.M.; Schoen, A.E.; Schweingruber, H.; Mylchreest, I.; Rudewicz, P.J. "Investigation of an Enhanced Resolution Triple Quadrupole Mass Spectrometer for High-Throughput Liquid Chromatography/Tandem Mass Spectrometry Assays," *Rapid Commun. Mass Spectrom.* **16,** 2060–2066 (2002).

33. Needham, S.; Jeanville, P.; Cole, M. "The Role of ESI-TOF in Drug Discovery Metabolism," in *Proceedings of the 47th ASMS Conference on Mass Spectrometry and Allied Topics*, Dallas, TX, June 13–16,1999.

34. Zhang, H.; Henion, J. "Comparison Between Liquid Chromatography-Time-of-Flight Mass Spectrometry and Selected Reaction Monitoring Liquid

Chromatography-Mass Spectrometry for Quantitative Determination of Idoxifene in Human Plasma," *J. Chromatogr. B. Biomed. Sci. Appl.* **757,** 151–159 (2001).

35. Hager, J.W. "A New Linear Ion Trap Mass Spectrometer," *Rapid Commun. Mass. Spectrom.*" **16,** 512–526 (2002).
36. Hopfgartner, G.; Husser, C.; Zell, M. "Rapid Screening and Characterization of Drug Metabolites Using a New Quadrupole-Linear Ion Trap Mass Spectrometer," *J. Mass Spectrom.* **38,** 138–150 (2003).
37. Simpson, H.; Berthemy, A.; Buhrman, D.; Burton, R.; Newton, J.; Kealy, M.; Wells, D.; Wu, D. "High Throughput Liquid Chromatography/Mass Spectrometry Bioanalysis Using 96-Well Disk Solid Phase Extraction Plate for Sample Preparation," *Rapid Commun. Mass Spectrom.* **12,** 75–82 (1998).
38. Jemal, M.; Teitz, D.; Ouyang, Z.; Khan, S. "Comparison of Plasma Sample Purification by Manual Liquid-Liquid Extraction, Automated 96-Well Liquid-Liquid Extraction and Automated 96-Well Solid-Phase Extraction for Analysis by High-Performance Liquid Chromatography with Tandem Mass Spectrometry," *J. Chromatogr. B Biomed. Sci. Appl.* **732,** 501–508 (1999).
39. Watt, A.P.; Morrison, D.; Locker, K.L.; Evans, D.G. "Higher Throughput Bioanalysis by Automation of a Protein Precipitation Assay Using 96-Well Format with Detection by LC-MS/MS," *Anal. Chem.* **72**, 979–984 (2000).
40. Mullett, W.M.; Martin, P.; Pawliszyn, J. "In-Tube Molecularly Imprinted Polymer Solid-Phase Microextraction for the Selective Determination of Propranolol," *Anal. Chem.* **73,** 2383–2389 (2001).
41. Rule, G.S.; Henion, J.D. "Determination of Drugs from Urine by On-Line Immunoaffinity Chromatography–High-Performance Liquid Chromatography–Mass Spectrometry," *J. Chromatogr.* **582,** 103–112 (1992).
42. Rule, G.; Henion, J. "A 384-Well Solid-Phase Extraction for LC/MS/MS Determination of Methotrexate and its 7-Hydroxy Metabolite in Human Urine and Plasma," *Anal. Chem.* **73,** 439–443 (2001).
43. Biddlecombe, R.A.; Benevides, C.; Pleasance, S. "A Clinical Trial on a Plate? The Potential of 384-Well Format Solid Phase Extraction for High-Throughput Bioanalysis Using Liquid Chromatography/Tandem Mass Spectrometry," *Rapid Commun. Mass Spectrom.* **15,** 33–40 (2001).
44. Henion, J.; Brewer, E.; Rule, G. "Sample Preparation for LC/MS/MS: Knowing the Basic Requirements and the Big Picture of an LC/MS System can Ensure Success in Most Instances," *Anal. Chem.* **70,** 650A–656A (1998).
45. Ackermann, B.L.; Murphy, A.T.; Berna, M.J. "The Resurgence of Column Switching Techniques to Facilitate Rapid LC/MS/MS Based Bioanalysis in Drug Discovery," *Am. Pharm. Rev.* **5,** 54–63 (2002).
46. Allanson, J.P.; Biddlecombe, R.A.; Jones, A.E.; Pleasance, S. "The Use of Automated Solid-Phase Extraction in the 96-Well Format for High Throughput Bioanalysis Using Liquid Chromtography Coupled to Tandem Mass Spectrometry," *Rapid Commun. Mass Spectrom.* **10,** 811–816 (1996).

47. Janiszewski, J.; Schneider, R.; Hoffmaster, K.; Swyden, M.; Wells, D.; Fouda, H. "Automated Sample Preparation Using Membrane Microtiter Extraction for Bioanalytical Mass Spectrometry," *Rapid Commun. Mass Spectrom.* **11,** 1033–1037 (1997).
48. Zhang, N.; Hoffman, K.L.; Li, W.; Rossi, D.T. "Semi-Automated 96-Well Liquid-Liquid Extraction for Quantitation of Drugs in Biological Fluids," *J. Pharm. Biomed. Anal.* **22,** 131–138 (2000).
49. Murphy, A.T.; Johnson, J.T.; Ackermann, B.L.; Gillespie, T.A.; Garner, C.O. "Bioavailability of Zyprexa Lead Generation Formulations in Dog Plasma Using LC/MS/MS Detection of Protein Precipitation Supernatants Injected From 96-Well Microtiter Plates," in *Proceedings of the 46th ASMS Conference on Mass Spectrometry and Allied Topics*, Orlando, FL, May 31–June 5, 1998.
50. Biddlecombe, R.A.; Pleasance, S. "Automated Protein Precipitation by Filtration in the 96-Well Format," *J. Chromatogr. B Biomed. Sci. Appl.* **12,** 257–265 (1999).
51. Polson, C.; Sarkar, P.; Incledon, B.; Raguvaran, V.; Grant, R. "Optimization of Protein Precipitation Based upon Effectiveness of Protein Removal and Ionization Effect in Liquid Chromatography-Tandem Mass Spectrometry," *J. Chromatogr. B Analyt. Technol. Biomed. Life Sci.* **785,** 263–275 (2003).
52. Kaye, B.; Herron, W.J.; Macrae, P.V.; Robinson, S.; Stopher, D.A.; Venn, R.F.; Wild, W. "Rapid, Solid Phase Extraction Technique for the High-Throughput Assay of Darifenacin in Human Plasma," *Anal. Chem.* **68,** 1658–1660 (1996).
53. Lensmeyer, G.L.; Oroskar, A. "Evaluation Model for Identification of Distinctive Silanol Variations in SPE Sorbents Used for LC-MS/MS Bioanalytical Sample Preparation," in *Proceedings of the 50th ASMS Conference on Mass Spectrometry and Allied Topics*, Orlando, FL, June 2–6, 2002.
54. Wells, D.A.; Lensmeyer, G.L.; Wiebe, D.A. "Particle-Loaded Membranes as an Alternative to Traditional Packed-Column Sorbents for Drug Extraction—In Depth Comparative Study," *J. Chromatogr. Sci.* **33,** 386–392 (1995).
55. Cheng, Y.F.; Neue, U.D.; Bean, L. "Straightforward Solid-Phase Extraction Method for the Determination of Verapamil and its Metabolite in Plasma in a 96-Well Extraction Plate," *J. Chromatogr. A* **828,** 273–281 (1998).
56. Mallet, C.R.; Lu, Z.; Fisk, R.; Mazzeo, J.R.; Neue, U.D. "Performance of an Ultra-Low Elution-Volume 96-Well Plate: Drug Discovery and Development Applications," *Rapid Commun. Mass Spectrom.* **17,** 163–170 (2003).
57. Harvey, M.C.; Stearns, S.D. "HPLC Sample Injection and Column-Switching Using a 10-Port Multifunctional Valve," *Am. Lab.* **14,** 68 (1982).
58. Needham, S.R.; Cole, M.J.; Fouda, H.G. "Direct Plasma Injection for High-Performance Liquid Chromatographic-Mass Spectrometric Quantitation of the Anxiolytic Agent CP-93393," *J. Chromatogr. B* **718,** 87–94 (1998).
59. Schellen, A.; Ooms, B.; van Gils, M.; Halmingh, O.; van der Vlis, E.; van de Lagemaat, D.; Verheij, E. "High Throughput On-Line Solid Phase Extraction/Tandem Mass Spectrometric Determination of Paclitaxel in Human Serum," *Rapid Commun. Mass Spectrom.* **14,** 230–233 (2000).

60. Ackermann, B.L.; Ruterbories, K.J.; Hanssen, B.R.; Lindstrom, T.D. "Increasing the Throughput of Microsomal Stability Screening Using Fast Gradient Elution LC/MS," in *Proceedings of the 46th ASMS Conference on Mass Spectrometry and Allied Topics*, Orlando, FL, May 31–June 5, 1998.

61. McLoughlin, D.A.; Olah, T.V.; Gilbert, J.D. "A Direct Technique for the Simultaneous Determination of 10 Drug Candidates in Plasma by Liquid Chromatography–Atmospheric Pressure Chemical Ionization Mass Spectrometry Interfaced to a Prospekt Solid-Phase Extraction System," *J. Pharm. Biomed. Anal.* **15,** 1893–1901 (1997).

62. Beaudry, F.; LeBlanc, J.Y.C.; Coutu, M.; Brown, N. "In Vivo Pharmcokinetic Screening in Cassette Dosing Experiments: The Use of On-Line Prospekt Liquid Chromatography/Atmospheric Pressure Chemical Ionization Tandem Mass Spectrometry Technology in Drug Discovery," *Rapid Commun. Mass Spectrom.* **12,** 1216–1222 (1998).

63. Gao, V.C.X.; Luo, W.C.; Ye Q.; Thoolen, M. "Column-Switching in High-Performance Liquid Chromatography with Tandem Mass Spectrometric Detection for High-Throughput Preclinical Pharmacokinetic Studies," *J. Chromatogr. A.* **828,** 141–148 (1998).

64. Zell, M.; Husser, C.; Hopfgartner, G. "Column-Switching High-Performance Liquid Chromatography Combined with Ionspray Tandem Mass Spectrometry for the Simultaneous Determination of the Platelet Inhibitor Ro 44-3888 and Its Pro-Drug and Precursor Metabolite in Plasma," *J. Mass Spectrom.* **32**, 23–32 (1997).

65. Boos, K.-S.; Rudolphi, A. "The Use of Restricted-Access Media in HPLC, Part I—Classification and Review," *LC-GC* **15,** 602–611 (1997).

66. Hagestam, I.H.; and Pinkerton, T.C. "Internal Surface Reversed-Phase Silica Supports for Liquid Chromatography," *Anal. Chem.* **57,** 1757–1763 (1985).

67. Boppana, V.K.; Miller-Stein, C.; Schaefer, W.H. "Direct Plasma Liquid Chromatographic-Tandem Mass Spectrometric Analysis of Ganisetron and its 7-Hydroxy Metabolite Utilizing Internal Surface Reversed-Phase Guard Columns and Automated Column-switching Devices," *J. Chromatogr. B.* **678,** 227–236 (1996).

68. Boos, K.-S.; Walfort, A.; Eisenbeiss, F.; Lubda, D. German Patent DE 4130475 (1991).

69. Boos, K.-S.; Rudolphi, A.; Vielhauer, S.; Walfort, A.; Lubda, D.; Eisenbeiss, F. "Alkyl-Diol Silica (ADS): Restricted Access Precolumn Packings for Direct Injection and Coupled-Column Chromatography of Biofluids," *Fresenius J. Anal. Chem.* **352**, 684–690 (1995).

70. Hsieh, Y.; Bryant, M.S.; Gruela, G.; Brisson, J.-M; Korfmacher, W.A. "Direct Analysis of Plasma Samples for Drug Discovery Compounds Using Mixed-Function Column Liquid Chromatography Tandem Mass Spectrometry," *Rapid Commun. Mass Spectrom.* **14,** 1384–1390 (2000).

71. Quin, H.M.; Takarewski, J.J. Int. Patent Number WO 97/16724 (1997).

72. Ayrton, J.; Dear, G.J.; Leavens, W.J.; Mallett, D.N.; Plumb, R.S. "The Use of Turbulent Flow Chromatography/Mass Spectrometry for the Rapid, Direct Analysis of a Novel Pharmaceutical Compound in Plasma," *Rapid Commun. Mass Spectrom.* **11,** 1953–1958 (1997).
73. Jemal, M.; Qing, Y.; Whigan, D.B. "The Use of High-Flow High Performance Liquid Chromatography Coupled with Positive and Negative Ion Electrospray Tandem Mass Spectrometry for Quantitative Bioanalysis via Direct Injection of the Plasma/Serum Samples," *Rapid Commun. Mass Spectrom.* **12,** 1389–1399 (1998).
74. Jemal, M.; Huang, M.; Jiang, X.; Mao, Y.; Powell, M.L. "Direct Injection Versus Liquid–Liquid Extraction for Plasma Sample Analysis by High Performance Liquid Chromatography with Tandem Mass Spectrometry," *Rapid Commun. Mass Spectrom.* **13,** 2125–2132 (1999).
75. Wu, J.-T.; Zeng, H.; Qian, M.; Brogdon, B.; Unger, S.E. "Direct Plasma Sample Injection in Multiple-Component LC-MS-MS Assays for High-Throughput Pharmacokinetic Screening," *Anal. Chem.* **72,** 61–67 (2000).
76. Ayrton, J.; Clare, R.A.; Dear, G.J.; Mallett, D.N.; Plumb, R.S. "Ultra-High Flow Rate Capillary Liquid Chromatography with Mass Spectrometric Detection for the Direct Analysis of Pharmaceuticals in Plasma at Sub-Nanogram per Milliliter Concentrations," *Rapid Commun. Mass Spectrom.* **13,** 1657–1662 (1999).
77. Bayliss, M.K.; Little, D.; Mallett, D.N.; Plumb, R.S. "Parallel Ultra-High Flow Rate Liquid Chromatography with Mass Spectrometric Detection Using a Multiplex Electrospray Source for Direct, Sensitive Determination of Pharmaceuticals in Plasma at Extremely High Throughput," *Rapid Commun. Mass Spectrom.* **14,** 2039–2045 (2000).
78. King, R.C.; Miller-Stein, C.; Magiera, D.J.; Brann, J. "Description and Validation of a Staggered Parallel High Performance Liquid Chromatography System for Good Laboratory Practice Level Quantitative Analysis by Liquid Chromatography/Tandem Mass Spectrometry," *Rapid Commun. Mass Spectrom.* **16,** 43–52 (2002).
79. Grant, R.P.; Cameron, C.; Mackenzie-McMurter, S. "Generic Serial and Parallel On-Line Direct-Injection Using Turbulent Flow Liquid Chromatography/Tandem Mass Spectrometry," *Rapid Commun. Mass Spectrom.* **16,** 1785–1792 (2002).
80. Zhang, H.; Heinig, K.; Henion, J. "Atmospheric Pressure Ionization Time-of-Flight Mass Spectrometry Coupled with Fast Liquid Chromatography for Quantitation and Accurate Mass Measurement of Five Pharmaceutical Drugs," *J. Mass Spectrom.* **35,** 423–431 (2000).
81. Warner, D.L.; Dorsey, J.G. "Reduction of Total Analysis Time in Gradient Elution, Reversed-Phase Liquid Chromatography," *LC-GC* **15,** 254–262 (1997).
82. Hopfgartner, G.; Bean, K.; Henion, J.; Henry, R. "Ion Spray Mass Spectrometric Detection for Liquid Chromatography: A Concentration—or a Mass-Flow-Sensitive Device?" *J. Chromatogr.* **647,** 51–61 (1993).

83. Plumb, R.S.; Dear, G.J.; Mallett, D.; Fraser, I.J.; Ayrton, J.; Ioannou, C. "The Application of Fast Gradient Capillary Liquid Chromatography/Mass Spectrometry to the Analysis of Pharmaceuticals in Biofluids," *Rapid Commun. Mass Spectrom.* **13,** 865–872 (1999).

84. Romanyshyn, L.; Tiller, P.R.; Hop, E.C.A. "Bioanalytical Applications of 'Fast Chromatography' to High-Throughput Liquid Chromatography/Tandem Mass Spectrometric Quantitation," *Rapid Commun. Mass Spectrom.* **14**, 1662–1668 (2000).

85. Romanyshyn, L.; Tiller, P.R.; Alvaro, R.; Pereira, A.; Hop, C.E.C.A "Ultra-Fast Gradient vs. Fast Isocratic Chromatography in Bioanalytical Quantification by Liquid Chromatography/Tandem Mass Spectrometry," *Rapid Commun. Mass Spectrom.* **15,** 313–319 (2001).

86. Cheng, Y.-F.; Lu, Z.; Neue, U. "Ultrafast Liquid Chromatography/Ultraviolet and Liquid Chromatography/Tandem Mass Spectrometric Analysis," *Rapid Commun. Mass Spectrom.* **15,** 141–151 (2001).

87. Hayward, M.J.; Munson, J.L.; Conneely, G.; Hargiss, L.O. "What Limits Productivity in High Flow Fast LC/MS? " in *Proceedings of the 49th ASMS Conference on Mass Spectrometry and Allied Topics*, Chicago, IL, May 27–31, 2001.

88. Murphy, A.T.; Berna, M.J.; Holsapple, J.L.; Ackermann, B.L. "Effects of Flow Rate on High-Throughput Quantitative Analysis of Protein-Precipitated Plasma Using Liquid Chromatography/Tandem Mass Spectrometry," *Rapid Commun. Mass Spectrom.* **16,** 537–543 (2002).

89. Covey, T.; Stott, B.; Joliffe, C.; Thomson, B. "Development of a High Sensitivity Triple Quadrupole Instrument for Quantitative Analysis," in *Proceedings of the 46th ASMS Conference on Mass Spectrometry and Allied Topics*, Orlando, FL, May 31–June 5, 1998.

90. Minakuchi, H.; Nakanishi, K.; Soga, N.; Ishizuka, N.; Tanaka, N. "Octadecylsilylated Porous Silica Rods as Separation Media for Reversed-Phase Liquid Chromatography," *Anal. Chem.* **68,** 3498–3501 (1996).

91. Tanaka, N.; Nagayama, H.; Kobayashi, H.; Ikegami, T.; Hosoya, K.; Ishizuka, N.; Minakuchi, H.; Nakanishi, K.; Cabrera, K.; Lubda, D. "Monolithic Silica Columns for HPLC, micro-HPLC and CEC," *J. High Resol. Chromatogr.* **23,** 111–116 (2000).

92. Wu, J.-T.; Zeng, H.; Deng, Y.; Unger, S.E. "High-Speed Liquid Chromatography/Tandem Mass Spectrometry Using a Monolithic Column for High-Throughput Bioanalysis," *Rapid Commun. Mass Spectrom.* **15,** 1113–1119 (2001).

93. Barbarin, N.; Mawhinney, D.B.; Black, R.; Henion, J.; "High-Throughput Selected Reaction Monitoring Liquid Chromatography-Mass Spectrometry Determination Of Methylphenidate and its Major Metabolite, Ritalinic Acid, in Rat Plasma Employing Monolithic Columns," *J. Chromatogr. B Anal. Technol. Biomed. Life Sci.* **783,** 73–83 (2003).

94. Przybyciel, M.; Majors, R.E. “Phase Collapse in Reversed-Phase Liquid Chromatography,” *LC–GC* **20,** 516–523 (2002).
95. Majors, R.E.; Przybyciel, M. “Columns for Reversed-Phase LC Separations in Highly Aqueous Mobile Phases,” *LC-GC* **20,** 584–593 (2002).
96. Miller-Stein, C.; Fernandez-Metzler, C. “Determination of Chiral Sulfoxides in Plasma by Normal-Phase Liquid Chromatography-Atmospheric Pressure Chemical Ionization Mass Spectrometry,” *J. Chromatogr. A* **964,** 161–168 (2002).
97. Strege, M.A.; Stevenson, S.; Lawrence, S.M. “Mixed-Mode Anion-Cation Exchange/Hydrophilic Interaction Liquid Chromatography-Electrospray Mass Spectrometry as an Alternative to Reversed Phase for Small Molecule Drug Discovery,” *Anal. Chem.* **72,** 4629–4633 (2000).
98. Naidong, W.; Lee, J.W.; Jiang, X.; Wehling, M.; Hulse, J.D.; Lin, P.P. “Simultaneous Assay of Morphine, Morphine-3-Glucuronide and Morphine-6-Glucuronide in Human Plasma Using Normal-Phase Liquid Chromatography-Tandem Mass Spectrometry with a Silica Column and an Aqueous Organic Mobile Phase,” *J. Chromatogr. B Biomed. Sci Appl.* **735,** 255–269 (1999).
99. Shou, W.Z.; Chen, Y.L.; Eerkes, A.; Tang, Y.Q.; Magis, L.; Jiang, X.; Naidong, W. “Ultrafast Liquid Chromatography/Tandem Mass Spectrometry Bioanalysis of Polar Analytes Using Packed Silica Columns,” *Rapid Commun. Mass Spectrom.* **16,** 1613–1621 (2002).
100. Korfmacher, W.A.; Veals, J.; Dunn-Meynell, K.; Zhang, X.P.; Tucker, G.; Cox, K.A.; Lin, C.C. “Demonstration of the Capabilities of a Parallel High Performance Liquid Chromatography Tandem Mass Spectrometry System for Use in the Analysis of Drug Discovery Samples,” *Rapid Commun. Mass Spectrom.* **13,** 1991–1999 (1999).
101. Van Pelt, C.K.; Corso, T.N.; Schultz, G.A.; Lowes, S.; Henion, J. “A Four- Column Parallel Chromatography System for Isocratic or Gradient LC/MS Analyses,” *Anal. Chem.* **73,** 582–588 (2001).
102. Yang, L.; Mann, T.D.; Little, D.; Wu, N.; Clement, R.P.; Rudewicz, P.J. “Evaluation of a Four-Channel Multiplexed Electrospray Triple Quadrupole Mass Spectrometer for the Simultaneous Validation of LC/MS/MS Methods in Four Different Preclinical Matrices,” *Anal. Chem.* **73,** 1740–1747 (2001).
103. Deng, T.; Zeng, H.; Unger, S.E.; Wu, J.T. “Multiple-Sprayer Tandem Mass Spectrometry with Parallel High Flow Extraction and Parallel Separation for High-Throughput Quantitation in Biological Fluids,” *Rapid Commun. Mass Spectrom.* **15,** 1634–1640 (2001).
104. Zeng, H.; Wu, J.T.; Unger, S.E. “The Investigation and the Use of High Flow Column-Switching LC/MS/MS as a High-Throughput Approach for Direct Plasma Sample Analysis of Single and Multiple Components in Pharmacokinetic Studies,” *J. Pharm. Biomed. Anal.* **27,** 967–982 (2002).
105. Deng, Y.; Wu, J.T.; Lloyd, T.L.; Chi, C.L.; Olah, T.V.; Unger, S.E. “High-Speed Gradient Parallel Liquid Chromatography/Tandem Mass Spectrometry

with Fully Automated Sample Preparation for Bioanalysis: 30 Seconds per Sample from Plasma," *Rapid Commun. Mass Spectrom.* **16,** 1116–1123 (2002).

106. Whalen, K.M.; Rogers, K.J.; Cole, M.J.; Janiszewski, J.S. "AutoScan: An Automated Workstation for Rapid Determination of Mass and Tandem Mass Spectrometry Conditions for Quantitative Bioanalytical Mass Spectrometry," *Rapid Commun. Mass Spectrom.* **14,** 2074–2079 (2000).
107. Chen, S.; Carvey, R. "Validation of Liquid-Liquid Extraction Followed by Flow-Injection Negative Ion Electrospray Mass Spectrometry Assay to Topiramate in Human Plasma," *Rapid Commun. Mass Spectrom.* **15,** 159–163 (2001).
108. Zheng, J.J.; Lynch, E.D.; Unger, S.E. "Comparison of SPE and Fast LC to Eliminate Mass Spectrometric Matrix Effects from Microsomal Incubation Products," *J. Pharm. Biomed. Anal.* **28,** 279–285 (2002).
109. Olech, R.M.; Pranis, R.A.; Jacobson, J.R.; Perman, C.A.; Boman, B.A.; Soldo, J.; Speziale, R.; Astle, T.W.; Cole, M.J.; Janiszewski, J.S.; Whalen, K.M. "A Novel SPE System for High-Throughput Quantification by ESI-LC/MS/MS Utilizing 96 Discrete Zones in a Disposable Card Format," in *Proceedings of the 49th Conference on Mass Spectrometry and Allied Topics*, Chicago, IL, May 27–31, 2001.
110. Dethy, J.M.; Ackermann, B.L.; Delatour, C.; Henion, J.D.; Schultz, G.A. "Demonstration of Direct Bioanalysis of Drugs in Plasma Using Nanoelectrospray Infusion From a Silicon Chip Coupled with Tandem Mass Spectrometry," *Anal. Chem.* **75,** 805–811 (2003).
111. Berman, J.; Halm, K.; Adkison, K.; Shaffer, J. "Simultaneous Pharmacokinetic Screening of a Mixture of Compounds in the Dog Using API LC/MS/MS Analysis for Increased Throughput," *J. Med. Chem.* **40,** 3–5 (1997).
112. Olah, T.V.; McLoughlin, D.A.; Gilbert, J.D. "The Simultaneous Determination of Mixtures of Drug Candidates by Liquid Chromatography/Atmospheric Pressure Chemical Ionization Mass Spectrometry as an In Vivo Drug Screening Procedure," *Rapid Commun. Mass Spectrom.* **11,** 17–23 (1997).
113. Tong, X.; Ita, I.E.; Wang, J.; Pivnichny, J.V. "Characterization of a Technique for Rapid Pharmacokinetic Studies of Multiple Co-Eluting Compounds by LC/MS/MS," *J. Pharm. Biomed. Anal.* **29,** 773–784 (1999).
114. White, R.E.; Manitpisitkul, P. "Pharmacokinetic Theory of Cassette Dosing in Drug Discovery Screening," *Drug Metab. Dispos.* **29,** 957–966 (2001).
115. Hop, C.E.C.A; Wang, Z.; Chen, Q.; Kwei, G. "Plasma-Pooling Methods to Increase Throughput for In Vivo Pharmacokinetic Screening," *J. Pharm. Sci.* **87,** 901–903 (1998).
116. Zhang, N.; Rogers, K.; Gajda, K.; Kagel, J.R.; Rossi, D.T. "Integrated Sample Collection and Handling for Drug Discovery Bioanalysis," *J. Pharm. Biomed. Anal.* **23,** 551–560 (2000).
117. Sadagopan, N.P.; Li, W.; Cook, J.A.; Galvan, B.; Weller, D.L.; Fountain, S.T.; Cohen, L.H. "Investigation of EDTA Anticoagulant in Plasma to Improve the

Throughput of Liquid Chromatography/Tandem Mass Spectrometric Assays," *Rapid Commun. Mass Spectrom.* **17,** 1065–1070 (2003).

118. Berna, M.J.; Murphy, A.T.; Wilken, B.F.; Ackermann, B.L. "Collection, Storage and Filtration of In-Vivo Study Samples Using 96-Well Filter Plates to Facilitate Automated Sample Preparation and LC/MS/MS Analysis," *Anal. Chem.* **74,** 1197–1201 (2002).
119. Korfmacher, W.A.; Cox, K.A.; Ng, K.J.; Veals, J.; Hsieh, Y.; Wainhaus, S.; Broske, L.; Prelusky, D.; Nomeir, A.; White, R.E. "Cassette-Accelerated Rapid Rat Screen: A Systematic Procedure for the Dosing and Liquid Chromatography/ Atmospheric Pressure Ionization Tandem Mass Spectrometric Analysis of New Chemical Entities as Part of New Drug Discovery," *Rapid Commun. Mass Spectrom.* **15,** 335–340 (2001).
120. Buhrman, D.L.; Price, P.I.; Rudewicz, P.J. "Quantitation of SR 27417 in Human Plasma Using Electrospray Liquid Chromatography-Tandem Mass Spectrometry: A Study of Ion Suppression," *J. Am. Soc. Mass Spectrom.* **7,** 1099–1105 (1996).
121. LeBlanc, J.C.Y.; Liu, C.C.; Covey, T.; Kovarik, P.; Thompson, C.; Javahari, H.; Jong, R. "Ion Suppression: Parameters That Influence the Ionization Efficiency in APCI-LC/MS/MS," in *Proceedings of the 49th ASMS Conference on Mass Spectrometry and Allied Topics*, Chicago, IL, May 27–31, 2001.
122. Shang, J.X.; Zhong, W.-Z.; Heath, T.G. "Examination of Matrix-Induced Ionization Suppression in Atmospheric Pressure Chemical Ionization," in *Proceedings of the 49th ASMS Conference on Mass Spectrometry and Allied Topics*, Chicago, IL, May 27–31, 2001.
123. Bonfiglio, R.; King, R.C.; Olah, T.V.; Merkle, K. "The Effect of Sample Preparation Methods on the Variability of the Electrospray Ionization Response for Model Drug Compounds," *Rapid Commun. Mass Spectrom.* **13,** 1175–1185 (1999).
124. Mei, H.; Hsieh, Y.; Nardo, C.; Xu, X.; Wang, S.; Ng, K.; Korfmacher, W.A. "Investigation of Matrix Effects in Bioanalytical High-Performance Liquid Chromatography/Tandem Mass Spectrometric Assays: Application to Drug Discovery," *Rapid Commun. Mass Spectrom.* **17,** 97–103 (2003).
125. Tong, X.S.; Wang, J.; Zheng, S.; Pivnichny, J.V.; Griffin, P.R.; Shen, X.; Donnelly, M.; Vakerich, K.; Nunes, C.; Fenyk-Melody, J. "Effect of Signal Interference from Dosing Excipients on Pharmacokinetic Screening of Drug Candidates by Liquid Chromatography/Mass Spectrometry," *Anal. Chem.* **74,** 6305–6313 (2002).
126. Shou, W.Z.; Naidong, W. "Post-Column Infusion Study of the 'Dosing Vehicle Effect' in the Liquid Chromatography/Tandem Mass Spectrometric Analysis of Discovery Pharmacokinetic Samples," *Rapid Commun. Mass Spectrom.* **17,** 589–597 (2003).
127. Avery, M.J. "Quantitative Characterization of Differential Ion Suppression on Liquid Chromatography/Atmospheric Pressure Ionization Mass Spectrometric Bioanalytical Methods," *Rapid Commun. Mass Spectrom.* **17,** 197–201 (2003).

128. Ramanathan, R.; Su, A.D.; Alvarez, N.; Blumenkrantz, N.; Chowdhury, S.K.; Alton, K.; Patrick, J. "Liquid Chromatography/Mass Spectrometry Methods for Distinguishing N-Oxides from Hydroxylated Compounds," *Anal. Chem.* **72,** 1352–1359 (2000).

129. Tong, W.; Chowdhury, S.K.; Chen, J.C.; Zhong, R.; Alton, K.B.; Patrick, J.E. "Fragmentation of N-Oxides (Deoxygenation) in Atmospheric Pressure Ionization: Investigation of the Activation Process," *Rapid. Commun. Mass Spectrom.* **15,** 2085–2090 (2001).

130. Jemal, M.; Quyang, Z. "The Need for Chromatographic and Mass Resolution in Liquid Chromatography/Tandem Mass Spectrometric Methods Used for Quantitation of Lactones and Corresponding Hydroxy Acids in Biological Samples," *Rapid Commun. Mass Spectrom.* **14,** 1757–1765 (2000).

131. Gobey, J.S.; Avery, M.J. "The Impact of Metabolites on LC/MS/MS Assays: A Real Life Example," in *Proceedings of the 49th ASMS Conference on Mass Spectrometry and Allied Topics*, Chicago, IL, May 27–31, 2001.

132. Jemal, M.; Ouyang, Z.; Powell, M.L. "A Strategy for a Post-Method-Validation use of Incurred Biological Samples for Establishing the Acceptability of a Liquid Chromatography/Tandem Mass-Spectrometric Method for Quantitation of Drugs in Biological Samples," *Rapid Commun. Mass Spectrom.* **16,** 1538–1547 (2002).

133. Schultz, G.A.; Corso, T.N.; Prosser, S.J.; Zhang, S. "A Fully Integrated Monolithic Microchip Electrospray Device for Mass Spectrometry," *Anal. Chem.* **72,** 4058–4063 (2000).

12

NEW STRATEGIES FOR THE IMPLEMENTATION AND SUPPORT OF BIOANALYSIS IN A DRUG METABOLISM ENVIRONMENT

WALTER A. KORFMACHER

INTRODUCTION

There are many challenges that one must overcome in order to achieve the goal of producing high-quality pharmacokinetic (PK) data in a higher throughput manner when working in a drug-metabolism (DM) environment in a pharmaceutical industry. The chapter in this book by Ackermann et al. provides an excellent overview of current practices and capabilities for high throughput bioanalytical support for drug-discovery samples (see also the recent review articles by Ackermann et al. [1] and Jemal [2]). This chapter focuses more on recent applications and strategies that can be implemented as needed when using mass spectrometry for lead optimization in a DM environment.

In 1997, Korfmacher et al. [3], described the basic strategies for using high-performance liquid chromatography–tandem mass spectrometry (HPLC-MS/MS) systems to support drug discovery in a DM environment. These strategies include the use of the triple-quadrupole HPLC-MS/MS system as a tool for providing the ability to perform rapid method development for PK sample assays as well as providing excellent sensitivity and selectivity for the compound (dosed drug) of interest. Due to the improved sensitivity

Integrated Strategies for Drug Discovery Using Mass Spectrometry, Edited by Mike S. Lee

over HPLC-ultraviolet (UV), smaller sample volumes were required, which allowed researchers to have a paradigm shift in the way rat PK studies were performed. Before the capabilities of the HPLC-MS/MS system were realized, HPLC-UV was the assay method and it was common to use individual rats for each concentration-time point when planning a PK study in the rat. This was due to the analytical need when using HPLC-UV to have 0.5–1.0 mL of plasma for the analysis. The HPLC-MS/MS systems allowed for 40 μL of plasma to be sufficient for an assay; this change allowed the use of serial bleeding in the rat. Serial bleeding allowed six rats to be used for an intravenous (IV) and oral (PO)-dosed PK study ($N = 3$ each dose route); previously, 48 rats were required for this same study, based on the need for 0.5–1.0 mL of plasma for the analysis. Furthermore, method development using HPLC-UV would have required two to four weeks; with HPLC-MS/MS, method development can be performed in one day or less. These basic concepts are still in use today.

What has changed since 1997 is that HPLC-MS/MS hardware and software have continually improved and that new strategies have developed to uniquely take advantage of these improvements. This chapter discusses some of these new strategies for using HPLC-MS/MS systems for various types of bioanalytical assays, as well as describing some unique challenges that have arisen when implementing these assays and some of their solutions.

Working in a Drug-Metabolism Environment

Several review articles have been written in the last few years that discuss the role that DM can play as part of a drug-discovery effort [4–10]. The importance of DM involvement in a drug-discovery effort was discussed in a review article by Kennedy [11] in 1997, in which historical data were used to show that a significant percentage of failures in the clinical phase were due to poor human PK properties of the investigational drug. The need for DM support of new drug-discovery efforts is now well accepted and most major pharmaceutical companies include DM input as part of their discovery strategy. What is needed from DM specialists is three types of information: (1) the PK properties for the compound in laboratory test animals (typically rats, dogs, and monkeys); (2) the potential human drug–drug interactions for the compound; and (3) the predicted human PK properties for the test compound.

To answer these needs, a series of in vitro and in vivo assays have been developed with the goal of providing these data in a time frame that is compatible with the requirements of medicinal chemists who are making the new chemical entities (NCEs). Most of the in vitro and in vivo drug metabolism and pharmacokinetic (DMPK) screens use LC-MS or LC-MS/MS for the analytical step. As shown in Figure 12.1, one approach for using these assays is to view the process of lead optimization as passing the lead

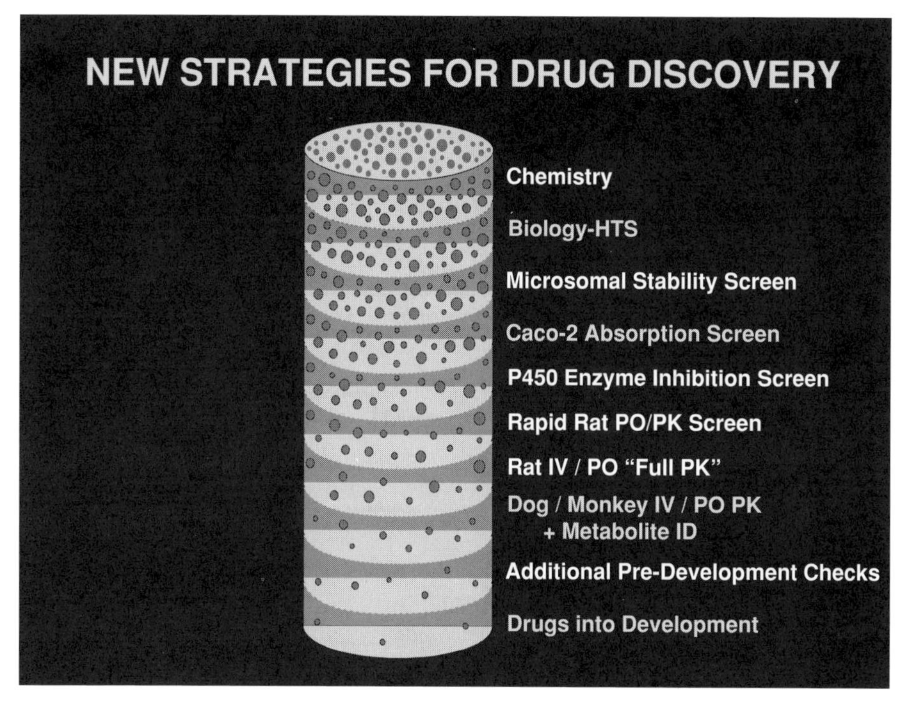

FIGURE 12.1 Schematic showing a series of screens that can be used for "sifting out" the problematic compounds (using DMPK criteria) with higher throughput screens and finding the smaller number of compounds that are suitable for more extensive testing in the lower throughput screens. (Adapted from W. A. Korfmacher, "Bioanalytical Assays in a Drug Discovery Environment" in "Using Mass Spectrometry for Drug Metabolism Studies", 2004, W. Korfmacher, Ed., CRC Press. Copyright © 2004, with permission from Taylor and Francis Group, LLC.)

compounds through a series of "sieves" designed to "weed out" compounds that "fail" in a test while allowing those that "pass" to proceed to the next assay. This screening process is useful if the assays are used in the correct sequence so that higher throughput assays are used first when there are larger numbers of compounds, followed by lower throughput assays which can be used when the number of compounds has decreased. While a significant rate of "false-positives" is acceptable, these assays should have a low rate of "false-negatives"; i.e., compounds should not be improperly deselected for further work by the assay that is being used to screen for a given DMPK property.

In a review article on high throughput screening approaches for evaluating NCEs for their DMPK properties [12], Roberts discusses another model for DMPK support of the lead optimization process. As shown in Figure 12.2, Roberts describes the DMPK effort as a combination of providing various in vitro and in vivo test results for lead compounds combined with suggestions for medicinal chemistry on how to improve the current lead molecule in order

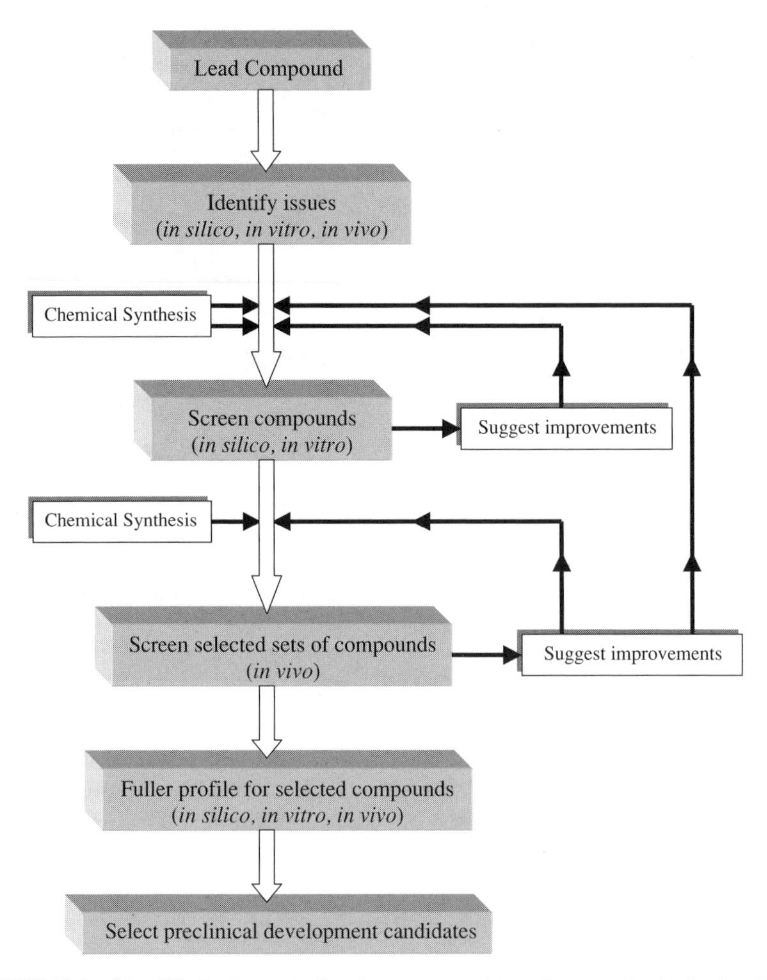

FIGURE 12.2 Simplified schematic showing stages and iterative steps in the lead optimization process from a DMPK perspective. (Reprinted from Roberts [12]. Copyright © 2001, with permission from Taylor & Francis Ltd.)

to improve its DMPK properties. The result of this iterative process is the selection of a suitable compound for preclinical development.

IN VITRO ASSAYS

In two review articles, Thompson [5,6] describes in detail the importance of metabolic stability as a DM parameter that needs to be optimized as part of the lead optimization paradigm. Thompson [5] discusses three ways to use a metabolic stability screen: (1) to solve a particular PK problem that was found in a lead; (2) as part of a screening strategy (Figure 12.1) where compounds that fail would be "filtered out"; (3) as a way to optimize a lead series by using the metabolic stability information to improve newly synthesized compounds. Another excellent review article by Masimirembwa et al. [13] describes how various higher throughput in vitro metabolic screens can be used in various stages of new drug discovery and drug development.

The ability to obtain microsomal stability data for NCEs in a higher throughput manner has continued to evolve. In 1999, Korfmacher et al. [14] described a microsomal stability assay based on an automated LC-MS system that was able to handle 75 compounds per week. More recently, Xu, Kassell et al. [15] have described a highly automated microsomal stability screen using an eight-channel parallel LC-MS system for the assay step; this system has the capacity to provide microsomal stability data for 176 compounds per day!

Another use of in vitro systems is to identify metabolites that will be produced in vivo. Several authors have described procedures for the identification of metabolites produced by in vitro systems [16–22]. Tiller and Romanyshyn [18] describe the use of LC-MS/MS procedures along with metabolite screening in vitro as part of a lead optimization process. Hop et al. [17] discuss the use of fast-gradient procedures specifically for in vitro metabolite identification as part of the drug-discovery process; their article suggests that rapid metabolite screening data can be obtained from HPLC gradients that last for as little as 2 minutes.

In vitro systems combined with LC-MS/MS can also be used for the prediction of problematic or reactive metabolites. As shown in Figure 12.3, Bertrand et al. [16] have used an LC-MS/MS assay to look for glutathione (GSH) adducts of a test compound. As they point out this technique can be successful without the need for a radio-labeled compound and is therefore applicable in the early drug-discovery process.

Another important use of LC-MS/MS is for the analysis of human colon adenocarcinoma (Caco-2) samples that are used as an in vitro assessment of drug absorption [4,8,10,12,16,23–28]. As shown in Figure 12.4, in order for a drug to provide efficacy after oral dosing, it must show a significant level of oral bioavailability (F). Oral bioavailability can be simply defined as the fraction of the dose that is given orally that makes it into the circulatory system. In the article by Chatuverdi et al. [8], F is defined as:

$$F = Fa \cdot Fg \cdot Fh \cdot Fl$$

where Fa is fraction of drug absorbed across the intestinal wall, and Fg, Fh, and Fl represent the fraction of the drug that gets through (is not cleared by) three additional barriers: the gastrointestinal tract, the liver, and the lung,

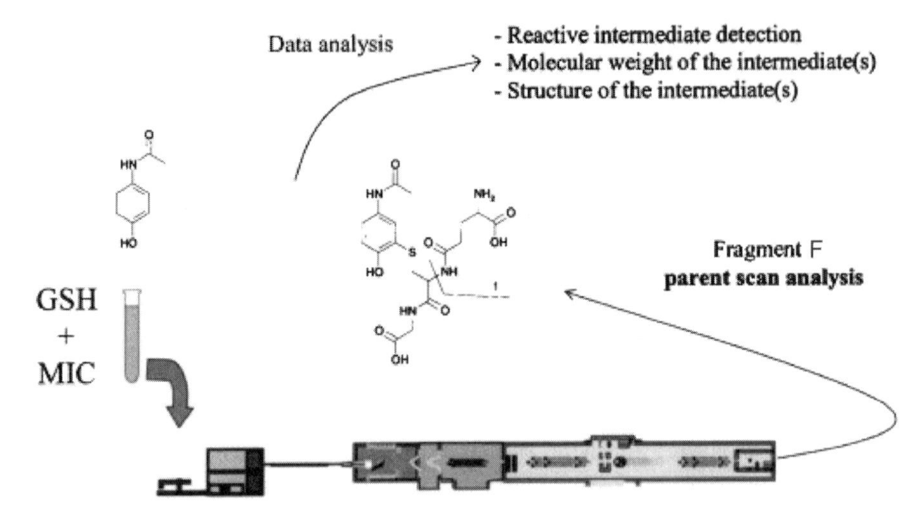

FIGURE 12.3 Reactive intermediate detection system. (Reprinted from Bertrand et al. [16]. Copyright © 2000, with permission from Elsevier.)

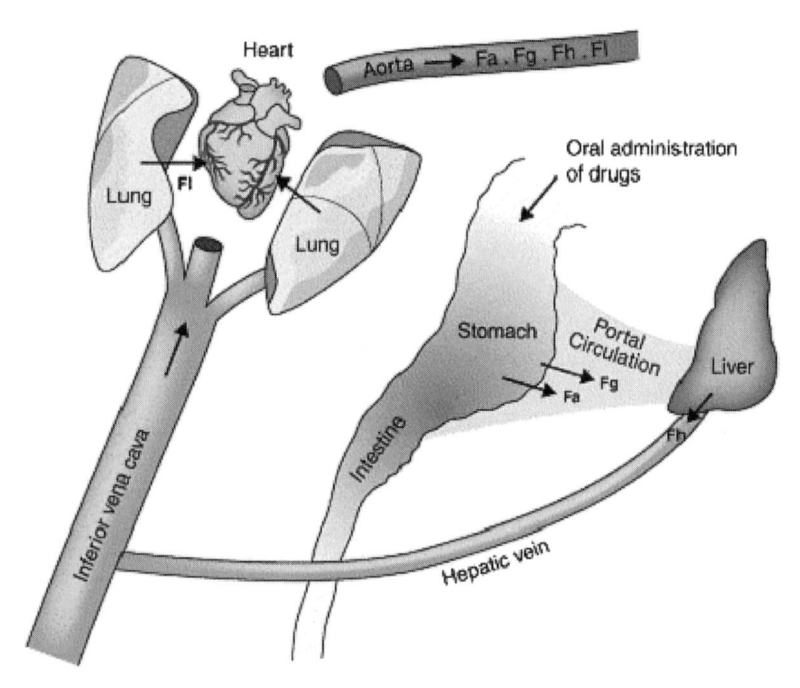

FIGURE 12.4 Schematic representation of the process of drug absorption following oral administration of drugs. (Reprinted from Chaturvedi et al. [8]. Copyright © 2001, with permission from Elsevier.)

respectively. Therefore, Fa can be properly modeled by the Caco-2 assay, which is simply a measure of a compound's permeability. As discussed by Burton et al. [24], predicting a compound's potential for absorption after oral dosing is difficult due to the multiple factors involved. Burton et al. list not only permeability and first-pass metabolism as factors in determining oral bioavailability but also the solubility of the compound in the intestinal lumen. Lipinski et al. [29,30] have discussed poor solubility as one of the hurdles that must be overcome in order to find new drugs that will have good oral bioavailability.

The prediction of human PK generally relies on a combination of human in vitro assays and various allometric scaling techniques based on PK data from animal models; a discussion of how this can be done can be found in the article by Naritomi et al. [31]. In a recent article, Mandagere et al. [32] describe the use of a graphical model for estimating the oral bioavailability of drugs in laboratory animals and humans by simply combining Caco-2 permeability data with microsomal stability data. As shown in Figure 12.5, the prediction can be used to categorize compounds into a low, medium, or high "bin" for predicted oral bioavailabilty.

Another type of in vitro assay is the measurement of plasma stability. As with other in vitro procedures, there has been a need to increase the

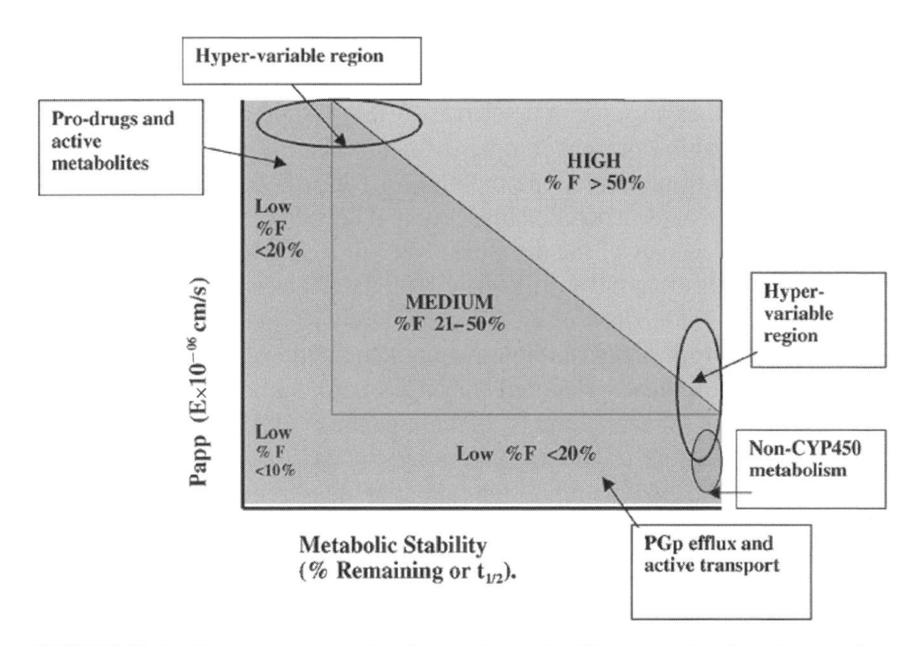

FIGURE 12.5 Graphical presentation for oral bioavailability estimation from in vitro data. (Reprinted from Mandagere [32]. Copyright © 2002, with permission from the American Chemical Society.)

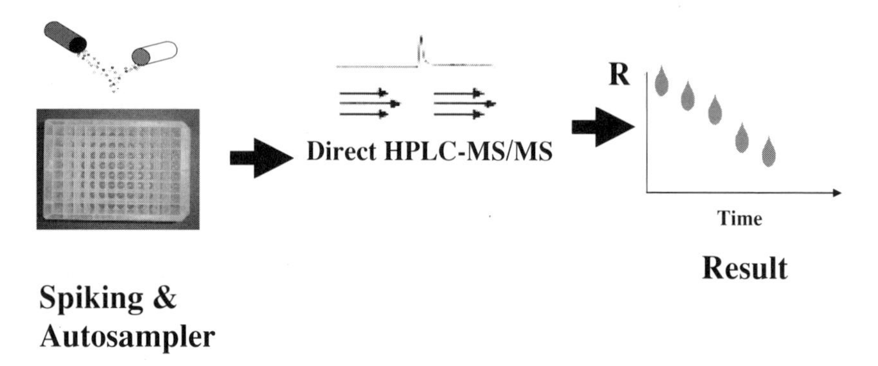

FIGURE 12.6 Schematic showing the semiautomated procedure for measuring plasma stability of test compounds. (Adapted from Wang and Hsieh [34]. Copyright © 2002, with permission from International Scientific Communications, Inc.)

throughput of the plasma stability assay in order to screen compounds in the drug-discovery arena. Recently, Wang and Hsieh have described a semi-automated procedure for measuring plasma stability of NCEs by using the autosampler as the incubation device [33,34]. As shown in Figure 12.6, the samples are incubated in a 96-well plate, and by using a direct plasma-injection procedure, the samples are assayed without the need for an off-line sample preparation step [34]. As shown in Figures 12.7 and 12.8, the method can also be used to monitor the appearance of a metabolite from a compound that shows a plasma stability liability [33].

The other higher throughput in vitro assay that is supported by LC-MS/MS is the enzyme inhibition potential assessment, which is used as a predictor of human drug–drug interaction potential for NCEs [7,9,12,35]. Kumar and Surapaneni [7] provide a detailed discussion of how enzyme inhibition potential can be used as part of the drug-discovery process and provide examples of marketed drugs that have drug–drug interactions as a liability. As shown in Figure 12.9, Eddershaw et al. [9] show how enzyme inhibition fits into the overall absorption, distribution, metabolism, and excretion (ADME) assessment that is typically required for NCEs before a compound can move into development. Several authors, including Chu et al. [36] and Testino and Patonay [37], have described higher throughput LC-MS/MS assays that are used as part of the enzyme inhibition potential measurement.

IN VIVO ASSAYS

In the review article by Riley et al. [4], the authors discuss parameters and a flow chart (see Figure 12.10) that describes the data that would be needed to provide a human dose projection. As this chart shows, a combination of in

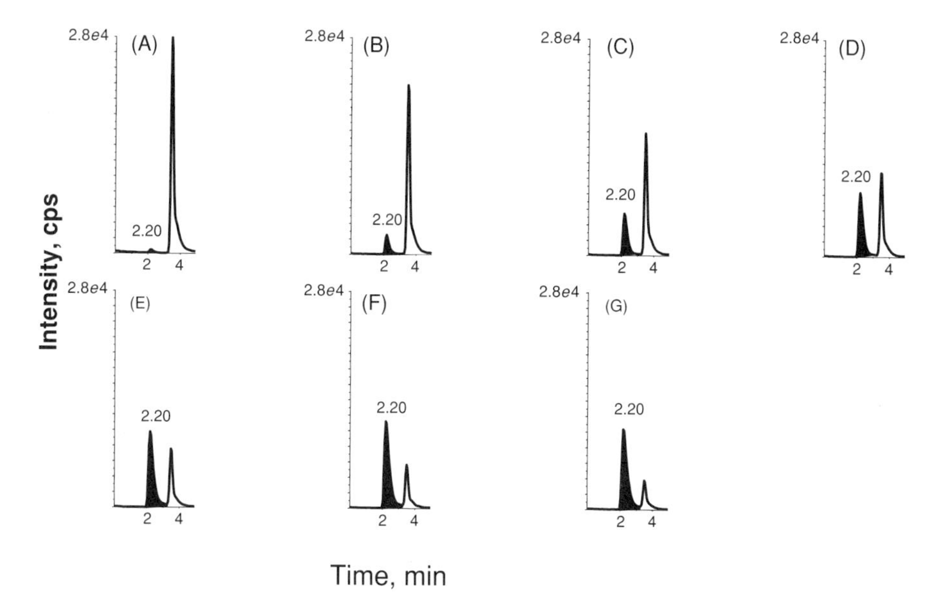

FIGURE 12.7 HLPC-MS/MS chromatograms of the carboxylic acid metabolite (retention time 2.20 min) in rat plasma after (A) 5 minutes, (B) 29 minutes, (C) 53 minutes, (D) 77 minutes, (E) 125 minutes, (F) 149 minutes, and (G) 173 minutes of incubation at 37°C. (Reprinted from Wang et al. [33]. Copyright © 2002, with permission from Elsevier.)

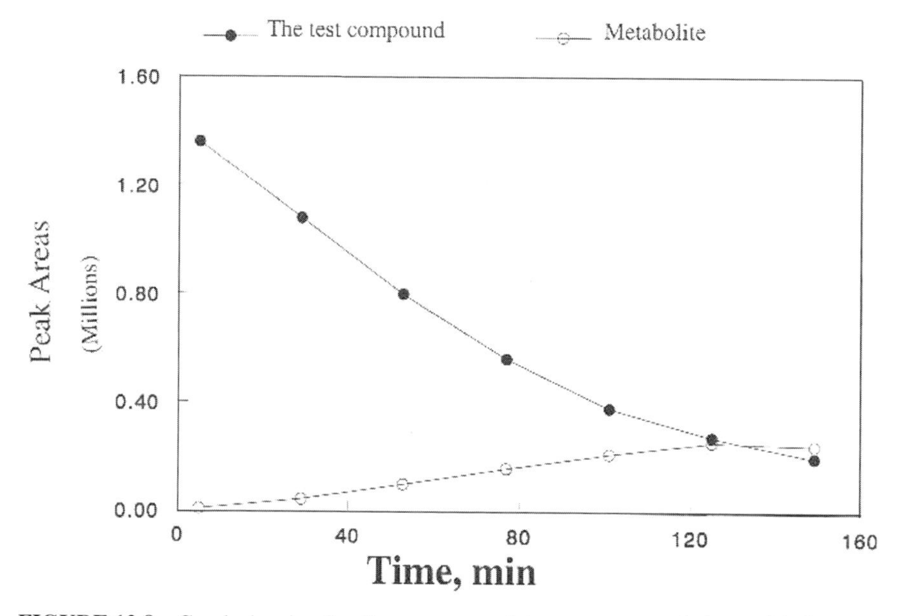

FIGURE 12.8 Graph showing the disappearance of the test compound along with the emergence of the metabolite (degradation product). (Reprinted from Wang et al. [33]. Copyright © 2002, with permission from Elsevier.)

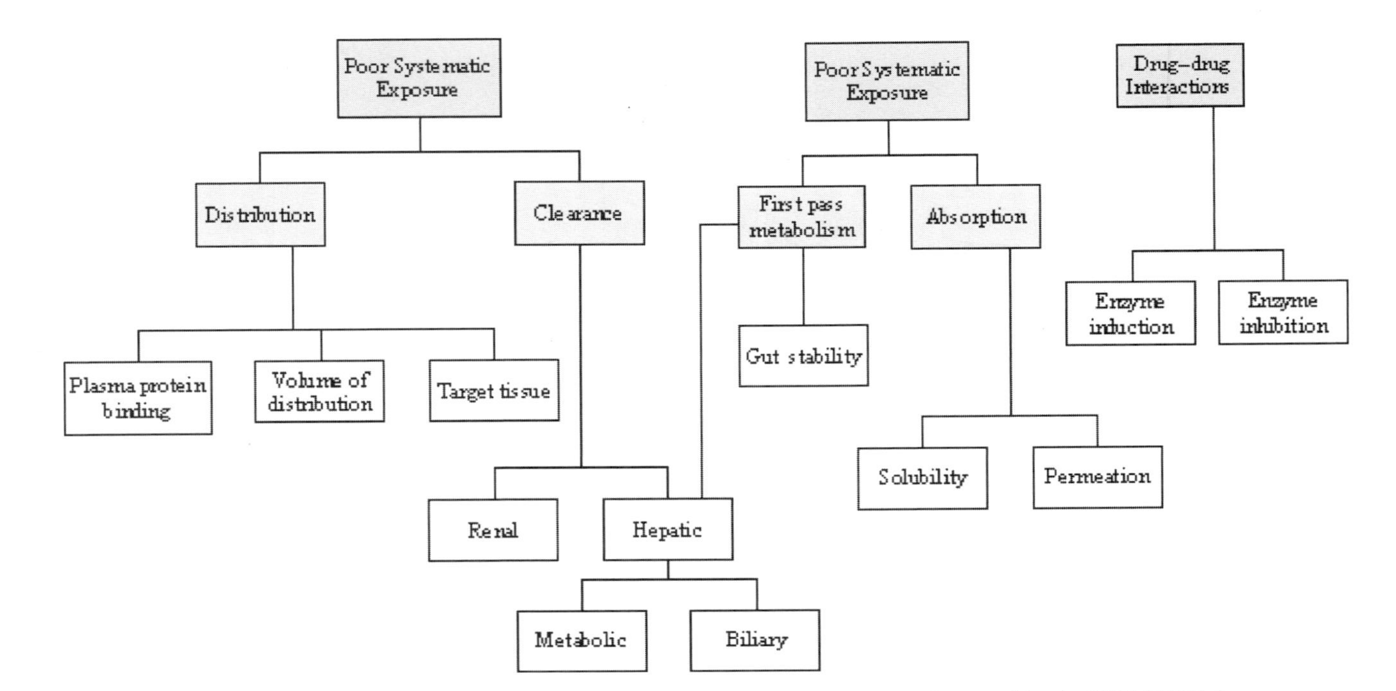

FIGURE 12.9 Schematic showing major absorption, distribution, metabolism, excretion and pharmacokinetic (ADME/PK) issues encountered in the lead optimization process as well as possible sources for the issue. (Adapted from Eddershaw et al. [9]. Copyright © 2000, with permission from Elsevier.)

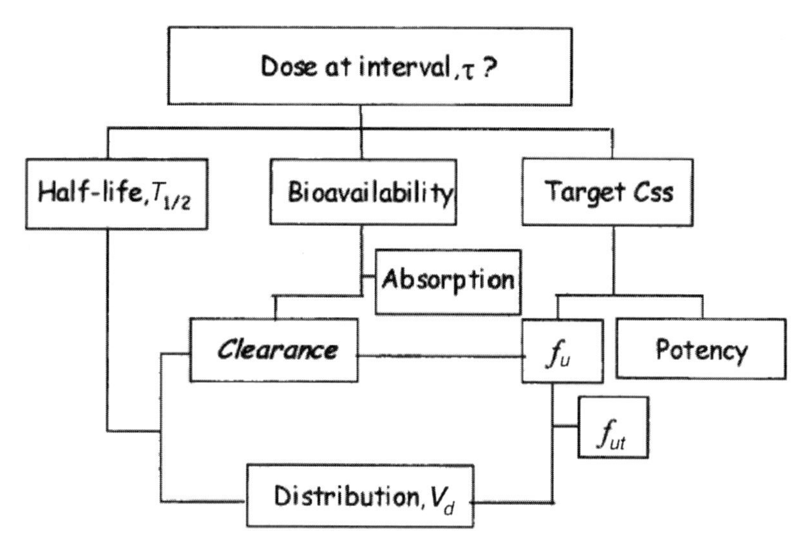

FIGURE 12.10 A general paradigm for estimating and optimizing the projected human dose. (Reprinted from Riley et al. [4]. Copyright © 2002, with permission from Bentham Science Publishers, Ltd.)

vitro and in vivo assays are needed in order to bring a compound forward into development. In this section, we discuss some of the strategies that have been used to increase the throughput of in vivo assays. As discussed in a review of this topic by Cox et al. [38], there are two basic strategies for increasing the throughput of assays used for various PK studies performed in a discovery environment: assay enhancement and sample reduction.

Assay-enhancement strategies are typically procedures that speed up the rate that samples can be processed and analyzed. Examples of this include various on-line sample preparation techniques that minimize the effort needed for manual sample preparation. One common technique is based on column switching (CS); in this procedure a short column is used to extract the analyte from the plasma matrix, then a second column is used to provide the sample chromatography [39–43]. As discussed by Hopfgartner et al. [44], CS can also be used along with protein precipitation in a combined sample cleanup approach. Another technique for direct plasma injection is based on the use of a single HPLC column, which serves not only as the means for analyte extraction, but provides chromatographic separation as well. Hsieh et al. [45–49] have provided several examples of how this technique can be used for drug-discovery applications. For most of the applications described by these authors, a mixed-functional column was used; in a recent article, Hsieh et al. [50] described the use of a monolithic silica column for direct plasma injection of discovery plasma samples into an HPLC-MS/MS system.

The various methods for sample reduction include cassette dosing [38,51,52], sample pooling strategies [38,53,54], and cassette-accelerated rapid rat screen (CARRS) [55]. As discussed recently in a review article by Papac and Shahrokh [56], the cassette-dosing approach has the disadvantage that drug–drug interactions can be an issue. As discussed by White and Manitpisitkul [57], these drug–drug interactions can lead to both false-positive and false-negative results. Multiple-compound sample pooling without cassette dosing has also been used to reduce the number of samples to be assayed, but as discussed by Papac and Shahrokh [56], this process has disadvantages in that careful attention has to be taken to ensure that isomeric compounds are not mixed and that possible metabolites of compounds do not interfere with dosed compounds.

The CARRS system described by Korfmacher et al. [55] avoids all of these problems while still providing the capability of a higher throughput in vivo rat PK screen. In the CARRS assay, compounds are grouped into cassettes of six selected from one drug-discovery team's new compounds. Each compound is dosed orally into two rats. Six plasma samples are taken from each rat at six time points (0.5, 1, 2, 3, 4, 6 hours postdose), and samples from the two rats dosed with one compound are pooled across the two rats to give a total of six samples (one at each time point) to assay per each compound that was dosed. Multiple cassettes are dosed each week (week 1). In the second week, the plasma samples are assayed in cassettes of six compounds each. Due to the standardization of the dosing and assay regimes, various steps have been streamlined to improve their efficiency. For example, as shown in Figure 12.11, all the samples and standards needed to assay one set of six compounds can fit into one 96-well plate [55]. Sample preparation consists

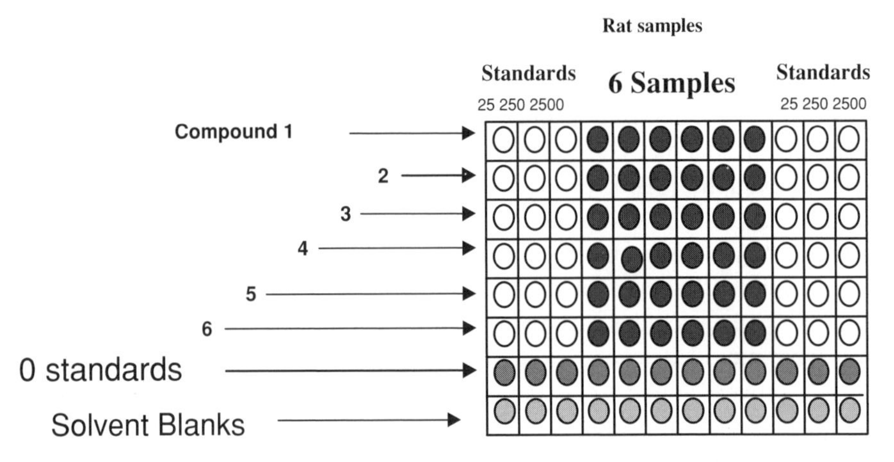

FIGURE 12.11 A schematic diagram of a 96-well plate used for holding all the samples and standards required for one set of six compounds as part of the CARRS assay. (Reprinted from Korfmacher et al. [55]. Copyright © 2001, with permission from John Wiley & Sons, Ltd.)

of protein precipitation, and can be performed using robotics such as the Tomtec Quadra 96 system. Sample injection into the LC-MS/MS system is also facilitated by the 96-well plate system by using smart autosamplers such as the LEAP CTC Pal system. Using this strategy, the compounds are dosed and assayed and the PK report is issued within 2 weeks of the compound delivery date [55].

MATRIX-ION SUPPRESSION

One of the challenges that has arisen in the last 6 years is the problem of matrix-ion suppression. This problem was not well known in the early days of using HPLC-MS/MS for PK assays, but has become a widely recognized issue that must be addressed not only as part of a validation package for a method used for good laboratory practices (GLP) studies, but also one that must be considered when developing and using a method for PK sample analysis in the drug-discovery arena. In a recent article, King et al. [58] described some of the possible causes of the matrix-effect problem. Unger [59] studied matrix effects caused by microsomal incubation samples on 27 pharmaceutical compounds; they found that either protein precipitation followed by fast LC or solid-phase extraction (SPE) could be used to eliminate the matrix effect in these samples. Ackermann et al. [60] recently discussed matrix effects versus flow rate for rapid PK assays; they concluded that the matrix effects were minor when the LC system provided good separation of the analyte of interest and highly polar matrix compounds. Weinmann et al. [61] studied the matrix effect that was observed after various sample preparation systems for two test compounds in serum; their conclusion was that as long as good chromatography is used, the matrix-effect problem can be minimized.

Two additional reports on matrix effects from unexpected sources should also be discussed. Tong et al. [62] describe the potential of matrix effects from the dosing formulation that may be used when dosing laboratory animals. These authors found that PEG 400 (a compound that is sometimes used in dosing formulations) could cause significant ion-suppression for polar compounds assayed using fast LC-MS systems; the ion-suppression problem was diminished when longer run times were used to assay the same samples. As shown in Figure 12.12, PEG 400 was found to cause significant matrix ion suppression in samples that were taken from not only IV-dosed animals, but even orally dosed animals [62]. The solution in this case was to use better chromatography in the analysis. Mei et al. [63], described the potential for matrix effects arising from exogenous material in certain tubes that are commonly used for plasma sample collection. In this case, there was a polymeric material that was leached from the sample tube and caused significant matrix-ion suppression problems. One unique feature of this problem was

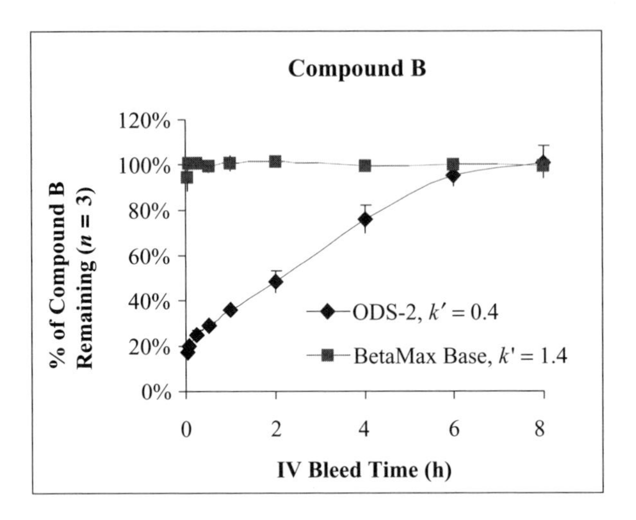

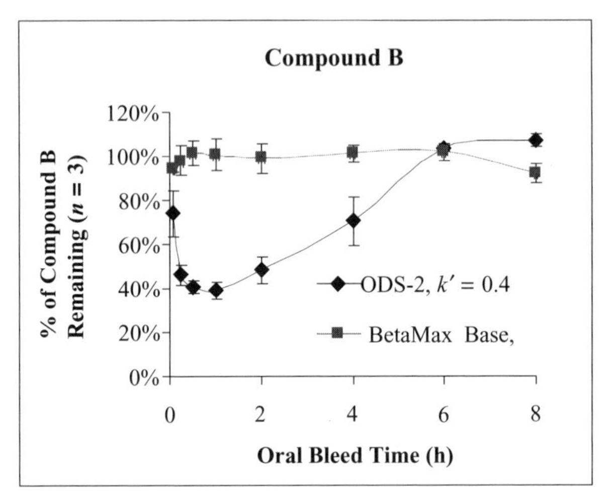

FIGURE 12.12 Graphical results showing the effect of PEG 400 on the peak intensity of compound B in electrospray LC/MS. For this study, rats were dosed IV and PO with a formulation that contained 40% PEG 400, and compound B was added to the extracted plasma samples obtained from the rats at various time points. (Reprinted from Tong et al. [62]. Copyright © 2002, with permission from the American Chemical Society.)

that the matrix effect was found in the nonpolar region of the chromatogram, making it much harder to solve using chromatographic optimization [63].

CONCLUSIONS

There continues to be an increased need for HPLC-MS/MS systems for the analysis of multiple types of samples that are generated as part of the drug-metabolism support of new drug discovery. In addition, there are new user strategies that continue to develop for increasing the throughput of these various assays. Finally, there are multiple strategies being tested to identify the new compounds for development and to ensure that selected compounds have the appropriate DMPK properties to succeed during clinical studies.

REFERENCES

1. Ackermann, B.L.; Berna, M.J.; Murphy, A.T. "Recent Advances in Use of LC/MS/MS for Quantitative High-Throughput Bioanalytical Support of Drug Discovery," *Curr. Top. Med. Chem.* **2**(1), 53–66 (2002).
2. Jemal, M. "High-Throughput Quantitative Bioanalysis by LC/MS/MS," *Biomed. Chromatogr.* **14**(6), 422–429 (2000).
3. Korfmacher, W.A.; Cox, K.A.; Bryant, M.S.; Veals, J.; Ng, K.; Lin, C.C. "HPLC-API/MS/MS: A Powerful Tool for Integrating Drug Metabolism into the Drug Discovery Process," *Drug Discov. Today* **2**, 532–537 (1997).
4. Riley, R.J.; Martin, I.J.; Cooper, A.E. "The Influence of DMPK as an Integrated Partner in Modern Drug Discovery," *Curr. Drug Metab.* **3**(5), 527–550 (2002).
5. Thompson, T.N. "Early ADME in Support of Drug Discovery: The Role of Metabolic Stability Studies," *Curr. Drug Metab.* **1**(3), 215–241 (2000).
6. Thompson, T.N. "Optimization of Metabolic Stability as a Goal of Modern Drug Design," *Med. Res. Rev.* **21**(5), 412–449 (2001).
7. Kumar, G.N.; Surapaneni, S. "Role of Drug Metabolism in Drug Discovery and Development," *Med. Res. Rev.* **21**(5), 397–411 (2001).
8. Chaturvedi, P.R.; Decker, C.J.; Odinecs, A. "Prediction of Pharmacokinetic Properties Using Experimental Approaches During Early Drug Discovery," *Curr. Opin. Chem. Biol.* **5**(4), 452–463 (2001).
9. Eddershaw, P.J.; Beresford, A.P.; Bayliss, M.K. "ADME/PK as Part of a Rational Approach to Drug Discovery," *Drug Discov. Today* **5**(9), 409–414 (2000).
10. Tarbit, M.H.; Berman, J. "High-Throughput Approaches for Evaluating Absorption, Metabolism and Excretion Properties of Lead Compounds," *Curr. Opin. Chem. Biol.* **2**, 411–416 (1998).
11. Kennedy, T. "Managing the Drug Discovery/Developmement Interface," *Drug Dis. Today* **2**(10), 436–444 (1997).
12. Roberts, S.A. "High-Throughput Screening Approaches for Investigating Drug Metabolism and Pharmacokinetics," *Xenobiotica* **31**(8–9), 557–589 (2001).
13. Masimirembwa, C.M.; Thompson, R.; Andersson, T.B. "In Vitro High Throughput Screening of Compounds for Favorable Metabolic Properties in Drug Discovery," *Comb. Chem. High Throughput Screen.* **4**(3), 245–263 (2001).
14. Korfmacher, W.A.; Palmer, C.A.; Nardo, C.; Dunn-Meynell, K.; Grotz, D.; Cox, K.; Lin, C.C.; Elicone, C.; Liu, C.; Duchoslav, E. "Development of an Automated Mass Spectrometry System for the Quantitative Analysis of Liver Microsomal Incubation Samples: A Tool for Rapid Screening of New Compounds for Metabolic Stability," *Rapid Commun. Mass Spectrom.* **13**(10), 901–907 (1999).

15. Xu, R.; Nemes, C.; Jenkins, K.M.; Rourick, R.A.; Kassel, D.B.; Liu, C.Z. "Application of Parallel Liquid Chromatography/Mass Spectrometry for High Throughput Microsomal Stability Screening of Compound Libraries," *J. Am. Soc. Mass Spectrom.* **13**(2), 155–165 (2002).
16. Bertrand, M.; Jackson, P.; Walther, B. "Rapid Assessment of Drug Metabolism in the Drug Discovery Process," *Eur. J. Pharm. Sci.* **11 (Suppl 2)**, S61–S72 (2000).
17. Hop, C.E.; Tiller, P.R.; Romanyshyn, L. "In Vitro Metabolite Identification Using Fast Gradient High Performance Liquid Chromatography Combined with Tandem Mass Spectrometry," *Rapid Commun. Mass Spectrom.* **16**(3), 212–219 (2002).
18. Tiller, P.R.; Romanyshyn, L.A. "Liquid Chromatography/Tandem Mass Spectrometric Quantification with Metabolite Screening as a Strategy to Enhance the Early Drug Discovery Process," *Rapid Commun. Mass Spectrom.* **16**(12), 1225–1231 (2002).
19. Clarke, N.J.; Rindgen, D.; Korfmacher, W.A.; Cox, K.A. "Systematic LC/MS Metabolite Identification in Drug Discovery," *Anal. Chem.* **73**(15), 430A–439A (2001).
20. Gu, M.; Lim, H.K. "An Intelligent Data Acquisition System for Simultaneous Screening of Microsomal Stability and Metabolite Profiling by Liquid Chromatography/Mass Spectrometry," *J. Mass Spectrom.* **36**(9), 1053–1061 (2001).
21. Lim, H.K.; Chan, K.W.; Sisenwine, S.; Scatina, J.A. "Simultaneous Screen for Microsomal Stability and Metabolite Profile by Direct Injection Turbulent-Laminar Flow LC-LC and Automated Tandem Mass Spectrometry," *Anal. Chem.* **73**(9), 2140–2146 (2001).
22. Cox, K.A.; Clarke, N.J.; Rindgen, D.; Korfmacher, W.A. "Higher Throughput Metabolite Identification in Drug Discovery: Current Capabilities and Future Trends," *Amer. Pharm. Rev.* **4**(1), 45–52 (2001).
23. Hidalgo, I.J. "Assessing the Absorption of New Pharmaceuticals," *Curr. Top. Med. Chem.* **1**(5), 385–401 (2001).
24. Burton, P.S.; Goodwin, J.T.; Vidmar, T.J.; Amore, B.M. "Predicting Drug Absorption: How Nature Made It a Difficult Problem," *J. Pharmacol. Exp. Ther.* **303**(3), 889–895 (2002).
25. Wang, Z.; Hop, C.E.; Leung, K.H.; Pang, J. "Determination of In Vitro Permeability of Drug Candidates through a Caco-2 Cell Monolayer by Liquid Chromatography/Tandem Mass Spectrometry," *J. Mass Spectrom.* **35**(1), 71–76 (2000).
26. Larger, P.; Altamura, M.; Catalioto, R.M.; Giuliani, S.; Maggi, C.A.; Valenti, C.; Triolo, A. "Simultaneous LC-MS/MS Determination of Reference Pharmaceuticals as a Method for the Characterization of the Caco-2 Cell Monolayer Absorption Properties," *Anal. Chem.* **74**(20), 5273–5281 (2002).
27. Markowska, M.; Oberle, R.; Juzwin, S.; Hsu, C.P.; Gryszkiewicz, M.; Streeter, A.J. "Optimizing Caco-2 Cell Monolayers to Increase Throughput in Drug Intestinal Absorption Analysis," *J. Pharmacol. Toxicol. Methods* **46**(1), 51–55 (2001).

28. Tannergren, C.; Langguth, P.; Hoffmann, K.J. "Compound Mixtures in Caco-2 Cell Permeability Screens as a Means to Increase Screening Capacity," *Pharmazie* **56**(4), 337–342 (2001).
29. Lipinski, C.A. "Drug-Like Properties and the Causes of Poor Solubility and Poor Permeability," *J. Pharmacol. Toxicol. Methods* **44**(1), 235–249 (2000).
30. Lipinski, C.A.; Lombardo, F.; Dominy, B.W.; Feeney, P.J. "Experimental and Computational Approaches to Estimate Solubility and Permeability in Drug Discovery and Development Settings," *Adv. Drug Deliv. Rev.* **46**(1–3), 3–26 (2001).
31. Naritomi, Y.; Terashita, S.; Kimura, S.; Suzuki, A.; Kagayama, A.; Sugiyama, Y. "Prediction of Human Hepatic Clearance from In Vivo Animal Experiments and In Vitro Metabolic Studies with Liver Microsomes from Animals and Humans," *Drug Metab. Dispos.* **29**(10), 1316–1324 (2001).
32. Mandagere, A.K.; Thompson, T.N.; Hwang, K.K. "Graphical Model for Estimating Oral Bioavailability of Drugs in Humans and Other Species from Their Caco-2 Permeability and In Vitro Liver Enzyme Metabolic Stability Rates," *J. Med. Chem.* **45**(2), 304–311 (2002).
33. Wang, G.; Hsieh, Y.; Lau, Y.; Cheng, K.C.; Ng, K.; Korfmacher, W.A.; White, R.E. "Semi-Automated Determination of Plasma Stability of Drug Discovery Compounds Using Liquid Chromatography-Tandem Mass Spectrometry," *J. Chromatogr. B Anal. Technol. Biomed. Life Sci.* **780**(2), 451–457 (2002).
34. Wang, G.; Hsieh, Y. "Utilization of Direct HPLC-MS-MS for Drug Stability Measurement," *Am. Lab.* **34**(24), 24–27 (2002).
35. Yan, Z.; Caldwell, G.W. "Metabolism Profiling, and Cytochrome P450 Inhibition and Induction in Drug Discovery," *Curr. Top. Med. Chem.* **1**(5), 403–425 (2001).
36. Chu, I.; Favreau, L.; Soares, T.; Lin, C.; Nomeir, A.A. "Validation of Higher-Throughput High-Performance Liquid Chromatography/Atmospheric Pressure Chemical Ionization Tandem Mass Spectrometry Assays to Conduct Cytochrome P450s CYP2D6 and CYP3A4 Enzyme Inhibition Studies in Human Liver Microsomes," *Rapid Commun. Mass Spectrom.* **14**(4), 207–214 (2000).
37. Testino, S.A., Jr.; Patonay, G. "High-Throughput Inhibition Screening of Major Human Cytochrome P450 Enzymes Using an In Vitro Cocktail and Liquid Chromatography-Tandem Mass Spectrometry," *J. Pharm. Biomed. Anal.* **30**(5), 1459–1467 (2003).
38. Cox, K.A.; White, R.E.; Korfmacher, W.A. "Rapid Determination of Pharmacokinetic Properties of New Chemical Entities: In Vivo Approaches," *Comb. Chem. High Throughput Screen.* **5**(1), 29–37 (2002).
39. O'Connor, D. "Automated Sample Preparation and LC-MS for High-Throughput ADME Quantification," *Curr. Opin. Drug Disc. Devel.* **5**(1), 52–58 (2002).
40. Zell, M.; Husser, C.; Hopfgartner, G. "Column-Switching High-Performance Liquid Chromatography Combined with Ionspray Tandem Mass Spectrometry for the Simultaneous Determination of the Platelet Inhibitor Ro 44-3888 and Its Pro-Drug and Precursor Metabolite in Plasma," *J. Mass Spectrom.* **32**(1), 23–32 (1997).

41. Zeng, H.; Wu, J.T.; Unger, S.E. "The Investigation and the Use of High Flow Column-Switching LC/MS/MS as a High-Throughput Approach for Direct Plasma Sample Analysis of Single and Multiple Components in Pharmacokinetic Studies," *J. Pharm. Biomed. Anal.* **27**(6), 967–982 (2002).
42. Powell, M.L.; Jemal, M. "Rapid Chromatography Coupled with Direct Injection Lc/Ms/Ms for Quantitative Bioanalysis," *Am. Pharm. Rev.* **4**(3), 63–69 (2001).
43. Ackermann, B.L.; Murphy, A.T.; Berna, M.J. "The Resurgence of Column Switching Techniques to Facilitate Rapid LC/MS/MS Based Bioanalysis in Drug Discovery," *Am. Pharm. Rev.* **5**(1), 54–63 (2002).
44. Hopfgartner, G.; Husser, C.; Zell, M. "High-Throughput Quantification of Drugs and Their Metabolites in Biosamples by LC-MS/MS and CE-MS/MS: Possibilities and Limitations," *Ther. Drug Monit.* **24**(1), 134–143 (2002).
45. Hsieh, Y.; Brisson, J.M.; Ng, K.; Korfmacher, W.A. "Direct Simultaneous Determination of Drug Discovery Compounds in Monkey Plasma Using Mixed-Function Column Liquid Chromatography/Tandem Mass Spectrometry," *J. Pharm. Biomed. Anal.* **27**(1–2), 285–293 (2002).
46. Hsieh, Y.; Brisson, J.M.; Ng, K.; White, R.E.; Korfmacher, W.A. "Direct Simultaneous Analysis of Plasma Samples for a Drug Discovery Compound and Its Hydroxyl Metabolite Using Mixed-Function Column Liquid Chromatography-Tandem Mass Spectrometry," *Analyst* **126**(12), 2139–2143 (2001).
47. Hsieh, Y.; Bryant, M.S.; Brisson, J.M.; Ng, K.; Korfmacher, W.A. "Direct Cocktail Analysis of Drug Discovery Compounds in Pooled Plasma Samples Using Liquid Chromatography-Tandem Mass Spectrometry," *J. Chromatogr. B Anal. Technol. Biomed. Life Sci.* **767**(2), 353–362 (2002).
48. Hsieh, Y.; Bryant, M.S.; Gruela, G.; Brisson, J.M.; Korfmacher, W.A. "Direct Analysis of Plasma Samples for Drug Discovery Compounds Using Mixed-Function Column Liquid Chromatography Tandem Mass Spectrometry," *Rapid Commun. Mass Spectrom.* **14**(15), 1384–1390 (2000).
49. Hsieh, Y.; Ng, K.; Korfmacher, W. "Development and Application of Single-Column Direct Plasma Injection Procedures for Drug Candidate Assays Using HPLC-MS/MS," *Am. Pharm. Rev.* **5**(4), 88–93 (2002).
50. Hsieh, Y.; Wang, G.; Wang, Y.; Chackalamannil, S.; Korfmacher, W.A. "Direct Plasma Analysis of Drug Compounds Using Monolithic Column Liquid Chromatography and Tandem Mass Spectrometry," *Anal. Chem.* **75**(8), 1812–1818 (2003).
51. Olah, T.V.; McLoughlin, D.A.; Gilbert, J.D. "The Simultaneous Determination of Mixtures of Drug Candidates by Liquid Chromatography/Atmospheric Pressure Chemical Ionization Mass Spectrometry as an In Vivo Drug Screening Procedure," *Rapid Commun. Mass Spectrom.* **11**(1), 17–23 (1997).
52. Berman, J.; Halm, K.; Adkison, K.; Shaffer, J. "Simultaneous Pharmacokinetic Screening of a Mixture of Compounds in the Dog Using API LC/MS/MS Analysis for Increased Throughput," *J. Med. Chem.* **40**(6), 827–829 (1997).
53. Cox, K.A.; Dunn-Meynell, K.; Korfmacher, W.A.; Broske, L.; Nomeir, A.A.; Lin, C.C.; Cayen, M.N.; Barr, W.H. "Novel Procedure for Rapid Pharmacokinetic

Screening of Discovery Compounds in Rats," *Drug Disc. Today* **4**(5), 232–237 (1999).

54. Hop, C.E.; Wang, Z.; Chen, Q.; Kwei, G. "Plasma-Pooling Methods to Increase Throughput for In Vivo Pharmacokinetic Screening," *J. Pharm. Sci.* **87**(7), 901–903 (1998).
55. Korfmacher, W.A.; Cox, K.A.; Ng, K.J.; Veals, J.; Hsieh, Y.; Wainhaus, S.; Broske, L.; Prelusky, D.; Nomeir, A.; White, R.E. "Cassette-Accelerated Rapid Rat Screen: A Systematic Procedure for the Dosing and Liquid Chromatography/Atmospheric Pressure Ionization Tandem Mass Spectrometric Analysis of New Chemical Entities as Part of New Drug Discovery," *Rapid Commun. Mass Spectrom.* **15**(5), 335–340 (2001).
56. Papac, D.I.; Shahrokh, Z. "Mass Spectrometry Innovations in Drug Discovery and Development," *Pharm. Res.* **18**(2), 131–145 (2001).
57. White, R.E.; Manitpisitkul, P. "Pharmacokinetic Theory of Cassette Dosing in Drug Discovery Screening," *Drug Metab. Dispos.* **29**(7), 957–966 (2001).
58. King, R.; Bonfiglio, R.; Fernandez-Metzler, C.; Miller-Stein, C.; Olah, T. "Mechanistic Investigation of Ionization Suppression in Electrospray Ionization," *J. Am. Soc. Mass Spectrom.* **11**(11), 942–950 (2000).
59. Zheng, J.J.; Lynch, E.D.; Unger, S.E. "Comparison of SPE and Fast LC to Eliminate Mass Spectrometric Matrix Effects from Microsomal Incubation Products," *J. Pharm. Biomed. Anal.* **28**(2), 279–285 (2002).
60. Murphy, A.T.; Berna, M.J.; Holsapple, J.L.; Ackermann, B.L., "Effects of Flow Rate on High-Throughput Quantitative Analysis of Protein-Precipitated Plasma Using Liquid Chromatography/Tandem Mass Spectrometry," *Rapid Commun. Mass Spectrom.* **16**(6), 537–543 (2002).
61. Muller, C.; Schafer, P.; Stortzel, M.; Vogt, S.; Weinmann, W. "Ion Suppression Effects in Liquid Chromatography-Electrospray-Ionisation Transport-Region Collision Induced Dissociation Mass Spectrometry with Different Serum Extraction Methods for Systematic Toxicological Analysis with Mass Spectra Libraries," *J. Chromatogr. B Anal. Technol. Biomed. Life Sci.* **773**(1), 47–52 (2002).
62. Tong, X.S.; Wang, J.; Zheng, S.; Pivnichny, J.V.; Griffin, P.R.; Shen, X.; Donnelly, M.; Vakerich, K.; Nunes, C.; Fenyk-Melody, J. "Effect of Signal Interference from Dosing Excipients on Pharmacokinetic Screening of Drug Candidates by Liquid Chromatography/Mass Spectrometry," *Anal. Chem.* **74**(24), 6305–6313 (2002).
63. Mei, H.; Hsieh, Y.; Nardo, C.; Xu, X.; Wang, S.; Ng, K.; Korfmacher, W.A. "Investigation of Matrix Effects in Bioanalytical High-Performance Liquid Chromatography/Tandem Mass Spectrometric Assays: Application to Drug Discovery," *Rapid Commun. Mass Spectrom.* **17**(1), 97–103 (2003).

13

AN INTEGRATED LC-MS STRATEGY FOR PRECLINICAL CANDIDATE OPTIMIZATION

JONATHAN L. JOSEPHS AND MARK SANDERS

INTRODUCTION

The process of successfully discovering and developing novel pharmaceutical agents has become increasingly challenging. New drugs must provide better therapies than existing treatments or address completely unmet medical needs. As pharmaceutical scientists understand more about drug toxicity and drug–drug interactions, the standards have been raised for what is considered to be a safe drug. The discovery of new targets and the development of technologies to synthesize and screen large numbers of compounds have provided greater numbers of potential lead compounds that enter into drug discovery. This situation is juxtaposed to the increased cost of drug development and the rising standards of safety, efficacy, and once-daily dosing. The logical response to this challenge has been to develop approaches to screen out those characteristics that cause attrition in development, namely administration, distribution, metabolism, and excretion (ADME), preclinical toxicology, and human safety profiles at an early stage. This response involves the determination those physiochemical, ADME, and toxicological properties of drug candidates that historically have been investigated in drug development [1]. The traditional approaches employed in development are not suited to the demands of discovery, namely high throughput and low sample consumption.

Integrated Strategies for Drug Discovery Using Mass Spectrometry, Edited by Mike S. Lee

This drug-discovery approach is addressed by a new area of pharmaceutical research known as profiling [2–5].

Profiling is employed subsequent to screening compound repositories and combinatorial libraries for structures with desired potency at the biological target of interest and continues through the lead optimization stage. The goal is to optimize physicochemical, ADME, and safety attributes of molecules in parallel with efficacy, at all stages of the discovery process, from virtual space to *in vivo* models, and from screening hits to drug candidates [6]. Therefore, to effectively guide medicinal chemistry programs with profiling data, results must be provided on a timeframe equal to that of efficacy screening. This coordination presents many analytical challenges as well as significant hurdles in logistics and data management that need to be overcome, especially when one considers the ease and speed with which many of the fluorescent-plate-based efficacy assays are performed [7].

The high throughput biological target screening assays are by their design intended to identify compounds that embrace a wide range of chemical diversity. This process results in large numbers of chemically diverse compounds that are introduced into the profiling assays. Compound selection is driven by the biological activity of these compounds in the profiling assays. Correlation of biological activity to chemical structure will only be as good as our knowledge of the identity and purity of the compounds under investigation, which are typically available as solutions that have been stored frozen in dimethyl sulfoxide (DMSO). Ideally, this determination should be made not just at the time a compound is submitted but each time it is assayed. This process allows for tracking chemical degradation and misplating or mislabeling that inevitably occur in a high-volume process where human input is required.

Many of the assays employed in profiling have the determination of concentration of the drug in a matrix as their common analytical endpoint. For example, metabolic stability is determined by comparing the concentration of the drug candidate at the end of an incubation compared to the concentration at the initial time point. The drug concentration may be determined by a number of techniques, including ultraviolet (UV) plate readers and high-performance liquid chromatography with in-line UV detection (HPLC-UV), but liquid chromatography with in-line mass spectrometry detection (LC-MS) has become the method of choice due to its sensitivity and selectivity, which allows for the rapid development of high throughput methods [8–11]. LC-MS using selected ion monitoring (SIM) has been used on single-quadrupole mass spectrometers due to its simplicity in method development [12]. Multiplexed systems such as those that employ selected gating of eluents from multiple chromatography columns into a single mass spectrometer offer the promise of extremely high throughput [13,14]. Selected reaction monitoring (SRM) on triple-quadrupole instrumentation is considered to be the "gold standard" in small molecule quantitation. This technique is highly sensitive and selective, but requires that individual methods be custom-made for each compound. Automated approaches to method development have been reported [15–18].

ANALYSIS OF ANALYTICAL PROBLEM

We reasoned that our biological screening and profiling assays would provide the most value when the scientists using the results had the highest confidence in their output. We therefore set out to obtain analytical data of very high quality. The overall philosophy was to perform the profiling analytical work with the same flexibility and quality as one would for a single analysis. Of course, some compromises would need to be made; however, the goal was to keep these to a minimum. The challenges of work flow, logistics, and data handling would be overcome by designing "smart" software that queues instruments for analyses, interprets data, and determines the next experiment to be performed. The confidence in the data from the profiling assays can be no greater than the confidence in the integrity of the compounds being assayed. Analytical-method development in support of combinatorial chemistry [19,20] has resulted in very high throughput methods that determine purity and identity for compounds as separate events [21]. More recently multiplexed (MUX) technology has been used to provide purity and identity in a single analysis [22–24]. The MUX technology is limited to electrospray ionization and is often implemented in positive-ion mode only, although pos/neg switching has been reported [25]. Recent experiences with LC-MS in support of traditional drug-discovery programs indicate that the diversity of chemotypes encountered across therapeutic areas can be best analyzed when a full range of ionization modes and polarities are available [26]. To avoid restricting the number of compounds that could be successfully run through our profiling assays or skewing our chemotype selection to those that work only by positive ion electrospray ionization (ESI), a structural integrity (SI) assay that employed the four common modes of ionization available for LC-MS (ESI positive, ESI negative, atmospheric-pressure chemical ionization (APCI) positive, APCI negative) was required. To avoid the need for all compounds to be run with each ionization mode, only those compounds that require the application of additional ionization modes are run. A tiered approach is applied, whereby results from previous ionization modes are used to determine the need for additional analyses. In a traditional set of analyses for a large number of compounds, this approach would be prohibitive due to the extensive amount of time that it would take to review all the data from each analysis and requeue the instrument to rerun the failures. There was clearly a need to automate this process and have the data presented to the analyst in a clear, rapidly interpretable format with a mechanism to efficiently access the raw data.

In keeping with the strategy of obtaining high-quality data as if one were analyzing compounds one at a time, quantitative determinations using liquid chromatography/tandem mass spectrometry (LC-MS/MS) in SRM mode was desired. This is the established method of choice, as it provides the required sensitivity, speed, and specificity. The downside of this approach is the cost of triple-quadrupole instrumentation and the time needed to develop the individual LC-MS/MS conditions required for each compound. A potential solution to both of these problems lay in the use of quadrupole ion-trap instrumentation. Ion traps provide access to MS/MS at lower cost, and unlike on triple quads, a single set of MS/MS conditions can be chosen that will be suitable for a diverse range of compounds [27]. The ease with which MS/MS spectra can be acquired in a data-dependent [28] manner provides the possibility of acquiring an MS/MS spectrum of the compound under investigation during the structural-integrity run without incurring any further penalty in instrument time or compound usage. The applicability of a single set of conditions for acquiring MS/MS spectra and the demonstrated spectral reproducibility of those spectra over sample concentration, time, and instrument platform has been well demonstrated [29,30]. The value that could be gained from the building of an MS/MS library [31–34] that contains all compounds that enter into the profiling assays has clear applicability to the processes of metabolite and degradant identification that take place during subsequent drug discovery and development investigations [35].

The tasks of creating instrument run lists, interpreting data, rerunning failures, creating an MS/MS database, and extracting the information required to build quantitative methods could not be performed manually. Automated software was needed; however, a commercial solution did not exist. Therefore, an in-house custom solution was developed. Automated, paperless data processing was required to maintain high-quality data and provide a mechanism to identify suspect data and allow rapid manual review. In addition, assays would be designed to give few to no false-positives. Information about compounds entering the assays would need to be obtained from the corporate database, used to set up experiment sample lists, and build custom instrument methods. After automated processing of data and any manual review, results would be entered back into the corporate database for retrieval with all other biological data. Since the chromatograms and mass spectra of the structural integrity results were of direct interest to medicinal chemists, these data were made available via the intranet.

As the number of assays within the profiling suite and the required throughput of those assays would change over time, a scalable system was developed. The LCQ family of ion traps from ThermoFinnigan met our requirements in terms of MS hardware and allowed external software control via the programmable component object model (COM) objects of the Xcalibur Development Kit (XDK).

INTEGRATED LC-MS SOLUTION

Integrated Approach

All assay plates for profiling assays are aliquoted from a master plate that contains the compounds to be profiled as solutions in DMSO. One of the assay plates is run through the structural-integrity assay, thereby ensuring the integrity of the linked profiling assays. The structural-integrity assay provides confirmation of compound identification along with a purity determination. All compounds are initially run by ESI using positive-ion–negative-ion switching on alternate scans with in-line UV detection. When the expected $[M + H]^+$ ion in positive-ion mode or $[M - H]^-$ ion in negative-ion mode is observed, the mass spectrometer is programmed to automatically obtain an MS/MS spectrum under standard conditions. Automated software protocols are used to integrate the UV peaks and analyze the corresponding mass spectra. Each sample is automatically assigned to one of eight color-coded conditions. The operator has the option of choosing which samples require an APCI positive-ion–negative-ion rerun based on their color code. The APCI data are then processed as previously shown. The identity and purity are reported based on the combined data from all ionization modes employed. Based on the mass and intensity of product ions present in the data-dependent MS/MS spectra from the different ionization modes, SRM quantitation methods are created for each compound. These quantitation methods are used to determine the concentration of drug in the various analytical samples generated by the profiling assays (Figure 13.1). Quantitation data are automatically processed and summary data are displayed in a color-coded graphical fashion. Suspect results are identified and can be reviewed and manually integrated in a custom view that shows all data related to a compound data set in a single view for rapid visual recognition [36,37].

Structural Integrity

UV at 220 nm was chosen as the method for purity determination. UV detectors interface seamlessly as in-line detectors with LC-MS systems, are extremely robust, reproducibile and sensitive at the required sample concentrations. No HPLC detector has uniform response across all compounds [38–41]. Evaporative light-scattering detection (ELSD) is reported to offer a more universal response factor [38], but the response factors compared to UV at 220 nm were

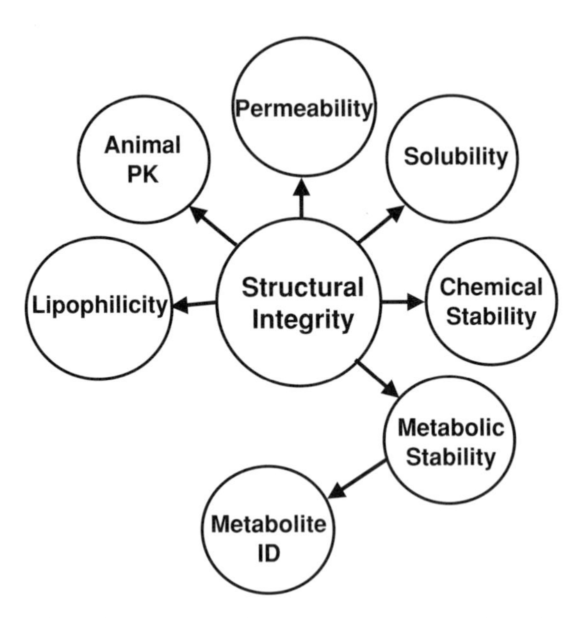

FIGURE 13.1 Structural integrity as the core of a suite of profiling assays.

found in our laboratory to be no more uniform across a range of compounds [42]. As ELSD was also observed to be less sensitive than UV at lower analyte concentrations, this lower sensitivity could lead to integration of the largest components of a mixture, but failure to detect impurities at less than 20% of the largest peak. Failure to detect all components in a mixture results in a falsely high-purity determination for the largest component in the sample. Our requirement was to determine the relative purity of the expected component rather than its quantitative purity; therefore, the uniform responses of components within the sample is more important than intersample comparisons. The U.S. Food and Drug Administration (FDA) is comfortable with assuming identical UV response factors of impurities in active pharmaceutical ingredient, as this is the usual way of reporting impurities in the chemistry and manufacturing controls (CMC) section of regulatory filings. This assumption was considered to be valid in the less rigorous drug-discovery environment. The wavelength of 220 nm was chosen as it is compatible with the mobile phases used, and the large majority of compounds that are of interest in drug discovery have absorption at this wavelength. A 2.1 × 30 mm, 5 μm C-18 column with a flow rate of 1 mL/min, which is five times the linear velocity for conventional chromatography, allows for a linear gradient from 2% organic to 90% organic in 3.5 minutes with acceptable peak shape and resolution for a diverse range of compounds. A water/acetonitrile mobile phase that containins 10 mM ammonium acetate at neutral pH is compatible with obtaining high-quality spectra in ESI and APCI in both positive- and negative-ion modes. The samples, supplied as 3 mM solutions in DMSO, are diluted 13:1 with a solvent system consisting of a 1:1:1 mixture of water/acetonitrile/isopropyl alcohol. A 4 μL injection was made onto the LC/UV/MS system. This dilution helps to minimize the effects of DMSO on the chromatography system while providing a solvent system that is compatible with the solubility characteristics of a wide range of compounds.

The mass spectrometer is set up to run in a positive-ion–negative-ion switching mode on alternate scans. The method is programmed to obtain an MS/MS spectra in a data-dependent manner if the expected $[M + H]^+$ or $[M - H]^-$ ion is observed in the preceding MS spectrum above a predetermined threshold. Previous experience in building libraries of MS/MS spectra for spectral searching had shown that a single collision energy of 45% relative collision energy using wide-band activation would be suitable for a wide range of compounds.

Compound name and location in a 96-well plate are provided in the form of an Excel spreadsheet. The software then accesses the corporate database and retrieves the empirical formula for each compound (Figure 13.2). Expected mass-to-charge ratio (m/z) values for the $[M + H]^+$ or $[M - H]^-$ ions are calculated and take into account that the formula may represent a salt form. Sample information along with the list of expected m/z values

	A	B	C	D
17	BATCH	Vial	Subname	Molecular Formula
18	Std	A01	(alfa)-methyl dopa	C10H13NO4 1.5H2O
19	Std	B01	Atenolol	C8 H9 N O2
20	Std	C01	Cimetidine	C10H13N02
21	Std	D01	Diclofenac Na (salt)	C9 H8 O4
22	Std	E01	Fluoxetine HCl (salt)	C16 H19 N3 O5 S
23	Std	F01	lactulose	C16H19N3O4S
24	Std	G01	Nifedapine	C11 H12 N2 O
25	Std	H01	Propronolol (salt)	C14H18N2O5
26	Std	A02	Acetamidophen	C14 H22 N2 O3
27	Std	B02	Metronidazole	C10H12Cl NO2
28	Std	C02	Cortisone	C18H24O2
29	Std	D02	Digoxin	C17H20N2O5S

FIGURE 13.2 Importing sample information.

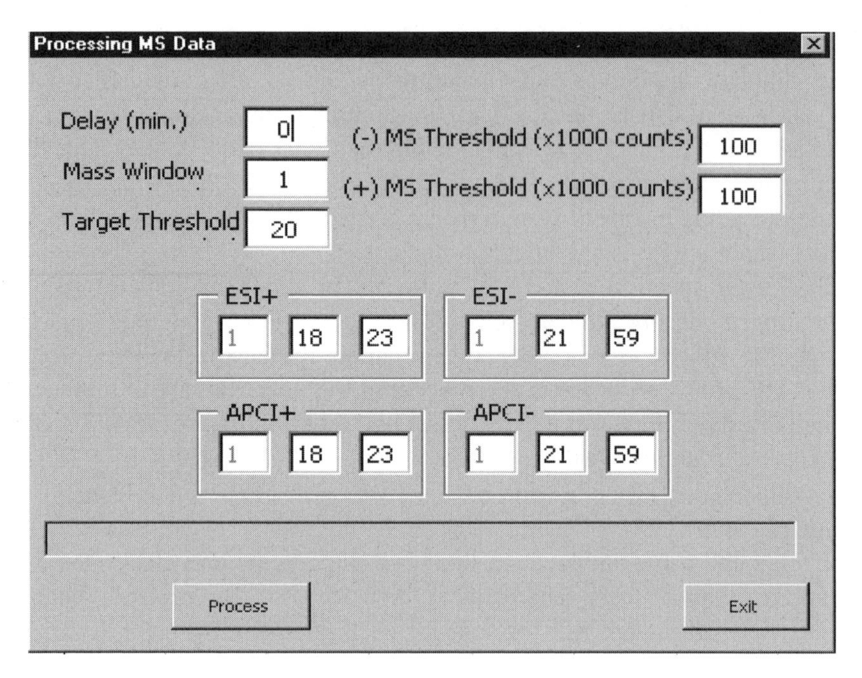

FIGURE 13.3 Dialogue box for setting expected adducts and intensity thresholds.

for data-dependent MS/MS are used to automatically build the sample sequence run list for the mass spectrometer. After the samples are run, the data are processed in a batch mode. The UV trace is integrated for each data file. A background-subtracted mass spectrum is obtained for each peak. Each spectrum is analyzed for ions at m/z values that would be expected from the compound molecular weight (MW) and the user-defined list of adducts (Figure 13.3) using an algorithm developed in-house. An analysis of the distribution of MWs of the compounds in our compound collection revealed that greater than 75% of the compounds fall within a MW range of 300 (Figure 13.4). Therefore, the chances of two random compounds with the same nominal MW are around 300:1. If the masses of protonated, ammoniated, and sodiated species are considered for confirmation of the expected compound, then the chances of two random compounds with the same MW increases to 100:1. When dimers and isotopes are considered, the chances of a random match increase further. An algorithm that incorporates a number of features to reduce the possibility of false positives is used. The intensity of an ion of a given m/z to be considered for MW confirmation must exceed both an absolute intensity threshold and a relative intensity threshold. These thresholds are relative to the base peak in the spectrum and are user-defined in the software. Isotopic pattern criteria and the nitrogen rule must be met. If the

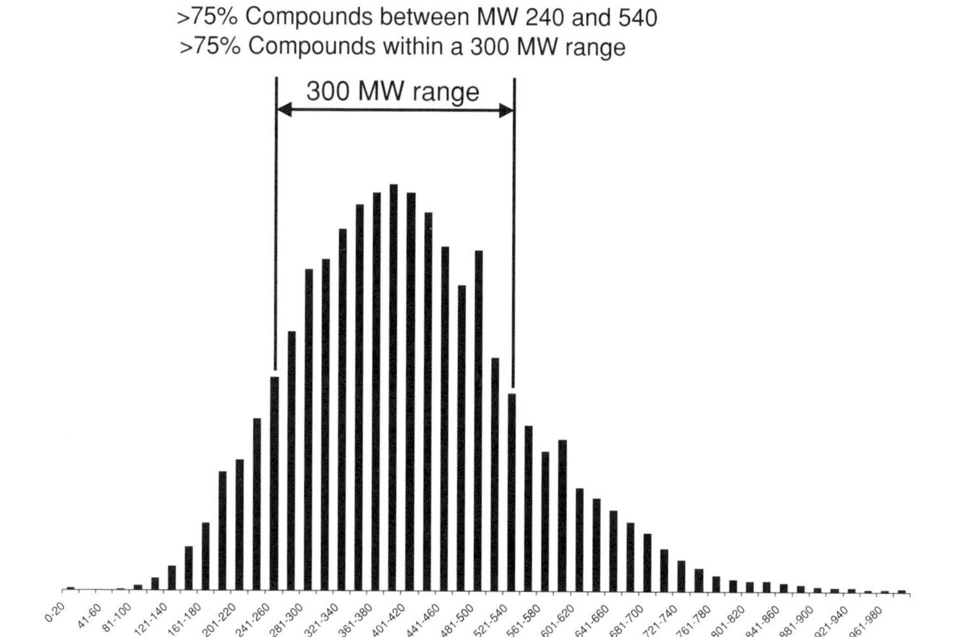

FIGURE 13.4 Distribution of compound MWs in compound collection.

confirmation is being made based on anything other than the expected $[M + H]^+$ or $[M - H]^-$ ion (depending on polarity), then the software identifies these data for manual review. Based on a combination of the purity by UV at 220 nm and analysis of the mass spectrum as described previously, the sample is described by one of the color-classified criteria (Figure 13.5) and reported in an Excel spreadsheet (Figure 13.6). The analyst chooses which of these color-coded criteria require a reanalysis (Figure 13.7). A rerun sequence list is generated and downloaded to the instrument for analysis. After analysis is complete, the samples are processed as described earlier, and the color-coded results are displayed in the summary sheet (Figure 13.8). The left-hand column shows the best result from the combined positive-ion–negative-ion ESI and positive-ion–negative-ion APCI runs. If the analyst wishes to manually review these data, then clicking on the individual results launches via a hyperlink the selected raw data file using the vendor-supplied LC-MS browser. After manual review, the analyst can update the summary sheet based on their judgment.

When any manual review of the data is complete, the data are ready to be reported into the central database. The multiple color-coded criteria are a very valuable aid and assist with the selection of compounds to rerun or manually review. However, this detailed information can be confusing to the end user.

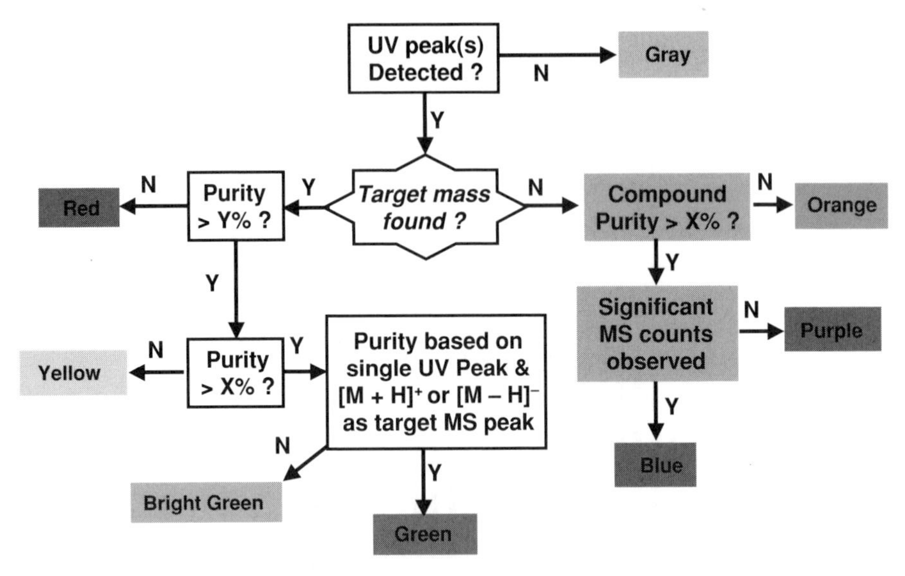

FIGURE 13.5 Automated integrity reporting.

Therefore, these eight criteria are condensed by condensing to four criteria (Figure 13.9), which are reported into the database.

To provide more detail to the end user beyond a purity number and a color code, the UV chromatogram and the background-subtracted spectra that correspond to the integrated peaks are written into a format that can be read by a custom browser that resides within the corporate analytical request-and-tracking system. In this way, the end user can view via the intranet the associated LC-MS data by simply clicking on the numeric value for SI, which is reported via the Web page with the other associated profiling data. There is no requirement to have any vendor software on their personal computer (PC).

Rerun

Name	% Area (Target)	UV Peaks	ESI+ link	ESI- link	APCI+ link	APCI- link
Cyclosporin A	No UV Dat	0	NO	NO		
Enalapril (Male	100	2	YES	YES		
Ampicillin	72	3	YES	YES		
Desipramine (sa	100	1	YES	NO		
Etoposide	100	2	YES	YES		
Nadolol	100	1	YES	YES		
Hydralazine HCl	0	3	NO	NO		
Reserpine	96	5	YES	YES		
beta-estradiol	0	1	NO	NO		

FIGURE 13.6 Color-coded summary of results from processing of pos/neg ESI data.

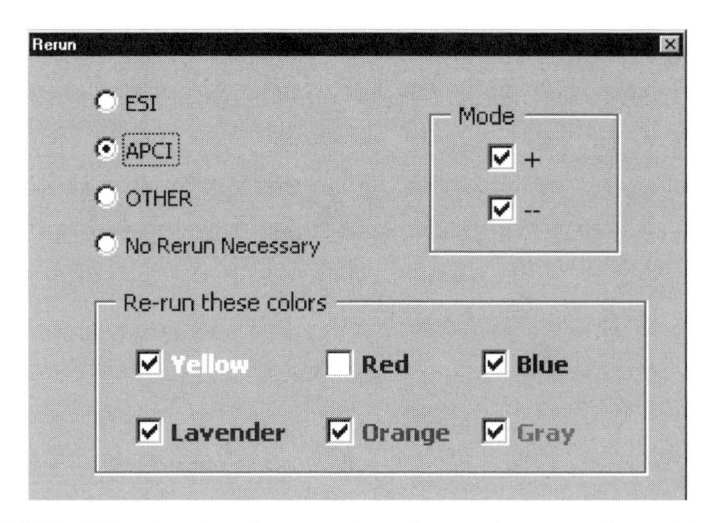

FIGURE 13.7 Dialog box that allows samples to be rerun based on color-coded summary of results from previous run.

The MS/MS spectra that are obtained in a data-dependent fashion during the structural-integrity analyses are extracted from the individual data files along with sample and data file information and written into a single file in a form that is readable by the National Institute of Standards and Technology (NIST) library searching program. These files are then imported into the library building and searching tool within the Xcalibur software package. This MS/MS database also serves as the basis for automatically generating the quantitative SRM methods for these compounds.

Review

Name	% Area (Target)	UV Peaks	ESI+ link	ESI- link	APCI+ link	APCI- link
Cyclosporin A	No UV Dat	0	NO	NO	NO	NO
Enalapril (Male	100	2	YES	YES		
Ampicillin	72	3	YES	YES	YES	YES
Desipramine (sa	100	1	YES	NO		
Etoposide	100	2	YES	YES		
Nadolol	100	1	YES	YES		
Hydralazine HCl	0	3	NO	NO	YES	NO
Reserpine	96	5	YES	YES		
beta-estradiol	0	1	NO	NO	NO	YES
Ketoprofen	0	1	NO	NO	NO	NO
Naproxen-03**	97	5	NO	YES		

FIGURE 13.8 Color-coded summary of results from processing of positive-ion–negative-ion ESI and positive-ion–negative-ion APCI data.

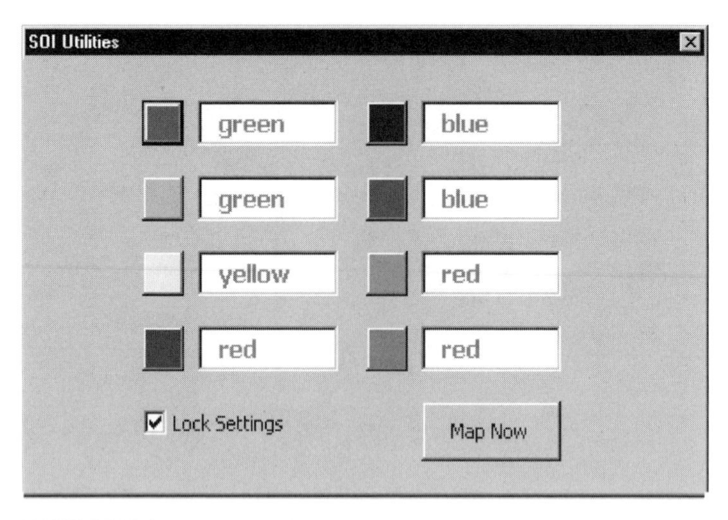

FIGURE 13.9 Mapping of analyst color codes to reporting color codes.

Quantitative Assays

Overview of Quantitative Assays

There are a number of assays within the profiling suite that require quantitation of the compound by LC-MS. The nature of the incubation and sample preparation methodologies of these assays may have considerable variation, but the analytical LC-MS determination is essentially the same. By using the same column type and mobile phase as the SI assay, the compounds can be expected to behave in the same fashion from both a chromatographic and MS standpoint. Typically, the column is shortened to 2×20 mm and the gradient time is reduced to provide a 1–1.5-minute sample-to-sample turnaround.

The MS/MS spectra extracted from the SI run just described also serve as the basis for automatically generating the quantitative SRM methods for these compounds. The most intense product ion from the full-scan MS/MS spectra obtained under the different ionization modes is selected for the SRM transition, and the method is automatically written in that ionization mode. The software approaches this method development procedure as an actual analyst would. SRM transitions that involve a simple loss of water or ammonia are not favored. If an SRM transition of suitable intensity cannot be obtained, then a SIM method is created instead. The $[M + H]^+$ or $[M - H]^-$ ion is favored for SIM. If these ions are not present above a user-defined threshold, then the software can create a SIM method that monitors an adduct. The SI method incorporates multiple scan types across LC peaks where the concentration of the analyte of interest rapidly changes. This rapid change in concentration presents a problem of how to determine which mode of operation and ions would provide the best quantitative method. It would not be correct to directly compare any two scans on an absolute scale for determining the best transition. To account for the changing concentration, an algorithm was developed that considers the intensity of the UV trace at the exact point that the MS scan was obtained. The MS intensities are normalized based on the ratio of UV intensity to that at the apex of the chromatographic peak.

Most quantitative LC-MS determinations employ an internal standard. Typically a closely related or stable label analog of the analyte is used. The use of a suitable analog allows for the correction of errors that are caused by sample-to-sample variability, such as extraction efficiency, injection errors, ion-suppression effects, and ionization-source fouling. However, when dealing with a profiling assay designed to run many different compounds in batch mode, the selection of an internal standard for each individual compound becomes impractical. Here, the solution has been to pick an unrelated compound as an internal standard. As a result, the generic MS internal standard is not a close chemical relative of the analyte and can no longer serve to correct for extraction efficiency or ion-suppression effects. The use of the internal standard is restricted to monitoring injection error and ion-source performance degradation. With this in mind a UV internal standard was selected to monitor injection performance. The fact that all analyses for any given compound are completed within 10–15 minutes makes the effects of ion-source degradation a minimal concern. The noninvolvement of ion-source degradation has been borne out by numerous and extensive validation sets; however, an external MS standard is monitored at regular intervals to verify MS performance. The use of a UV internal standard rather than a MS internal standard approach, was to compensate for the relatively long cycle time of the ion trap (relative to SRM on a triple quad) for doing an MS/MS experiment. Using this approach it was possible to maximize the amount of time spent analyzing the component of interest, and consequently, the precision improved. The UV internal standard with a large extinction coefficient but a weak MS response was selected to avoid ion suppression or possibility of interference if the internal standard and analyte happened to coelute.

The three-dimensional (3D) ion trap has sufficient sensitivity and precision for determining analyte concentrations in *in vitro* profiling assays where concentrations are typically in the 100-nM to 10 μM range. Recent work on the emerging technology of linear ion traps suggests that this process may be extended to animal PK work where sensitivity requirements are more rigorous [43]. This instrument demonstrates increased sensitivity and duty cycle over the 3D trap, but exhibits the same characteristic as the 3D trap in allowing universal conditions to be used for obtaining product-ion spectra [43].

As an extension to the approaches used in the SI assay, software that automatically processes the data files from the individual quantitation runs and

Compound	Integration	Mass Balance	Donor time 0	Permeability
acetophenetidine	0,0,0,0,0	92.9%	116.2%	81.04
amiloride	0,0,1,0,0	119.1%	70.2%	2.80
antipyrine	0,0,0,0,0	106.2%	98.5%	16.67
atenolol	0,0,0,0,0	121.9%	190.5%	2.10
cimetidine	0,0,0,0,1	87.3%	91.2%	17.36
desipramine salt	0,0,0,0,0	49.8%	85.1%	46.84
diltiazen HCl	0,0,0,0,1	61.1%	91.8%	100.70
famotidine	0,0,0,0,0	97.9%	104.1%	8.63
fenoterol	0,0,0,0,0	136.0%	90.0%	3.13
ketoconazole	0,0,0,0,0	18.5%	123.8%	13.74
ketoprofen	0,0,0,0,1	106.1%	104.4%	44.23
metoprolol salt	0,0,0,0,0	103.8%	96.7%	25.08
norfloxacin	0,0,0,0,0	90.5%	118.5%	6.52
pindolol	0,0,0,0,0	95.4%	98.1%	24.67

FIGURE 13.10 Summary page of PAMPA quantitation experiment.

presents the data in a summary page has been developed (Figure 13.10). In keeping with our approach to give no false-positives, the software automatically identifies suspect data, and hyperlinks are provided to allow manual review of the data. Clicking on the embedded hyperlinks in the Excel summary launches a custom browser that was built to allow the analyst to see all the relevant data from a single compound on the screen simultaneously (Figure 13.11). As manual integration changes are made, the results are updated in realtime. When all data have been reviewed and approved, uploads are made in batch mode to the corporate database via Activity Base software [44] for viewing by the end user along with the data from the other profiling assays.

Examples of how this integrated approach is applied to assays in the profiling suite are illustrated for determining permeability via the parallel artificial-membrane permeabilty assay (PAMPA) [45] and metabolic stability via a microsomal stability assay.

Parallel Artificial Membrane Permeability Assay

The use of PAMPA to provide a predictive measure of passive intestinal permeability was introduced by Kansy [45], and has been extensively employed in the pharmaceutical industry [46,47] for this purpose and extended for use in predicting permeability of the blood–brain barrier [48]. The original analytical output employed a 96-well photometer with simultaneous measurement of six different wavelengths and comparison to reference spectra. Others have used both the plate reading method and HPLC/UV for determining compound concentration [47,48]. We felt that UV plate-reading methods could be subject to interferences, particularly if the compound being assayed was not 100% pure, as is often the case with compounds that are stored as solutions in DMSO. HPLC/UV methods may be able to minimize the effect of interferences, but at the expense of longer analysis times. Automated LC-MS/MS methods seem

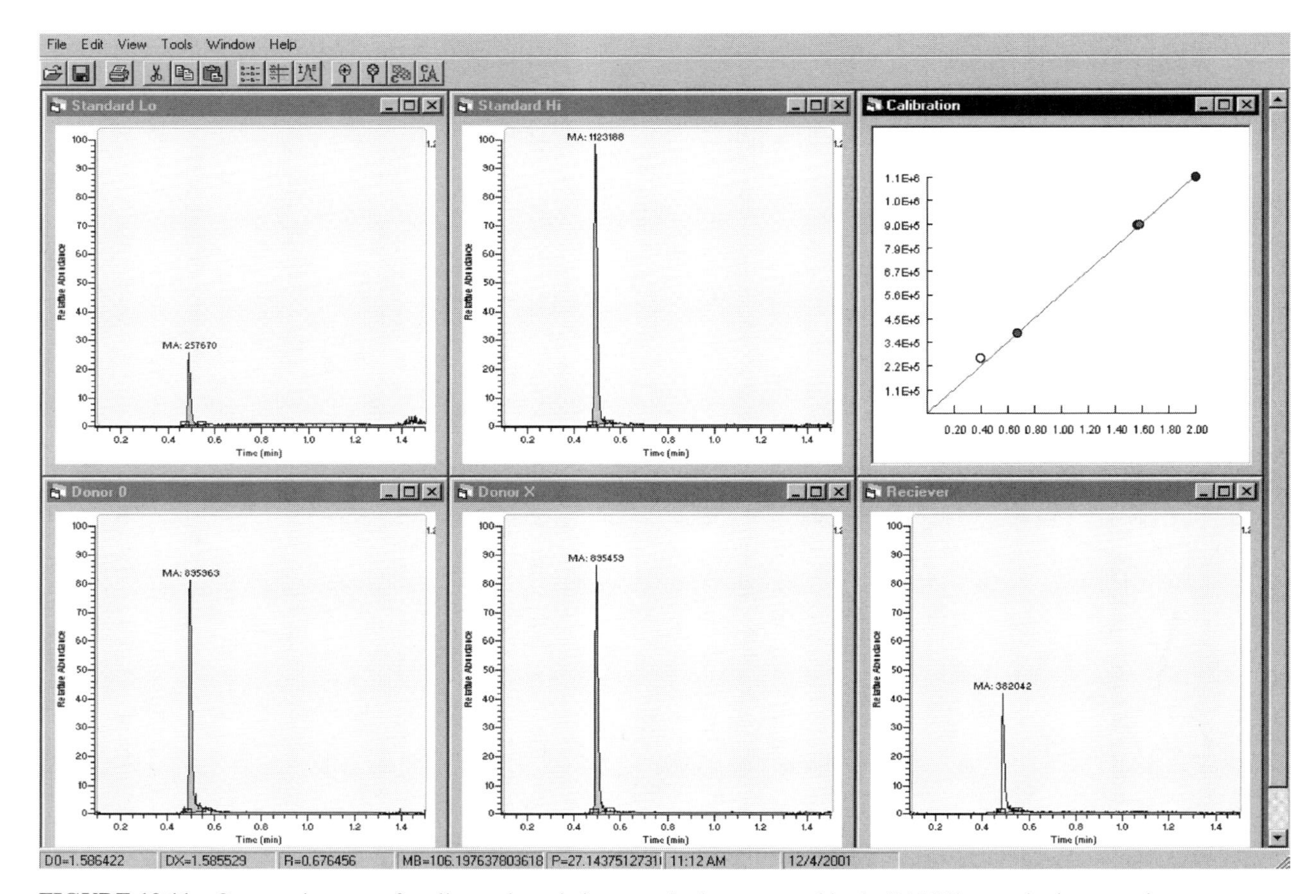

FIGURE 13.11 Quan review page for all samples relating to a single compound in the PAMPA quantitation experiment.

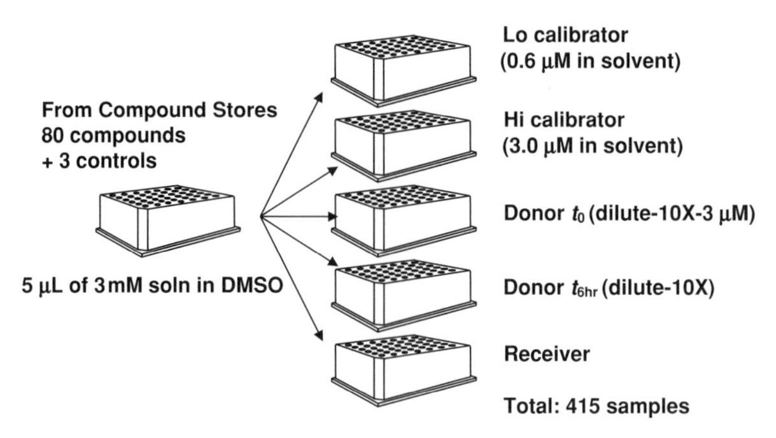

FIGURE 13.12 Sample generation for PAMPA assay.

well suited to providing the required selectivity, sensitivity, and throughput, and would afford the opportunity to incorporate controls to ensure the quality of the data. Rather than simply measuring the MS responses for donor solution before incubation and the receiver solution after incubation for determining permeability, the determination of actual concentrations from a two-point calibration curve made by dilution into organic solvent was more reliable. The permeability measurement is calculated from the concentrations of donor and receiver solutions at the incubation endpoint (6 hours). A donor at time zero solution is also measured and compared to the high standard to assess adequate solubility in the assay buffer and to monitor mass balance. The work flow for preparing the sample for analysis is shown in Figure 13.12. The incubation is performed in duplicate to provide an indication of well-to-well variability. The entire process, including preparation of the lipid bilayer, has been automated on a CyBio liquid handler. Each 96-well plate of compounds to be analyzed results in a 96-well plate containing all the low standards for the compounds, a 96-well plate containing all the high standards for the compounds, and 96-well plates for donor at time zero, donor at time 6 hours, and receiver at time 6 hours. From an automation perspective, it would appear to be simplest to run one full plate at a time. However, this procedure would separate in time the various measurements for individual compounds that will be used to calculate the permeability readout. The software prepares the instrument run lists to acquire all related analytical measurements sequentially. A Leap HTS Pal autosampler equipped with multiple trays is used to select the samples from the various plates without impacting the injection-to-injection time. The software then automatically integrates the analytical runs and provides the data in an Excel summary (Figure 13.10) as previously described. This assay has a capacity of 1000 compounds/month, with a turnaround time of two days.

Metabolic Stability

The use of liver microsomal assays prepared from various species for the early prediction of stability of compounds to first pass phase-one metabolism is well documented [49]. A number of approaches that use LC-MS for the analytical portion of the assay have been reported [12–18]. Automated LC-MS/MS methods appear to be well suited to provide the necessary selectivity, sensitivity, and throughput and may incorporate controls to ensure the quality of the data [50]. Automated methods are created for each compound as described earlier. The percent of parent drug remaining and rate of clearance are calculated. Comparison of the parent-drug concentration in the incubation at time-zero solution to the high standard allows for the calculation of recovery. The work flow for preparing the samples for analysis is shown (Figure 13.13). Incubations with rat, mouse, and human microsomes are routinely performed; other species are run by request. The entire process of preparation of standards in organic solvent, serial dilutions, and incubations in duplicate is carried out on a Tecan liquid handler. As described previously for the PAMPA assay, each plate contains all compounds to be analyzed, with each plate consisting of a particular species and time point, duplicate, or analytical standard. The software prepares the instrument run lists to acquire all related analytical measurements, then automatically integrates the analytical runs and provides the data in an Excel summary (Figure 13.14) that reports percent recovery, percent remaining, and rate for each individual compound. Red flags indicate data that are outside the range of values set by the operator, alerting to manual review. The data are then uploaded into the corporate database, as described for the PAMPA assay. This assay is currently configured with a capacity of 720 compounds/month with a turnaround of 4 days.

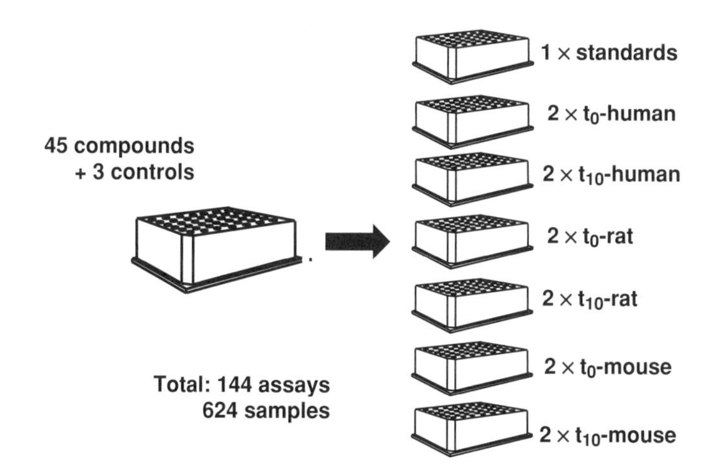

FIGURE 13.13 Sample generation for metabolic-stability assay.

			%Recovery		%Remaining		
Form-Lot	Compound	Species	Average	SD	Average	SD	Rate (avg)
000-00	Nefazadone	human	110.7%	3.8%	0.4%	0.1%	0.299
		mouse	120.1%	6.1%	0.1%	0.0%	0.300
		rat	99.2%	7.0%	0.3%	0.1%	0.299
000-00	Verapamil	human	105.1%	3.1%	32.2%	3.8%	0.204
		mouse	112.7%	1.5%	30.0%	0.1%	0.210
		rat	108.7%	0.4%	28.9%	5.1%	0.213
000-00	Tolbutamide	human	126.0%	9.9%	97.8%	0.8%	0.007
		mouse	[illegible]	1.0%	85.9%	14.6%	0.042
		rat	147.3%	6.0%	99.9%	1.8%	0.000
01-003	BMS_597367	human	106.9%	5.2%	80.2%	3.1%	0.059
		mouse	110.8%	4.7%	10.3%	1.3%	0.269
		rat	101.2%	[illegible]	6.6%	0.5%	0.280
01-001	BMS_604729	human	126.1%	[illegible]	0.0%	0.0%	0.300
		rat	131.7%	9.9%	0.0%	0.0%	0.300
01-001	BMS_604730	human	82.4%	1.1%	1.2%	1.4%	0.296
		rat	97.6%	4.7%	23.5%	2.1%	0.229
01-001	BMS_604721	human	111.1%	5.8%	0.0%	0.0%	0.300
		rat	119.4%	[illegible]	0.0%	0.0%	0.300
01-001	BMS_604088	human	83.3%	2.4%	73.0%	7.4%	0.081
		rat	91.1%	9.6%	78.2%	4.2%	0.065
01-001	BMS_595752	human	113.1%	4.5%	100.6%	6.4%	-0.002
		rat	133.4%	[illegible]	90.8%	7.8%	0.028
01-001	BMS_589543	human	107.6%	7.7%	91.2%	6.2%	0.026
		rat	123.2%	[illegible]	84.4%	5.7%	0.047
01-001	BMS_592671	human	115.8%	2.5%	60.6%	4.5%	0.118
		rat	114.0%	1.5%	95.7%	2.3%	0.013
01-001	BMS_593010	human	99.6%	0.3%	51.3%	8.5%	0.146
		rat	101.9%	3.7%	76.0%	5.0%	0.072
01-001	BMS_593212	human	96.4%	2.5%	91.0%	4.3%	0.027
		rat	99.1%	7.5%	82.2%	2.1%	0.053
01-001	BMS_593763	human	98.8%	5.3%	89.1%	4.2%	0.033
		rat	97.1%	3.5%	97.4%	3.9%	0.008

Detail | Results

FIGURE 13.14 Summary page of metabolic-stability quantitation experiment.

CONCLUSIONS AND FUTURE DIRECTIONS

An integrated LC-MS/MS infrastructure has been conceived and built that emphasizes quality in concert with throughput and turnaround for the high throughput profiling assays that are used to support candidate optimization in drug discovery. Quality, in terms of the breadth of chemical space covered, is achieved via the orchestrated application of tiered multiple ionization modes. The process of checking the fidelity of samples used in the profiling assays at the time and place of use ensures the highest correlation between biological readout and compound structure. The traditional dogma that high-quality quantitative LC-MS requires triple-quadrupole instrumentation (and the associated costs of custom method development for each compound) has been challenged. Ion-trap instrumentation provides near universal collision-induced dissociation (CID) conditions, while the ease of recording product ion spectra via data-dependent MS/MS acquisition provides access to these spectra at no additional cost. The facile access to high-quality product-ion spectra of all compounds under investigation in the profiling assays provides an immediate return, since this approach allows all compounds to be quantitated with the speed, sensitivity, and selectivity that LC-MS/MS affords. Archiving of these spectra into a library (currently $>$ 72,000) that can be searched by both spectra and structure provides a powerful internal tool for impurity and degradant studies. The promise of combining this structure/spectra database with cross-correlation algorithms moves us toward the promise of an empirically based structural-elucidation tool [35].

The key driver for this automated integrated approach has been the access to processing and instrument control via the COM objects of the Finnigan Xcalibur software package. This access to the programming objects of the vendor software has allowed us to focus our programming efforts on those aspects that allow us to meet the specific challenges of the profiling assays by creating new tailored analytical approaches. Without access to these objects we would have had to produce an entire hardware/software solution from the ground up, which would have been impractical. The ability to customize vendor software provides freedom from any vendor preconceptions of how specific analytical challenges should be solved. For example, the development of decision algorithms that extend beyond the simplistic "MW $+$ 1 $=$ compound found" genera of algorithms used for identifying the presence of an expected compound. Recognition that no algorithm can be perfect and an appreciation of the human visual process led us to develop a low false-positive automation approach coupled with single-page color-coded rapid manual review tools.

ACKNOWLEDGMENT

The authors gratefully acknowledge the considerable contributions of their present and former colleagues at Bristol-Myers Squibb, who have contributed to the successful execution of the vision of an integrated LC-MS/MS platform described here.

REFERENCES

1. Venkatesh, S.; Lipper, R.A. "Role of the Development Scientist in Compound Lead Selection and Optimization," *J. Pharm. Sci.* **89**, 145–154 (2000).
2. Di, L.; Kerns, E.H. "Profiling Drug-like Properties in Discovery Research," *Curr. Opin. Chem. Biol.* **7**, 402–408 (2003).
3. Kerns, E.H.; Di, L. "Pharmaceutical Profiling in Drug Discovery," *Drug Disc. Today* **8**, 316–323 (2003).

4. Kerns, E.H. "High Throughput Physicochemical Profiling for Drug Discovery," *J. Pharm. Sci.* **90**, 1838–1858 (2001).
5. Kerns, E.H.; Di, L. "Multivariate Pharmaceutical Profiling in Drug Discovery," *Curr. Topics Med. Chem.* **2**, 87–98 (2002).
6. Biller, S.A.; Custer, L.; Dickinson, K.E.; Durham, S.K.; Gavai, A.V.; Hamann, L.G.; Josephs, J.L.; Moulin, F.; Pearl, G.M.; Flint, O.P.; Sanders, M.; Tymiak, A.A.; Vaz, R. "The Challenge of Quality in Candidate Optimization," In Borchardt, R.T.; Thakker, D.R.; Kerns, E.H.; Wang, B.; Lipenski, C.A., eds., *Pharmaceutical Profiling in Drug Discovery for Lead Selection*, AAPS Press, Arlington, VA 413–430 (2004).
7. Hertzberg, R.P.; Pope, A.J. "High-throughput Screening: New Technology for the 21st Century," *Curr. Opin. Chem. Biol.* **4**, 445–451 (2000).
8. Lee, M.S.; Kerns, E.H. "LC/MS Applications in Drug Development," *Mass Spectrom. Rev.* **18**, 187–279 (1999).
9. Hoke, S.H.; Morand, K.L.; Greis, K.D.; Baker, T.R.; Harbol, K.L.; Dobson, R.L.M. "Transformations in Pharmaceutical Research and Development, Driven by Innovations in Multidimensional Mass Spectrometry-based Technologies," *Int. J. Mass Spectrom.* **212**, 135–196 (2001).
10. Cole, M.J.; Janiszewski, J.S.; Fouda, H.G. pp. 211–249. In Pramanik, B.N.; Ganguly, A.K.; Gross, M.L., eds., "Electrospray Mass Spectrometry in Contemporary Drug Metabolism and Pharmacokinetics," *Applied Electrospray Mass Spectrometry*, Vol. 32, Dekker, New York (2002).
11. Lee, M.S. *LC/MS Applications in Drug Development*, Wiley, New York (2002).
12. Korfmacher, W.A.; Palmer, C.A.; Nardo, C.; Dunn-Meynell, K.; Grotz, D.; Cox, K.; Lin, C.-C.; Elicone, C.; Liu, C.; Duchoslav, E. "Development of an Automated Mass Spectrometry System for the Quantitative Analysis of Liver Microsomal Incubation Samples: A Tool for Rapid Screening of New Compounds for Metabolic Stability," *Rapid Commun. Mass Spectrom.* **13**, 901–907 (1999).
13. Xu, R.; Nemes, C.; Jenkins, K.M.; Rourick, R.A.; Kassel, D.B.; Liu, C.Z.C. "Application of Parallel Liquid Chromatography/Mass Spectrometry for High Throughput Microsomal Stability Screening of Compound Libraries," *J. Am. Soc. Mass Spectrom.* **13**, 155–165 (2002).
14. Xu, R.; Nemes, C.; Liu, C.Z.C.; Kassel, D.B. "A High Throughput Parallel LC/MS Method for Assessing Metabolic Stability and Permeability of Compound Libraries," in *Proceedings of the 47th ASMS Conference on Mass Spectrometry and Allied Topics*, Dallas, TX, 1999.
15. Whalen, K.M.; Rogers, K.J.; Cole, M.J.; Janiszewski, J.S. "AutoScan: An Automated Workstation for Rapid Determination of Mass and Tandem Mass Spectrometry Conditions for Quantitative Bioanalytical Mass Spectrometry," *Rapid Commun. Mass Spectrom.* **14**, 2074–2079 (2000).
16. Rogers, K.J.; Cole, M.J.; Fouda, H.G.; Janiszewski, J.S.; Liston, T.E.; Whalen, K.M. "High-Throughput LC/MS Analysis of ADME Screen Samples: An

Integrated Module Approach," in *Proceedings of the 48th ASMS Conference on Mass Spectrometry and Allied Topics*, Long Beach, CA, June 11–15, 2000.

17. Janiszewski, J.S.; Rogers, K.J.; Whalen, K.M.; Cole, M.J.; Liston, T.E.; Duchoslav, E.; Fouda, H.G. "A High-Capacity LC/MS System for the Bioanalysis of Samples Generated from Plate-Based Metabolic Screening," *Anal. Chem.* **73**, 1495–1501 (2001).
18. Di, L.; Kerns, E.H.; Hong, Y.; Kleintop, T.A.; McConnell, O.J.; Huryn, D.M. "Optimization of a Higher Throughput Microsomal Stability Screening Assay for Profiling Drug Discovery Candidates," *J. Biomol. Screening* **8**, 453–462 (2003).
19. Kassel, D.B. "Combinatorial Chemistry and Mass Spectrometry in the 21st Century Drug Discovery Laboratory," *Chem. Rev.* **101**, 255–267 (2001).
20. Triolo, A.; Altamura, M.; Cardinali, F.; Sisto, A.; Maggi, C.A. "Mass Spectrometry and Combinatorial Chemistry: A Short Outline," *J. Mass Spectrom.* **36**, 1249–1259 (2001).
21. Morand, K.L.; Burt, T.M.; Regg, B.T.; Chester, T.L. "Techniques for Increasing the Throughput of Flow Injection Mass Spectrometry," *Anal. Chem.* **73**, 247–252 (2001).
22. De Biasi, V.; Haskins, N.; Organ, A.; Bateman, R.; Giles, K.; Jarvis, S. "High Throughput Liquid Chromatography/Mass Spectrometric Analyses Using a Novel Multiplexed Electrospray Interface," *Rapid Commun. Mass Spectrom.* **13**, 1165–1168 (1999).
23. Xu, R.; Wang, T.; Isbell, J.; Cai, Z.; Sykes, C.; Brailsford, A.; Kassel, D.B. "High-Throughput Mass-Directed Parallel Purification Incorporating a Multiplexed Single Quadrupole Mass Spectrometer," *Anal. Chem.* **74**, 3055–3062 (2002).
24. Edwards, C.; Liu, J.; Smith, T.J.; Brooke, D.; Hunter, D.J.; Organ, A.; Coffey, P. "Parallel Preparative High-performance Liquid Chromatography with On-line Molecular Mass Characterization," *Rapid Commun. Mass Spectrom.* **17**, 2027–2033 (2003).
25. Shipkova, P.; Salyan, M.E.; Josephs, J.L.; Langish, R.A.; Sanders, M.; Pragnell, C. "High Throughput LC/MS Analysis Using an Eight-Way Multiplexed Electrospray Source with Switching between Positive and Negative Ion Modess," in *Proceedings of the 50th ASMS Conference on Mass Spectrometry and Allied Topics*, Orlando, FL, 2002.
26. Flynn, M.J.; Josephs, J.L.; Sanders, M. "High-Throughput Automated Quantitation using a Quadrupole Ion Trap Mass Spectrometer," Unpublished results, 2001.
27. Cole, M.J.; Needham, S.R.; Jeanville, P.; Hemenway, E.; Cunniff, J. in *Proceedings of the 47th ASMS Conference on Mass Spectrometry and Allied Topics*, Dallas, TX, 1999.
28. Josephs, J.L. "Detection and Characterization of Fumonisin Mycotoxins by Liquid Chromatography/Electrospray-ionization Using Ion Trap and Triple

Quadrupole Mass Spectrometry," *Rapid Commun. Mass Spectrom.* **10**, 1333–1344 (1996).

29. Josephs, J.L.; Sanders, M. "Creation and Comparison of MS/MS Spectral Libraries Using Quadrupole Ion Trap and Triple Quadrupole Mass Spectrometers," *Rapid Commun. Mass Spectrom.* **17**, 743–759 (2004).
30. Lopez, L.L.; Tiller, P.R.; Senko, M.W.; Schwartz, J.C. "Automated Strategies for Obtaining Standardized Collisionally Induced Dissociation Spectra on a Benchtop Ion Trap Mass Spectrometer," *Rapid Commun. Mass Spectrom.* **13**, 663–668 (1999).
31. Josephs, J.L. "Characterization of Over-the-Counter Cough/Cold Medications by Liquid Chromatography/Electrospray Mass Spectrometry," *Rapid Commun. Mass Spectrom.* **9**, 1270–1274 (1995).
32. Josephs, J.L.; Sanders, M.; DiDonato, G.; Hail, M.; Kerns, E.; Volk, K.; Lee, M.S. "Automated LC/MS/MS and CID Spectral Library Searching of Pharmaceutical Impurilios Using Ion Trups and Triple Quads," in *Proceedings of the 45th ASMS Conference on Mass Spectrometry and Allied Topics*, Palm Springs, CA, 1997.
33. Josephs, J.L.; Sanders, M.; Bolgar, M.; DiDonato, G.; Hail, M.; Kerns, E.; Lee, M.S. "The role of LC/MS/MS Spectral Library Searching for impurity Analysis Using Ion Trap and Triple Quadrupole Mass Spectrometers," in *Proceedings of the 46th ASMS Conference on Mass Spectrometry and Allied Topics*, Orlando, FL, 1998.
34. Baumann, C.; Cintora, M.A.; Eichler, M.; Lifante, E.; Cooke, M.; Przyborowska, A.; Halket, J.M. "A Library of Atmospheric Pressure Ionization Daughter Ion Mass Spectra Based on Wideband Excitation in an Ion Trap Mass Spectrometer," *Rapid Commun. Mass Spectrom.* **14**, 349–356 (2000).
35. Josephs, J.L.; Sanders, M.; Langish, R.A.; Whitney, J.; Phillips, J.J. " Detection and Characterization of Pharmaceutical Metabolites, Degradants and Impurities by the Application of MS/MS Software Algorithms," in *Proceedings of the 51st ASMS Conference on Mass Spectrometry and Allied Topics*, Montreal, Canada, June 8–12, 2003.
36. Josephs, J.L.; Sanders, M.; Langish, R.A.; Hnatyshyn, S.Y.; Salyan, M.E.; Shipkova, P.; Drexler, D.; Balimane, P.; Zvyaga, T.A.; Chong, S. "A High Quality High Throughput LC/MS Strategy for profiling of Drug Candidates with Applications to Structural Integrity, Permeability, Stability and related Assays," in *Proceedings of the 50th ASMS Conference on Mass Spectrometry and Allied Topics*, Orlando, FL, 2002.
37. Sanders, M.; Josephs, J.L.; Langish, R.A.; Hnatyshyn, S.Y.; Salyan, M.E.; Shipkova, P.; Drexler, D.; Flynn, M.J.; Burdette, H.L. "An Integrated LC/MS Strategy for Preclinical Candidate Optimization from compound Integrity to the High Throughput Quantitative Assays of Permeability, Metabolic and Chemical Stability," in *Proceedings of the 50th ASMS Conference on Mass Spectrometry and Allied Topics*, Orlando, FL, 2002.
38. Hsu, B.H.; Orton, E.; Tang, S.-Y.; Carlton, R.A. "Application of Evaporative Light Scattering Detection to the Characterization of Combinatorial and Parallel

Synthesis Libraries for Pharmaceutical Drug Discovery," *J. Chromatogr. B* **725**, 103–112 (1999).

39. Fang, L.; Wan, M.; Pennacchio, M.; Pan, J. "Evaluation of Evaporative Light-scattering Detector for Combinatorial Library Quantitation by Reversed Phase HPLC," *J. Comb. Chem.* **2**, 254–257 (2000).
40. Yan, B.; Fang, L.; Irving, M.; Zhang, S.; Boldi, A.M.; Woolard, F.; Johnson, C.R.; Kshirsagar, T.; Figliozzi, G.M.; Krueger, C.A.; Collins, N. "Quality Control in Combinatorial Chemistry: Determination of the Quantity, Purity, and Quantitative Purity of Compounds in Combinatorial Libraries," *J. Comb. Chem.* **5**, 547–559 (2003).
41. Cremin, P.A.; Zeng, L. "High-throughput Analysis of Natural Product Compound Libraries by Parallel LC-MS Evaporative Light Scattering Detection," *Anal. Chem.* **74**, 5492–5500 (2002).
42. Shipkova, P.; Josephs, J.L. Unpublished results, 2002.
43. Sanders, M.; Josephs, J.L.; Wang, J.; Phillips, J.J.; Mylchreest, I. "Highly Automated Process for the Quantitation of Samples from Animal Pharmacokinetic Studies using a Linear Ion Trap Mass Spectrometer," in *Proceedings of the 51st ASMS Conference on Mass Spectrometry and Allied Topics*, Montreal, Canada, June 8–12, 2003.
44. *Activity Base Software*, 5.0, ID Business Solutions.
45. Kansy, M.; Senner, F.; Gubernator, K. "Physicochemical High Throughput Screening: Parallel Artificial Membrane Permeation Assay in the Description of Passive Absorption Processes," *J. Med. Chem.* **41**, 1007–1010 (1998).
46. Wohnsland, F.; Faller, B. "High-Throughput Permeability pH Profile and High-Throughput Alkane/Water log P with Artificial Membranes," *J. Med. Chem.* **44**, 923–930 (2001).
47. Zhu, C.; Jiang, L.; Chen, T.-M.; Hwang, K.-K. "A Comparative Study of Artificial Membrane Permeability Assay for High Throughput Profiling of Drug Absorption Potential," *Eur. J. Med. Chem.* **37**, 399–407 (2002).
48. Di, L.; Kerns, E.H.; Fan, K.; McConnell, O.J.; Carter, G.T. "High Throughput Artificial Membrane Permeability Assay for Blood–Brain Barrier," *Eur. J. Med. Chem.* **38**, 223–232 (2003).
49. Thompson, T.N. "Optimization of Metabolic Stability as a Goal of Modern Drug Design," *Med. Res. Rev.* **21**, 412–449 (2001).
50. Drexler, D.; Edinger, K.; Hnatyshyn, S.Y.; Josephs, J.L.; Langish, R.A.; McNaney, C.; Sanders, M. "An Automated High Quality High Throughput LC/MS Process for the Analysis of Metabolic Stability Samples in Support of Drug Discovery," in *Proceedings of the 51st ASMS Conference on Mass Spectrometry and Allied Topics*, Montreal, Canada, June 8–12, 2003.

14

NEW APPROACHES FOR METHOD DEVELOPMENT AND PURIFICATION IN LEAD OPTIMIZATION

YINING ZHAO AND DAVID J. SEMIN

INTRODUCTION

Evolution of the Discovery Analytical Model

The traditional discovery analytical model in the pharmaceutical and biotechnology industries in the 1990s was basically set up to support combinatorial chemistry and the open-access analytical environment. One of the initial tasks for discovery analytical scientists was to develop high throughput techniques to ensure the purity and structure confirmation of compounds that were used in high throughput screening (HTS) campaigns. Nevertheless, compromises were made in the quality-assessment phase because of the sheer number of compounds.

Due to the large HTS campaigns over the last decade, scientists faced great difficulties during the hit-validation phase. For example, many of the original libraries were of poor quality and resulted in an abundance of false-positives. Most significantly, despite the promise and the widespread adoption of HTS approaches throughout the pharmaceutical industry, combinatorial chemistry has yet to yield any significant clinical candidates [1]. Due to these drawbacks and concerns, the drug-discovery paradigm has shifted toward the improved quality of libraries with the development of smaller, more focused libraries that are derived from specific pharmacophores. Additionally, more rational design

Integrated Strategies for Drug Discovery Using Mass Spectrometry, Edited by Mike S. Lee

tools are applied at an earlier stage of drug discovery to enhance physicochemical, potency, selectivity, as well as absorption, distribution, metabolism, and excretion (ADME) parameters. Under this new paradigm, medicinal chemists apply parallel syntheses techniques to generate targeted libraries that consist of hundreds of compounds in the 10–100-mg range. Compounds are contained in separate vessels that can be used for various screening activities during the lead optimization stage.

As a result, the analytical philosophy has changed from a model that was based on raw capacity to a model that features capacity coupled with quality. To achieve this transition, high-quality information must be effectively delivered within a high throughput environment. Nowadays, the enhancement of analytical capacity and efficiency with the aid of automation technology is no longer a top priority. The goal is to generate high-quality data and maximize the content of analytical information within a minimal time frame. To fulfill this requirement set by the current discovery paradigm, some new analytical concepts will need to be established.

Traditional Versus New Concepts

First, the concept of "stage-appropriate analysis" needs to be introduced. With the intense competition in drug discovery, and the drive to compress the timeframe, comes an evaluation of what resources and requirements are needed to maximize return on investments. The mere increase in automation technology and human resources does not necessarily equate to higher efficiency and success. Moreover, the main question becomes, What techniques and methodologies should be used to obtain the necessary information to drive the decision-making process? The philosophy of stage-appropriate analysis attempts to answer this question by the selection of the appropriate analytical technique at the appropriate stage of drug discovery to enable either the hit-to-lead or lead optimization stages. Table 14.1 illustrates the stage-appropriate analysis model between these two stages. The hit-to-lead stage is simply defined as the stage where initial hits from a high throughput screening campaign are assessed from a qualitative and biological standpoint to provide tractable and viable leads. At this stage, high throughput techniques are widely used for the qualitative assessment. In the lead optimization stage, significant medicinal chemistry resources are used to find safe and efficacious compounds that could be selected as clinical candidates. It is at this stage where multidisciplinary approaches and multidimensional separation techniques are critically important. The increased purity requirements at this stage necessitate detailed analysis to ensure that the lead compounds are fully characterized.

The second concept to be addressed is how to *rapidly* develop methods [2] to enable structure characterization, purification, quantification, and chirality

TABLE 14.1 The Stage-Appropriate Analysis Model Applied for Hit-to-Lead and Lead Optimization Stages

	Hit-to-lead	Lead Optimization
Strategy	• Combichem-oriented support • Open-access support • Ensure the compound purity and integrity • Quick-and-dirty assay	• Medchem-oriented support • Open-access support • Perform stage-appropriate analysis but maximize the analytical information • Quick-and-detailed assay
Methodology	• High throughput techniques with the standardized platform • Fast generic method with "one-size-fits-all" approach • Mostly LC-MS based techniques together with FIA-NMR • Automation and miniaturization	• Multidisciplinary approaches (using LC, pSFC, CE, GC and their hyphenated techniques) • Multidimensional separation (Orthogonal fashions) • Multiple on-line detections (MS, NMR, ELSD and CLND, etc.) • Automation
Major Applications	• Structure confirmation • Purity assessment • Small scale purification (<50 mg) • ADME profiling	• Structure characterization and elucidation • Purity assessment • Small-to-medium scale purification (<50g) • ADME profiling • Special analysis (e.g. chirality separation)
Major Clients	• Combichem • HTS	• Medchem • HTS • Process chemistry

determination. For more than a decade, the traditional approach to high throughput analyses for the assessment of compound purity and structure confirmation has been to apply fast liquid chromatography–mass spectrometry (LC-MS) methods with very steep gradients, high flow rates, high temperature, short columns, and acidic volatile mobile phases. Despite the popularity and quite acceptable separation performance of this approach, which is supported by sound theoretical explanations [3–5], only so-called "quick-and-dirty" types of method development could be handled. This approach trades selectivity and resolution for high speed with the use of MS for highly selective analysis. The downside of this approach is experienced with the analysis of focused libraries that contain structurally related impurities with small differences in polarity, lipophilicity, or any other physicochemical property from the compound of interest. As a result, sample components that closely elute are observed. These components can compromise resolution during the preparative scale-up stage.

The third concept is to generate as much useful information as possible from a single analysis (e.g., to obtain both structural and quantitative information from a single run without tedious external or internal calibrations) [39].

Finally, one of the most difficult bottlenecks to overcome in drug discovery is the purification process. The successful introduction of an automated preparative LC-MS (prepLC-MS)-based high throughput purification (HTP) platform in the late 1990s [6,7] has enabled purification to be achieved quickly and easily. The introduction of this technology, however, only moves the bottleneck to the next step of the process [8]. The biggest bottleneck still remains with the tedious postpurification process that involves multiple steps (e.g., transferring fractions, solvent evaporation, purity assessment, weighing, and reformatting the fractions for HTS) to achieve the final products. Under the traditional model, laboratories focus on improvements with the purification or solvent-evaporation technologies per se instead of the entire purification platform. Therefore, the overall efficiency of HTP platforms has yet to be realized. A new concept is required to develop the next generation of purification infrastructure. This approach aims to integrate all fragmented purification-related platforms, such as open-access, HTP, large-scale, and specialized purification, into one highly efficient and functional process based upon a stage-appropriate analysis concept.

HIGH THROUGHPUT TECHNIQUES FOR STRUCTURE CHARACTERIZATION

How Fast Is Fast Enough?

Since the early 1990s, numerous research groups have reported on the use of automated flow-injection-analysis MS (FIA-MS) [9–10] and automated LC-MS [11] for high throughput analysis. The demands for rapid analytical tools that provide purity and identity information have pushed specialized coupled chromatography-MS techniques (e.g., LC-MS, supercritical fluid chromatography SFC-MS, capillary electrophoresis CE-MS) to the forefront, and have dramatically changed the traditional structure characterization protocols that use melting points, LC-UV, elemental analysis, and nuclear magnetic resonance (NMR). Efforts have been focused mainly on increased speed of analysis by the introduction of new column technologies, improved automation, as well as advanced hardware and software. However, the goal is to maintain good quality of data with the development of diversified high throughput techniques and implementation based on stage-appropriate analysis concepts.

Table 14.2 presents a high throughput analysis platform that consists of five major techniques, and indicates why and where specific analytical tools

TABLE 14.2 Summary of Major high Throughput Analytical Techniques in Support of Medchem

Technique	Speed (min/96-well plate)	Operation	Comments	Employed area
FIA-MS	**10**	• Needs re-plumbing the system to reduce the injection cycle	• Fastest way to confirm compound identity but no purity information	• Open-access
Single channel LC-UV-MS/ELSD	**200–300**	• Uses short column small particle size (typically 10–30 × 2.0 mm i.d., 2–3 u), steep gradient, high temperature (up to 60 C) and high flow rate (1 ml/min) • Uses monolithic column technology (50 × 4.6 mm i.d.)	• Most acceptable for both qualitative and quantitative analysis • Most flexible and easily automated • Smaller i.d. monolithic column is not yet available, and the column efficiency needs improving	• Open-access • Hit validation • Combichem and parallel synthesis QC • Lead optimization • Process development • Preclinical PKDM
Parallel LC-MS	**20–40**	• Fully automated system with 8 columns in parallel (MUX™ system)	• Best for high throughput analysis • Lack of flexibility	• Hit validation • Combichem and parallel synthesis QC
pSFC-MS	**10–200**	• Normal phase-like separation mechanism • Uses CO_2-MeOH-additive as mobile phase mostly coupled to APCI-MS	• Higher efficiency than LC • Orthogonal separation to LC • Needs more studies on separation and MS-ionization mechanisms	• Combichem and parallel synthesis • Lead optimization • Process development
CE and CE-MS	**300**	• A simple device requires small sample and buffer size • Orthogonal separation to LC	• Use for general purity screening but rarely employed in reality • Difficult to be coupled to MS	• Lead optimization • Process development

should be used. In an open-access environment, where a synthetic organic chemist conducts the high throughput analysis, fast LC-MS techniques (ca. 2 min/sample) have begun to replace flow injection analysis FIA-MS, even though ultrahigh speeds (ca. 10 s/sample) [12] have been achieved. LC-MS offers purity information and less matrix interference. Additionally, LC-MS is a more flexible and standard format that covers a wider range of applications as opposed to the customized, replumbed, ultrahigh speed FIA-MS setup. However, to assess the purity and integrity of large libraries or compound collections, the use of parallel LC-MS is considered to be the best choice. Due to its unsurpassed high throughput capacity, parallel LC-MS formats achieve good separation through the use of longer columns and gradients. With the emergence of SFC technology for purification, SFC-MS has become more popular for pre- and postpurification quality control (QC). The advantage of SFC-MS within the current discovery analytical arena is to introduce an orthogonal separation to LC-MS. Both techniques are complementary.

Parallel LC-MS

One of the most important recent innovations for discovery analytical support is parallel LC-MS with a multiplexed (MUX™) electrospray interface on an orthogonal time-of-flight (TOF) or a triple-quadrupole MS platform. The commercially available interface consists of four or eight multi-indexed rotating sprayers plus an extra fixed probe that is used to introduce an internal standard for exact mass measurement. Although the eluent from LC columns are simultaneously sprayed into the MS, the position of the sampling rotor allows the spray from only one probe to be admitted into the sampling cone [13]. An increasing number of applications have been reported due to its four- or eightfold increase of throughput for both qualitative and quantitative tasks, such as ADME analysis, structure characterization of proteins, peptides, and combinatorial libraries [14–17]. Recently, Eldridge et al. added eight evaporative light-scattering detectors (ELSD) onto this parallel LC-MS system for natural-product-libraries analysis to capture the structure information of those components that have no UV chromophores [18]. Two major concerns with this system are reduced sensitivity because of the limited cycle time per channel and the interchannel cross contamination. However, these concerns have been reported to be almost negligible for most applications [19,20].

A fully automated high throughput analysis platform that has been successfully established is shown in Figure 14.1. This platform is able to provide unattended, continuous operation, by control of the entire process (i.e., load plates, scan bar codes, sample injection, analysis, report generation, and data archival) with a single computer workstation [21]. The throughput is 1200 samples per day with acceptable chromatographic resolution. Moreover, when

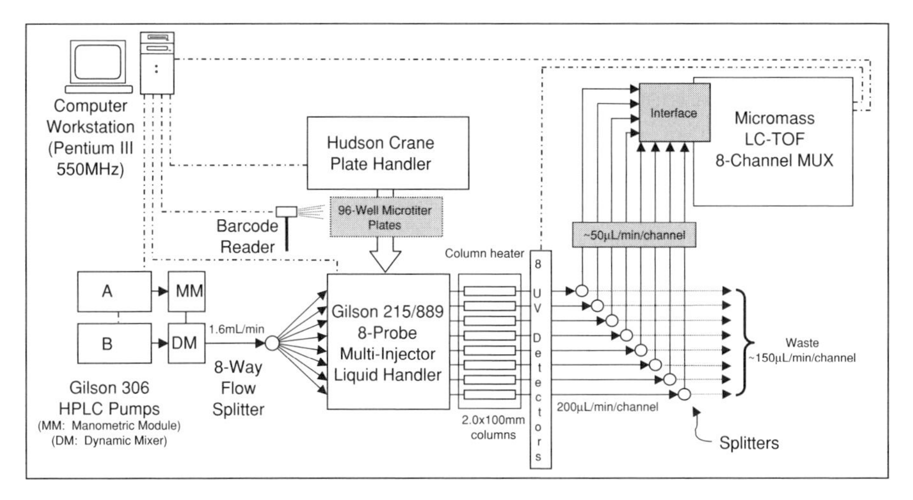

FIGURE 14.1 Scheme of a fully automated high throughput analysis platform based on a MUX system.

combined with a sample pooling strategy, this system has also been used to determine lipophilicity (logP) [22] and protein-compound binding affinity [23]. Both systems have been validated for routine operation.

SFC-MS

Since its introduction in the 1960s, SFC has experienced several ups and downs in its development. Either a gas or a liquid above its critical temperature and pressure is used as the mobile phase for SFC. In most cases, CO_2 is used because of its favorable critical parameters (i.e., a critical temperature of 31°C and a critical pressure of 7.3 MPa). Moreover, CO_2 is cheap, nontoxic, and nonflammable. A high-pressure pump delivers the mobile phase through either a packed (pSFC) or capillary column (cSFC) to the detector. The mobile phase is maintained under supercritical or subcritical conditions via an electronic controlled variable restrictor that is positioned after detection (pSFC) or via a fixed restrictor positioned before a gas-phase detector (cSFC). The retention characteristics of the analytes are influenced by the properties of the stationary phase and by the polarity, selectivity, and density of the CO_2 mobile phase. The density is controlled by variation of the temperature and pressure of the supercritical medium. Furthermore, the elution of very polar compounds under high densities can be achieved with a precolumn addition of polar modifiers such as methanol. Nowadays, pSFC formats use the same injector and column configurations as LC methods. Consequently, pSFC formats are considered to be more useful for routine operation than cSFC. The most remarkable advances of pSFC in drug discovery are the application of pSFC-MS for high throughput analysis and the use of automatic preparative SFC for achiral/chiral compound purification.

Commercial pSFC instruments are easily coupled to electrospray ionization (ESI) or atmospheric pressure chemical ionization (APCI) mass spectrometry with only a very minor modification. Pure CO_2 does not produce ions under ESI or APCI conditions. Furthermore, CO_2 does not play any direct role in ion formation due to the lack of CO_2-derived primary ions. Therefore, it is essential to add postcolumn modifiers (e.g., MeOH or other solvents) into the CO_2 stream. This is especially true with pSFC/ESI-MS to introduce a solution chemistry mechanism where (de)protonated or adduct-based molecular ions of the analytes are produced.

Because of its much lower viscosity, higher diffusion coefficients, and lower column pressure drop than LC, pSFC offers a number of significant advantages. For example, pSFC methods can provide a 5–10 times faster analysis, threefold higher efficiencies, and facilitate the use of longer columns [24]. The shorter run times of pSFC-MS, coupled with the greater resolving power of SFC relative to LC makes this technique a very powerful tool for high throughput analysis for both qualitative [25] and quantitative tasks [26]. Generally, pSFC-MS, is used with LC-MS to provide two sets of separation profiles for each sample. This feature is extremely helpful for purity assessment. Figure 14.2 shows the separation profiles of two samples obtained from parallel synthesis. The impurity in sample A is separated by LC, but co-elutes with the parent compound in SFC, as opposed to sample B. Therefore, two different methodologies are often required to maximize the selectivity to assess compound purity.

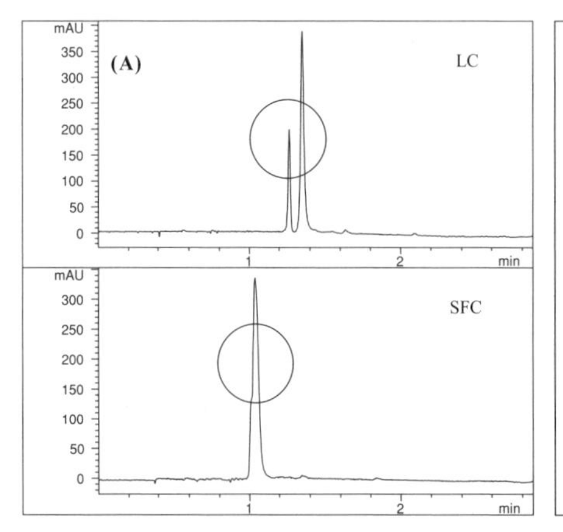

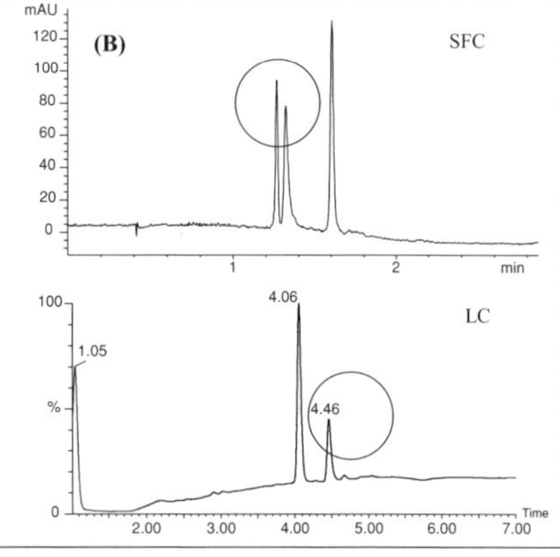

FIGURE 14.2 High throughput analyses of samples A and B using LC-MS vs. SFC-MS.

NOVEL APPROACHES TO RAPID METHOD DEVELOPMENT

The conventional "one-at-a-time" and "one-size-fits-all" approaches of method development are inefficient and are unable to satisfy increased demands of drug discovery. Therefore, some novel approaches have been effectively and extensively implemented, as shown in Figure 14.3.

Multidisciplinary Techniques

The traditional one-at-a-time fashion of method development is usually based on one technique (mostly LC) and one mobile phase system with very long gradients to gain the maximum resolution and selectivity. However, the main challenge in drug discovery is to find the best solution with minimum turnaround time. An effective approach to method development is to perform different techniques on different platforms in parallel. As a result, a series of methods are screened with different separation techniques (e.g., LC, SFC and CE) to effectively and efficiently obtain optimal separation conditions. Two novel parallel-method screen strategies for achiral (Figure 14.4A) and chiral (Figure 14.4B) separation have been constructed. The entire method screen system is automated with a simple column- and buffer-switching device. The

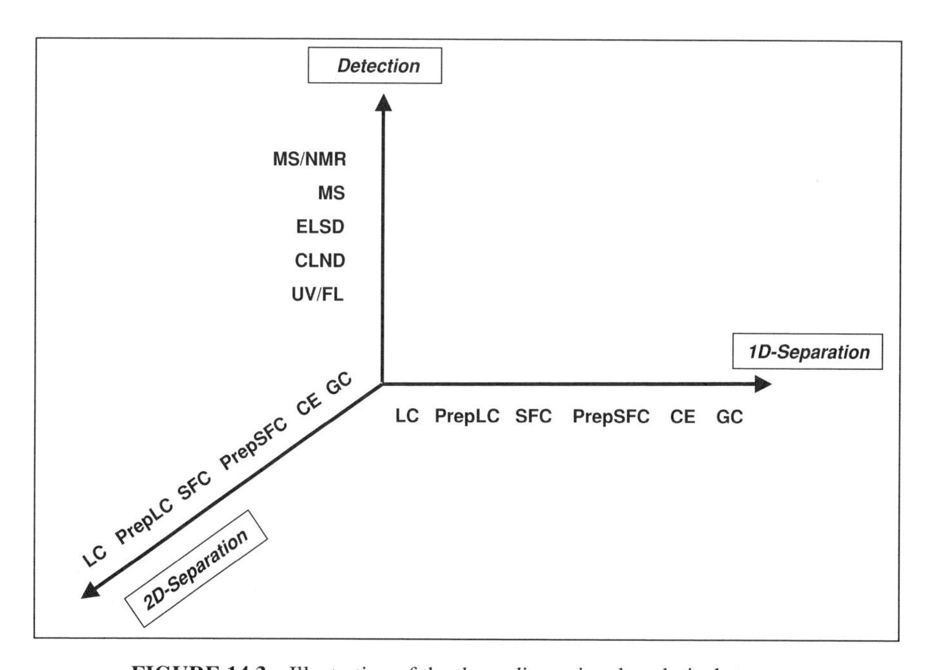

FIGURE 14.3 Illustration of the three-dimensional analytical strategy.

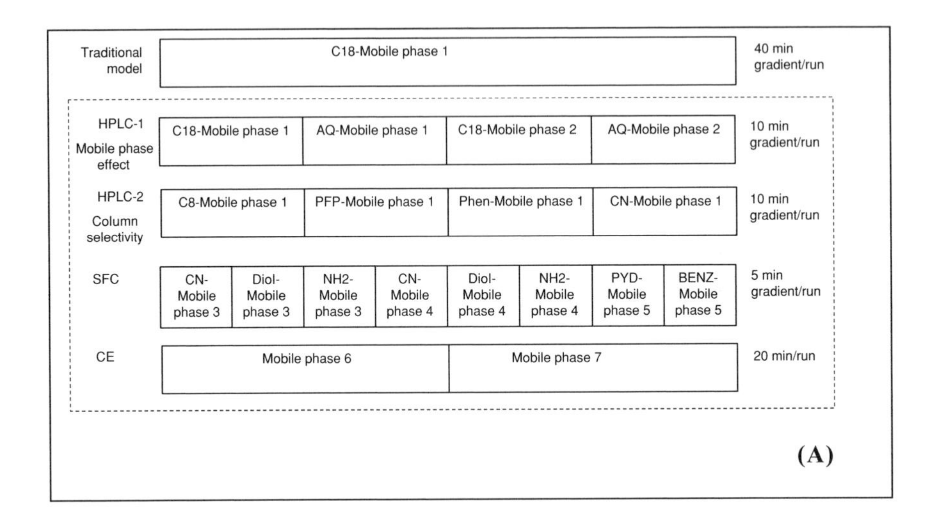

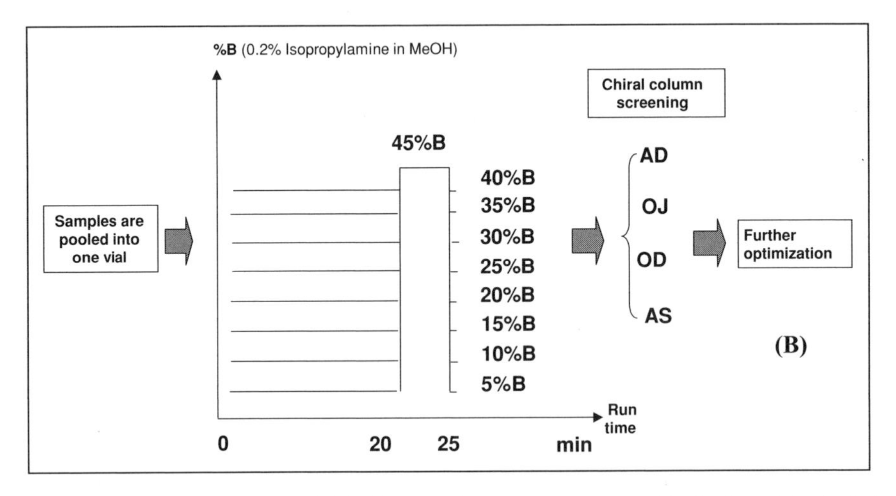

FIGURE 14.4 (A) Achiral- and (B) chiral-method screen strategies. (A) Platform 1 (HPLC-1) uses two different mobile phases, one is acidic (mobile phase 1) and another is neutral-to-basic (mobile phase 2) to screen on two columns, a classic reverse-phase C-18 and a special polar group–embedded C-18 (AQ). Platform-2 (HPLC-2) uses one mobile phase (pH is usually acidic) to screen four different columns of wide range of polarity (PFP: fluorinated, and Phen: phenyl-hexyl phases). Platform-3 (SFC) uses three different mobile phases and five different columns (PYD: 2-ethylpyridine, BENZ: benzamide). Platform-4 (CE) uses two different mobile phases. (B): AD, OJ, OD, AS are different polysaccharide-based chiral stationary phases.

technical details are not discussed in this chapter, but have been published elsewhere [27,28].

As shown in Figure 14.4A, a traditional approach can only generate one achiral method within 40 minutes versus 18 methods from four technique platforms previously described. The purpose of this approach is to examine the pH effect on the selectivity for relatively polar and apolar analytes and to identify the most appropriate stationary phases. pSFC and CE are also used as a method screen due to their orthogonal separation mechanism to the reverse-phase liquid chromatography (RP-LC). Figure 14.5A illustrates different impurity profiles of the same sample generated from LC, pSFC, and CE method screening platforms. Some critical issues such as purification scale-up, buffer restriction for some bioassays, method transfer, and method validation are also taken into early consideration in this method screening stage. Based on the state-appropriate analysis concept and knowledge of the chemistry, one or two platforms can be selected instead of application of the whole system to avoid data redundancy.

There are three major challenges associated with the rapid development of chiral methods: (1) the significant number of compounds requested for chiral separation on a daily basis; (2) the wide variety of compounds that correspond to starting materials, intermediates, or drug substances; and (3) the achiral purity of the compounds can often be less than 90%, sometimes barely over 70%. Achiral impurities can severely interfere with the chiral-method development. pSFC is currently considered the primary choice over LC and CE for chiral separation. Previous approaches with SFC have not taken advantage of the MS selectivity and sensitivity to help develop chiral methods or to determine enantiomeric excess (EE%) in a complex mixture. Therefore, Zhao et al. [28] have developed a chiral screening strategy, which is illustrated in Figure 14.4B. The extracted ion chromatogram (EIC) function of the MS is used to distinguish the enantiomers. Sample pooling substantially increases the throughput without fear of interferences from achiral impurities, with the exception of structural isomers. This screen is performed under eight gradients on four chiral stationary phases (32 conditions). As an example, Figure 14.5B shows the chiral separation profiles of six pooled racemic samples obtained on a chiralpak-AD column at 20% B.

The downside of this multidisciplinary approach, however, is that a tremendous amount of information is produced, and consequently, data processing and interpretation are complicated. Therefore, another approach that can be used to maximize the resolving power of chromatography is to introduce a multidimensional separation with column-coupling or column-switching (CS) as an alternative procedure, especially for purity assessment.

Multidimensional Separation

Multidimensional separation approaches can be incorporated on-line via CS or off-line via fraction collection. Developed in the early 1980s, this technique has been mainly used to isolate multiple components that are contained in a complex mixture [29,30], particularly for natural product, petrochemical, and environmental analyses. To date, the most important applications involve trace enrichment and sample cleanup with environmental and pharmaceutical

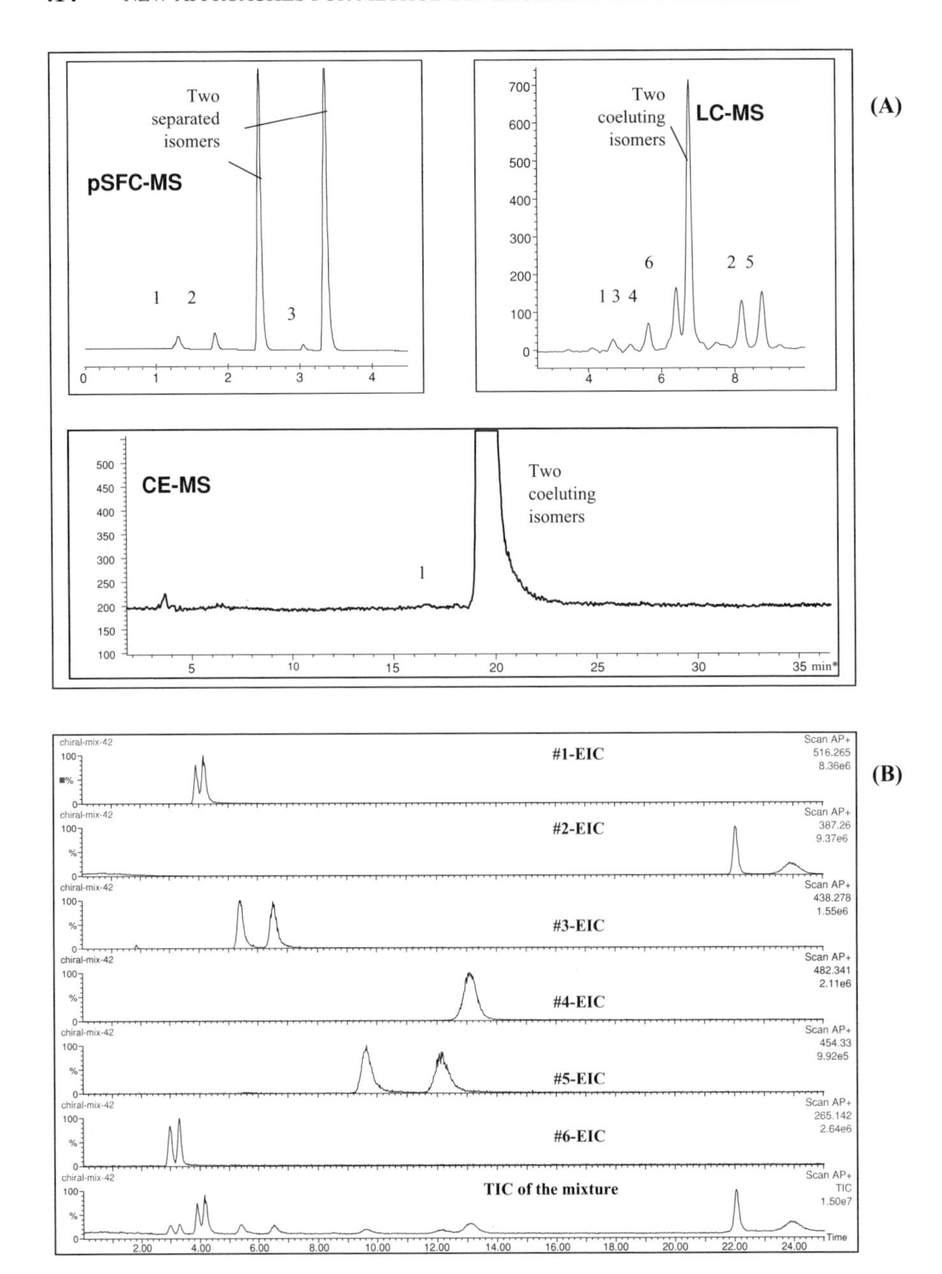

FIGURE 14.5 (A) Different separation profiles of impurities generated from the LC, pSFC, and CE parallel-method development platforms [66]. (B) Chiral separation profiles of six pooled racemic samples on a chiralpak-AD column at 20% B run as one method using the strategy presented in Figure 14.4B [28].

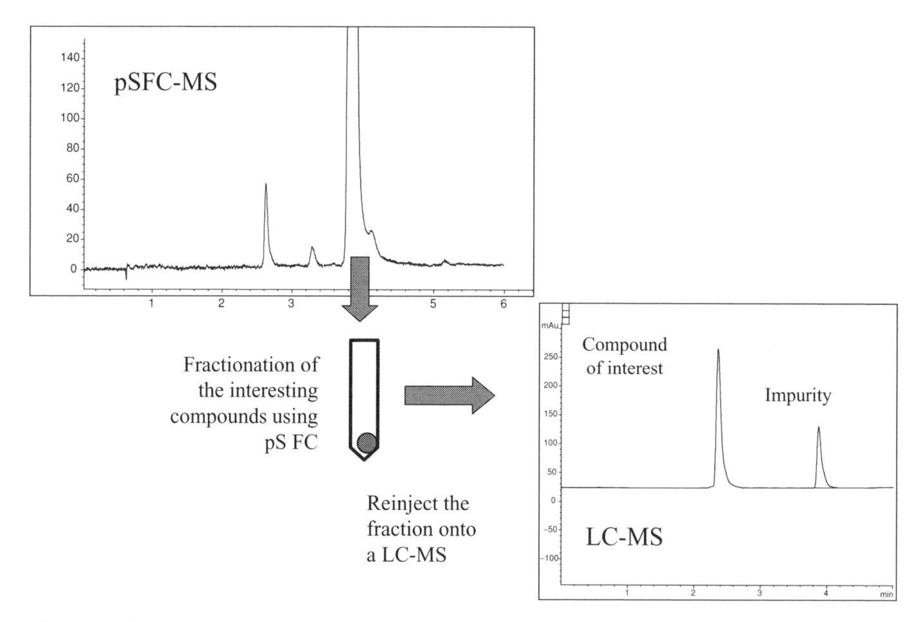

FIGURE 14.6 An example of performing an off-line pSFC/LC-MS for purity assessment and structure analysis [66].

analyses that demand accurate determination of analytes at very low concentrations in complex matrices [31]. The recent interest with LC-LC-MS for proteomic studies has increased rapidly and been extensively reported [32,33].

The purpose of multidimensional chromatography approaches is to provide optimal efficiency and selectivity for separation of the component of interest, while the analysis time is minimized. Generally, this approach helps to validate that no other compounds co-elute with the compound of interest. Krstulovic et al. [34] and Venkatramani and Zelechonok [35] have recently proposed several simple on-line LC-LC CS designs for the separation of co-eluting impurities. Sandra et al. [36] have introduced an on-line and off-line SFC-SFC system for both analytical and semipreparative purposes. Figure 14.6 shows the separation of minor impurities (<2% in total) along with the compound of interest with pSFC-MS followed by reinjection of the fraction that contains the major peak onto a second LC-MS. The final result shows a clear separation of a significant amount of impurity (30%) that coelutes with the major peak on LC. Since fractionation by pSFC has a significantly smaller fraction size than LC due to the use of CO_2 and methanol, sample preconcentration on the secondary column becomes more efficient.

Multiple On-Line Detection

To maximize the information that can be obtained from each analysis, two multifunctional analytical platforms that use multiple detectors were established:

(1) LC-UV-ELSD/chemiluminescence nitrogen detector (CLND)/MS [2] and 2) LC/UV/MS/NMR [37]. The LC/UV/ELSD/CLND/MS platform allows for structure identification and quantitative analysis. The biggest advantage of the LC/UV/ELSD/CLND/MS platform is that the four detectors can complement each other to provide both qualitative and quantitative information. Furthermore, synthesis of each analyte standard to prepare a calibration curve for quantification is not required. Several groups have reported the use of this platform for quality assessment of combinatorial chemistry libraries [38], hit validation [38], and metabolite characterization and quantification [39]. CLND detection systems are restricted to solvents that do not contain nitrogen, such as methanol, which can affect the chromatographic separation. To circumvent this limitation, some groups have attempted to implement pSFC/UV/ELSD/CLND/MS with mostly alcohols with CO_2 as the eluent [25]. However, these early attempts were not really practical until a new generation of pSFC columns was introduced [40]. The new columns allow for improved selectivity and resolution without the use of basic additives such as diethylamine (DEA).

LC-MS does not always provide unambiguous structure identification even when the molecular and fragment ions are identified from MS spectra. In many cases, the direct coupling of LC/NMR is required to obtain detailed structural information, especially when structural, conformational, and optical isomers need to be identified. Conversely, NMR provides only a partial solution when the molecule under investigation contains structures that lack NMR resonances [41]. Nonetheless, neither LC-MS nor LC/NMR alone can provide complete structure information for all compounds. Therefore, it is logical to integrate LC-MS and LC/NMR into a single combined LC-MS/NMR platform for unequivocal structure elucidation. Since the first published results of LC-MS/NMR appeared in 1995 [42], the groups of Bayer-Albert et al. [43] and Wilson et al. [41,44] have made tremendous contributions to the hardware improvement and pharmaceutical application for the structure determination of metabolites, degradants, and impurities. The integration of LC-MS and LC/NMR platforms into the daily structural analysis work flow is illustrated in Figure 14.7. Figure 14.7 shows an example of both platforms used to help identify and quantify the impurities present in a batch submitted for a toxicology study. Four major impurities (each at ca. the 0.2–0.3% level by UV response) were first detected and characterized by LC/UV-ELSD/CLND/MS with $MeOH/H_2O/0.1\%$ trifuroroacetic acid (TFA) as eluent (Figure 14.7A). It was observed from the MS spectra that all impurities contained the same number of nitrogen atoms, which led to the determination of their concentration based on the CLND response. Further structure elucidation of the four impurities was achieved with LC-MS/NMR (Figure 14.7B); as a 500-μL injection, a loop-storage mode was featured. The method is transferred between the two

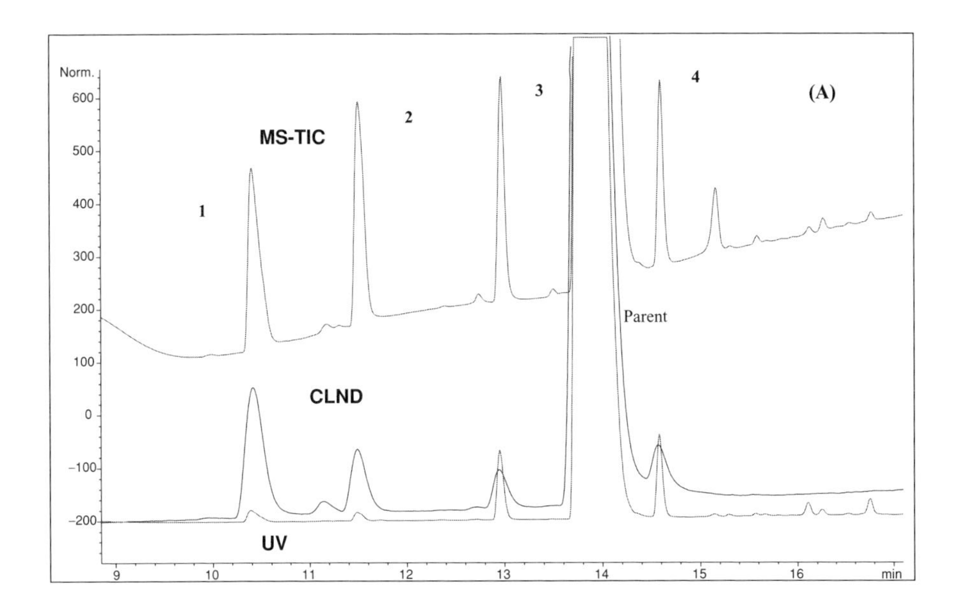

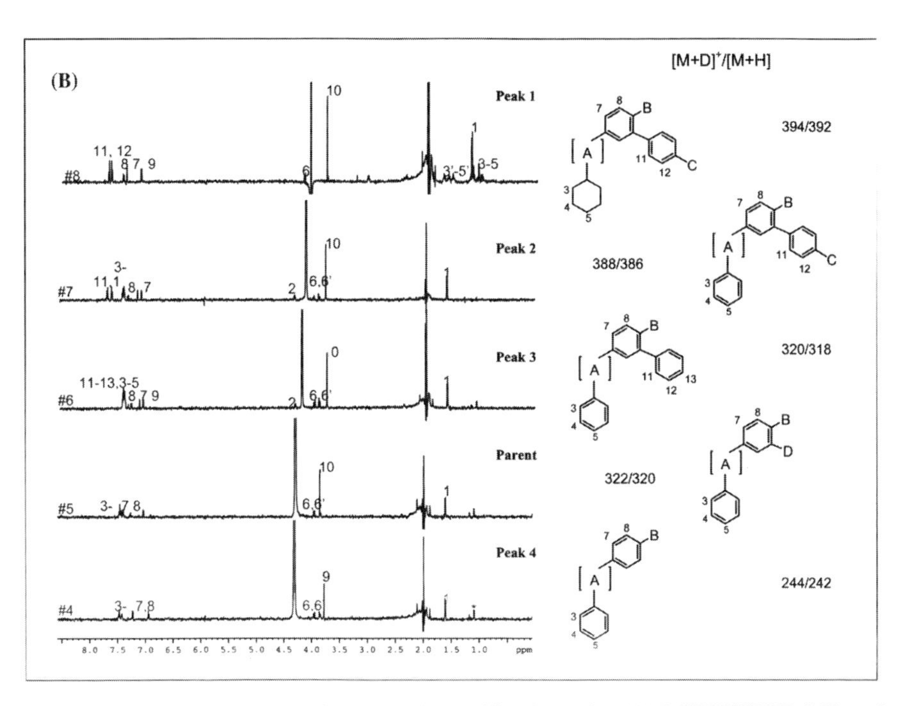

FIGURE 14.7 Impurity identification and quantification using (A) LC/UV/CLND/MS, and (B) LC/UV/MS/NMR.

platforms and the previous eluent was replaced by MeCN/D_2O/0.1%TFA, which in this case causes only minor changes in resolution and peak shape, while the elution order is maintained.

INTEGRATED PURIFICATION PLATFORMS

Integrated purification platforms are simply defined as the informatics and compound work-flow continuum from the purification system to the subsequent assay. Figure 14.8 depicts a modular approach for an integrated purification platform [45]. These modules generally contain both compound and information-processing components. The emergence of high throughput UV- and MS-triggered purification platforms in the late 1990s [46–49] led to an increased burden on the postpurification processes. A postpurification process requires integrated work-flow solutions to track, reformat, or rearray crude and purified compounds. Reynolds et al. [50] have developed an entire postsynthesis work flow that accommodates different purification techniques based on different purity requirements. A postpurification process based on prepLC-MS and stand-alone liquid handlers and robotic weigh stations [45] has been developed with pre- and postpurification QC conducted on a parallel LC-MS.

There are many requirements for an integrated purification platform. First, the platform must accommodate single as well as multiple compounds for either high throughput or late-stage analyses. Second, the platform must be scalable to accommodate milligram to kilogram amounts of material. Third, the platform must be able to deliver multiple formats to multiple groups. Fourth, the platform must be able to process a range of required purities based

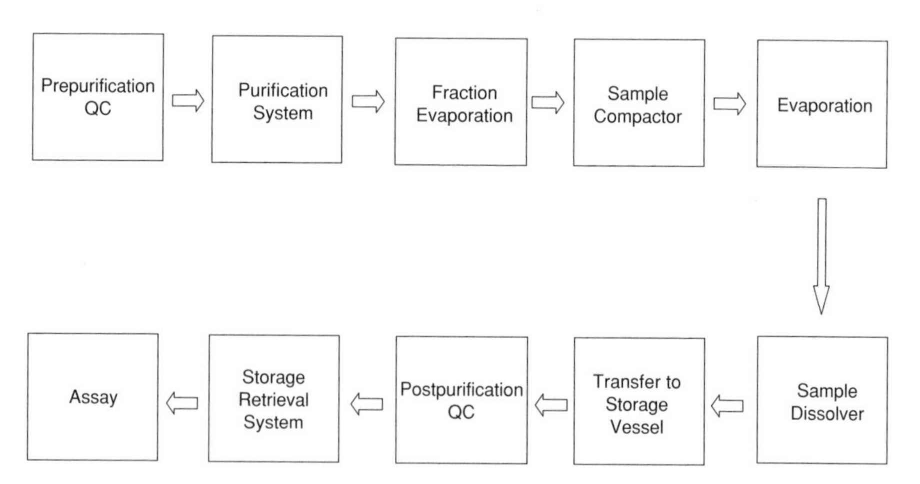

FIGURE 14.8 A high-level purification platform architecture diagram. The modular components represent compound and informatics-processing routines.

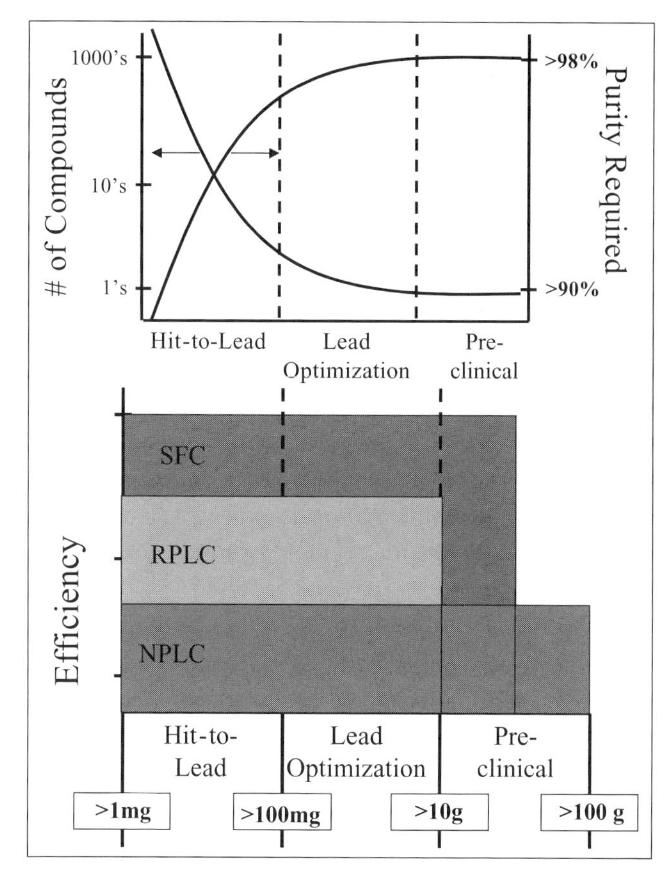

FIGURE 14.9 Stage-appropriate purification.

on the stage-appropriate purification philosophy. As shown in Figure 14.9, the approximate purity that is required from the hit-to-lead stage to the preclinical stage generally increases from >90% to >98%. A representation of purification technique that is based on efficiency and quantity of material is also depicted in Figure 14.9. Some recent technologies, such as preparative SFC (prepSFC) and the automated workstations from Symyx Technologies, have provided increased capability and efficiency. These technologies are discussed in the following section.

Preparative SFC

As illustrated in Figure 14.9, prepSFC has demonstrated a wider coverage of purification and higher efficiency than LC in the lead optimization stage. Compared to analytical-scale pSFC-MS, prepSFC has long been adopted to purify fatty acids esters, synthesis intermediates, steroids, fullerences, and most importantly, chiral compounds [51,52]. Most systems have been designed for use with pure CO_2 as the mobile phase and with a cyclone separator to collect a single or a few components from a feedstock that remains unchanged. Although there are numerous examples of "semiprep" scale pSFC (tens of milligrams up to grams) with CO_2 and a polar modifier (e.g., MeOH) for complex mixtures and enantiomeric purification [36], these methods suffer from the loss of 30–50% of the analytes, as aerosols, during fraction collection. These losses are due to the difficulty associated with the separation and methods used to trap the sample quantitatively from the eluent. The sample is micronized by rapid depressurization at the system outlet, and the sample tends to be carried away with the large volume of the high-speed gas phase [51]. Only until after Berger et al. [53] and Perrut et al. [52] developed new hardware designs to effectively trap the samples on-line for semi- and large-scale purification, did prepSFC become useful for medicinal chemistry compound purifications.

The critical problem with this approach, however, is with the integration of SFC technology into the existing HTP platform, especially since the current prepSFC purification is a UV-triggered system. Although some exciting breakthroughs recently have been reported on the development of a parallel mass-triggered prepSFC-MS purification system to maximize the purification throughput [54,55], the commercialization of these techniques and improvement of hardware/software reliability have a long way to go. Due to the lack of adequate attention and research resources from industry and academia, there remains a significant gap between theoretical studies and real-world applications.

Recent efforts focus on the investigation of new pSFC column technologies to achieve good selectivity and peak symmetry with no additives (mainly basic additives such as DEA) in pSFC. The purified fractions will remain in pure methanol solution before the dry-down stage and offers a significant advantage over prepLC, which uses TFA as a common additive. Figure 14.10A presents some new column stationary phases [56]. The ethyl-pyridine–type stationary phase has proved to be able to replace traditional phases such as CN and diol for HTP [57]. The ethyl-pyridine stationary phase shows no tailing effect and even better selectivity (Figure 14.10B), with only methanol as the modifier. More theoretical studies are needed to determine the separation mechanisms on those stationary phases. These studies are expected to provide an in-depth understanding of solute–stationary-phase interaction under pSFC conditions.

Symyx™ Automated Workstation

A prerequisite for compound scale-up in clinical trials is to avoid the time-consuming and costly step of chromatographic-based purification. An efficient

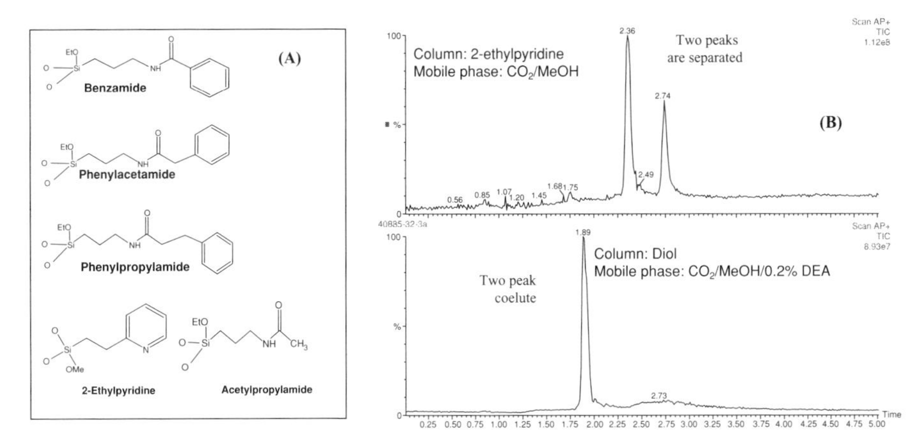

FIGURE 14.10 (A) Some new pSFC stationary phases under investigation. (B) An example of using 2-ethylpyridine vs. diol columns for basic compounds separation [66].

transition from medicinal chemistry to process chemistry requires an understanding of the solubility of impurities, key intermediates, and final product. The goal of process chemistry is to develop a robust and reliable streamlined process from the starting materials that are easily obtained to the final product. The final-product polymorph form and particle-size properties are dictated by the final step of product isolation. The final-product isolation step often involves recrystallization as the workup conditions. The workup conditions and subsequent fine-tuning of the conditions must be completely understood for material scale-up. In addition to the issue of scalability, efficient method transfer processes are involved with facilities responsible for scale-up (i.e., kilo labs, pilot plants). And now, with the need to understand the workup conditions earlier in the discovery cycle, analytical chemists are responsible for the decision-support process with medicinal and process chemistry. The batch-to-batch variability that is often observed with the scale-up from a medicinal chemistry bench to process chemistry can significantly affect the polymorphic forms and inherent solubility. As a result, the salt-selection process is often used as a tool to increase the solubility of the drug substance.

The Symyx^{TM} technology [58] allows a greater number of experiments to be conducted earlier in the process and with smaller compound amounts of material compared to manual methods [59]. Symyx has been able to significantly increase the number of experiments and solvent combinations to investigate solubility profiles, salt formations, and polymorph forms with the integration of commercial and sometimes slightly modified hardware components (e.g., powder and liquid dispensers, LC, X-ray, Raman, and birefringence microscopy as well as through client-server applications. Figure 14.11 depicts

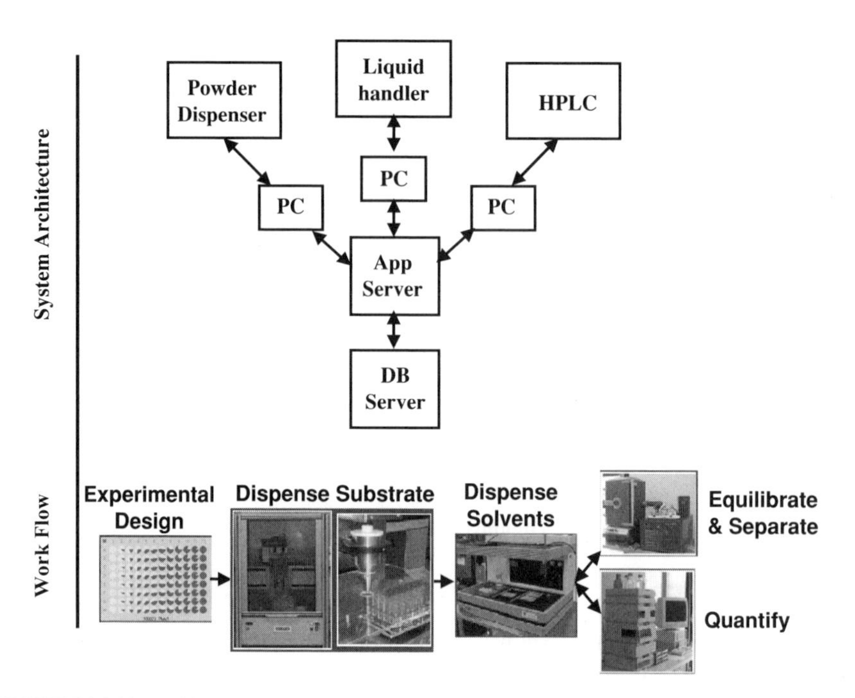

FIGURE 14.11 A high-level representation of the Symyx solubility system architecture. The work flow from experimental design in Library Studio through quantitation by LC is shown in the bottom panel.

the high-level instrument architecture for the Symyx solubility workstation [58]. The application server acts as the hub and distributor of a coordinated informatics flow that use independent client computers to control the commercial hardware. The information is passed and received among the hardware devices through extensible markup language (XML) documents. Symyx has developed a software package, Library Studio®, which is user-friendly and powerful for experimental design. The input parameters to Library Studio are used to drive the hardware devices.

The input parameters define the compound quantities and positions, solvents, temperatures, and equilibration time. The first step of the work flow is to dispense the compound with an automated powder dispenser. The parameters from Library Studio are used to drive a powder dispenser from Autodose (Geneva, Switzerland). The next step of the work flow requires the liquid handler to take the parameters outlined in Library Studio and to dispense the solvents across the plate. Besides the routines to aspirate and dispense, the liquid-handler platform has the capability to stir and control temperature. Concurrently, while this part of the work flow is executed, the compound standards, LC methods, and calibration curves are generated for quantification purposes. After equilibration on the liquid handler, the next step in the work flow requires a separation procedure. Separation procedures can be either centrifugal or filter based. The separation procedure is used to ensure that the excess compound is separated from the solvent under saturated conditions to avoid contamination with the subsequent quantitation step. The final step in the work flow is the quantitation-based analysis with LC. The parameters from Library Studio are transferred to the LC and resultant peak areas, based on experimental conditions (e.g., reaction volumes, dilution ratios, injection volumes), and are used to calculate the solubility of the compound or compounds as a function of the solvent(s), temperature, and time.

This system architecture enables a streamlined work flow that significantly decreases the amount of time needed for an experiment. The technology enables order of magnitudes more experiments (i.e., combination of conditions) to be conducted in the same period as done manually. The power of the technology is the ease at which design, execution, and analysis experiments can be performed across a platform of hardware components. If these experiments were conducted manually, then these processes would be extremely time-consuming and laborious. Furthermore, fewer combinations would be investigated.

The Symyx solubility workstation has been integrated into the medicinal and process chemistry work flows at Amgen [60]. The technology has allowed for the investigation of numerous parameters that would otherwise not be assessed during a first-pass analysis. Experiments on lead optimization candidates to study single- and multiple-solvent effects (e.g., co-, tri-, antisolvent, etc.) as a function of temperature, time, and pH have been conducted.

CONCLUSIONS AND FUTURE OUTLOOK

Emerging Technology

The future technological advances in drug discovery will likely involve separation sciences, MS, and hyphenated techniques. More detailed discussions on the technology development and future trends in bioanalysis, laboratory automation, and MS instrumentation have been elaborated in other chapters. Therefore, only three technologies are specifically addressed with regard to their potential application in drug discovery: pSFC; monolithic column technologies; and chip-based separations.

As previously outlined in this chapter, pSFC and its related techniques have become more widely accepted over the last few years. However, for pSFC to become a more widely used and universal technique such as LC, will likely require collaborative efforts among instrument vendors, academia, and industrial research laboratories. These efforts must address two main issues. First, the redesign of the hardware infrastructure of pSFC is required. Second, the fundamentals of the separation mechanism will need to be understood to fully exploit the advantages of pSFC over LC.

In recent years, there have been a number of reports by Svec et al. [61] and Iberer et al. [62] on the monolithic columns in terms of monolithic materials preparation, properties, and applications. There are several advantages of monolithic column technologies over common columns packed with silica particles. First, the higher column permeability of monolithic columns accelerates the rate of mass transfer and enables a substantial increase in the separation speed [63]. Second, polymer-based monolithic columns can be fabricated into different formats such as disks and tubes. These unique formats can be used for many separation modes, such as ion exchange, hydrophobic interaction, affinity, and reversed phase, which has been proved very useful for the significant enhancement of analysis speed of large molecules. Another essential prospect is associated with developments of capillary electrochromatography (CEC) and chip-based separation devices. Monolithic material can be easily fabricated in situ without the use of retaining frits.

Significant academic research on miniaturized separations has been carried out since the early 1980s. Although miniaturized separation techniques have been one of the hottest topics in the industry for the past 10 years [64], the format has yet to be accepted in drug discovery. Because the separation format does not change the separation mechanism, there is a long way to go in terms cost reduction, column-packing improvements, and the prospects of hyphenated techniques that feature miniaturized separations. A novel product from Nanostream, Inc. (Pasadena, California) has recently become commercially available. This device can be regarded as a prototype model with desirable features (e.g., high parallel capability, etc.). A diagram of the Nanostream™ microfluidic cartridge [65] (160 × 270 mm L) with 24 LC separation channels (80 × 0.5-mm ID per channel) is depicted in Figure 14.12A and 14.12B. Good column-to-column reproducibility is demonstrated with an injection of a mixture that contains five standards.

Infrastructure

Analytical techniques and methodologies will no doubt constantly evolve. The real challenge in the future will not come exclusively from the implementation of technology into the work flow, but rather will be the design of an overall analytical laboratory architecture that can support, drive, and exploit a diverse array of enabling technologies. An example of a high-level system architecture that can be used to enable the previously discussed philosophies is shown in Figure 14.13. This architecture is designed to enable automated systems to

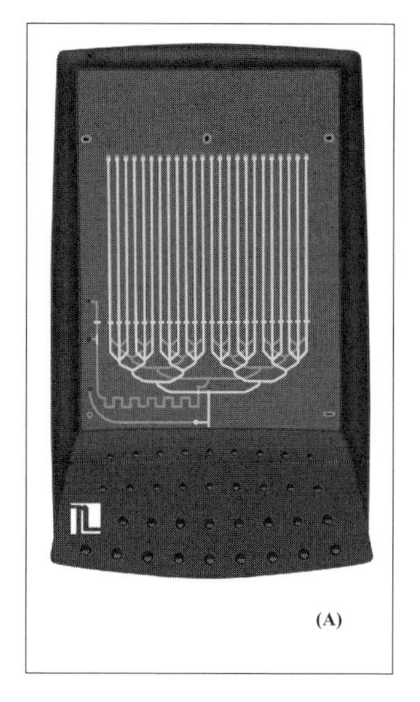

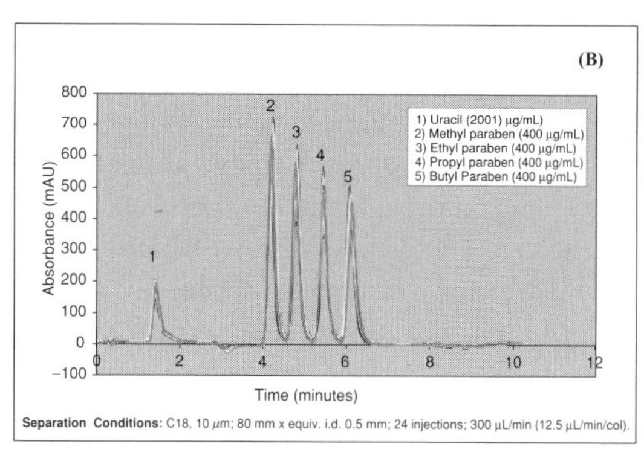

FIGURE 14.12 (A) A picture of the Nanostream microfluidic cartridge with 24 LC separation channels. (B) A demonstration of good column-to-column reproducibility by injecting a five-standard mixture. (Reprinted with permission from Nanostream Inc.)

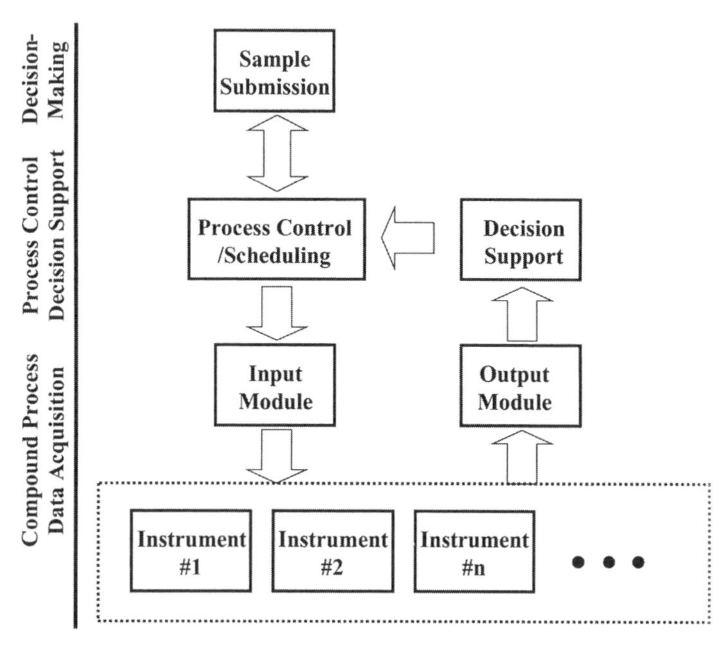

FIGURE 14.13 A high-level analytical laboratory system architecture diagram. The modular components represent compound and informatics-processing routines.

be fully used through informatics work-flow solutions. Three key advantages include: (1) a modular approach that enables a laboratory to leverage the incorporation of new technologies without compromising core business practices; (2) the scalable approach that is aligned to the stage-appropriate philosophy; and (3) the approach enables focus-specific skill sets to be applied to the different layers and modules. To benefit from these advantages, an analytical chemist in the future must be highly skilled not only with the comprehension of multidisciplinary analytical techniques but also with the development of work flows through integrated software, hardware, and decision support tools. The successful groups will be able to reflect on the advancements made over the past decade to adopt new ways to think or conceptualize a solution and effectively change the way analytical support is provided to medicinal chemistry in the future.

ACKNOWLEDGMENT

We thank D. T. Chow, Dr. B. Marquez, Dr. L. Poppe, and D. Smith for their contributions to LC-MS/NMR; Dr. K. Gahm, Dr. P. Grandsard, and S. Thomas for their support on the pSFC-MS and HTP platforms; H. Tan for his contribution to the Symyx™ platform; Dr. P. Schnier and Z. Hua for their help on LC/CLND/MS. Special thanks go to Greg Woo for his tremendous contribution to the IT infrastructure, and to Dr. J. Cheetham for the review of this manuscript. Finally, one of the authors (Y. Z.) thanks Prof. Dr. P. Sandra (University of Ghent, Belgium) for his valuable discussion on SFC-MS and Dr. H. Rong (Merck Research Laboratory) for her inspiration and valuable input on bioanalysis.

REFERENCES

1. Minter, B.; Bolten, B.M. “Combinatorial Chemistry Industry Trends,” *Spectrum, Drug Disc. Design* **8,** 1–14 (2001).
2. Zhao, Y. “New Strategies and Methodologies for Rapid Method Development in Support of Lead Optimization,” oral presentation at *5th Annual Symposium on Chemical and Pharmaceutical Structure Analysis (CPSA2002)*, Princeton, NJ, October 8–10, 2002.
3. Wehr, T. “Fast LC for High-throughput LC-MS,” *LC-GC North America* **20**(1) 40–47 (2001).
4. Mutton, I.M. “Fast Generic HPLC Methods,” pp.73–85 in Valko, K., ed., *Separation Methods in Drug Synthesis and Purification*, Elsevier Science, Amsterdam, (2000).

5. Sandra, P.; Vanhoenacker, G.; Lynen, F.; Li, L.; Schelfaut, M. "Considerations on Column Selection and Operating Conditions for LC-MS," *LC-GC Europe: Guide to LC-MS* December 2001, pp. 8–21.
6. Kassel, D.B. "Combinatorial Chemistry and Mass Spectrometry in the 21st Century Drug Discovery," *Chem. Rev.* **101**(2), 255–268 (2001).
7. Kyranos, J.N.; Cai, H.; Wei, D.; Goetzinger, K.W. "High-throughput Techniques for Compound Characterization and Purification," *Drug Disc. Today* **6**(9), 471–477 (2001).
8. Underwood, T.; Boughflower, R.J.; Brinded, K.A. "The Development and Industrial Application of Automated Preparative HPLC," in Valko, K., ed., *Separation Methods in Drug Synthesis and Purification*, Elsevier Science, Amsterdam, 2000, pp. 293–335.
9. Taylor, L.C.E.; Johnson, R.L.; Raso, R. "Open Access Atomospheric Pressure Chemical Ionization Mass Spectrometry for Routine Sample Analysis," *J. Am. Soc. Mass Spectrom.* **6**(5), 387–393 (1995).
10. Hayward, M.J.; Snodgrass, J.T.; Thompson, M.L. "Flow Injection Thermospray Mass Spectrometry for the Automated Analysis of Potential Agricultural Chemicals," *Rapid Commun. Mass Spectrom.* **7**(1), 85–91 (1993).
11. Pullen, F.S.; Richards, D. "Automated Liquid Chromatography/Mass Spectrometry for Chromatographers," *Rapid Commun. Mass Spectrom.* **9**(2), 188–190 (1995).
12. Morand, K.; Burt, T.M.; Regg, B.T.; Chester, T.L. "Techniques for Increasing the Throughput for Flow Injection Mass Spectrometry," *Anal. Chem.* **73**(2), 247–252 (2001).
13. Sage, A.B.; Little, D.; Giles, K. "Using Parallel LC-MS and LC-MS-MS to Increase Sample Throughput," *LC-GC* **18,** S20–S29 (2000).
14. Yang, L.; Wu, N.; Rudewicz, P.J. "Application of New Liquid Chromatography–Tandem Mass Spectrometry Technologies for Drug Development Support," *J. Chromatogr. A* **926,** 43–55 (2001).
15. De Biasi, V.; Haskins, N.; Organ, A.; Bateman, R.; Giles, K.; Jarvis, S. "High Throughput Liquid Chromatography/Mass Spectrometric Analyses Using a Novel Multiplexed Electrospray Interface," *Rapid Commun. Mass Spectrom.* **13**(12), 1165–1168 (1999).
16. Rudewicz, R.J.; Yang, L. "Novel Approaches to High-Throughput Quantitative LC-MS/MS in a Regulated Environment," *Am. Pharm. Rev.* **4**(2), 64–70 (2001).
17. Feng, B.; Patel, A.; Keller, P.M.; Slemmon, J.R. "Fast Characterization of Intact Proteins Using a High-throughput Eight-channel Parallel Liquid Chromatography/Mass Spectrometry System," *Rapid Commun. Mass Spectrom.* **15**(10), 821–826 (2001).
18. Eldridge, G.R.; Vervoort, H.C.; Lee, C.M.; Cremin, P.A.; Williams, C.T.; Hart, S.M.; Goering, M.G.; O'Neil-Johnson, M.; Zeng, L. "High-throughput Method for the Production and Analysis of Large Natural Product Libraries for Drug Discovery," *Anal. Chem.* **74,** 3963–3971 (2002).

19. Corens, D. "The Role of LC-MS in Drug Discovery," *LC-GC Europe: Recent Applications in LC-MS*, November, p. 5. (2002).
20. Zhao, Y.; Chow, D.T.; Wilson, J.; Woo, G.; Semin, D.J. "Investigation of a High Throughput Parallel LC-UV-TOF-MS for Small Molecule Drug Discovery," in *Proceedings of the HPLC2000 Symposium*, Seattle, WA, June 24–30, 2000, p. 1541.
21. Chatwin, S.; Chow, D.T.; Maliski, E.; Talen, M.; Woo, G.; Zhao, Y.; Semin, D.J. "System Aids Analysis of Compound Collections," *Drug Disc. Devel.*, May 2001, p. 50–52.
22. Zhao, Y.; Jona, J.; Chow, D.T.; Rong, H.; Semin, D.; Xia, X.; Zanon, R.; Spancake, C.; Maliski, E. "High-throughput logP Measurement Using Parallel Liquid Chromatography/Ultraviolet/Mass Spectrometry and Sample-pooling," *Rapid Commun. Mass Spectrom.* **16,** 1548–1555 (2002).
23. Schnier, P.; DeBlonc, R.; Woo, G.; Gigante, W.; Cheetham, J. "Ultra-high Throughput Affinity Mass Spectrometry for Screening Protein Receptors," in *Proceedings of the 51th ASMS Conference on Mass Spectrometry and Allied Topics*, Montreal, Quebec, Canada, June 8–12, 2003.
24. Berger, T.A.; Wilson, W.H. "High-speed Screening of Combinatorial Libraries by Gradient Packed-column Supercritical Fluid Chromatography," *J. Biochem. Biophys. Methods* **43,** 77–85 (2000).
25. Ventura, M.C.; Farrell, W.P.; Aurigemma, C.M.; Greig, M.J. "Packed Column Supercritical Fluid Chromatography/Mass Spectrometry for High-throughput Analysis," *Anal. Chem.* **71,** 2410–2416, 4223–4231 (part 2) (1999).
26. Hoke, S.H. II.; Tomlinson, J.A.; Bolden, R.D.; Morand, K.L.; Pinkston, J.D.; Wehmeyer, K.R. "Increasing Bioanalytical Throughput Using pcSFC-MS/MS: 10 Minutes per 96-Well Plate," *Anal. Chem.* **73,** 3083–3088 (2001).
27. Zhao, Y.; Chow, D.T.; Thomas, S.; Yin, L.; Semin, D. "Rapid Analytical Method Development in Support of Medicinal Chemistry Using Parallel and Multidimensional Separation Techniques," in *Proceedings of the HPLC2003 Symposium*, Nice, France, June 15–19, 2003, p. 149.
28. Zhao, Y.; Woo, G.; Thomas, S.; Semin, D.; Sandra, P. "Rapid Method Development for Chiral Separation in Drug Discovery Using Sample-pooling and Supercritical Fluid Chromatography/Mass Spectrometry," *J. Chromatogr.* **1003,** 157–165 (2003).
29. Davis, J.M.; Giddings, J.C. "Statistical Theory of Component Overlap in Multicomponent Chromatograms," *Anal. Chem.* **55,** 418–424 (1983).
30. Giddings, J.C. "Two-dimensional Separation: Concepts and Promise" *Anal. Chem.* **56,** 1258A–1270A (1984).
31. Corradini, C. "Coupled-column Liquid Chromatography," pp. 17–42 in Mondello, L.; Lewis A.C.; Bartle, K.D. eds., *Multidimensional Chromatography*, Wiley, UK (2002).

32. Wolters, D.A.; Washburn, M.P.; Yates, J.R. "An Automated Multidimensional Protein Identification Technology for Shotgun Proteomic," Anal. Chem. **73**(23), 5683–5690 (2001).
33. Wehr, T. "Multimensional Liquid Chromatography in Proteomic Studies," *LC-GC Europe* **16**(3), 154–162 (2003).
34. Krstulovic, A.M.; Lee, C.R.; Firmin, S.; Jacquet, G.; Van Dau, C.; Tessier, D. "Applications of LC-MS Methodology in the Development of Pharamaceuticals," *LC-GC Europe* **15**(1), 31–41 (2002).
35. Venkatramani, C.J.; Zelechonok, Y. "An Automated Orthogonal Two-dimensional Liquid Chromatography," *Anal. Chem.* **75**(14), 3484–3494 (2003).
36. Sandra, P.; Medvedovici, A.; Kot, A.; David, F. "Selectivity Tuning in Packed Column Supercritical Fluid Chromatography," pp. 177–194 in Anton, K.; Berger, C., eds., *Supercritical Fluid Chromatography with Packed Columns*. Dekker, New York (1998).
37. Chow, D.T.; Poppe, L.; Zhao, Y.; Cheetham, J. "LC-MS/NMR in Drug Discovery," in *Proceedings of the SMASH 2003 Conference*, Verona, Italy, Sept. 14–17, 2003.
38. Yurek, D.A.; Branch, D.L.; Kuo M-S. "Development of a System to Evaluate Compound Identity, Purity, and Concentration in a Single Experiment and Its Application in Quality Assessment of Combinatorial Libraries and Screen Hits," *J. Comb. Chem.* **4**(2), 138–148 (2002).
39. Taylor, E.W.; Jia, W.; Bush, M.; Dollinger, G.D. "Accelerating the Drug Optimization Process: Identification, Structure Elucidation, and Quantification of In Vivo Metabolites using stable isotopes with LC/MSn and the chemiluminescent nitrogen detector," *Anal. Chem.* **74**(13), 3232–3238 (2002).
40. Zhao, Y.; Sandra, P.; Thomas, S.; Semin, D. "SFC and SFC-MS Applications in Modern Drug Discovery," submitted for publication in *LC-GC* 2004.
41. Wilson, I. "Advancing Hyphenated Chromatographic Systems," *Anal. Chem.* 534A–542A (2000).
42. Pullen, F.S.; Swanson, A.G.; Newman, M.J.; Richards, D.S. "On-line Liquid Chromatography/Nuclear Magnetic Resonance Mass Spectrometry—A Powerful Spectroscopic Tool for the Analysis of Mixtures of Pharmaceutical Interest," *Rapid Comm. Mass Spectrom.* **9**(11), 1003–1006 (1995).
43. Albert, K., ed., *On-line LC-NMR and Related Techniques*, Wiley, West Sussex, UK (2002).
44. Lindon, J.C.; Nicholson, J.K.; Wilson, I.D. "Direct Coupled HPLC-NMR and HPLC-NMR-MS in Pharmaceutical Research and Development," *J. Chromatogr. B* **748,** 233–258 (2000).
45. Thomas, S.B.; Hua, Z.; Chow, D.; Zhao, Y.; Jones, W.; Overland, D.; Petersen, J.; Woo, G.; Hulme, C.; Semin, D. "High-Throughput Purification Platform and Post Purification Processes at Amgen," Prep2001, Washington, DC, May 2001.

46. Kibbey, C.E. "An Automated System for the Purification of Combinatorial Libraries by Preparative LC/MS," *Lab. Rab. Autom.* **9,** 309–321 (1997).
47. Zeng, L.; Kassel, D.B. "Developments of a Fully Automated Parallel HPLC/Mass Spectrometry System for the Analytical Characterization and Preparative Purification of Combinatorial Libraries," *Anal. Chem.* **70,** 4380–4388 (1998).
48. Zeng, L.; Burton, L.; Yung, K.; Shushan, B.; Kassel, D.B. "Automated Analytical/Preparative High-performance Liquid Chromatography–Mass Spectrometry System for the Rapid Characterization and Purification of Compound Libraries," *J. Chromatogr. A* **794,** 3–13 (1998).
49. Zeng, L. et al. "New Developments in Automated PrepLCMS Extends the Robustness and Utility of the Method for Compound Library Analysis and Purification," *Comb. Chem. High Throughput Screen.* **1,** 101–111 (1998).
50. Reynolds, D.P.; Brinded, K.; Hollerton, J.; Lane, S.J.; Seffler, A.; Lewis, K. "High-Throughput Analysis, Purification, and Quantification of Combinatorial Libraries of Single Compounds," *Chimia*, **54** (ILMAC Congress Focal Point: Medicinal Chemistry), 37–40 (2000).
51. Jusforgues, P.; Shaimi, M.; Barth, D. "Preparative Supercritical Fluid Chromatography: Grams, Kilograms and Tons," pp. 403–427 in Anton, K.; Berger, C., eds., *Supercritical Fluid Chromatography with Packed Columns*, Dekker, New York, (1998).
52. Nicoud, R.M.; Clavier, J.Y.; Perrut, M. "Preparative SFC: Basics and Applications," pp. 397–429 in Caude, M.; Thiebaut, D., eds., *Practical Supercritical Fluid Chromatography and Extraction*, Harwood, Amsterdam, The Netherlands (1999).
53. Berger, T.A.; Fogleman, K.; Staats, T.; Bente, P.; Crocket, I.; Farrell, W.; Osonubi, M. "The Development of a Semi-preparatory Scale Supercritical-Fluid Chromatograph for High-throughput Purification of Combi-chem Libraries," *J. Biochem. Biophys. Methods* **43,** 87–111 (2000).
54. Wang, T.; Barber, M.; Hardt, I.; Kassel, D.B. "Mass-directed Fractionation and Isolation of Pharmaceutical Compounds by Packed-column Supercritical Fluid Chromatography/Mass Spectrometry," *Rapid Commun. Mass Spectrom.* **15,** 2067–2075 (2001).
55. Barker, G.E.; Romano, S.J.; Short, K.M.; Slee, D.; Park, W. "Purification of Combinatorial Libraries by OntoCHROMTM: A Novel High-throughput SFC Prep Technique," in *ACS Proceedings*, San Diego, CA, USA, (2001) 221-ORGN-046.
56. Caldwell, J.; Caldwell, W. "Selectivity of Polar Bonded Phases for SFC and New Developments," *Poster presentation on the HPLC2003 Symposium*, Nice, France, June 15–19 (2003).
57. Farrell, W.P. "Implementation of Supercritical Fluid Chromatography for High-throughput Purification," *Abstracts of papers on 226th ACS national meeting*, New York, NY, September 7–11, 2003 (2003).
58. http://www.symyx.com.

59. Semin, D.S., unpublished internal results.

60. Semin, D.S., unpublished internal results.

61. Svec, F.; Tennikova, T.B.; Deyl, Z., eds., *Monolithic Materials—Preparation, Properties and Applications*, Elsevier, Amsterdam (2003).

62. Iberer, G.; Hahn, R.; Jungbauer, A. "Column Watch: Monoliths as Stationary Phase for Separating Biopolymers—Fourth-generation Chromatography Sorbents," *LC-GC Europe* **17,** 88–95 (2000).

63. Svec, F. "Porous Monoliths: The Newest Generation of Stationary Phases for HPLC and Related Methods," *LC-GC Europe* **16**(6a), 24–28 (2003).

64. Heller, M.J.; Guttman, A. eds., *Integrated Microfabricated Biodevices—Advanced Technologies for Genomics, Drug Discovery, Bioanalysis and Clinical Diagnostics*, Dekker, New York (2002).

65. http://www.nanostream.com.

66. Zhao, Y.; Sandra, P.; Woo, G.; Thomas, S.; Gahm, K.; Semin, D. "Perspectives of Applying Packed-column Supercritical Fluid Chromatography-Mass Spectrometry (pSFC-MS) to Drug Discovery," submitted for publication in *LC-GC Europe*, April 2004, pp. 224–237.

15

HIGH THROUGHPUT METABOLIC AND PHYSICOCHEMICAL PROPERTY ANALYSIS: STRATEGIES AND TACTICS FOR IMPACT ON DRUG DISCOVERY SUCCESS

EDWARD H. KERNS AND LI DI

INTRODUCTION

Pharmaceutical profiling is a new activity in drug discovery that has appeared during the past few years. It involves the assessment of physicochemical and physiological properties of compounds in order to predict the performance of drug-discovery compounds against the hurdles faced by a pharmaceutical agent [1]. Several reviews have described the attrition of development candidates due to inadequate biopharmaceutical properties and the need for optimization of pharmaceutical properties during drug discovery [2–11]. Pharmaceutical profiling provides the data and strategies necessary to optimize the properties of compounds.

The development of pharmaceutical profiling follows the trend to enhance drug discovery with new technologies, in order to more rapidly produce development candidates with a greater likelihood of surviving the hurdles of pharmaceutical development to be approved for clinical therapy. These technologies include combinatorial chemistry, high throughput screening, computer-aided drug design, genomics, proteomics, and metabonomics. Each of these has emerged from existing techniques through the application of a new

Integrated Strategies for Drug Discovery Using Mass Spectrometry, Edited by Mike S. Lee

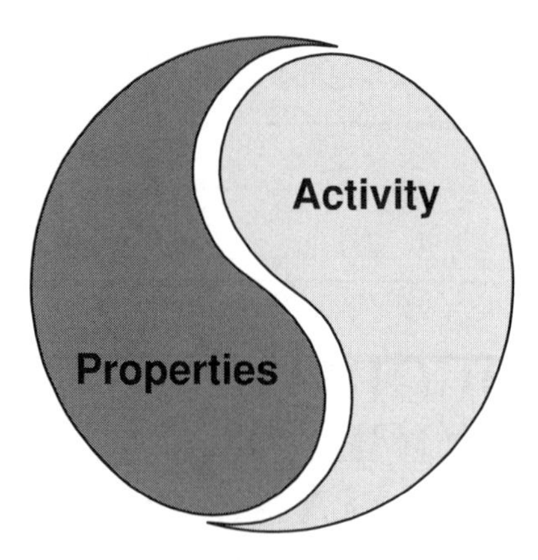

FIGURE 15.1 A pharmaceutical must balance both selective activity for the target and adequate properties for safe delivery of the drug to the target tissue.

strategy, in order to meet a specified need in the discovery process. Drug discovery is an incubator for new technologies, due to its intensely investigative nature. Early successes with nascent technologies lead to greater utilization by mainstream discovery scientists. Eventually, these technologies become "disciplines" in their own right, which contribute as an organizational unit to discovery teams and the processes of drug discovery. Pharmaceutical profiling has progressed to mainstream usage by discovery scientists.

A holistic model for a pharmaceutical agent is shown in Figure 15.1. Pharmaceuticals must balance activity and selectivity for the target with adequate properties for safe delivery of sufficient drug concentration to the target tissue. Inadequate properties, that limit delivery of drug to the target, place increased burdens on the development process. For example, poor delivery to the target results in higher dosing levels, which causes increased toxicity risk.

Early pharmaceutical profiling work grew from metabolism, pharmacokinetics (PK), physical chemistry, pharmaceutics, and toxicology disciplines that aimed to identify and screen the critical pharmaceutical properties at early stages of discovery. As a result of successes, the field has become formalized in many pharmaceutical organizations as a collaborative activity among departments that each perform one or more of the assays in their field of expertise. More recently, pharmaceutical organizations have formalized these activities by centralization of pharmaceutical profiling into one group. This trend toward centralization offers several advantages: increases the efficiency of coordination and model building; provides one group for discovery teams to interact with; and provides clear goals, priorities, and commitments by the pharmaceutical profiling group toward rapid and efficient drug-discovery support without being distracted by other responsibilities.

This field became more formally recognized as a result of the First AAPS Frontiers Symposium, "From Good Ligands to Good Drugs: Optimizating Pharmaceutical Properties by Accelerated Screening," February 19–21, 1998, in Bethesda, Maryland. More recently, the AAPS Workshop, "Pharmaceutical Profiling for Lead Selection," May 19–21, 2003, in Whippany, New Jersey, focused on the in silico; physicochemical; in vitro absorption, distribution, metabolizing, excretion (ADME); and permeability aspects of pharmaceutical profiling and case studies of their application in medicinal chemistry. These influential congresses assembled leaders in the field, promoted discussion, and helped many pharmaceutical research organizations crystallize their goals and tactics in this critical emerging area.

Therefore, a key strategy in drug discovery is to profile compound properties to select and optimize leads for safety and efficacy. Techniques and data analysis that contribute to tactics supporting this strategy are discussed in this chapter.

As property information has become more available, we have recognized that these data can assist with many areas of drug-discovery research. In the same way that physicochemical and physiological properties affect delivery of drug to the therapeutic target tissue in vivo, the same properties also affect delivery of drug to the target for in vitro bioactivity experiments performed in drug discovery. Lack of understanding of these properties and their possible effect on discovery experimental data can lead to incorrect experimental planning and data interpretation, which can misdirect research teams with incorrect structure activity relationships (SAR). Exposure of the test compound to the therapeutic target protein is critical. For example, if the solubility of the test compound is lower than the concentration at which it is bioassayed, then a lower concentration will be present at the target protein and the assay results will be falsely low. Properties can also produce erroneous property assays. For example, if a compound has a solubility of 3 μM and it is tested for metabolic stability at 10 μM, then the stability of the compound will be judged to be falsely high.

Thus, a second key strategy in drug discovery is to improve the planning and interpretation of discovery experiments using pharmaceutical-profiling property data. Drug-discovery researchers of all disciplines should have access to compound property data and be aware of the potential effects on the experiments they are performing.

The scope, goals, and implementation of pharmaceutical profiling continue to be enhanced as drug discovery becomes more sophisticated. The central goal of pharmaceutical profiling will continue to be the delivery of useful property information and interpretations to discovery teams so that they can make "informed decisions."

Various tools and techniques continue to emerge to assist this goal. Not only can understanding of individual properties, such as solubility, be of use to teams, but tactics that combine data from more than one technique, are emerging. One tool, multivariate analysis (MVA), seeks to understand the relationships of properties and identify which properties are most critical for a desired result.

PHARMACEUTICAL PROFILING IN SELECTION AND OPTIMIZATION OF LEADS

Experience has shown that it is most efficient to intervene early in the discovery process. This is when discovery teams traditionally have the least amount of information available for making decisions. Informed decisions at this stage have the greatest leverage on productive downstream work. Correct choice of one or more lead series at this stage can make later work more successful. Conversely, lack of information at this stage can lead to investment of time and resources in a compound series that is later found to have a fatal flaw. This is demoralizing for the team, wasteful of resources, and leads to later entry into the clinical market.

Working at early stages can, however, impose additional constraints on the pharmaceutical profiling program. Typically, very little compound is available at this point and activity measurements usually have the highest priority. Thus, profiling methods must use only milligram-level quantity of each compound. Methods must also have rapid throughput, in order to provide the discovery team with data in a time frame that is consistent with the fast-paced decisions of early discovery. Profiling methods must be high capacity, because hundreds of compounds are often evaluated for each project team and most pharmaceutical organizations have many early-discovery projects underway at a given time. Finally, large amounts of data are generated and must be effectively delivered to discovery scientists and archived.

These requirements have lead to the following analytical solutions. Assays for low amounts of compounds must typically have highly sensitive detection schemes. Rapid throughput often relies on robotic methods. The large number of compounds has led to parallel high-density analysis formats, such as 96-well plates. To handle the large amounts of data, special software, databases, and visualization methods have been developed.

It is also advantageous to include assays that cover diverse physicochemical and physiological properties. This multivariate approach allows the survey of diverse property space to (1) measure the contribution of each property against the pharmaceutical hurdles, and (2) develop multivariate analysis models. These are discussed in later sections.

PHARMACEUTICAL PROFILING IN PLANNING AND INTERPRETING DISCOVERY EXPERIMENTS

Drug-discovery scientists have been accustomed to having limited knowledge about the hundreds or thousands of compounds that project teams consider during their research. Now that high throughput property assays are available, a new source of information for more informed decisions is available. Typically, pharmaceutical profiling data are used to predict absorption, distribution, metabolism, excretion, and toxicity (ADMET). However, it can be applied more broadly to better plan and interpret discovery experiments. A compound must successfully pass the battery of discovery experiments to be considered for human experiments. Property information can provide improved insights for these experiments.

Figure 15.2 illustrates some of the discovery bioactivity experiments in which a test compound must be successful to advance. If erroneous activity or selectivity data are generated or misinterpreted, the SAR will mislead the project team. SAR is a central strategy of drug-discovery research. If the activity assays are affected by properties in addition to target protein interaction, then the SAR will be a composite of multiple variables. Table 15.1 lists some of the potential effects on SAR from lack of property data application in planning and interpretation of drug-discovery bioassays.

Property data from pharmaceutical profiling is not exclusively for optimizing PK. It can be considered as part of the multivariate ensemble of data (e.g., MW, chirality, purity, IC_{50}, LD_{50}) that is available to research teams for application to any drug-discovery experiment.

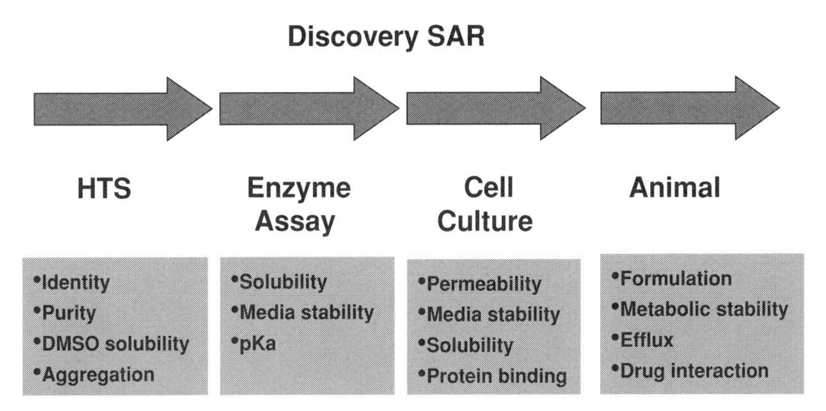

FIGURE 15.2 Many critical stages in drug discovery relate to bioassays. The stages of bioassay along the discovery time line are shown, along with potential hurdles related to pharmaceutical properties below. A candidate must succeed in all of these assays to reach development. Use of pharmaceutical profiling property data can assist the proper planning and interpretation of bioassays to obtain the highest value from drug-discovery investments.

TABLE 15.1 Examples of Application of Pharmaceutical Properties for Discovery Support

Property	Property Effect	Results	Erroneous Activity Result	Possible Action if Property Known
Integrity	Impurity or unknown structure in sample	Interference with activity or property assay	False high activity; Inaccurate SAR	Do not test unknown and impure compounds
Solubility	Low solubility in plate well causes lower than expected concentration	Low concentration in solution increases IC_{50}	False low activity	Test at lower concentration; select more soluble compounds in series
	Assay materials (e.g., protein) help suspend insoluble compound	Unrecognized solubilization is not reproducible and varies with assay	Variable activity results	Test at lower concentration
	Precipitate forms at cell surface	High concentration at cell surface causes high intracellular concentration	Falsely high activity	Test at lower concentration
	Compound is not soluble in dilutions before being added to assay plate well	Low concentration in plate well increases IC_{50}	Falsely low activity	Make dilutions with solubilizers (e.g., pH, co-solvent)
	Low solubility in NMR binding experiment	Concentration too low for experiment	Inability to measure or interpret binding	Do not test low solubility compounds

	Compounds have low solubility in DMSO [18]	Concentration lower than expected in assay	Falsely low activity	Ensure compound is dissolved initially; add co-solvent
	Aggregate formation in assay [53]	Interference with assay	Falsely high activity	Light-scattering assay for aggregates; nonionic detergent [53]
	Compound has low dissolution in vivo	Low concentration of compound for absorption	False low in vivo activity	Initiate formulation or salt form for in vivo study
Permeability	Compound has low permeability	Cell-based assay activity is much lower than expected from enzyme assay	False low activity	Synthesis to increase permeability saves series
Permeability and PGP[a]	Compound is not permeable to BBB due to lack of passive diffusion or PGP efflux	Low brain-to-plasma ratio causes low activity	False low activity	Select permeable compounds for in vivo studies
Stability	Compound is unstable in assay medium	Low bioassay results from removal of compound	False low activity	Identify unstable substructure and synthesize more stable analogs

[a] PGP = polyglycoproteins.

TABLE 15.2 Examples of Pharmaceutical Hurdles and Assays Used in the Pharmaceutical Industry to Predict the Performance of Compounds

Pharmaceutical Hurdle	Assays	Method	References
QSAR	Lipophilicity	Fast HPLC	[13]
		E log *P*	[14,15]
		Log $P_{\text{Octanol-Water}}$	[54,55]
Absorption	Solubility	Turbidimetry	[56]
		Nephelometry	[57]
		pH Metric	[58]
	Permeability	Caco-2	[59]
		MDCK Cells	[60]
		PAMPA	[19]
		HPLC-IAM	[61]
		Alkane artificial membrane	[62]
	pK_a	SGA	[20,63]
		Potentiometric	[58]
		Titration	
Distribution & Excretion	In vivo screen	Rapid Rat	[64]
		N-in-one/cassette	[23–25]
	Plasma–protein binding	Capillary	[65]
		Electrophoresis	[66]
		Equilibrium dialysis biacore	
Metabolism	Stability	Microsomal	[67]
		Hepatocytes	[68]
		Plasma	[69]
	Metabolite ID	LC-MS/MS	[70]
		LC/NMR/MS	[71]
Safety	CYP450 inhibition	Fluorescent Probes	[72]
		LC-MS/MS	[73]

SELECTION OF PHARMACEUTICAL PROFILING ASSAYS

Table 15.2 lists some pharmaceutical profiling assays in common use throughout the industry. High throughput physicochemical methods [10] and physiological methods [11] have been reviewed. Despite the great activity in the pharmaceutical-profiling field there are at this time no standard methods or experimental formats that are universal throughout the industry. Each laboratory tends to implement their own variations of general methods. The

TABLE 15.3 Multivairate Pharmaceutical Profiling Assays Used in One Pharmaceutical Research Organization

Pharmaceutical Hurdle	Assays	Method	References
Quality	Integrity and Purity	LC/UV/MS	[12]
QSAR	Lipophilicity	Rapid HPLC	[16]
		calculation	Prolog D Software
Absorption	Solubility	Direct UV	[17]
	Permeability	PAMPA	[21]
	pK_a	SGA	[59,20]
		Calculation	ACD[a] software
Distribution and excretion	In vivo Exposure/PK	LC-MS/MS of plasma	
	Permeability/blood–brain barrier	Proprietary assay	[21]
	Tissue penetration	In vivo tissue uptake and LC-MS/MS	[74]
Metabolism	Stability	Microsomal, plasma, light, heat, acid, base	[63,22]
	Metabolite ID	LC-MS/MS and LC/NMR	[66,67]
Safety	CYP450 inhibition	Fluorimetric probes	[68]

[a] ACD is Advanced Chemistry Development Inc.

standardization of protocols and biological reagents is most common within an organization to ensure comparability of data. Variations in the biological reagents and protocols between organizations make the data less comparable between organizations. Many analytical instrument vendors have developed common methods on their instruments and provide them as "solutions."

It is anticipated that integrated workstations will be developed that incorporate all of the steps of a method in one system, thus reducing the burden on the laboratory scientist to integrate the system and placing the emphasis on generating results, rather than developing methods.

Pharmaceutical research organizations select assays that meet their own expertise and goals in assessing the performance of their compounds against pharmaceutical hurdles. While it is not easy, from the literature, to assemble a list of what each organization is actively using, Table 15.3 and Figure 15.3 provide an example from the Wyeth Research organization.

Compound quality is addressed with a rapid LC/UV/MS assay [12]. Many groups use compounds that were synthesized several years before or obtained from an outside source. Structure assignment inaccuracies, impurities, and degradants can cause SAR and SPR confusion for the drug-discovery team.

Traditional medicinal chemistry quantitative structure-activity relationship (QSAR) studies often involve octanol–water distribution coefficients (Log D). These can be miniaturized and automated for profiling. The Log D experiment

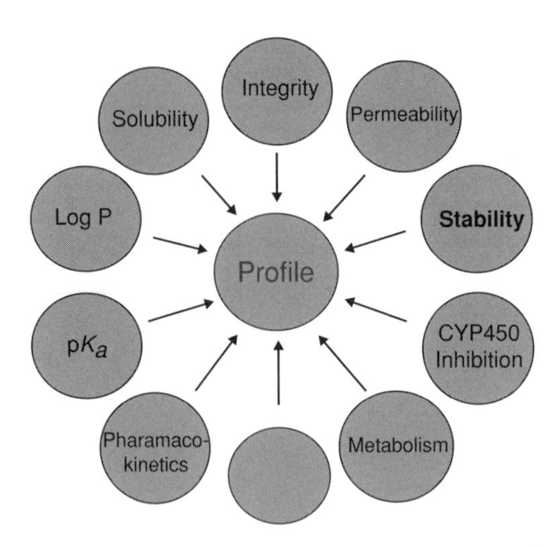

FIGURE 15.3 Example of a multivariate pharmaceutical profiling program.

may seem simple, but many variables (e.g., pH, ionic strength, counterions, co-solvents) have a major effect on the results. Log D is often measured by chromatographic techniques for higher throughput [13–16], or it may be calculated for efficiency.

Three common properties that affect intestinal absorption of drugs after oral administration are solubility, permeability, and pK_a. Traditional solubility experiments measure solubility of solids placed into aqueous phases (thermodynamic solubility), but these methods are too slow or they consume too much material for drug discovery. Higher throughput methods must be used. The direct ultraviolet (UV) method [17] adds compound dissolved in dimethyl sulfoxide (DMSO) to an aqueous buffer and measures the UV absorption of the aqueous phase using a 96-well plate reader after equilibration and filtration (kinetic solubility). Lipinski has discussed the pitfalls that inadequate solubility information can have for a drug-discovery organization [18].

Many organizations use colon adenocarcinoma (Caco-2) for detailed study of permeability; however, this method can be resource intensive. Parallel artificial-membrane permeability (PAMPA) [19] has proven to be a reliable predictor of passive transcellular permeability for intestinal absorption prediction. It is also useful to interpret results of cell-based discovery assays, in which cell-membrane permeability is limiting. Finally, pK_a provides insight into the pH dependence of solubility and permeability. It can be measured [20] or calculated to get an understanding of the regions of the intestine in which the compound will be best absorbed, as well as to anticipate the effect of pH on solubility and permeability. Permeability at the blood–brain barrier (BBB) also can be rapidly profiled [21].

Metabolism is also important to assess, and most organizations have implemented microsomal, hepatocyte, or other in vitro assays to assess relative physiological stability. Other destabilizing factors are plasma, acid (stomach), base (large intestine), light (laboratory and factory), and heat (storage) [22].

TACTICS FOR PHARMACEUTICAL PROFILING IN THE DISCOVERY PROCESS

We have found that experience and innovation lead to useful tactics for applying pharmaceutical-profiling assays. Applications tactics can be as important as the assays in determining the impact of pharmaceutical-profiling efforts for drug discovery. In the following discussion, some useful tactics are highlighted in italics. These have been developed based on observation of the needs of medicinal chemists and pharmacologists at various stages of drug discovery.

Exploration Stage

During the exploration stage, high throughput screening (HTS) reveals compounds ("hits") that are active in modifying the action of the target protein. Pharmaceutical profiling contributes by *confirmation of HTS hits* using an integrity and purity assay. Often, the majority of compounds in a HTS screening library were synthesized many years earlier, came from an extramural source, or were members of a combinatorial screening library that were not completely characterized. The integrity and purity assay start the discovery team off with known structures on which to base synthetic series. If the HTS hit is impure, due to degradation or reaction by-products, the integrity and purity assay will serve as the first stage of structure elucidation, if the team wants to identify the active component in the sample.

The integrity and purity assay also serves to *qualify compounds for nonspecific assays* of biological activity or pharmaceutical properties. For example, if an assay uses a 96-well plate UV spectrophotometer, the presence of an impurity can significantly change the assay results, especially if the impurity's properties or activity differ from the parent compound.

Also at this stage, pharmaceutical profiling assays, such as solubility, permeability, and metabolic stability provide a multivariate property profile of HTS hits that contribute to the team's *holistic selection of leads* for synthetic optimization. The pharmaceutical-property profile is considered in relation to the ensemble of activity properties for making informed decisions. Conversely, an *alert to poor properties* can save the team effort and time on problematic structures, or forewarn of issues that must later be solved via synthetic modification, salt form, or formulation.

In many organizations, advancement from the exploratory stage requires in vivo model. Some of the profiling assays can provide evidence of in vivo exposure to the blood and penetration into the target tissue. Thus, the assays can *provide evidence for meeting advancement criteria.*

Early-Discovery Stage

In early discovery, pharmaceutical-profiling assays can assist with *monitoring properties of series analogs as they are synthesized.* This can provide feedback about how synthetic modifications that are intended to explore SAR also change the properties of the analogs. Comparisons are made to the historical baseline of the HTS hits and leads. It is very common for synthetic modifications to add molecular mass and lipophilicity to the molecule, often resulting in higher lipophilicity that leads to lower solubility and permeability. Monitoring these property changes as series compounds are synthesized keeps the team informed of movement of the compounds away from optimum properties. The effective activity of the drug is a function of both its intrinsic activity with the target protein and the concentration of the drug that is delivered to the target.

As discovery research moves forward, great value is derived from dosing experiments with living animals. In vitro profiling assays are used to *predict the in vivo ADME performance*, in order to select compounds for animal studies.

Simple in vivo assays of blood or tissue provide *the first measurement of the ADME properties* of the compound. In early discovery, studies of mixtures of compounds [23–25], which are co-administered to a single animal, can efficiently provide this exposure and tissue-penetration information. Early in vivo exposure studies help to establish dosing levels for animal models. Blood samples can also be obtained from the animal activity model following dosing to *correlate in vivo pharmacology with plasma levels.* Lower than expected in vivo activity can often be correlated with low plasma or tissue levels.

When inadequate in vivo exposure or PK properties are observed, pharmaceutical profiles can be used to *troubleshoot the cause of poor* in vivo *exposure or PK* [26]. Assays for solubility, permeability, and stability (metabolic, plasma, acid) can help to *track down the inadequate properties responsible for poor in vivo performance.* Property optimization synthesis can then be initiated. Subsequent series analogs can be assayed to *rank order compounds by properties for subsequent in vivo tests*, in order to give the highest likelihood of success. Often animal studies are expensive and time-consuming, especially if they are performed using the animal activity model. Simple in vitro profiling assays can provide information for improved decisions and efficiency.

Early discovery also makes extensive use of in vitro and cell-culture activity models. Activity measurements can be inconsistent when the compound has low aqueous solubility. Thus, the solubility assay, can *assist with the interpretation of biological activity tests*. Another cause of poor bioactivity in vitro is poor stability in the assay medium. Also, there can be discrepancies with activity when compounds are moved from biological models in 96–384-well format to living-cell models. Differences in activity can often be traced to permeability of the compound through the cellular lipid membrane and correlate with results of the permeability assay.

Mid-Discovery Phase

As the optimization activities proceed, pharamceutical profiling *provides feedback for property optimization*. As the project moves ahead, a deeper understanding of individual compounds becomes more important. Pharmaceutical profiling data can then be used to *compare series analogs for selection*. Often this information can be used for decisions to narrow the selection field from thousands to hundreds or tens of compounds for in-depth study.

Profiling assays related to metabolism can also provide important benefits here. High levels of metabolism can lead to rapid removal from the living animal; thus, metabolite identification assays can *assist with synthetically blocking labile "hot spots" in the molecular structure* for prolonged in vivo action. Also, metabolism can generate active metabolites. When a compound is more active in vivo than is expected from in vitro tests, then active metabolites may be suspected. Profiling methods that deal with identification of metabolites can *enable the synthesis of metabolites for activity testing*. If metabolites are identified during the earlier phases of discovery, while chemists are still in the process of synthesizing series analogs, then the compounds can be synthesized and tested for activity to add quality to the optimization phase. The plasma stability assay can be used to *screen pro-drugs prior to in vivo dosing*.

In summary, pharmaceutical profiles can be applied using the following strategies to improve the efficiency, speed, and effectiveness of drug discovery:

- Confirm HTS "hits"
- Qualify compounds for assays with nonspecific detection
- Early evaluation of compounds
- Holistic selection of leads
- Alert to poor properties
- Evidence for meeting advancement criteria
- Monitor properties of series analogs as they are synthesized
- Use in vitro assays to predict in vivo ADME performance

- First in vivo screen of ADME properties
- Interpret in vivo pharmacology based on plasma levels
- Use in vitro assays to troubleshoot the cause of poor in vivo exposure or PK
- Track down inadequate properties responsible for poor in vivo performance
- Rank-order compounds by properties for subsequent in vivo tests
- Interpretation of biological activity tests
- Provide feedback for property optimization
- Assist with synthetically blocking labile "hot spots" in the molecular structure
- Enable the synthesis of metabolites for activity testing
- Screen prodrugs prior to in vivo dosing

Data Analysis

Data analysis and interpretation is an important and challenging part of the discovery process, due to the large volumes of in vitro/in vivo data generated from pharmacology and high throughput pharmaceutical-profiling programs. The goals of data analysis and interpretation for pharmaceutical profiling are:

1. *Model building*: correlation between different properties of a drug-candidate series; making useful predictions.
2. *Problem solving*: diagnosing and forecasting the issues/liabilities and advantages for each series; correlating these to the underlying properties that are responsible.
3. *Decision making*: providing data and interpretations for teams to make go/no go decisions and to plan optimization work to improve series liabilities.
4. *Result presentation*: communicating and relating the data in a format that medicinal chemists and pharmacologists can readily grasp and incorporate.

Caldwell [4] describes a useful method of data analysis and presentation of multiple variable data sets to provide discovery teams with predictions of oral bioavailability. By ranking the compounds as high, intermediate, and low for in vitro solubility, acidic stability (to simulate stomach), hepatocyte stability (to simulate liver), and Caco-2 permeability, reliable early predictions can be made for in vivo oral bioavailability.

Structure	MW	Permeability	Solubility	Log *D*	CYP 3A4	Microsome Stability	Plasma Stability	Bioavailability
		P_e (10^{-6} cm/s)	(ug/mL)	pH 7.4	% Inhib.	% Remaining	% Remaining	%
A	601	0.04	55	1.4	17	85	100	10
B	582	0.01	65	0.8	18	85	100	4
C	655	0.02	50	1.2	18	80	95	7
D	498	0.01	45	1.6	35	80	100	8
E	617	0.03	> 100	1.3	23	80	90	12
F	629	0.01	90	1.9	20	95	100	9
High		> 1	60		< 15	> 80	> 80	
Moderate		1–0.1	10–60		15–50	20–80	20–80	
Low		< 0.1	10		> 50	< 20	< 20	

FIGURE 15.4 Hypothetical compound series having good in vitro potency, but low in vivo activity after oral administration. Low permeability is apparently responsible for low bioavailability.

Presentation of the data with a ranking color scheme (e.g., green for high, yellow for moderate, red for low) can be useful for formatting pharmaceutical profiling data to provide increased insights for in vitro and in vivo properties. The results are easily visualized through color recognition. This information helps to diagnose the potential causes for undesirable properties and what can be done to improve these properties. For example, a hypothetical series of compounds has good in vitro potency, but have no activity in vivo after oral administration. Correlating the in vivo bioavailability data with in vitro solubility, permeability, microsomal stability, and plasma stability with color recognition in Figure 15.4 quickly indicates that permeability is the major cause of poor oral bioavailaility for the series. Strategies can then be formulated to overcome this hurdle early in discovery, such as increase lipophilicity or pro-drug approaches to enhance permeability. If the undesirable feature is due to some properties that cannot be fixed, a "go/no go" decision can be made quickly to avoid investment in a series that is not likely to succeed.

Multivariate Data Analysis for Pharmaceutical Profiling

A common method of data analysis is to examine changes in one variable at a time (e.g., solubility) and relate these to the structure of the compound. The large volumes of data that are now produced from high throughput methods, as well as their complexity and correlation, suggest that there are more productive ways to derive information from the data. One useful approach for analyzing large and complex data sets is multivariate analysis [27,28]. With this approach an entire data set is simultaneously analyzed to derive increased information.

Many biological activities and druglike properties are governed by a set of strongly intercorrelated molecular and physicochemical properties. Multivariate analysis allows the inclusion of all theoretically calculated and experimentally measured descriptors for model building, QSAR, and quantitative structure-property relationships (QSPR) [29,30]:

1. Computationally derived descriptors (e.g., molecular weight, size, shape, polarity, electrostatic interactions, charge, lipophilicity, hydrogen bonding capacity, polar surface area).
2. In vitro experimentally measured physicochemical properties (e.g., solubility, permeability, stability, Log P, pK_a) and biochemical characteristics (e.g., microsomal stability, CYP450 inhibition, plasma stability, plasma protein binding).
3. In vivo PK and tissue distribution (e.g., clearance, volume of distribution, half-life, mean residence time, fraction drug bound).
4. In vitro and in vivo biological data (e.g., enzyme or receptor assays, cell-based functional assays, efficacy animal models).

In general, experimentally measured descriptors provide more reliable data than calculated descriptors [31]. On the other hand, calculated parameters are advantageous for predicting the behavior of new molecules, even before they are actually synthesized.

A pharmaceutical-profiling program measures many diverse properties. Each compound is usually characterized by more than 10 different properties (descriptors/variables). Using multivariate analysis methods, these properties and their effect on compound performance can be correlated in order to address the goals for data analysis and interpretation in greater depth. Multivariate analysis treats the data as a multidimensional array. The data are mapped in multidimensional space to find the correlations and reduced to a two- or three-dimensional space format that is easier for people to visualize. Multivariate analyses methods, such as principal components analysis (PCA), multiple linear regression (MLR), and partial least-squares (PLS), provide low-dimensional and statistical representation of data points with a few information-rich parameters [28]. Multivariate analysis tools are powerful for extracting underlying information, to find regularities in the data, and separate them from the "noise" [27]. Data sets having a few errors, outliers, or missing data can be dealt with [32].

Both qualitative and quantitative information can be derived from multivariate analysis. A qualitative classification model provides overviews of how the properties relate to each other and how changes can be made to molecules in a qualitative sense. The main goal for qualitative analysis is to assist understanding and put all the pieces of the puzzle together. A quantitative model focuses on accurate prediction of various desirable properties for designing new molecules and combinatorial libraries with good potency and druglike properties.

Quantitative multivariate models provide the advantage of predicting the properties of new molecules, even before they are synthesized. This provides synthetic direction for improving pharmaceutical properties and helps to prioritize the synthetic efforts. Winiwarter et al. [33] used multivariate analysis to develop a model for predicting in vivo human jejunal permeability using experimentally and theoretically derived descriptors. Statistical software SIMCA from Umetrics AB (www.umetrics.com) was used. Its stepwise process for model building, which appears to be widely applicable, was as follows.

1. A critical but difficult to measure property (human jejunal permeability) was experimentally determined for a set of 22 compounds.
2. Several other properties or descriptors of each of these compounds were then experimentally measured or calculated.
3. PCA was performed on this compound set to demonstrate that the set was representative of drugs in general.
4. Compounds that were likely to have a different mechanism (paracellular or active transport) than the others (passive transcellular transport) were omitted.
5. The remaining compounds were divided into a training set and a test set. PLS models were generated on the training set.
6. Variables (measured properties and calculated descriptors) that were shown by these models not to correlate with the property of interest (human jejunal permeability) were dismissed as being of low relative importance, based on the method of variable importance (VIP).
7. A linear equation relating the important variables to the property of interest was derived. The models were evaluated using the test set. Once the model was verified, the final equation model was calculated using both the training and test sets.

Similar multivariate analysis approaches have been widely applied for model building and predicting druglike and physicochemical properties. For example, the following were predicted:

- Caco-2 permeability and intestinal drug absorption [31,34–41]
- Blood–brain barrier penetration [39,42,43]
- *P*-glycoprotein substrates [44,45]
- Log *P* and pK_a values [41]
- PK and metabolism [32,43]

Multivariate analysis can be used to predict multiple parameters at the same time and give insights into the overall profile of a compound series or library. It also allows the integration of QSAR and QSPR analysis, and thus simultaneous optimization of biological activity properties (in vitro and in vivo potency) and druglike properties (solubility, permeability, metabolism, PK) at early stages of drug discovery [44,45]. For example, the potency (IC_{50}) and pharmaceutical-property data (solubility and permeability) can be analyzed at the same time by using simple descriptors (H-bonding, log P, molecular weight (MW), etc.). The correlation provides insights for modification of the molecules to make them more potent and more druglike, or to find a compromise between the two. Molecular descriptors should be simple and intuitive, if possible, so medicinal chemists can easily incorporate them to make changes in their compounds. Simple answers and directions tend to be very effective. For instance, more lipophilicity may be needed to make a compound series permeable, or increasing MW may increase potency, but decrease solubility. With the dynamic correlation of property variables, finding a balance between the two properties is necessary.

Figure 15.5 shows a hypothetical model PLS model for a set of compounds. The analysis shows that the cell-based functional assays (isopropionic acid) (IPA)-rat, IPA-human, and bioassay) do not correlate well with the receptor-binding assay. On the other hand, the cell-based assays are correlated well with the pharmaceutical properties, such as solubility and permeability. The model shows that increased MW and hydrogen-bonding capacity will improve potency. That, however, will decrease solubility and permeability and make them less druglike. So, for this series of compounds, it is important to find a balance between potency and druglike properties.

As new molecules are made, in vitro, pharmaceutical properties may be rapidly determined. By applying predictive multivariate models, other properties (e.g., in vivo permeability) that are more difficult to measure may be predicted. As the database increases, models become more precise and general. This iterative process provides useful information for drug and property design. Follow-up compounds and libraries can then be designed to have higher potency and better properties.

Multivariate data analysis can be coupled with data visualization programs, such as Spotfire, to enhance data visualization [46]. Many of these visualization programs are currently being applied for HTS data to develop SAR and for data mining. Similar methodologies can also be used to visualize pharmaceutical-profiling data and for SPR. These visualization tools can be used interactively by medicinal chemists to help them look for the important interactions and provide conceptual understanding of structure–property relationships.

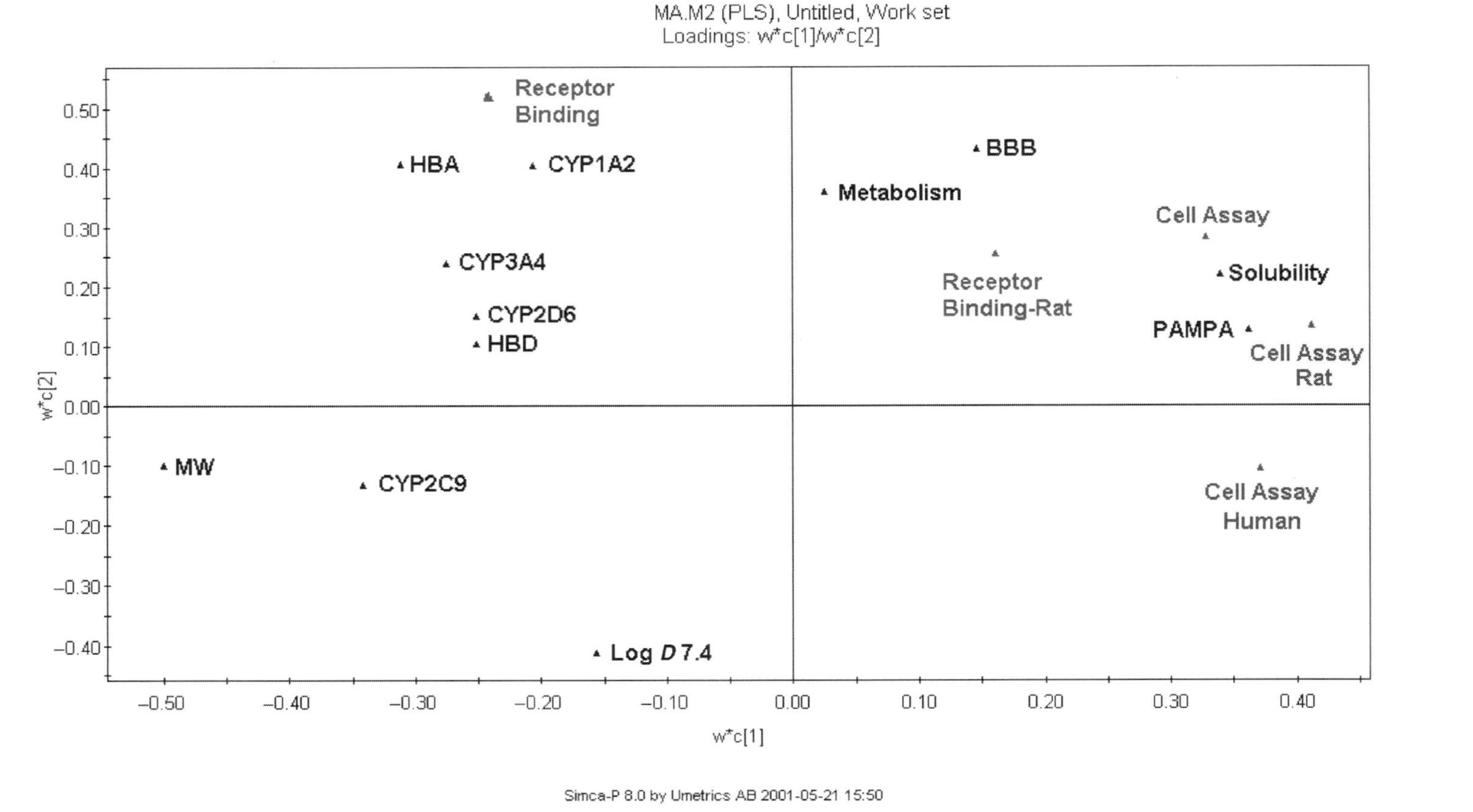

FIGURE 15.5 PLS model for a compound series indicating the correlation of various calculated and experimentally derived properties.

CONCLUSION

Pharmaceutical-profiling [47,48] programs provide multivariate property data, starting at the earliest phases of drug discovery. This information is used to enhance discovery-team information for effective decisions on selection and optimization of leads. A comprehensive program provides a diverse set of measurements of properties critical to successful discovery experiments, as well as the safe delivery of sufficient drug to the target-disease tissue. Multivariate analysis techniques are useful for developing models, making predictions, and effectively presenting data for discovery-team use. Pharmaceutical profiling has become an integral "technology" in drug discovery and is growing in its contribution to discovering new drugs.

ACKNOWLEDGMENTS

The authors thank Oliver McConnell, Guy Carter, and Magid Abou-Gharbia for their personal and professional support, encouragement, leadership, and enthusiasm for the pharmaceutical-profiling scientists and efforts at Wyeth Research. They also gratefully acknowledged the critical contributions of the pharmaceutical-profiling team to the development, implementation, and dedicated delivery of pharmaceutical profiling at Wyeth Research: Mark Tischler, Larry Mallis, Michael Chlenov, Marilyn Dar, Rachel Davis, Natasha Kagan, Teresa Kleintop, Susan Li, Hanlan Liu, Susan Petusky, Yelena Pyatski, Chantel Sabus, Ani Sarkahian, Deanna Yaczko, Mairead Young, and Mei-Yi Zhang.

REFERENCES

1. Venkatesh, S.; Lipper, R.A. "Role of the Development Scientist in Compound Lead Selection and Optimization," *J. Pharm. Sci.* **89,** 145–154 (2000).
2. Gaviraghi, G.; Barnaby, R.J.; Pellegatti, M. "Pharmacokinetic Challenges in Lead Optimization," pp. 3–14 in Testa, B.; van de Waterbeemd, H.; Folkers, G.; Guy, R. eds., *Pharmacokinetic Optimization in Drug Research. Biological, Physicochemical and Computational Strategies*, Verlag Helvetica Chimica Acta, Zurich (2001).
3. van de Waterbeemd, H.; Smith, D.A.; Beaumont, K.; Walker, D.K. "Property-based Design: Optimization of Drug Absorption and Pharmacokinetics," *J. Med. Chem.* **44,** 1313–1332 (2001).
4. Caldwell, G.W. "Compound Optimization in Early- and Late-Phase Drug Discovery: Acceptable Pharmacokinetic Properties Utilizing Combined Physicochemical, *In Vitro* and *In Vivo* Screens," *Curr. Opin. Drug Disc. Dev.* **3,** 30–41 (2000).

5. Smith, D.A. "High Throughput Drug Metabolism," pp. 137–143 in Gooderham, N., ed., *Drug Metabolism: Towards the Next Millennium*, IOS Press, London (1998).
6. Krämer, S.D. "Absorption Prediction From Physicochemical Parameters," *Pharm. Sci. Technol. Today* **2,** 373–380 (1999).
7. Abraham, M.H.; Gola, J.M.R.; Kumarsingh, R.; Cometto-Muniz, J.E.; Cain, W.S. "Connection Between Chromatographic Data and Biological Data," *J. Chromatogr. B* **745,** 102–115 (2000).
8. Sinko, P.J. "Drug Selection In Early Drug Development: Screening for Acceptable Pharmacokinetic Properties Using Combined *In Vitro* and Computational Approaches," *Curr. Opin. Drug Disc. Dev.* **2,** 42–48 (1999).
9. Barthe, L.; Woodley, J.; Houin, G. "Gastrointestinal Absorption of Drugs: Methods and Studies," *Fund. Clin. Pharmacol.* **13,** 154–168 (1999).
10. Kerns, E.H. "High Throughput Physicochemical Profiling for Drug Discovery," *J. Pharm. Sci.* **90,** 1838–1858 (2001).
11. Lin, J.H.; Rodrigues, A.D. "*In Vitro* Models for Early Studies of Drug Metabolism," pp. 217–243 in Testa, B.; van de Waterbeemd, H.; Folkers, G.; Guy, R., eds., *Pharmacokinetic Optimization in Drug Research. Biological, Physicochemical and Computational Strategies*, Verlag Helvetica Chimica Acta, Zurich (2001).
12. Kleintop, T.; Liu, H.; Sarkaihan, A.; Sabus, C.; Zhang, M. "LC/UV/MS Method for Integrity Evaluation of Phamraceutical Discovery Compounds," in *Proceedings of the Annual Conference of the American Association of Pharmaceutical Scientists*, Denver, CO, October 21–24, 2004; Page 2343.
13. Valko, K.; Du, C.M.; Bevan, C.D.; Reynolds, D.; Abraham, M.H. "High Throughput Lipophilicity Determination: Comparison With Measured and Calculated Log P/Log D Values," *Log P 2000 Conference*, Laussanne, Switzerland, March 5–9, 2000.
14. Lombardo, F.; Shalaeva, M.Y.; Tupper, K.A.; Gao, F.; Abraham, M.H. "ElogP_{oct}: A Tool for Lipophilicity Determination in Drug Discovery," *J. Med. Chem.* **43,** 2922–2928 (2000).
15. Lombardo, F.; Shalaeva, M.Y.; Tupper, K.A.; Gao, F. "ElogDoct: A Tool for Lipophilicity Determination in Drug Discovery. 2. Basic and Neutral Compounds." *J. Med. Chem.* **44,** 2490–2497 (2001).
16. Kerns, E.H.; Di, L.; Petusky, S.; Kleintop, T.; McConnell, O.J., Carter, G. "Pharmaceutical Profiling Method for Lipophilicity and Integrity Using LC/MS," *J. Chromatogr. B* **791,** 381–388 (2003).
17. Avdeef, A. "High Throughput Measurements of Solubility Profiles," pp. 305–326 in Testa, B.; van de Waterbeemd, H.; Folkers, G.; Guy, R., eds., *Pharmacokinetic Optimization in Drug Research: Biological, Physiological, and Computational Strategies*, Verlag Helvetica Chimica Acta, Zurich (2001).
18. Lipinski, C.A. "Avoiding Investment in Doomed Drugs," *Curr. Drug Disc.* **April**, 17–19 (2001).

19. Kansy, M.; Senner, F.; Gubernator, K. "Physicochemical High Throughput Screening: Parallel Artificial Membrane Permeability Assay in the Description of Passive Absorption Processes," *J. Med. Chem.* **41,** 1007–1010 (1998).
20. Box, K.J.; Comer, J.E.A.; Hosking, P.; Tam, K.Y.; Trowbridge, L. "Rapid Physicochemical Profiling as an Aid to Drug Candidate Selection," pp. 67–74 in Dixon, G.K.; Major, J.S.; Rice, M.J., eds., *High Throughput Screening: The Next Generation*, BIOS Scientific Publishers Ltd., Oxford, UK (2000).
21. Di, L.; Kerns, E.H.; McConnell, O.J.; Carter, G. "High Throughput Artificial Membrane Permeability Assay for Blood-Brain Barrier," *Eur. J. Med. Chem.* **38,** 223–232 (2003).
22. Kerns, E.H.; Di, L.; Kleintop, T.; Zhang, M.; Petusky, S.; McConnell, O. "High Throughput Stability Profiling for Drug Discovery," in *Presented at the Symposium on Chemical and Pharmaceutical Structure Analysis*, September, 2000.
23. Olah, T.V.; McLoughlin, D.A.; Gilbert, J.D. "The Simultaneous Determination of Mixtures of Drug Candidates by Liquid hromatography/Atmospheric Pressure Chemical Ionization Mass Spectrometry as an *In Vivo* Drug Screening Procedure," *Rapid Commun. Mass Spectrom.* **11,** 17–23 (1997).
24. Bayliss, M.K.; Frick, L.W. "High-Throughput Pharmacokinetics: Cassette Dosing," *Curr. Opin. Drug Discovery Dev.* **2,** 20–25 (1999).
25. Halm, K.A.; Adkison, K.K.; Berman, J.; Shaffer, J.E.; Tong, Q.W.; Lee, F.W.; Unger, S.E. "Rapid *In vivo* Pharmacokinetic Screening of Drug Discovery Mixtures Using APCI LC/MC," in *Proceedings of the 44th Annual Conference on Mass Spectrometry and Allied Topics*, American Society for Mass Spectrometry, Santa Fe, 392 (1996).
26. Gan, L.S.L.; Thakker, D.R. "Application of the Caco-2 Model in the Design and Development of Orally Active Drugs: Elucidation of Biochemical and Physical Barriers Posed by the Intestinal Epithelium," *Adv. Drug Deliv. Rev.* **23,** 77–98 (1997).
27. Wold, S.; Albano, C.; Dunn, W.J., III; Edlund, U.; Esbensen, K.; Geladi, P.; Hellberg, S.; Johansson, E.; Lindberg, W.; Sjöström, M. "Multivariate Analysis in Chemistry," pp. 17–95 in Kowalski, B.R., ed., *Chemometrics. Mathematics and Statistics in Chemistry*, D. Reidel, Dordrecht (Netherlands) (1984).
28. Eriksson, L.; Johnsson, J.; Sjöström, M.; Wold, S. "A Strategy for Ranking Enviromentally Occurring Chemicals. Part II. An Illustration With Two Data Sets of Chlorinated Aliphatics and Aliphatic Alcohols," *Chemom. Intell. Lab. Syst.* **7,** 131–141 (1989).
29. Grover, M.; Singh, B.; Bakshi, M.; Singh, S. "Quantitative Structure-Property Relationships in Pharmaceutical Research—Part 1," *Pharm. Sci. Technol. Today* **3,** 28–35 (2000).
30. Grover, M.; Bakshi, M. "Quantitative Structure-Property Relationships in Pharmaceutical Research—Part 2," *Pharm. Sci. Technol. Today* **3,** 50–57 (2000).
31. van de Waterbeemd, H.; Camenisch, G.; Folkers, G.; Raevsky, O.A. "Estimation

of Caco-2 Cell Permeability Using Calculated Molecular Descriptors," *Quant. Struct.-Act. Relat.* **15,** 480–490 (1996).

32. Spencer, R.W. "High Throughput Screening of Historic Collections: Observations on File Size, Biological Targets, and File Diversity,"*Biotechnol. Bioeng. (Comb. Chem.)* **61,** 61–67 (1998).
33. Winiwarter, S.; Bonham, N.M.; Fredrick, A.; Hallberg, A.; Lennernäs, H.; Karlén, A. "Correlation of Human Jejunal Permeability (In Vivo) of Drugs with Experimentally and Theoretically Derived Parameters. A Multivariate Data Analysis Approach," *J. Med. Chem.* **41,** 4939–4949 (1998).
34. Stenberg, P.; Norinder, U.; Luthman, K.; Artursson, P. "Virtual Screening of Intestinal Drug Permeability," *J. Controlled Release* **65,** 231–243 (2000).
35. Stenberg, P.; Norinder, U.; Luthman, K.; Artursson, P. "Experimental and Computational Screening Models for the Prediction of Intestinal Drug Absorption," *J. Med. Chem.* **44,** 1927–1937 (2001).
36. Egan, W.; Merz, K.M., Jr.; Baldwin, J.J. "Prediction of Drug Absorption Using Multivariate Statistics," *J. Med. Chem.* **43,** 3867–3877 (2000).
37. Norinder, U.; Österberg, T.; Artursson, P. "Theoretical Calculation and Prediction of Caco-2 Cell Permeability Using MolSurf Parametrization and PLS Statistics," *Pharm. Res.* **14,** 1786–1791 (1997).
38. Norinder, U.; Österberg, T.; Artursson, P. "Theoretical Calculation and Prediction of Intestinal Absorption of Drugs in Human Using MolSurf Parametrization and PLS Statistics," *Eur. J. Pharm. Sci.* **8,** 49–56 (1999).
39. Norinder, U.; Österberg, T. "Applicability of Computational Chemistry in the Evaluation and Prediction of Drug Transport Properties," *Perspect. Drug Discovery Design* **19,** 1–18 (2000).
40. Sugawara, M.; Takekuma, Y.; Yamada, H.; Kobayashi, M.; Iseki, K.; Miyazaki, K. "A General Approach for the Prediction of Intestinal Absorption of Drugs: Regression Analysis Using the Physicochemical Properties and Drug-Membrane Electrostatic Interaction," *J. Pharm. Sci.* **87,** 960–966 (1998).
41. Bravi, G.; Wikel, J.H. "Application of MS-WHIM Descriptors: 3. Prediction of Molecular Properties," *Quant. Struct.-Act. Relat.* **19,** 39–49 (2000).
42. Norinder, U.; Sjöberg, P.; Österberg, T. "Theoretical Calculation and Predition of Brain-Blood Partitioning of Organic Solutes Using Mol Surf Parametrization and PLS Statistics," *J. Pharm. Sci.* **87,** 952–959 (1998).
43. Luco, J.M. "Prediction of the Brain–Blood Distribution of a Large Set of Drugs from Structurally Derived Descriptors Using Partial Least-squares (PLS) Modeling." *J. Chem. Inf. Comput.* **39,** 396–404 (1999).
44. Österberg, T.; Norinder, U. "Theoretical Calculation and Prediction of P-glycoprotein-Interaction Drugs Using MolSurf Parametrization and PLS Statistics." *Eur. J. Pharm. Sci.* **10,** 295–303 (2000).
45. Österberg, T.; Norinder, U. "Prediction of Drug Transport Processes Using Simple Parameters and PLS Statistics: The Use of ACD/log P and ACD/ChemSketch Descriptors," *Eur. J. Pharm. Sci.* **12,** 327–337 (2001).

46. Schneider, G.; Coassolo, P.; Lave, T. "Combining *In Vitro* and *In Vivo* Pharmacokinetic Data for Prediction of Hepatic Drug Clearance in Humans by Artificial Neural Networks and Multivariate Statistical Techniques," *J. Med. Chem.* **42,** 5072–5076 (1999).

47. Guba, W.; Cruciani, G. "Molecular Field-Derived Descriptors for the Multivariate Modeling of Pharmacokinetic Data," K. Gundertofte; and F.S. Jorgensen, eds., in *Molecular Modeling and Prediction of Bioactivity*, Kluwer Academic / Plenum Publishers, New York, 89–94 (2000).

48. van der Graaf, P.H.; Nilsson, J.; van Schaick, E.A.; Danhof, M. "Multivariate Quantitation Structure—Pharmacokinetic Relationships (QSPKR) Analysis of Adenosine A_1 Receptor Agonists in Rat," *J. Pharm. Sci.* **88,** 306–312 (1999).

49. Pickett, S.; McLay, I.M.; Clark, D.E. "Enhancing the Hit-to-Lead Properties of Lead Optimization Libraries," *J. Chem. Inf. Comput. Sci.* **40,** 263–272 (2000).

50. Gedeck, P.; Willett, P. "Visual and Computational Analysis of Structure-Activity Relationships in High-Throughput Screening Data," *Curr. Opin. Chem. Biol.* **5,** 389–395 (2001).

51. Kerns, E.H.; Di, L. "Pharmaceutical Profiling in Drug Discovery," *Drug Disc. Today* **8,** 316–323 (2003).

52. Di, L.; Kerns, E.H. "Profiling Drug-like Properties in Discovery Research." *Curr. Opin. Chem. Biol.* **7,** 402–408 (2003).

53. McGovern, S.; Caselli, E.; Grigorieff, N.; Shoichet, B.K. "Common Mechanism Underlying Promiscuous Inhititors from Virtual and High-Throughput Screening," *J. Med. Chem.* **45,** 1712–1722 (2002).

54. Wilson, D.M.; Wang, X.; Walsh, E.; Rourick, R. "High Throughput log D Determination Using Liquid Chromatography-Mass Spectrometry," *Comb. Chem. High Throughput Screen* **4**, 511–519 (2001).

55. Avdeef, A.; Box, K.J.; Comer, J.E.A.; Hibbert, C.; Tam, K.Y. "pH-Metric LogP 10. Determination of Liposomal Membrane-Water Partition Coefficients of Ionizable Drugs," *Pharm. Res.* **15,** 209–215 (1998).

56. Lipiniski, C.A.; Lombardo, F.; Dominy, B.W.; Feeney, P.J. "Experimental and Computational Approaches to Estimate Solubility and Permeability in Drug Discovery and Development Settings," *Adv. Drug Delivery Rev.* **23,** 3–25 (1997).

57. Bevan, C.; Lloyd, R.S. "A High-Throughput Screening Method for the Determination of Aqueous Drug Solubility Using Laser Nephelometry in Microtiter Plates," *Anal. Chem.* **72,** 1781–1787 (2000).

58. Avdeef, A. "pH—Metric Log P: Refinement of Partition Coefficients and Ionization Constants of Multiprotic Substances," *J. Pharm. Sci.* **82,** 183–190 (1993).

59. Caldwell, G.W.; Easlick, S.M.; Gunnet, J.; Massucci, J.A.; Demarest, K. "*In vitro* Permeability of Eight β-Blockers Through Caco-2 Monolayers Utilizing Liquid Chromatography/Electrospray Ionization Mass Spectrometry," *J. Mass Spectrom.* **33,** 607–614 (1998).

60. Irvine, J. D.; Takahashi, L.; Lockhart, K.; Cheong, J.; Tolan, J.W.; Selick, H. E.; Grove, J.R. "MDCK (Madin-Darby Canine Kidney) Cells: A Tool for Membrane Permeability Screening," *J. Pharm. Sci.* **88,** 28–33 (1999).
61. Caldwell, G.W.; Masucci, J.A.; Evangelisto, M.; White, R. "Evaluation of the Immobilized Artificial Membrane Phosphatidylcholine Drug Discovery Column for High Performance Liquid Chromatographic Screening of Drug–Membrane Interactions," *J. Chromatogr. A* **800,** 161–169 (1998).
62. Wohnsland, F.; Faller, B. "High-Throughput Permeability pH Profile and High-Throughput Alkane/Water Log P With Artificial Membranes," *J. Med. Chem.* **44,** 923–930 (2001).
63. Hill, A.; Bevan, C.; Reynolds, D. United Kingdom Patent WO00/13328 (1999).
64. Korfmacher, W.A.; Cox, K.A.; Ng, K.J.; Veals, J.; Hsieh, Y.; Wainhaus, S.; Broske, L.; Prelusky, D.; Nomeir, A.; White, R.E. "Cassette-Accelerated Rapid Rat Screen: A Systematic Procedure for the Dosing and Liquid Chromatography/Atmospheric Pressure Ionization Tandem Mass Spectrometric Analysis of New Chemical Entities as Part of New Drug Discovery," *Rapid Commun. Mass Spectrom.* **15,** 335–340 (2001).
65. McDonnell, P.A.; Caldwell, G.W.; Masucci, J.A. "Using Capillary Electrophoresis/Frontal Analysis to Screen Drugs Interacting With Human Serum Proteins," *Electrophoresis* **19,** 448–454 (1998).
66. Kariv, I.; Cao, H.; Oldenberg, K.R. "Development of a High Throughput Equilibrium Dialysis Metho," *J. Pharm. Sci.* **90,** 580–587 (2001).
67. Di, L.; Kerns, E.H.; Hong, Y.; Kleintop, T.; McConnell, O.J. "Optimization of a Higher Throughput Microsomal Stability Screening Assay for Profiling Drug Discovery Candidates," *J. Biomol. Screening* **8,** 453–462 (2003).
68. Li, A.P.; Lu, C.; Brent, J.A.; Pham, C.; Fackett, A.; Ruegg, C.E.; Silber, P.M. "Cryopreserved Human Hepatocytes: Characterization of Drug-Metabolizing Enzyme Activities and Applications in Higher Throughput Screening Assays for Hepatotoxicity, Metabolic Stability, and Drug–Drug Interaction Potential," *Chem. Biol. Interact*. **121,** 17–35 (1999).
69. Linget, J.; du Vignaud, P. "Automation of Metabolic Stability Studies in Microsomes, Cytosol and Plasma Using a 215 Gilson Liquid Handler," *J. Pharm. Biomed. Anal.* **19,** 893–901 (1999).
70. Kerns, E.H.; Rourick, R.A.; Volk, K.J.; Lee, M.S. "Buspirone Metabolite Structure Profile Using a Standard Liquid Chromatographic–Mass Spectrometric Protocol," *J. Chromatog. B* **698,** 133–145 (1997).
71. Burton, K.I.; Everett, J.R.; Newman, M.J.; Pullen, F.S.; Richards, D.S.; Swanson, A.G. "On-Line Liquid Chromatography Coupled with High Field NMR and Mass Spectrometry (LC-NMR-MS): A New Technique for Drug Metabolite Structure Elucidation," *J. Pharm. Biomed. Anal.* **15,** 1903–1912 (1997).
72. Crespi, C.L. "Higher-Throughput Screening with Human Cytochromes P450," *Curr. Opin. Drug Disc. Dev.* **2,** 15–19 (1999).

73. Dierks, E.A.; Stams, K.R.; Lim, H.K.; Cornelius, G.; Zhang, H.; Ball, S.E.A. "Method for the Simultaneous Evaluation of the Activities of Seven Major Human Drug-Metabolizing Cytochrome P450s Using an *In Vitro* Cocktail of Probe Substrates and Fast Gradient Liquid Chromatography Tandem Mass Spectrometry," *Drug Metab. Dispos.* **29,** 23–29 (2001).
74. Zhang, M. "Brain Uptake Studies Using LC/MS/MS," in *Proceedings of the American Society for Mass Spectrometry. 49th Annual Conference on Mass Spectrometry and Allied Topics*, Santa Fe, NM (2001).

16

ORGAN PERFUSION AND MASS SPECTROMETRY

C. GERALD CURTIS, BEN CHIEN, DAVID BAR-OR, AND KUMAR RAMU

INTRODUCTION

How much does any single technique contribute to the process of drug development? The answer is, not surprisingly, not very much! The complex process from drug discovery to market requires many disciplines, techniques, flair, and intuition. Luck helps a great deal too! Ask project managers at the beginning of a development program which of the preceding requirements they would prefer and most would likely select "luck." Few, if any, would insist on a particular technique. Thus, when different techniques are compared for use in both drug-candidate selection and the subsequent drug-development programs, the critical issue is not whether the technique can produce data, but rather, whether the quality of the science is sufficient to allow "go/no-go" decisions to be made. It seems wasteful to undertake experiments in drug development, regardless of the results, that do not alter the next step in the program.

With this in mind, the two major themes of this chapter focus on (1) a comparison of the quality and usefulness of the data generated in ex vivo isolated blood-perfused organ models with the data obtained from in vivo and in vitro studies, and (2) the advantages of liquid chromatography with tandem mass spectrometry (LC-MS/MS) over other analytical techniques for the measurement of the fate and effects of drugs in perfused organ models.

Integrated Strategies for Drug Discovery Using Mass Spectrometry, Edited by Mike S. Lee

ORGAN-PERFUSION PLATFORM

The ability to remove or isolate an intact mammalian organ or tissue from a host and maintain the organ in a viable state by perfusion of the vascular bed with blood or a suitable substitute is not a new technique. Experiments of this nature were carried out in the seventeenth century. Over the last 50 years, many systems have been described for the perfusion of numerous organs in various species. The major differences between the published models deal with the types of perfusate, the mechanical setup used to deliver the perfusates, and the surgical protocols used to isolate the organs [1–4].

Despite the subtle and not so subtle differences in methodology, there is a general consensus that these models provide data that (1) cannot be obtained directly in vivo, and (2) depicts events in vivo more accurately than tissue slices, isolated cells, or subcellular particles. The latter has been attributed to the fact that intact organs contain a full complement of cell types with normal architecture, microcirculation, and physiological hemodynamics. The maintenance of enzyme activities and endogenous substrates are within normal limits, and the delivery of drugs are controlled via the routes expected in vivo.

Figure 16.1 summarizes the major components of the flexible perfusion platform used at Bowman Research that are representative of those used in

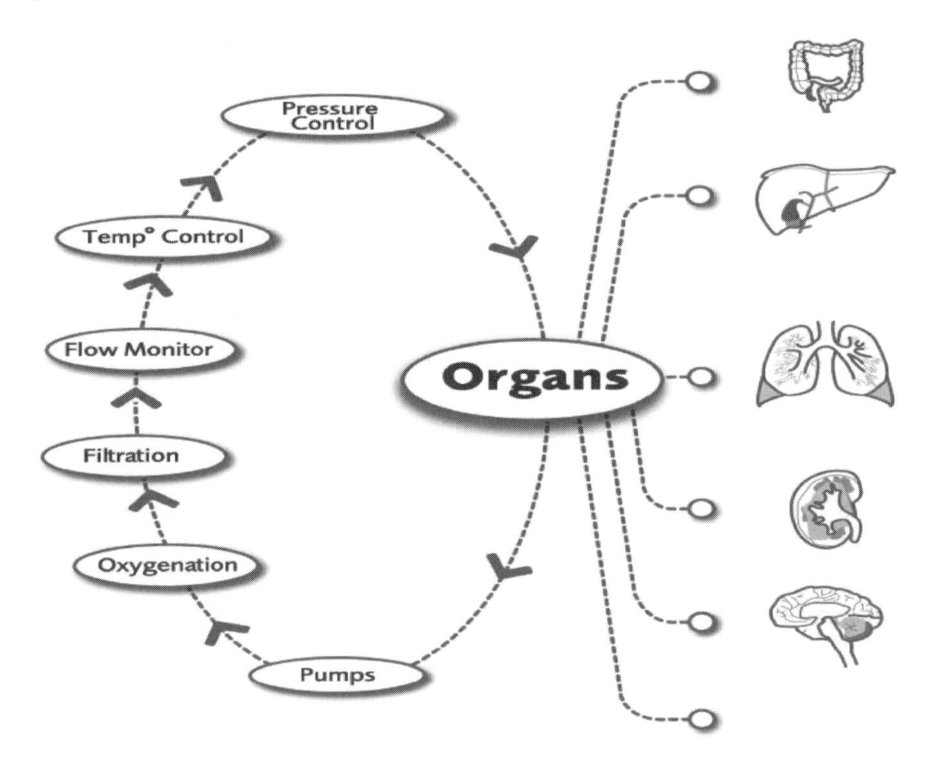

FIGURE 16.1 Organ perfusion setup.

a number of laboratories. The mechanics shown on the left side of Figure 16.1 provides oxygenated blood perfusate that contains the appropriate biochemicals for each organ at 37°C under physiological conditions of perfusion pressure and flow rate. Once perfused through the organ, the effluent blood can either be recirculated or collected first pass. In drug development, the perfused organs most commonly used are intestine, liver, lung, kidney, and brain.

Vascular Perfused Intestine

The gut model is used most often for direct measurement of drug transport from the gut lumen into blood, the nature and extent of gut metabolism, and the effects of different formulations on these processes.

The Gut Model

The intestine is isolated and perfused via the mesenteric artery with oxygenated perfusate, preferably, heparinized blood at 37°C and at a physiological pressure and blood flow. Venous blood from the portal vein is either collected first pass or returned to the reservoir for reoxygenation and recirculation. The effect of bile on absorption from the gut can be assessed by the infusion of bile via an indwelling cannula that is inserted into the bile duct in the direction of the gut.

For absorption studies, the drug(s) are administered directly into the gut lumen. In cases where the contribution of the gut to the whole body metabolism of blood-borne compounds is required, drugs are added to the blood. Several compounds can be dosed as a cassette for drug-screening applications. After dosing under recirculating conditions, blood samples (1–2 mL) are obtained at timed intervals for 3 to 4 hours (postdose) for quantitative analysis and metabolite profiling by LC-MS/MS. Under first-pass conditions, blood is collected at timed intervals up to 1 to 2 hours postdose for quantitative analysis and metabolic profiling. Routinely, the results are expressed as rate of drug absorption from the gut lumen into the circulating blood (or vice versa) as parent drug or metabolites. Typical examples are shown in Figures 16.2 and 16.3.

Assessment of the Model/Data for Drug Development

In the quest for quality data, the question inevitably arises as to how the isolated gut preparation compares with intravenous versus oral administration in vivo or the in situ cannulation of veins draining specific areas of the gut into which the compound(s) has been administrated or in vitro studies that use Ussing chambers or colon adenocarcinoma cells (Caco-2).

The decision as to which model should be used can depend on the question asked. If, for example, the question is "What is absorbed from the intestine into

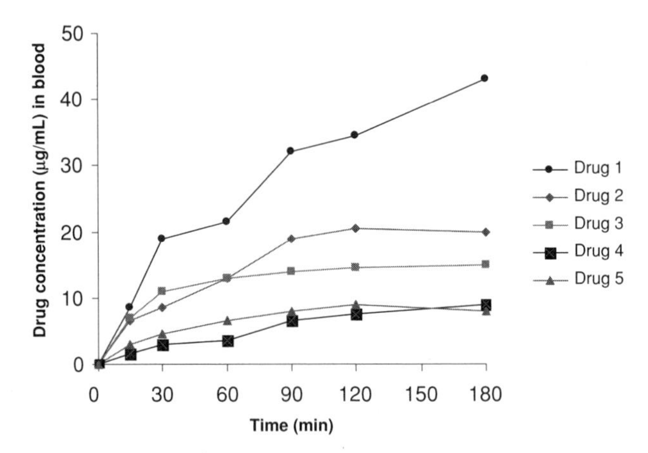

FIGURE 16.2 Relative rates of transport of five drug candidates (1–5) from the gut lumen into blood in the isolated ex vivo perfused-gut preparation.

the blood? how much? and in what form?," then any model that does not have a blood supply (i.e., all the in vitro models) will not give unequivocal data and must rely upon certain assumptions. For example, the transport across the entire gut wall from the serosal side of the Ussing chamber would be equivalent in vivo to transport from the gut lumen into the peritoneal cavity. It would be assumed that the data from this model are equivalent to transport from the lumen into the blood and that the blood and blood vessels have no role in this process! This assumption indeed may be valid for some compounds. However, is this assumption true for all compounds?

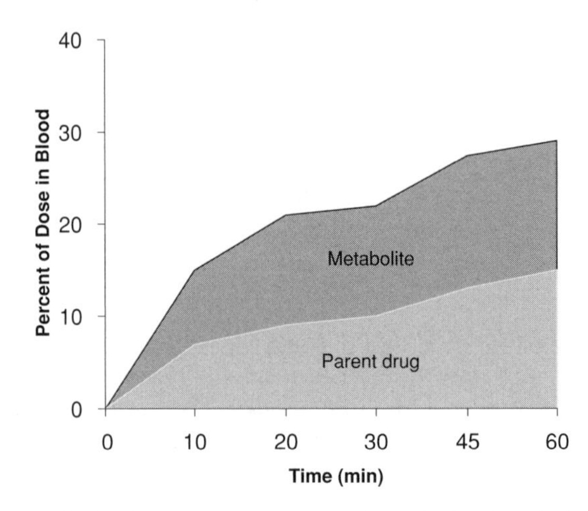

FIGURE 16.3 Metabolism in the gut and transport of metabolites into the blood in an isolated ex vivo perfused-gut preparation.

The use of Caco-2 cells to study both drug transport and metabolism has been intensive within the pharmaceutical industry. Much emphasis has been placed on the use of these cells to screen for absorption, because these cells are of human origin and the throughput is high. But what assumptions are made that influence the quality of the data? Clearly, these cells do not produce mucus; therefore, it must be assumed that this layer does not play a significant role in drug absorption! Despite the heterogeneity of the gut wall, it is assumed that this colonic epithelium adequately mimics the properties of the entire intestinal epithelium! The lack of certain key drug metabolizing, enzymes in Caco-2 cells can be overcome by replacement. But how much enzyme should be added? Furthermore, are the amounts of enzymes important? In short, while some studies with Caco-2 cells and closely related compounds seem to show good correlation with in vivo absorption data, other compounds do not. Therefore, if gut metabolism and transport into blood can be measured directly, then why take the risk with less physiological models?

In vivo comparison of plasma or blood profiles after intravenous and oral administration provide values for absolute bioavailability (or presystemic elimination). However, such studies do not distinguish between the contribution of the gut and liver and do not measure absorption directly. Low bioavailability can be attributed to either low absorption or extensive metabolism or a combination of both.

By contrast, the in situ cannulation of veins that drain specific areas of the gut and the collection of all the effluent blood from this area (which is replaced by continuous infusion of blood into the cannula) has all the advantages of the totally isolated model. The major difficulty with this model is the ability to control the blood pressure and flow from the dosed area of the gut.

Until recently, the reluctance to use the isolated perfused gut preparation for direct measurement of absorption and gut metabolism has been due in part to the resources and time required to set up the system, the high maintenance, and the limited number of compounds that could be examined simultaneously. However, this reluctance no longer exists, as throughput has increased dramatically with (1) cassette dosing and (2) LC-MS/MS methods. In this regard, a strategy has been developed whereby the absorption rate of up to 25,000 compounds can be evaluated per year. Therefore, measurement of absorption from the gut need not be rate limiting in drug optimization and drug development.

Mechanistically, the data from perfused gut experiments will distinguish unequivocally whether poor absorption is due to the inherent, physicochemical properties of the molecule, or to metabolism in the gut. In this way, the data clearly dictate the direction of subsequent attempts to improve absorption. If the compound is not metabolized, then changes in formulation may improve absorption. On the other hand, if gut metabolism is limiting absorption, then changes within the structure (at appropriate site(s) on the molecule) is more likely to improve absorption.

Once the gut model and the LC-MS/MS assay(s) have been set up, it is simply routine to extend the model to examine those factors that might affect absorption, such as formulation, dose, regional differences, diet, species, strains, gender, age, blood flow, and other drugs. Moreover, it is possible to define precisely what form(s) of the biliary products reenter the blood from experiments that involve the infusion of bile from donor animals that are dosed with drug(s) into the lumen of isolated gut preparations.

When compounds are added directly to the blood, rather than into the gut lumen, the perfused gut can be used to measure the contribution of the gut to the clearance/metabolism of blood-borne drugs. If (and only if) there is a significant quantitative efflux from the blood into the gut lumen, then it is worthwhile to pursue the role of transporter proteins (e.g., polyglycoproteins (PGP)), as well as the effects other substances may have with the same transporter system.

In conclusion, by dosing into the lumen or blood or via the bile, it is possible to define the ebb and flow of drug and drug-related products between the gut and its blood supply. The factors that can affect these processes and change the therapeutic (or toxic) profile in vivo can be quantified. Table 16.1 summarizes the major applications of the perfused-gut model in drug-candidate optimization and subsequent drug development.

Vascular Perfused Liver

The isolated vascular perfused liver preparation mimics the liver in vivo and has normal cell–cell topography, vasculature, blood, and bile flows. This model is used to directly measure the contribution of the liver in the presystemic elimination of drugs, and to evaluate factors that affect this process.

The Liver Model

The liver is routinely perfused with heparinized blood from the same species, or with a perfusate comprised of washed bovine erythrocytes that are suspended in a buffer that contains albumin. Perfusion is achieved via the portal vein (and via the hepatic artery, if necessary), and the venous effluent is collected either first pass or reperfused through the liver after reoxygenation. Drugs are administered either by pulse dosing (single or cassette) or by constant infusion into the perfusate. Routinely, perfusate samples are taken at timed intervals up to 6 hours after dosing for quantitative analysis of parent drug, metabolites, and markers (of liver damage) by LC-MS/MS. Bile is collected throughout the study and the parent drug or metabolites are analyzed. The results are usually expressed as the rate of hepatic uptake and biliary

TABLE 16.1 Major Applications of Vascular Perfused Organs in Drug-Candidate Optimization and Development

Perfused Organ	Screening Chemical Libraries For	SAR	Contribution to Whole Body Metabolism	Factors Affecting	Others
Gut	Absorption into blood	Yes	Regional differences	Dose formulation Bile flow Other drugs Gut damage	Role of transporter proteins
Liver	First-pass hepatic CL	Yes	Quantitative distribution of metabolites in liver, blood, and bile	Dose Blood flow Other drugs Cholestasis Liver toxicity	Role of transporter proteins Quantify free-radical production
Kidney	Renal CL	Yes	Quantitative distribution of metabolites in kidney, blood, and Urine	Dose Other drugs Protein binding Diuretics Kidney toxicity	Mechanisms of excretion
Lung	Absorption from airways	Yes	Metabolism via airways vs. blood Volatile products	Dose formulation Other drugs Respiratory rate Pulmonary toxicity	Pulmonary transit times
Brain	Uptake from blood	Yes	Quantitative distribution of metabolites in brain, blood, and CSF	Dose Blood flow Other drugs Ischemia TBI	Role of transporter proteins

Notes: SAR–Structure activity relationship; CL–clearance; CSF–cerebrospinal fluid; TBI–traumatic brain injury.

excretion of drug(s), as well as quantitative and qualitative partitioning of metabolites between blood and bile.

Assessment of the Model/Data for Drug Development

Scientists who have used a perfused liver preparation for any length of time are aware of the quality of the data in terms of hepatic clearance, contribution of the liver to presystemic elimination and whole-body metabolism, partitioning of drugs, and their metabolites between blood and bile. Yet despite the ability to quantify the time-dependant flux of drugs/metabolites between blood, liver, and bile with this ex vivo model over 6 hours and longer, alternative in vitro techniques are used more frequently in the pharmaceutical industry. These techniques include hepatocytes, S9 or microsome preparations, tissue slices, and homogenates. Hepatocyte and microsome preparations are more convenient to use and technically demand less time and fewer resources. It is certainly possible to screen many more compounds with these in vitro preparations. Sample throughput is clearly an important benchmark within the industry. But how good are the data and how are they used? In defense of these in vitro systems, especially systems that incorporate isolated hepatocytes and microsome preparations, there have been a number of elegant studies that show the metabolic data obtained in vitro can be scaled up to predicted hepatic clearance data that are consistent with those found in vivo.

On the other hand, it is also the case that more compounds are identified that are metabolized extensively in the intact liver, but only trace amounts of metabolites are produced in hepatocytes or microsome preparations. When this situation occurs, it is not difficult to conjure up a host of possible explanations. For example, the lack of metabolism in hepatocytes, but not in whole liver, may be due to metabolism in other liver cells such as Kupffer cells or cells of the biliary tree. The inability to remove metabolites via the bile may cause feedback inhibition/toxicity [5]. The cells may have lost key substrates or enzymes, and intermediary metabolism may be depleted despite an intact plasma membrane with high viability measurements. For example, protein synthesis may be zero, but the cell membrane is still intact and able to exclude dyes.

Most of the phase II conjugation reactions (with microsome preparation) are missing, so the assumption is made that these reactions and the removal of the conjugates do not affect the kinetics of metabolism in the liver. In contrast, extensive metabolism can be obtained with in vitro systems, but trace amounts or different metabolites can be obtained in whole livers. This situation could be due to metabolism in whole liver by other cells or pathways; factors such as blood or bile flow contribute in whole liver, but do not exist in in vitro systems. The inability of compounds that can penetrate to the site of metabolism during perfusion is observed in vitro.

As mentioned previously, the choice of the system or method employed ultimately depends on the question asked and how the data generated are subsequently used. Advocates of perfused-liver models would argue to determine first-pass hepatic clearance in preclinical development, then measure clearance directly, and avoid as many assumptions as possible. Table 16.1 summarizes the major applications of the perfused liver model in drug candidate optimization and subsequent development. As is the case with the gut, the contribution of the liver to the fate and effects of drugs can be influenced by numerous factors, many of which can be studied with the perfused-liver model. These effects include the blood flow, species and gender differences, and induction.

Vascular Perfused Kidney

While the liver is generally recognized as a major contributor to the PK and whole-body metabolism of many disparate compounds, this case is true for all drugs. A number of extrahepatic organs or tissues take up and metabolize a significant proportion of some drugs. The kidney, for example, is able to handle peptides and proteins, as well as a wide range of other compounds. The extent to which the organs are involved in metabolism and clearance can be measured directly in the isolated perfused-kidney model.

The Kidney Model

The isolated kidney receives a heparinized blood-based perfusate via the renal artery, at the appropriate temperature, pressure, and blood flow. Venous effluent from the renal vein is either collected first pass or recirculated. Urine is collected continuously during perfusion from a cannula that is inserted into the ureter.

Compounds are added directly to the perfusate as a pulse dose or constant infusion. Aliquots of the perfusate are taken for analysis at appropriate times, usually over 3 hours, and the urine is collected at timed intervals up to 3 hours. For renal binding studies, the kidney is first perfused with the drug and, after an appropriate time, perfusion is switched to fresh, drug-free perfusate. The subsequent release of drug-related material(s) into blood or urine can then be monitored. Table 16.1 summarizes the major applications of the perfused-kidney model in drug development.

Vascular Perfused Lungs

Inhalation into the lungs provides an important route of administration for some types of drugs. However, measurement of the extent of absorption via the lungs and the nature of the materials absorbed into the pulmonary vasculature is difficult in vivo because of the variable amounts swallowed and subsequently absorbed from the gut lumen. In the perfused-lung model, this problem does not arise.

The Lung Model

The surgically isolated respiring lung is perfused with heparinized blood via the pulmonary artery at physiological temperature pressure and blood flow. Effluent blood is either collected at timed intervals (first pass) or allowed to recirculate. Compounds are dosed via the trachea or added directly to the blood. After dosing, aliquots of blood are removed for quantitative analysis up to 3 hours postdose. Volatile compounds/metabolites eliminated via the expired air can be collected throughout the perfusing period with suitable trapping agents. Results are usually expressed as the percent of dose transported from the airways into blood (or vice versa) as parent drug or metabolites. Table 16.1 contains a summary of the major applications of perfused lungs in drug-candidate optimization and drug development.

Vascular Perfused CNS

The undeniable advantages that the isolated vascular perfused-brain model brings is probably due to the fact that the entire blood supply of the organ, its architecture, and especially the blood–brain barrier, remains intact. Thus, not only is the organ more viable, but the organ is exposed to chemicals as it would be in vivo and is therefore more likely to respond to chemical challenge in a way that mimics (and therefore predicts) events in vivo [6].

Many other alternative systems have been pursued with vigor and include the use of tissue biopsies, cells in culture, as well as subcellular particles, for screening compounds for metabolism, therapeutic activity, and toxicity. Unfortunately, the data from these simplified systems are often not sufficiently reliable to explain or predict either the fate or the effects of compounds in whole organs. For example, brain slices readily metabolize pyruvate, but the intact brain is unable to do so, probably because of restricted transport.

In terms of mechanics, perfusing the brain is not significantly different from any of the other organs previously described. There is a serious ethical issue, however, namely if the brain is perfused with oxygenated blood, would it "feel pain" and would it be "conscious." When asked this question for the first time, the initial answer of most scientists is "no," but soon after, it is clear that doubts and uncertainties begin to arise and invariably the issue is not resolved. Thus, while the use of perfusion models for most organs appears to hold the moral high ground as alternatives to whole-animal experiments, this is not the case with the intact brain. Until the issue of "pain and consciousness" is resolved, the only safe approach is to maintain an anesthetic in the perfusate at sufficient concentration for complete sedation. While this approach should satisfy the ethical issue, it is not ideal for determination of the fate of drugs, because of the possible modifications of uptake and modification to the blood–brain barrier and transport in and out of the brain by anesthetics.

The Brain Model

In the in situ rat-brain-perfusion model of Takasato et al. [7,8], or the modified Takasato model of Smith [9], the perfusion catheter is placed in the external carotid artery or the common carotid artery, respectively. In both cases, after ligature of appropriate vessels, one of the cerebral hemispheres is completely perfused via the internal carotid artery. Alternatively other groups insert the infusion cannula in the left cardiac ventricle or in the aorta. These systems ensure perfusion to both hemispheres of the brain [10–12].

The brain is perfused with heparinized blood or a suitable substitute containing plasma concentrations of anesthetic that are sufficient to maintain total anesthesia in vivo. These levels of anesthetics must be maintained throughout the perfusion period. The effluent blood from the internal jugular veins is collected first pass or recirculated.

Once the perfusion pressure and flow rate are stable, test compound(s) is (are) added directly to the perfusate. The initial concentration is invariably chosen on either PK data or plasma concentrations known to be associated with therapeutic or toxic effects in vivo. Table 16.1 summarizes the major applications of the perfused brain in lead candidate optimization and development.

A STRATEGY FOR THE PREDICTION OF THE NATURE AND EXTENT OF DRUG METABOLISM IN HUMANS

When individual organs are perfused with whole blood at flow rates that mimic those organs in vivo, then addition of drug(s) to the perfusate and measurement of time-dependent generation of metabolites establishes which of the major organs is primarily responsible for metabolism in vivo. It cannot be assumed that the liver will be the major site of metabolism. The kidney, for example, can be more important quantitatively than the liver or other tissues for hydrolytic reactions. In the hydrolysis of merepenem in the rat, for example, the rate of metabolism in the kidney is approximately three- to fourfold higher than the liver, regardless of the difference in size of the two organs. Cross-species comparisons of metabolism by liver tissue in vitro would produce results. However, it is unlikely that the data could be used to predict or explain species difference in the nature and extent of drug metabolism in vivo. In short, it is sensible to use the same tissue or organ in other species that include human. While it is possible to use the same perfused-organ techniques for cross-species comparisons in animals, this is clearly not possible in humans where in vitro systems are the only alternative.

TABLE 16.2 Plasma Clearance of Merepenem in Isolated Perfused Rat Liver, Kidneys, Gut, and Lung

Perfused Organ	$T\,^1\!/_2$ (min)
Liver	18
Kidneys	5
Gut	25
Lung	>60

This raises the question of what in vitro system to use and which system will provide real data on the relative rates of metabolism in different species. Should tissue slices, isolated cells, homogenates, or subcellular fractions be used? One approach to this difficulty is to use the rate of metabolism per gram of tissue (or mg protein) derived from the perfused organ studies in rat as the yardstick by which to validate the rate of metabolism in in vitro systems. Often the rate of metabolism in slices, isolated cells, homogenates, or subcellular fractions with an arbitrary incubation medium is well below the expected levels. However, by judicious manipulation of the incubation medium, it should be possible to improve the nature and rates of metabolism in vitro so that the rates approach the whole perfused organ. If the improved rates are 80% or more than that in the whole organ, then cross-species comparisons that use the same conditions should provide data that can be used for go/no-go decisions.

In summary, a recommended strategy to obtain reliable metabolic data in humans is as follows. First, identify the organ(s) that control metabolism in the rat. Second, validate the in vitro metabolism system with the rat tissue selected as stated earlier. Finally, investigate cross-species differences with the validated in vitro system.

With the metabolism of merepenem as an example, perfused organ studies in the rat identified the kidney as the major organ of metabolism (Table 16.2). Subsequently, it was established that fresh rat kidney slices and homogenates could be used in vitro. Cross-species studies showed that the rate of metabolism in rat kidney preparations was three- to fourfold higher than in human kidney (Table 16.3). These data predicted that the plasma clearance

TABLE 16.3 Relative Rates of Merepenem Metabolism in Rat and Human Kidney Preparations In Vitro

Tissue Preparation	Metabolism Ratio
Rat kidney slices	3.7
Human kidney slices	1.0
Rat kidney homogenates	2.9
Human kidney homogenates	1.0

of this compound in humans in vivo would be severalfold slower than in rat and similar to dog, which indeed was the case.

DRUG–DRUG INTERACTIONS

It is highly likely that many drugs in development will, if they go to market, be administered simultaneously with others. Surprisingly, despite all the polypharmacy, few (as a percent of total drugs administered to humans) drugs fail because of adverse interactions with others, although when this happens, the results can be devastating. In an attempt to be aware of potential interactions, it is common practice to incubate cassettes of drugs over a range of concentrations with hepatic microsomes in vitro. However, the lack of metabolic interactions in vitro does not exclude other possible mechanisms, such as competition for binding to plasma proteins, cellular uptake, intracellular binding, and secretory and excretory processes. Any of these mechanisms could influence the plasma concentration of free drug(s), and therefore, drug potency and toxicity.

These mechanisms of metabolic interactions pertain to the whole organ with the intact blood supply; thus, this model may well be a more comprehensive and reliable approach to test for drug–drug interactions. The use of perfusion techniques helps to establish which organ(s) is (are) responsible for the control of the plasma concentration in a class/family of structurally analogous drug candidates (whether by metabolism or clearance). This model should be a suitable substitute for whole-animal studies to evaluate potential drug–drug interactions. The advantages include tight control of blood levels of each drug; normal binding to plasma proteins; and the delivery of each drug to the organ under physiological conditions. When drugs are administered in combination, these drugs show extensive inhibition/activation with hepatic microsomes. However, the drugs do not affect the plasma concentration of each other in the perfused liver model or in vivo. In this way, the perfused organ models eliminate false-positives and identify interactions at extramicrosomal sites.

FURTHER APPLICATIONS OF PERFUSED ORGANS IN METABOLIC STUDIES

The direct measurement of the nature and extent of metabolism by individual organs over a range of drug–drug candidate doses provides only a snapshot of metabolic changes. It is well known that a number of factors influence metabolism in vivo, and therefore, these influences are observed at the organ level too. The effect of these factors on the potential therapeutic/toxic effects of drug candidates is not considered to be on the critical path of development.

These factors can haunt development teams before market launch. These factors include the effects of formulation; diet; species; age; gender; blood flow; polymorphism; and other drugs. These factors can be evaluated with perfused-organ models by comparison with baseline metabolism. In addition, an important area that has yet to be fully exploited is the effect(s) of specific organ/tissue damage on the nature and extent of drug metabolism. The purpose is to closely mimic the disease or damage that is likely to be found in the patients for whom the drug is intended. If, for example, the liver is the major site of metabolism for a particular compound, then there may be substantial differences in fatty, cirrhotic, neoplastic, viral, or bacterial infected or jaundiced livers. Interestingly, the effects of inflammation or gastroenteritis on gut absorption and gut metabolism would be valuable to monitor. The same rationale can be applied to any other organ that plays a major role in the fate of a compound in vivo.

LC-MS/MS PLATFORM

The throughput in organ perfusion studies has increased dramatically with the use of technologies for fast LC-MS separation, dual- or multiple-column LC-MS, turbulent-flow LC-MS, and cassette analysis [13–17].

LC-MS/MS is the current method of choice for the analysis of small molecules in biological fluids (i.e., perfusate plasma, urine, bile) from organ-perfusion studies. The speed, high specificity, and sensitivity of this analysis format are widely applicable to compounds with diverse structure. Simultaneous dosing of a number of compounds followed by analysis of multiple analytes with LC-MS/MS (*N*-in-1) has been developed and shown to be an effective approach to improve the throughput of pharmacokinetics (PK) and metabolite screening [18,19].

High Throughput Cassette Analysis

Biological matrices are not directly compatible with LC-MS analysis, since these samples tend to block LC columns and contaminate the ion source. Extraction of compounds of interest from biological fluids is required prior to LC-MS/MS analysis [20]. Sample extraction can be achieved off-line with protein precipitation (PP), liquid–liquid extraction (LLE), or solid-phase extraction (SPE) [21]. With the ease of use and sophistication of automated liquid-handling systems, sample extraction procedures in a 96-well format can handle microliter volumes with multiple sorbents per plate and can simplify and expedite SPE method development [22,23]. The technique can be used to routinely develop methods for multiple analytes and examine a set of eluent compositions for each analyte [16].

Direct plasma injection [24] with a combination of turbulent-flow and column-switching (CS) [25] systems have been used for on-line sample extraction in multiple-component LC-MS/MS analysis. This simple and reliable technique combines speed and ruggedness of turbulent chromatography with the focusing and resolving power of the CS format [26]. Manual protein precipitation by addition of an organic solvent along with an appropriate internal standard prior to sample loading is an indirect on-line SPE extraction technique [27].

High-pressure liquid chromatography (HPLC) formats that use reverse-phase analytical columns dominate the bioanalytical field. Isocratic elution formats are used for traditional LC-MS/MS bioanalysis, since these separations do not require additional time for column reequilibration. However, gradient elution is a more effective approach for cassette analysis in highthroughput screening methods for discovery applications because of its versatility for the separation of compounds that have a wide variety of lipophilicity.

The use of CS techniques reduces the column equilibration time and flushes the matrix components to waste (decreases source contamination), while narrow-bore LC columns help not only with flow rates that were amenable to most atmospheric- pressure ionization (API)- interfaces but also provide increased sensitivity [28]. A change from a serial to a parallel format further increases throughput for bioanalysis, with timed injections at intervals from multiple LC systems and staggered MS detection [29–31].

There are two modes of ionization: electrospray (ESI) and atmospheric-pressure chemical ionization (APCI). ESI is considered to be more applicable for the ionization a wide variety of molecules that include polar and non-volatile molecules [32]. Detection can be made in either positive- or negative-ion mode. The negative-ion mode of detection is generally used for acidic compounds, while the positive-ion mode of detection performs better for neutral or basic compounds.

The use of positive-ion ESI with reverse-phase liquid chromatography is common for most methods. Mobile phases are often a combination of either methanol or acetonitrile combined with water that contains additives such as acids or volatile ammonium buffers. Optimization between increased sensitivity with a higher percent of organic solvent in the mobile phase and adequate retention on the column is important for LC-MS/MS detection.

Bioanalysis has been successfully conducted using different mass analyzers such as single-quadrupole (SQ), triple-quadrupole (TQ), ion-trap (IT), time-of-flight (TOF), and quadrupole time-of-flight (Q-TOF) mass spectrometers. The use of multiple-reaction monitoring (MRM) with MS/MS detection on a TQ mass spectrometer offers the selectivity and dynamic range that are essential for high-throughput assays. Sensitivity, reproducibility, selectivity, and specificity are key to successful high throughput bioanalysis [27,31].

CONCLUSIONS

Automated high throughput screening and bioanalysis accompanied by fast and efficient LC-MS/MS separation with ESI interfaces is the current method of choice for the analysis of small molecules in biological fluids. The simultaneous dosing of numerous compounds followed by multiple-component analysis using LC-MS/MS is an effective way to improve the throughput of PK and metabolism screens. The application of these LC-MS–based analytical techniques to perfused-organ technology has resulted in high-quality data that can be produced at rates that are orders of magnitude greater than previously possible.

REFERENCES

1. Ross, B.D. *Perfusion Techniques in Biochemistry*, Clarendon Press, Oxford, 1972.
2. Ritchie, H.D.; Hardcastle, J.D. *Isolated Organ Perfusion*, Crosby Lockwood, Staples, London, 1973.
3. Borchardt, R.T.; Smith, P.L.; Wilson, G., eds. *Models for Assessing Drug Absorption and Metabolism*, Plenem Press, New York and London, 1996.
4. Powell, G.M.; Hughes, H.M.; Curtis, C.G. "Isolated Perfused Liver Technology for Studying Metabolic and Toxicological Problems," *Drug Metabol. Drug Intract.* **7,** 53–86 (1989).
5. Hughes, H.M.; George, I.M.; Evans, J.C.; Rowlands, C.C.; Powell, G.M.; Curtis, C.G. "The Role of the Liver in the Production of Free Radicals During Halothane Anaesthesia in the Rat," *Biochem. J.* 795–800 (1991).
6. Smith Q.R. "Brain Perfusion Systems for Studies of Drug Uptake and Metabolism in the Central Nervous System," pp. 285–307 in Borchardt, R.T.; Smith, P.L.; Wilson, G., eds., *Models for Assessing Drug Absorption and Metabolism*, Plenem Press, New York and London (1997).
7. Takasato, Y.; Rapoport, S.I.; Smith, Q.R. "A New Method to Determine Cerebrovascular Permeability in the Anesthetized Rat," *Soc. Neurosci. Abstr.* **8**:850 (1982).
8. Takasato, Y.; Rapoport, S.I.; Smith, Q.R. "An In Situ Brain Perfusion Technique to Study Cerebrovascular Transport in the Rat," *Am. J. Physiol.* **247,** H484–H493 (1984).
9. Smith, Q.R. "Brain Perfusion Systems of Drug Uptake and Metabolism in the Central Nervous System in Models for Assessing Drug Absorption and Metabolism," *Pharm. Biotech.* **8,** 285–307 (1996).
10. Thompson, A.M.; Robertson, R.C.; Bauer, T.A. "A Rat Head-Perfusion Technique Developed for the Study of Brain Uptake of Materials," *J. Appl. Physiol.* **24,** 407–411 (1968).

11. Greenwood, J.; Luthert, P.J.; Pratt, O.E.; Lantos, P.L. "Maintenance of the Integrity of the Blood–Brain Barrier in the Rat During an In Situ Saline-Based Perfusion," *Neurosci. Lett.* **56,** 223–227 (1985).
12. Luthert, P.J.; Greenwood, J.; Pratt, O.E.; Lantos, P.L. "The Effect of Metabolic Inhibitor upon the Properties of the Cerebral Vasculature During Whole-Head Saline Perfusion of the Rat," *Quart. J. Exp. Physiol.* **72,** 129–141 (1987).
13. Korfmacher, W.A; Cox, K.A; Ng, K.J; Veals, J.; Hsieh, Y.; Wainhaus, S.; Broske, L.; Prelusky, D. "Cassette-Accelerated Rapid Rat Screen: A Systematic Procedure for the Dosing and Liquid Chromatography/Atmospheric Pressure Ionization Tandem Mass Spectrometric Analysis of New Chemical Entities as Part of New Drug Discovery," *Rapid Commun. Mass Spectrom.* **15,** 335–340 (2001).
14. Cai, Z.; Han, C.; Harrelson, S.; Fung, E.; Sinhababu, A.K. "High-Throughput Analysis in Drug Discovery: Application of Liquid Chromatography/Ion-Trap Mass Spectrometry for Simultaneous Cassette Analysis of Alpha-1a Antagonists and Their Metabolites in Mouse Plasma," *Rapid Commun. Mass Spectrom.* **15,** 546–550 (2001).
15. Cai, Z.; Sinhababu, A.K.; Harrelson, S. "Simultaneous Quantitative Cassette Analysis of Drugs and Detection of Their Metabolites by High Performance Liquid Chromatography/Ion Trap Mass Spectrometry," *Rapid Commun. Mass Spectrom.* **14,** 1637–1643 (2001).
16. Janiszewski, J.S.; Swyden, M.C.; Fouda, H.G. "High-Throughput Method Development Approaches for Bioanalytical Mass Spectrometry," *J. Chromatogr. Sci.* **38,** 255–258 (2000).
17. Zweigenbaum J.; Heinig K.; Steinborner S.; Wachs T.; Henion J. "High-Throughput Bioanalytical LC/MS/MS Determination of Benzodiazepines in Human Urine: 1000 Samples per 12 Hours," *Anal. Chem.* **71,** 2294–2300 (1999).
18. Brewer, E.; Henion J. "Atmospheric Pressure Ionization LC/MS/MS Techniques for Drug Disposition Studies," *J. Pharm. Sci.* **87,** 395–402 (1998).
19. Jemal, M. "High-Throughput Quantitative Bioanalysis by LC/MS/MS," *Biomed. Chromatogr.* **14,** 422–429 (2000).
20. Henion, J.; Brewer, E.; Rule, G. "Sample Preparation for LC/MS/MS: Knowing the Basic Requirements and the Big Picture of an LC/MS System can Ensure Success in Most Instances," *Anal. Chem.* **70,** 650A–656A (1998).
21. Mallet C.R.; Mazzeo J.R.; Neue U. "Evaluation of Several Solid Phase Extraction Liquid Chromatography/Tandem Mass Spectrometry On-Line Configurations for High-Throughput Analysis of Acidic and Basic Drugs in Rat Plasma," *Rapid Commun. Mass Spectrom.* **15,** 1075–1083 (2001).
22. Jemal, M.; Teitz, D.; Ouyang, Z.; Khan, S. "Comparison of Plasma Sample Purification by Manual Liquid–Liquid Extraction, Automated 96-Well Liquid–Liquid Extraction and Automated 96-Well Solid-Phase Extraction for Analysis by High-Performance Liquid Chromatography with Tandem Mass Spectrometry," *J. Chromatogr. B Biomed. Sci. Appl.* **732,** 501–508 (1999).

23. Watt, A.P.; Morrison, D.; Locker, K.L.; Evans, D.G. "Higher Throughput Bioanalysis by Automation of a Protein Precipitation Assay Using 96-well Format with Detection by LC-MS/MS," *Anal. Chem.* **72,** 979–984 (2000).
24. Needham, S.R.; Cole, M.J.; Fouda, H.G. "Direct Plasma Injection for High-Performance Liquid Chromatographic-Mass Spectrometric Quantitation of the Anxiolytic Agent CP-93393," *J. Chromatogr. B.* **718,** 87–94 (1998).
25. Gao, V.C.; Luo, W.C.; Ye, Q.; Thoolen, M. "Column Switching in High-Performance Liquid Chromatography with Tandem Mass Spectrometric Detection for High-Throughput Preclinical Pharmacokinetic Studies," *J. Chromatogr. A.* **18,** 141–148 (1998).
26. Wu, J.T.; Zeng, H.; Qian, M.; Brogdon, B.L.; Unger, S.E. "Direct Plasma Sample Injection in Multiple-Component LC-MS-MS Assays for High-Throughput Pharmacokinetic Screening," *Anal. Chem.* **72,** 61–67 (2000).
27. Ackermann, B.L.; Ruterbories, K.J.; Hanssen, B.R.; Lindstrom, T.D. "Increasing the Throughput of Microsomal Stability Screening Using Fast Gradient Elution LC/MS," in *Proceedings of the 46th ASMS Conference on Mass Spectrometry and Allied Topics*, Orlando, FL, 1998.
28. Warner, D.L.; Dorsey, J.G. "Reduction of Total Analysis Time in Gradient Elution, Reversed-Phase Liquid Chromatography," *LC–GC* **15,** 254–262 (1997).
29. Korfmacher, W.A.; Veals, J.; Dunn-Meynell, K.; Zhang, X.P.; Tucker, G.; Cox, K.A.; Lin, C.C. "Demonstration of the Capabilities of a Parallel High Performance Liquid Chromatography Tandem Mass Spectrometry System for use in the Analysis of Drug Discovery Samples," *Rapid Commun. Mass Spectrom.* **13,** 1991–1999 (1999).
30. Wu, J.T. "The Development of a Staggered Parallel Separation Liquid Chromatography/Tandem Mass Spectrometry System with On-Line Extraction for High-Throughout Screening of Drug Candidates in Biological Fluids," *Rapid Commun. Mass Spectrom.* **15,** 73–81 (2001).
31. Yang, L.; Mann, T.D.; Little, D.; Wu, N.; Clement, R.P.; Rudewicz, P.J. "Evaluation of a Four-Channel Multiplexed Electrospray Triple Quadrupole Mass Spectrometer for the Simultaneous Validation of LC/MS/MS Methods in Four Different Preclinical Matrices," *Anal. Chem.* **73,** 1740–1747 (2001).
32. Bruins, A.P. "Mechanistic Aspects of Electrospray Ionization," *J. Chromatogr.* **794,** 345–357 (1998).

17

SAMPLE PREPARATION FOR DRUG DISCOVERY BIOANALYSIS

DAVID A. WELLS

INTRODUCTION

Pharmaceutical drug discovery and development laboratories are under increased demands to perform their work in a rapid yet accurate manner to meet shortened industry timelines. Within this constraint, it can be a challenge to quickly develop bioanalytical methods that selectively separate drugs and metabolites from endogenous materials in the sample matrix. Fortunately, many different sample preparation techniques are available to meet the desired objectives for an individual assay. This chapter discusses the objectives for bioanalytical sample preparation prior to LC-MS/MS analysis and provides an overview of the most widely used sample preparation methods. Each technique is described in terms of its fundamentals, overall consideration for usage, and high throughput methodologies and applications.

SAMPLE-PREPARATION PERSPECTIVES

Importance of Sample Preparation

The determination of drug concentrations in biological samples yields the data necessary to understand the metabolism and time course of drug action (pharmacokinetics (PK)) in animals, and is an essential component of the

Integrated Strategies for Drug Discovery Using Mass Spectrometry, Edited by Mike S. Lee

drug-discovery process. To obtain these data, each sample of fluid obtained from an in vitro or in vivo study is subjected to sample preparation, chromatographic separation, and analytical detection. There have been tremendous advances in liquid chromatography mass spectrometry (LC-MS) and LC tandem mass spectrometry (LC-MS/MS) detection systems in recent times, and other chapters in this book discuss these enhanced capabilities. The focus of this chapter is to introduce the goals for sample preparation and the many choices available to the drug-discovery scientist. Applications that demonstrate high throughput sample preparation are of particular interest since they help meet the rapid turnaround needs required within the drug-discovery research environment.

Sample preparation is an important and necessary step in the overall analytical process, because most analytical instruments cannot accept the sample matrix directly. If these materials were to be injected, then the consequences may include a rapid deterioration in the separation performance of the chromatographic column, clogged frits or lines that result in an increased system back pressure, and detector fouling that may reduce system performance and result in downtime for cleaning the ionization source.

The major goals for sample preparation are to remove unwanted endogenous matrix components (e.g., protein, salts, metabolic by-products) that can cause interferences upon analysis, thus improving method specificity; concentrate an analyte to improve its limit of detection; and exchange the analyte from a totally aqueous environment into a chosen percentage of organic/aqueous solvent appropriate for injection into a chromatographic system. Some additional goals for the sample-preparation step may include the removal of material that could block chromatographic tubing, solubilization of analytes, and dilution to reduce solvent strength or avoid solvent incompatibility [1]. When biological fluid samples such as plasma, serum, bile, or urine are used, this technique is described as *bioanalytical sample preparation*. Samples obtained from in vitro studies are also common in bioanalysis, such as aliquots from a tissue homogenate, perfusate, dialysate, and hepatocyte or microsomal incubation.

Considerations

Many different sample-preparation techniques are available to a drug-discovery scientist performing bioanalytical sample preparation. These techniques vary in several regards, such as simplicity (number of steps), time requirement, materials cost per sample, expertise of the analyst, ease of method development, ability to be automated, concentration factor attained, and cleanliness of the final extract. It is important to note that the sample-preparation needs within drug discovery are different from those in other phases of drug development. In drug discovery, new compounds are introduced as often as weekly, high quantitation limits are acceptable, rapid data generation is demanded, and thus there is little time for method development. The drug-discovery scientist favors a quick and generic sample-preparation method that can be applied to a range of compounds under study. In contrast, once a drug candidate has been selected, studied in animals, and approved for further study with human subjects, clinical assays then require ultrasensitivity and a fully validated, rugged method. At this time, a very selective sample-preparation method is chosen that can also be automated and performed in high throughput.

The distinctions within the drug-discovery environment previously noted naturally favor the use of simpler and faster sample-preparation techniques. However, most all sample-preparation methods are potentially viable choices, as recent applications have demonstrated method-development techniques that are both rapid and selective. Also, there is value for the quick development of a bioanalytical method with selectivity for a series of similarly related structures, as that one method can often meet the needs for PK and toxicokinetic (TK) studies well into the discovery process, rather than just for one initial study. The variety of sample-preparation methodologies for bioanalysis are now detailed.

PROTEIN PRECIPITATION

Fundamentals

Most biological matrices contain protein to varying extents. Protein-binding phenomena are known to influence drug–drug interactions in the clinical setting. Among the various plasma proteins, serum albumin is the most widely studied and is regarded as the most important carrier for drugs. The presence of these materials for bioanalysis is problematic, and they must be removed before chromatographic separation and detection. When injected into a chromatographic system, proteins will precipitate upon contact with the organic solvents used in LC mobile phases. The precipitated mass of protein builds up within the column inlet, reduces the column lifetime and increases the system back pressure. When protein is carried through the analytical system, it may reach the mass spectrometer to foul the interface and require frequent cleaning; the result is system downtime.

A common approach to remove proteins from an injected sample is precipitation with an organic solvent (methanol, ethanol, or acetonitrile). Typically, a volume of sample matrix (1 part) is diluted with a volume of precipitating agent (3 or 4 parts), followed by vortex mixing, and then centrifugation or filtration to isolate or remove the mass of precipitated protein (Figure 17.1). The supernatant or filtrate is then analyzed directly. Note that ionic salts (saturated

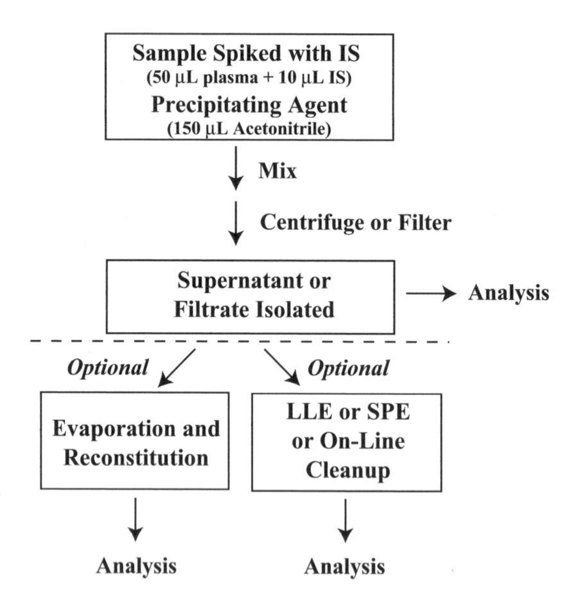

FIGURE 17.1 Schematic diagram of the protein-precipitation technique for bioanalysis. (Reprinted from Wells [54]. Copyright © 2003, with permission from Elsevier Science.)

aqueous ammonium sulfate or zinc sulfate heptahydrate (10%, w/v)) and inorganic acids (trichloroacetic acid (10%, w/v), perchloric acid (6%, w/v), and metaphosphoric acid (5% w/v)) are also effective protein precipitating agents [2], but are not commonly used in bioanalysis, since these materials are not compatible for direct injection with LC-MS techniques.

Considerations

Protein precipitation is often used as the initial sample-preparation scheme in the analysis of small drug molecules, since one universal procedure is followed for all compounds and method development is unnecessary. The speed of this technique presents a real time savings. The high resolving power of LC-MS/MS analytical methods accommodates this nonselective cleanup procedure. This mode of sample preparation is also inexpensive and does not chemically alter the analyte. Typical sample matrices that are used with protein-precipitation techniques are plasma, serum, tissue homogenates, and in vitro incubation mixtures. In general terms, a sensitivity of 1–10 ng/mL is often achieved from 50-µL sample volumes with this technique with LC-MS/MS analysis.

As with any methodology, there are some disadvantages to consider. Protein precipitation typically dilutes the sample by a factor of 3 or more, so it is a useful technique only when analyte concentrations are relatively high and the detection limits allow adequate quantitation. However, the supernatant can be evaporated with nitrogen and heat to concentrate the sample. This evaporation step requires an additional transfer step (with possible transfer loss) and added time for the dry-down and reconstitution procedures. The analyte volatility for newly synthesized compounds is often unknown, and lower recovery may result by introducing a dry-down step for an analyte that is labile to heat.

Matrix components are not efficiently removed with protein precipitation and will be contained in the isolated supernatant or filtrate. In MS/MS detection systems, matrix contaminants have been shown to reduce the efficiency of the ionization process with atmospheric-pressure ionization (API) techniques [3–12]. The observation seen is a loss in response, and this phenomenon is referred to as *ionization suppression*. This effect can lead to decreased reproducibility and accuracy for an assay and failure to reach the desired limit of quantitation. Additionally, the efficiency of protein removal with organic solvents is not complete and typically ranges from 98.7% to 99.8% [2] to leave residual amounts of protein that carry over into the analytical system and foul the ionization source of a mass spectrometer after repeated injections.

High Throughput Methods

Protein precipitation traditionally has been performed with individual tubes, but this approach is labor intensive with its required labeling and frequent manipulations and does not satisfy the needs for high throughput. Since the microplate format is commonly used in autosamplers for injection into the chromatographic system, it is desirable to retain that format and perform the sample-preparation procedure in a microplate. Two general approaches are common for performing protein precipitation in the high throughput microplate format:

1. Use a collection plate or microtube rack as the source container, pellet the precipitated protein at the bottom of the wells by centrifugation, and aspirate the supernatant for analysis.
2. Use a filtration microplate to trap the precipitated protein on top of a filter and collect and analyze an aliquot of the filtrate contained within a mated collection plate.

Each of these approaches is practical for performing high throughput protein-precipitation methods. The determining factors for selection of the collection-plate format over the filtration plate, or vice versa, include the extent of available hardware and automation accommodating the microplate format, total cost of materials, and thus the cost per sample, number of physical manipulations, and the degree of transfer loss deemed acceptable.

It is common for bioanalytical support laboratories in drug-discovery programs to choose protein precipitation in the 96-well plate as their preferred sample-preparation procedure prior to LC-MS/MS analysis, as described in two chapters within this volume ("Recent Advances in High Throughput Quantitative Drug-Discovery Bioanalysis," by Bradley L. Ackermann, Michael J. Berna, and Anthony T. Murphy, and "New Strategies for the Implementation and Support of Bioanalysis in a Drug Metabolism Environment," by Walter A. Korfmacher). Additional reports of protein-precipitation applications are available in the literature with either the collection plate [13–22] or the filtration plate [23–27], and several of these papers discuss the automation of these formats to achieve high throughput.

An example of the successful use of rapid sample-preparation procedures in bioanalysis is the fast-track status of Gleevec™ conferred by the Food and Drug Administration (FDA). The duration from first dose in man to completion of the New Drug Application filing was about 2.6 years. The method chosen, developed, and validated to complete the pharmacokinetic studies with the speed required to meet target dates was a semiautomated protein-precipitation procedure with deep well collection plates [19]. The method described used an LC-MS/MS run time of 2.5 min (injection-to-injection cycle). Typical batch sizes were two to four plates a day. Linearity and reproducibility of calibration curves for Gleevec in human plasma were acceptable between 4 and 10,000 ng/mL. The mean accuracy of standard calibration samples that cover the preceding concentration ranged from 98.0% to 102% with %CV values from 1.66% to 5.97%. Automation in the form of a 96-tip liquid-handling workstation (Quadra® 96; Tomtec Inc., Hamden, Connecticut) was used to transfer supernatant (250 μL) from plates following centrifugation into clean collection plates for injection (10 μL). Filtration plates were used on occasion to filter large aggregates from the supernatant solutions upon transfer into the microplates used for injection.

FILTRATION

Fundamentals and Considerations

Filtration is used as a sample-preparation method to remove particulates and debris that can potentially foul the LC lines, column frits, or mass spectrometer interface. Also, it is generally accepted that all workstations and pipetting systems can benefit from sample filtration because of the universal issues related to plasma clot formation that introduce pipetting challenges. Applications that use filtration include the removal of a mass of precipitated protein or of debris from raw plasma before use with any of the traditional sample-preparation techniques, as well as direct injection techniques (turbulent flow chromatography, restricted-access media (RAM), and on-line solid-phase extraction). Also, filtration can effectively clarify eluates and reconstituted extracts before chromatographic analysis [28], as well as remove particulates from urine before analysis by nuclear magnetic resonance (NMR) techniques, as performed for metabonomics research in drug discovery [29].

Vacuum is a practical approach to filtration. However, this approach cannot always achieve enough force to completely pull liquids through a fine porosity membrane; residual liquid can be left above the membrane or adsorbed onto the plastic inside a well below the membrane. Centrifugation can achieve these needed higher forces and is preferred to pass the full volume of liquid through a fine porosity filter. The filter plate is mated on top of a deep-well collection plate for this procedure and then is placed into a centrifuge for processing at about 3000 rpm for 10 minutes.

High Throughput Methods

Filtration in the microplate format allows high throughput sample preparation that is automatable with liquid-handling workstations. Generally, plasma samples are frozen before analysis and the freeze/thaw process can introduce or promote clot formation as the proteins separate from the water. The prevalence of clot formation and its magnitude become greater with repeated freeze/thaw cycles [30]. These clots are capable of plugging pipet tips and create pipetting errors with automated systems. Berna et al. [16] describe an application in which 20-μm polypropylene filter plates (Captiva™; Varian Inc., Harbor City, California) were used as the source container for a set of plasma samples. The top of the filter plate is sealed, as well as the bottom. Prior to analysis, the filter plate containing frozen plasma samples is allowed to thaw; the bottom cover is removed and the filter plate is placed over a deep-well microplate to collect the plasma filtrate. The filter plate/collection plate combination can be placed into a centrifuge to complete the filtration process, if required, or a vacuum can be applied with a vacuum collar or manifold.

DILUTION FOLLOWED BY INJECTION

Sample dilution is used to reduce the concentration of salts and endogenous materials in a sample matrix and is commonly applied to urine. Drug concentrations in urine are usually fairly high and allow this dilution without an adverse effect on sensitivity. While the amount of protein in a sample is usually a concern, protein concentrations in urine are negligible under normal physiological conditions. An example of a urine dilution procedure is reported for indinavir in which 1 mL urine was diluted with 650 μL acetonitrile so that the resulting concentration of organic in the sample was equal to or less than that of the mobile phase. An aliquot of 6 μL was injected into an LC-MS/MS system [7]. Although co-eluting endogenous species in urine were not seen in the selected ion-monitoring mode, their presence did adversely affect the ionization of analytes, leading to increased variation in MS/MS responses.

Dilution is not typically used for plasma due to the high amounts of protein present and the greater effect that dilution has on sensitivity. However, dilution can be very attractive for the minimal effort and time involved. In one report [31], dog-plasma samples were centrifuged, pipetted into wells of a microplate, and then placed on a pipetting workstation. A volume of 15 μL plasma was diluted with 485 μL of a solution of water/methanol/formic acid (70:30:0.1, v/v/v) containing internal standard. The samples were sealed and mixed; the dilution resulted in a slightly viscous solution with no observed precipitation. An aliquot of 5 μL was injected into an LC-MS/MS system. The limit of quantitation for the dilution assay (2 ng/mL) was 400 times higher than that of a more selective procedure that also concentrates the analyte (liquid–liquid extraction; 5-pg/mL limit of quantitation). However, the advantage offered by the dilution procedure was fifty times greater throughput. In this case, throughput was a more important consideration than analyte sensitivity.

LIQUID–LIQUID EXTRACTION

Fundamentals

Liquid–liquid extraction (LLE) is a technique used to separate analytes from interferences in the sample matrix by partitioning between two immiscible liquids. A given volume of aqueous sample solution (e.g., plasma) containing analytes is mixed with an internal standard in solution. A volume of buffer at a known pH (or a strongly acidic or basic solution that adjusts pH) is then added to maintain the analytes in their unionized (uncharged) state. The resulting solution is then vigorously mixed with several ratio volumes of a water-immiscible organic solvent or mixtures of two or more solvents such as hexane, diethyl ether, methyl *tert*-butyl ether (MTBE), or ethyl acetate. Substances in the sample matrix distribute between these two liquid phases (aqueous and organic) and partition preferentially into the organic phase when analytes are unionized (uncharged) and demonstrate solubility in that organic solvent. Isolation of the organic phase, followed by evaporation and reconstitution in a mobile-phase–compatible solvent, yields a sample ready for injection (Figure 17.2). The method provides efficient sample cleanup, as well as sample enrichment.

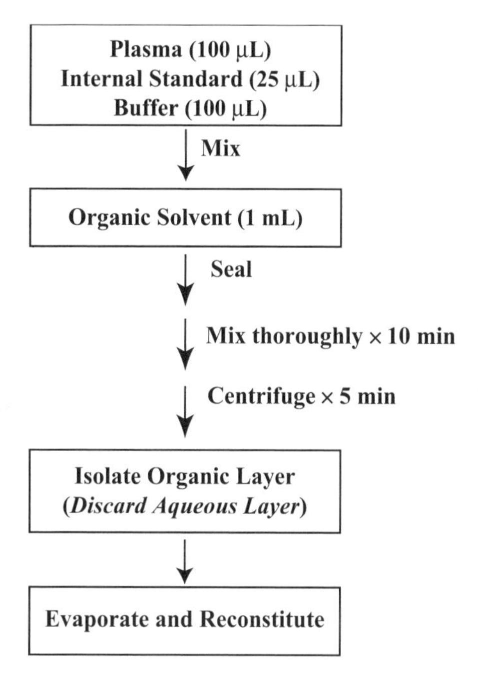

FIGURE 17.2 Schematic diagram of the liquid–liquid extraction procedure in which analyte partitions from an aqueous phase into an organic phase. (Reprinted from Wells [54]. Copyright © 2003, with permission from Elsevier Science.)

Considerations

LLE is a popular technique for sample preparation because it is widely applicable for many drug compounds and is a relatively inexpensive procedure. With proper selection of organic solvent and adjustment of sample pH, very clean extracts can be obtained with good selectivity for the target analytes. Inorganic salts are insoluble in the solvents commonly used for LLE and remain behind in the aqueous phase along with proteins and water-soluble endogenous components. These interferences are excluded from the chromatographic system and a clean sample is prepared for analysis. However, some disadvantages of LLE include its labor-intensive nature with several disjointed steps such as vortex, mix, and centrifugation. The organic solvents used are volatile and present hazards to worker safety. Also, emulsions can be formed without warning and can result in loss of sample.

High Throughput Methods

LLE has been demonstrated in collection microplates to provide for high throughput sample preparation [32–42]. The use of liquid-handling workstations reduces the hands-on analyst time required for this technique and offers semiautomation to an otherwise labor-intensive task. The plate sealing and vortex mixing steps still require manual intervention, but the pipetting steps are all performed by the workstation. A relevant example of a high throughput LLE procedure in microplates is described by Steinborner and Henion [43]. They reported a semiautomated LLE procedure (following protein precipitation with acetonitrile) for the quantitative analysis of the anticancer drug methotrexate (MTX) and its major metabolite from human plasma. The calibration curves for MTX and its metabolite were linear over the ranges 0.5 to 250 ng/mL and 0.75 to 100 ng/mL, respectively. Sample preparation throughput was reported as four sample plates (384 samples) processed in 90 minutes by one person. The analytical throughput was reported as 768 samples analyzed within 22 hours (maximum 820 samples per 24 hours) with a selected reaction monitoring mass spectrometry method with a 1.2 minute analysis time per sample.

SOLID-SUPPORTED LIQUID–LIQUID EXTRACTION

Fundamentals

The many disjointed mixing and centrifugation steps of traditional LLE can be eliminated by performing LLE in a flowthrough column filled with inert diatomaceous earth particles. This technique is referred to as solid-supported LLE (SS-LLE). The high surface area of the diatomaceous earth particle facilitates efficient, emulsion-free interactions between the aqueous sample and the organic solvent. Essentially, the diatomaceous earth with its treated aqueous phase behaves as the aqueous phase of a traditional LLE, yet it has the characteristics of a solid support.

The procedure for use of solid-supported particles to perform LLE follows. A mixture of sample (e.g., plasma), internal standard, and buffer solution to adjust pH is prepared. This aqueous mixture is then added directly to the dry particle bed without a conditioning or pretreatment step. The mixture is allowed to partition for about 3–5 minutes on the particle surface via gravity flow. The analyte in aqueous solution is now spread among the particles in a high surface area. A hydrophobic filter on the bottom of each column or well prevents the aqueous phase from breaking through into the collection vessel placed underneath. Organic solvent is then added to the column, and as it slowly flows through the particle bed via gravity the analyte partitions from the adsorbed aqueous phase into the organic solvent, the eluate is collected into an appropriately sized receiving column, tube, or microplate. A second addition of organic solvent is performed and the combined eluates are evaporated to dryness. The residue is reconstituted in a mobile-phase–compatible solution and an aliquot is injected into the chromatographic system for analysis.

Considerations and High Throughput Methods

SS-LLE can be performed in microplates for high throughput and is fully automatable with liquid-handling workstations, since each step is simply a liquid transfer or addition. The many capping and mixing steps as used for traditional LLE are unnecessary, so less hands-on analyst time is required. However, there are a few precautions with SS-LLE that deserve mention. The diatomaceous earth particles used in these procedures demonstrate a capacity limit for adsorption of aqueous sample matrix. It is possible to overload the particle with too much sample volume; the end result of overloading the column is incomplete recovery. Also, method transfer from an established LLE procedure to one that uses SS-LLE is not always straightforward. An investment in time may be required to optimize the method in terms of sample volume loaded onto the particle bed, the mass of particle in each well or column, the volume of organic solvent required, and the mechanics of collecting and working with the isolated organic solvent.

Reported bioanalytical applications that use SS-LLE products include both the cartridge (tube) format and the microplate format. Some of these applications include the determination of mexiletine [44], amiodarone [45] and other antiarrhythmic drugs [46], proxyphylline [47], 16β-hydroxystanozolol [48], dextromethorphan [49], and simvastatin [50] from biological fluids. Microplate applications for SS-LLE include a crude purification of crude combinatorial library samples [51], carboxylic acid–based matrix metalloprotease inhibitors [52], and a β3-adrenergic receptor agonist [53].

SOLID-PHASE EXTRACTION (OFF-LINE)

Fundamentals

Solid-phase extraction (SPE) is a procedure in which an analyte, contained in a liquid phase, comes in contact with a solid phase (sorbent particles in a column or disk) and is selectively attracted to the surface of that solid phase. All other materials not adsorbed by chemical attraction or affinity remain in the liquid phase and go to waste. A wash solution is then usually passed through the sorbent bed to remove any loosely adsorbed contaminants from the sample matrix, yet retain the analyte of interest on the solid phase. Finally, an eluting solvent (usually an organic solvent such as methanol or acetonitrile that may be modified with acid or base) is added to the sorbent bed. This solvent disrupts the attraction between the analyte and solid phase that cause desorption,

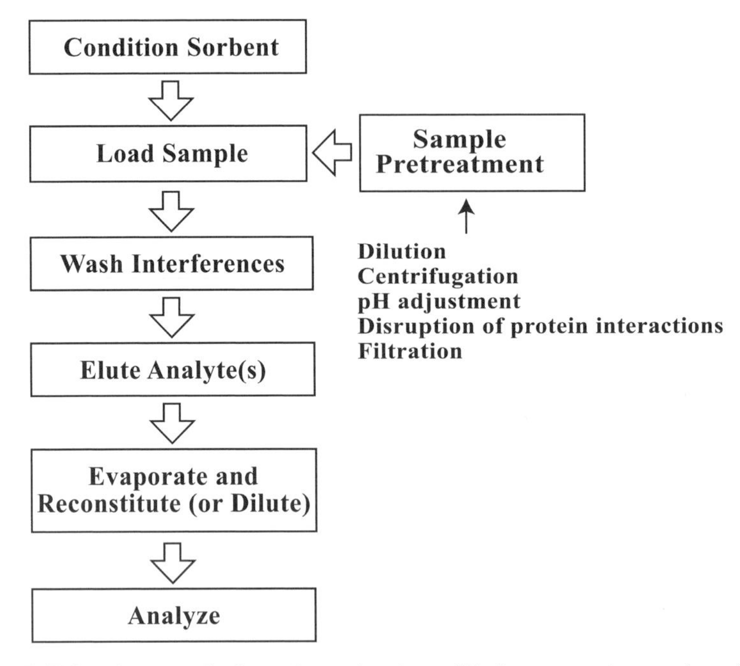

FIGURE 17.3 The general scheme for performing solid-phase extraction consists of several sequential steps. Sample pretreatment is important to ensure analyte retention. (Reprinted from Wells [54]. Copyright © 2003, with permission from Elsevier Science.)

or elution, from the sorbent. Liquid processing through the sorbent bed can be accomplished via vacuum or positive displacement. Solvent exchange (evaporation followed by reconstitution) or dilution is followed by analysis on a chromatographic system. The general scheme for performing SPE is shown in Figure 17.3. The term off-line refers to performing a sample preparation procedure independently of the chosen analytical detection system.

Considerations

SPE is an attractive sample-preparation technique for these reasons: very selective extracts can be obtained (reducing the potential for ionization suppression from matrix materials); wide variety of sample matrices accepted; analytes can be concentrated; high recoveries with good reproducibility; improved throughput via parallel processing; low solvent volumes; suitable for full automation; and no emulsion formation as seen with LLE. In addition, so many product formats and chemistry choices are available that nearly all extraction requirements can be met.

In spite of its great advantages, SPE may not be the preferred sample-preparation technique in a laboratory supporting drug-discovery bioanalysis, primarily for its higher cost per extraction compared with other methods such as protein precipitation and LLE. Additional considerations against the use of SPE may be that it consists of several sequential steps, a perceived difficulty to master use of the many sorbent chemistries and a historical perception that method development takes too much time.

There have been many advances in SPE over the years that have made it a more attractive choice for sample preparation in drug discovery. The technology has been improved with the introduction of new solid sorbent chemistries that allow for generic methods, multiple types of disk-based SPE devices, smaller bed mass sorbent loading, and a multitude of 96-well plate formats to accommodate a range of sample volumes from microliters to milliliters. Liquid-handling workstations have also been improved to offer completely unattended automation for performing the SPE procedure in microplates.

The amount of published information available on SPE is numerous, and further coverage of its theory and chemistry is outside the scope of this text. The reader is referred to a comprehensive book on the subject by this author [54] that reviews the theory of SPE, sorbent chemistries, product formats, high throughput applications, method development and optimization procedures, and automation methodologies. In addition, Hennion provides an informative overview of the established solid-phase extraction sorbents, their modes of interaction with analytes, on-line coupling with liquid chromatography, and the method development process (including discussion of predictive models) [55]. Hennion et al. also present strategies for the prediction, rapid selection, and optimization of SPE parameters with emphasis on the extraction of polar analytes [56].

Sorbent Chemistries Allowing for Generic Methods

Advantage for Drug Discovery

The use of a single sorbent chemistry that involves a generic or universal set of extraction conditions for the majority of analytes encountered is of great utility for drug-discovery bioanalysis, since the time for method development is eliminated or at least greatly reduced. Additionally, a generic sorbent is most useful for a parent drug and its metabolites that demonstrate differences in polarity. The development process then consists simply of running the method with several analyte concentrations with multiple replicates and then assaying for recovery and performance (and documenting precision and matrix effect). The chemistries chosen for such a generic method are commonly reverse-phase polymer sorbents and mixed-mode polymer chemistries, each of which uses a single suggested extraction protocol.

Polymer Sorbents

The great appeal of a reverse-phase polymer sorbent is its ability to successfully extract analytes with a generic methodology (condition with methanol then water, load sample, wash with 5% methanol in water, and then elute with methanol). Summaries of the characterization of a hydrophilic lipophilic balanced (HLB) polymer sorbent have been published by Bouvier et al. [57–58], and a discussion of this same polymer sorbent is contained within a review of sample-preparation methods by Gilar et al. [59]. A review of polymer-based sorbents for SPE has been prepared by Huck and Bonn [60]. Many published research reports have shown the utility of such a generic sorbent for SPE methods in both individual columns [3,61,62] and 96-well plates [11,63–70]. Some examples of polymer choices from manufacturers include Oasis® (Waters Corporation, Milford, Massachusetts), Focus™ and Nexus™ (Varian Inc., Harbor City, California), Strata™ X (Phenomenex, Torrance, California), Isolute® 101 and ENV+ (Argonaut Technologies, Foster City, California), Universal Resin (3M, St. Paul, Minnesota) and Polycrom™ (CERA Inc., Baldwin Park, California).

An example of the universal approach to SPE is described in a report by Cheng et al. [71] for the extraction of verapamil and its metabolite from a plasma matrix with Oasis HLB in a 96-well plate format. As executed, this method is a "catch-all" approach that aims to capture or adsorb as many analytes as possible, each displaying different degrees of hydrophobicity and polarity. The wash step in this procedure is 95% aqueous, which is effective to displace proteins from the sorbent bed; lipophilic interfering components remain adsorbed, along with the analytes of interest. Elution with the polar solvent methanol is expected to disrupt the attraction of analyte to sorbent bed, eluting the analyte. Since no selectivity is developed within this method, however, unwanted interfering components are also eluted. A generic set of conditions is intended simply to extract the analytes of interest. The selectivity of the mass spectrometer is used to distinguish the analytes from all other components in the injected sample.

Mixed-Mode Chemistry

The sorbent chemistry chosen for a universal method is not required to be a polymer. When the analyte contains at least one ionizable functional group, a mixed-mode interaction is extremely useful and relies on dual mechanisms of attraction—ion exchange (cation or anion exchange) and reverse-phase (hydrophobic) affinities. When the analyte contains an amine group, protonation of the amine is exploited for cationic attraction; likewise, an acid group is ionized for anionic attraction. The elution solvent shown to work optimally for mixed-mode cation exchange is organic modified with base (e.g., methanol containing 2–5% NH_4OH); likewise, for mixed-mode anion, the elution solvent is organic modified with acid (e.g., methanol containing 2–5% HCOOH). An important result of a mixed-mode chemistry is that very selective extracts can be obtained; because the analyte is held by an ionic attraction (which is much stronger than a reverse-phase attraction, as with a polymer), a 100% organic solvent can be used as a powerful wash to completely remove matrix interferences, yet retain the analyte of interest [72]. Many published reports detail the successful use of mixed-mode chemistry for SPE [73–77].

Disk and Small Bed Mass Product Formats

A recent shift toward the discovery and development of new chemical entities that have greater potency has required their dosing at lower levels; popular sample-preparation techniques such as protein precipitation are less useful for analyte concentrations below 1 ng/mL. Additionally, the frequent use of mouse plasma necessitates the use of sample volumes $\leq$50 μL and effectively miniaturizes the sample-preparation process. In these situations, SPE is especially appealing due to packaging of sorbents within small-diameter, thin disks requiring dramatically lower solvent and elution volumes [54,78–80]. Also, the benefits of disks are often attainable with small sorbent bed masses packed in columns that are now available in bed masses as low as 2 mg [81].

High Throughput Methods

High throughput SPE uses the 96-well microplate format, and reviews of the historical development of this product format and the accessory products required for successful use are available [82,83]. Many sorbent chemistries (>45) and formats are available to meet most all high throughput sample-preparation needs. The large number of published applications for quantitative bioanalysis with 96-well plates confirms the wide acceptance of this technique, and some pertinent applications are documented [84–91]. Some reports have even used a higher-density 384-well format [92,93]. Although SPE microplates can be processed manually, liquid-handling workstations are preferred; the use of 96-tip semiautomated workstations are popular for high throughput, while 8- and 12-tip workstations are able to perform additional functions, such as reformat samples from tubes to plates and offer full automation. The automation of SPE in microplates is covered in detail within several book chapters [54,94,95] and reviews in the literature [96,97].

Generally, in terms of throughput for semiautomated SPE procedures that use 96-well plates, from 2.0 to 2.5 plates per day (192–240 samples) can be prepared and analyzed per analyst [98]. Extraction times with eight-tip workstations are reported generally as from 50 to 60 minutes, and for 96-tip workstations are from 10 to 15 minutes. The practical limitations to improving throughput one step further with 96-tip automation were evaluated by Rule and Henion [86]. One analyst was able to prepare and analyze 384 samples within a 24-hour period. The interday and intraday accuracy and precision obtained over the course of these samples was within 8% coefficient of variation when analyzed by atmospheric-pressure chemical ionization (APCI) mass spectrometry with positive-ion detection.

The combination of automated sample preparation with high-speed chromatography with a monolithic column provides tremendous throughput and is very useful in drug-discovery applications. Wu et al. [63] developed a monolithic chromatography method for three analytes with a run time of only one minute. Sample preparation was performed with a 96-tip workstation. Three components were measured in 600 plasma extracts during an overnight run. In an extension of this work by the same laboratory, Deng et al. [68] described a component system capable of extracting and analyzing 1152 plasma samples (12 plates) within 10 hours. A Zymark track robot (Zymark Corporation, Hopkinton, Massachusetts) was interfaced with a Tecan Genesis (Tecan US, Research Triangle Park, North Carolina) liquid-handling workstation for the simultaneous and fully automated SPE of four 96-well plates. The extracts were injected onto four parallel monolithic columns for separation via a four-injector autosampler. The effluent from each of the four columns was directed into an LC-MS/MS system equipped with an indexed four-probe electrospray ionization source (MUX®, Micromass, Beverly, Massachusetts). An overall throughput was reported as 30 seconds per sample with this novel four-channel parallel format.

SOLID-PHASE EXTRACTION (ON-LINE)

Fundamentals

SPE performed in the off-line mode (separate from the analytical detection system) with individual cartridges or microplates was discussed in the previous section. However, the technique is also performed in automated fashion with disposable extraction cartridges that are processed on-line with the chromatographic system. A typical example of on-line SPE with disposable extraction cartridges is demonstrated by the Symbiosis™ System (Spark Holland, Emmen, The Netherlands); earlier versions of this successful technology were named Prospekt-2™ and Prospekt™. The SPE cartridge used for on-line applications has a standard dimension of 10 × 2 mm and can withstand LC system pressures to 300 bar; note that column dimensions of 10 × 3 mm and 10 × 1 mm are also available. A complete set of sorbent chemistries (>38) is available in this format. Typical particle sizes used are 40 μm, although the HySphere™ cartridges (Spark Holland) specifically use a smaller particle size, <10 μm. Sorbent mass per cartridge is typically between 20 and 45 mg, depending on particle chemistry. The standard capacity of the system is 192 cartridges (two trays containing 96 cartridges each); an optional feeder mechanism allows access up to 960 cartridges (10 trays). Note that the Waters Corporation provides their Oasis HLB chemistry in a disposable column for on-line SPE through Spark Holland, while this same chemistry is also available as a reusable column (25 μm) directly from Waters (2.1 mm × 20 mm).

The use of on-line SPE with LC analysis in a serial mode is straightforward; one sample is processed after the previous one has finished. With the advent of parallel and staggered parallel LC systems, throughput of on-line SPE can similarly be increased. The use of two SPE cartridges in staggered parallel fashion, one being eluted while the other is performing extraction, becomes an important feature in helping to meet throughput needs in drug-discovery laboratories. Note that to achieve maximum throughput, the cycle time for on-line SPE must be faster than the LC run time.

Considerations

An important feature and advantage of on-line SPE, compared with off-line SPE, is direct elution of the analyte from the SPE cartridge into the mobile phase of the LC system. The time-consuming off-line steps of evaporation, reconstitution, and preparation for injection are eliminated and make the on-line SPE more efficient and fully automated. Since the entire volume of eluate is analyzed, maximum sensitivity for detection is obtained. Some other advantages of this on-line approach include:

1. Samples and SPE cartridges are processed in a completely enclosed system, protected against light and air.
2. The operator is protected from working with hazardous or volatile organic solvents.
3. Less handling and manipulation are involved, with no transfer loss of analyte.

The method development process for on-line SPE is ideal for its ability to offer full automation. Method development involves the examination of several extraction variables with subsequent optimization. These parameters include the sorbent chemistry, the composition of the load, wash and elution solvents, sorbent particle size, solvent volumes, and flow rates. A general overview of the method development process for on-line SPE coupled to LC and LC-MS has been published by Ooms et al. [99]. A unique feature of on-line SPE is its ability to perform thermally assisted SPE; temperatures for loading, washing, and eluting (desorption) can be manipulated. Typically, raw sample matrix (combined with internal standard) is used for injection onto the SPE cartridge, although sometimes a pretreatment sample preparation step, such as filtration, may precede this on-line analysis.

Applications

A full array of applications has been reported with the Symbiosis on-line SPE technique. Its usefulness has been documented across many clinical and research disciplines, including bioanalytical support to drug discovery. McLoughlin et al. [100] reported the simultaneous determination of 10 proprietary discovery compounds in plasma (cassette dosing) by APCI MS/MS with a cyano SPE cartridge; limits of quantitation ranged from 2.5 ng/mL to 5 ng/mL for these compounds. A C-18 on-line extraction with LC-MS/MS detection was demonstrated as useful for the determination of 64 generic pharmaceutical compounds in support of drug-discovery cassette dosing techniques [101]. The extracted limit of detection for 41 of these analytes was 25 pg, and for 57 others was 250 pg, based on a 0.1-mL sample volume. Some staggered parallel applications for higher throughput have also been demonstrated [102–105]. A general overview that highlights the use of this system for high throughput analysis with MS techniques has been published by Koster and Ooms [106]. The applications described include a 2-minute cycle time for 11 compounds extracted from serum with HySphere Resin GP cartridges; detection was by LC-MS/MS.

TURBULENT FLOW CHROMATOGRAPHY (ON-LINE)

Fundamentals and Considerations

Turbulent flow liquid chromatography (TFC) has been shown to eliminate the need for traditional off-line sample preparation, as it allows for the direct injection of plasma or serum samples following only a filtration pretreatment step. TFC is different from traditional laminar flow chromatography in that it is performed at higher flow rates and with a single short column (1 × 50 mm) containing particles of a larger size (30–50 μm). When a sample in a biological fluid is injected into such a system, the liquid chromatograph delivers a high flow rate (~5 mL/min) of a highly aqueous solvent (typically 100% water with pH or salt additives as needed). A divert valve is set to waste for the load- and-wash period, which normally lasts for less than a minute.

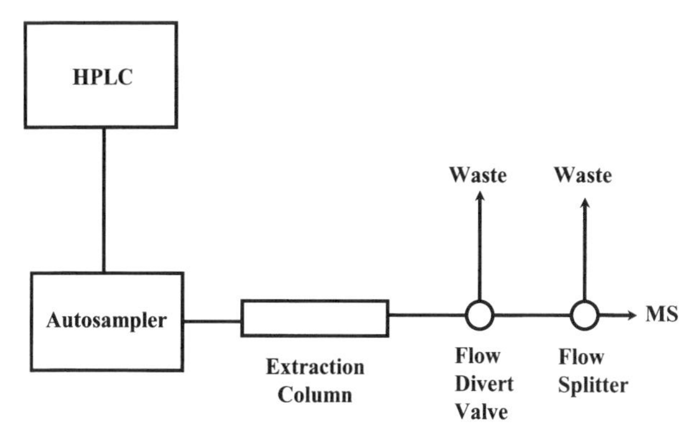

FIGURE 17.4 Schematic diagram of a single-column turbulent flow on-line extraction system. (Reprinted from Wells [54]. Copyright © 2003, with permission from Elsevier Science.)

When a reverse-phase sorbent is used in the extraction column, the analyte is retained via partitioning, while hydrophilic components in the sample matrix, including proteins, are washed off the column and go to waste (Figure 17.4).

After the load and wash steps are completed, the flow is changed to a high-percentage organic solvent and the divert valve is switched to direct the flow into a splitter; a portion of the split flow enters the mass spectrometer. Depending on the type of ion source and the ionization mode used, the flow is usually split from ~5 mL/min down to 0.3–1.0 mL/min. The retained analyte is eluted from the column with some limited separation and is detected by the mass spectrometer. The elution step usually takes less than one minute. Following this elution, the flow composition is changed back to high aqueous to equilibrate the extraction column in preparation for the next sample. Again, this equilibration step can be completed in less than a minute. Therefore, the entire on-line extraction and analysis procedure can usually be accomplished in about 3 minutes for each sample. This single-column mode provides the distinct advantage of high throughput and simplicity. However, carryover between samples can be a potential concern if the equilibration procedure is not adequate.

The single-column mode of turbulent flow chromatography is the simplest and yields the shortest injection cycle. It is often the first approach in the development of a new assay. However, when the single-column mode does not provide sufficient cleanup, an analytical column can be placed in series downstream from the first column. This dual-column mode provides the advantage of improved separation performance, as well as detection sensitivity that results from chromatographic focusing. Another advantage of the dual-column configuration mode is that the extraction and separation processes can be performed simultaneously, since they are now driven by two separate

LC pumps. While the sample is running on the analytical separation column, the extraction column is in use with the next sample. Many applications have been reported with TFC in both the single-column mode [107,108] and the dual-column mode [109–111].

Applications

In a drug discovery absorption, distribution, metabolism, and excretion (ADME) laboratory providing bioanalytical support, new analytes and compound series are frequently encountered. The time required developing new methods for each analyte or series of analytes needs to be reduced to move toward true high throughput ADME screening. It is not possible to develop a unique method for each new compound being screened; rather, the approach taken is to develop one method that will work for >95% of compounds. Recent development of a polymeric stationary phase for the TFC extraction column has made this technique viable. A sound and proved generic method that meets this objective was developed by Herman [112] with TFC with fast LC-MS/MS detection techniques. Herman's method (Cyclone™ polymeric extraction column with an Eclipse XDB C-18 analytical column) has been used with great success to screen over 1000 compounds in many different biological matrices. Grant et al. also report a generic serial and parallel on-line direct injection with TFC and LC-MS/MS [113].

The turbulent flow column switching (CS) approach has also been used for the bioanalysis of a single analyte, as well as for mixtures of multiple analytes. Jemal et al. used this scheme to effectively separate and quantify two positional isomers in plasma [114]. Multiple components in plasma were simultaneously quantified with good chromatographic separation and accuracy by Wu et al. [65]. Lim et al. used this approach to simultaneously screen for metabolic stability and profiling [115].

RESTRICTED-ACCESS MEDIA (ON-LINE)

Fundamentals

Another technique that allows for the direct injection of plasma or serum on-line with the chromatographic system uses an analytical column containing particles referred to as restricted-access media (RAM); traditional laminar flow liquid chromatography is employed. These RAM particles are designed to prevent or restrict large macromolecules from accessing the inner adsorption sites of the bonded phase. Commercially available RAM columns, all silica based, include internal surface reverse phase (ISRP), semipermeable surface (SPS), and a hydrophobic shielded phase named Hisep™ (Supelco, Bellefonte, Pennsylvania). The most popular column variety used in bioanalysis is the ISRP column; the internal surface is covered with a bonded reverse-phase material and the external surface is covered with a nonadsorptive but hydrophilic material. This dual-phase column permits effective separation of the analyte of interest from macromolecules in the sample matrix; drugs and other small molecules enter the pores of the hydrophobic reverse phase to partition and be retained, while proteins and larger matrix components are excluded. Essentially, a combination of size exclusion chromatography and partition chromatography is observed. When serum or plasma is injected onto a RAM column, the proteins are excluded by the outer, hydrophilic layer and pass through to waste. Note that conventional polymers (e.g., Oasis HLB) are subject to protein adsorption from nonspecific binding; column lifetimes are reduced tenfold compared with RAM columns [116].

Considerations

The dual-phase nature of these RAM materials allows the direct injection of the biological sample matrix onto the column without pretreatment. Some disadvantages with the use of RAM columns are that retention times can be long (>10 minutes); the column must be washed between injections and the mobile phases are not always compatible with some ionization techniques used in LC-MS/MS. Dual-column RAM techniques are also used. These methods use an analytical separation column placed in series downstream from the RAM column. A general overview of the use of RAMs in LC has been published in two parts [117,118].

The LiChrospher™ ADS (Alkyl-Diol-Silica, Merck kGaA, Darmstadt, Germany) RAM column, for example, can be operated in either single-column or dual-column mode. In single-column mode, cleanup, extraction, and separation occur simultaneously; there is no sample enrichment. Sample volumes for direct injection are small (<100 μL) in single-column mode; column overloading is a potential problem. There is limited peak capacity for analytes, and the limit of detection is high. In contrast, for ADS used in dual-column mode, cleanup, extraction, and separation occur sequentially, and sample enrichment is realized. Sample volumes are large (100–500 μL) and limits of detection are much lower.

The ADS column lifetime in single-column mode is short: 100–500 injections of 10–20-μL sample volumes. Flow is performed via one direction only, so the total mass injected goes toward the detector; regeneration requires additional time. In coupled column mode, about 2000 injections of 50 μL can be made. A backflush mode cleans the precolumn; regeneration of the precolumn is included within the analytical cycle.

Some disadvantages regarding RAM column usage include long retention times (>10 minutes), washing of the column is required between injections, and the mobile phases are not always compatible with the ionization techniques used in LC-MS/MS. However, in the case of LiChrospher ADS, the extraction column conditioning step is part of the entire process, so that while separation and quantification occur, the extraction column is conditioned and prepared for the next injection. Other general disadvantages noted for RAM columns are low chromatographic efficiency, limited loading capacity, and low extraction efficiency for highly (99.9%) protein-bound analytes [119]. Some possible solutions to lessen the influence of protein binding include the addition of a dilute solution of trichloroacetic acid; dilution of the sample with an aqueous solution; adjustment of the solution pH; or the addition of organic modifiers. The overall approach with RAM extraction columns is automated, however. There is a low capital outlay for automation equipment, since this technique uses common high-performance liquid chromatography (HPLC) hardware.

Applications

In addition to many demonstrated applications of ADS with C-4, C-8, and C-18 bonded-silica reverse-phase materials within the inner pores of the particles, a mixed-mode column packing is also available that relies on affinity and size-exclusion principles. The CAT-PBA™ column (Iris Technologies, Lawrence, Kansas) is filled with a porous copolymer that contains specially modified phenylboronic acid (PBA) as an affinity ligand. Low molecular-weight analytes such as catecholamines (e.g., epinephrine, norepinephrine, and dopamine) reach the inner pore surface. Under slightly alkaline load conditions (pH 8.7), these catecholamines are extracted onto the PBA; selective retention occurs due to the formation of a cyclic ester between the boronic acid bound to the stationary phase and the 1,2-dihydric alcohol (diol) functional group on the catecholamines. Following a wash step to remove residual components, the pH is changed to acidic, which causes the cyclic boronic acid ester to hydrolyze, and the catecholamines are eluted onto an analytical column via a switching valve. Separation is achieved under reverse-phase C-18 conditions [120]. As an LC separation with electrochemical detection is taking place, the CAT-PBA SPE column is automatically regenerated and prepared for another extraction. The lifetime of the CAT-PBA SPE column has been reported by the manufacturer to exceed 2000 untreated urine samples (100 μL each) and 1000 untreated plasma samples (500 μL each).

COMBINATIONS OF TECHNIQUES

Although many individual choices for sample preparation exist, sometimes the combination of two techniques yields a more desirable result. Within a drug discovery environment, however, throughput and cost are important factors that influence the choice of sample-preparation methods. The protein-precipitation procedure can quickly yield a sample that is ready for analysis, but its potential for carryover of matrix interferences is problematic. The isolated supernatant can be filtered or centrifuged before injection, but these procedures remove only particulates or proteinaceous material, not the materials that cause matrix interferences. Thus, protein precipitation is sometimes followed by LLE, SPE, or an on-line technique for a more selective cleanup before analysis. A particularly challenging sample-preparation requirement is the analysis of animal tissues, as is commonly performed in drug-discovery support laboratories that perform drug uptake studies. Here, a series of sequential sample-preparation steps are common [121].

CONCLUSION

The need for sample preparation within a drug discovery research environment is essential with the use of LC/MS-MS detection systems. The removal of unwanted endogenous matrix components (e.g., protein, salts and metabolic byproducts) from the sample must be rapid and as complete as possible to avoid the introduction of interferences upon analysis. Many different techniques are available to a drug discovery scientist performing bioanalytical sample preparation. Off-line procedures, independent of the analysis, include protein precipitation and removal, filtration, dilution followed by injection, liquid-liquid extraction and solid-phase extraction. These methodologies are performed in high throughput mode using the 96-well microplate format with suitable automation. On-line sample prep procedures are performed in sync with the analysis and include solid-phase extraction as well as turbulent flow chromatography using classic chromatographic media or restricted access media. High throughput using on-line procedures is achieved via parallel analyses and/or detection systems. The choice of which sample preparation methodology is most suitable for an assay involves a consideration of simplicity, time requirement, materials cost per sample, expertise of the analyst, ease of method development, ability to be automated, concentration factor attained and cleanliness of the final extract. The rapid turnaround needs within the drug discovery environment naturally favor the use of simpler and faster sample preparation techniques. However, most all sample preparation methods are potentially viable choices as recent applications have demonstrated method development techniques that are both rapid and selective.

REFERENCES

1. Kataoka, H.; Lord, H. "Sampling and Sample Preparation for Clinical and Pharmaceutical Analysis," in: Pawliszyn, J. ed., *Sampling and Sample Preparation for Field and Laboratory: Fundamentals and New Directions in Sample Preparation*, Elsevier Science, Amsterdam, Chapter 23, 779–836 (2002).
2. Blanchard, J. "Evaluation of the Relative Efficacy of Various Techniques for Deproteinizing Plasma Samples Prior to High-Performance Liquid Chromatographic Analysis," *J. Chromatogr.* **226,** 455–460 (1981).
3. Bonfiglio, R.; King, R.C.; Olah, T.V.; Merkle, K. "The Effects of Sample Preparation Methods on the Variability of the Electrospray Ionization Response for Model Drug Compounds," *Rapid Commun. Mass Spectrom.* **13**(12), 1175–1185 (1999).
4. King, R.; Bonfiglio, R.; Fernandez-Metzler, C.; Miller-Stein, C.; Olah, T. "Mechanistic Investigation of Ionization Suppression in Electrospray Ionization," *J. Am. Soc. Mass. Spectrom.* **11,** 942–950 (2000).
5. Constanzer, M.; Chavez-Eng, C.; Matuszewski, B. "Determination of a Thermally Labile Metabolite of a Novel Growth Hormone Secretagogue in Human and Dog Plasma by Liquid Chromatography with Ion Spray Tandem Mass Spectrometric Detection," *J. Chromatogr. B* **760,** 45–53 (2001).
6. Matuszewski, B.K.; Costanzer, M.L.; Chavez-Eng, C.M. "Matrix Effect in Quantitative LC/MS/MS Analyses of Biological Fluids: A Method for Determination of Finasteride in Human Plasma at Picogram Per Milliliter Concentration," *Anal. Chem.* **70,** 882–889 (1998).
7. Fu, I.; Woolf, E.J.; Matuszewski, B.K. "Effect of the Sample Matrix on the Determination of Indinavir in Human Urine by HPLC with Turbo Ion Spray Tandem Mass Spectrometric Detection," *J. Pharm. Biomed. Anal.* **18,** 347–357 (1998).
8. Buhrman, D.L.; Price, P.I.; Rudewicz, P.J., "Quantitation of SR 27417 in Human Plasma using Electrospray Liquid Chromatography–Tandem Mass Spectrometry: A Study of Ion Suppression," *J. Am. Soc. Mass. Spectrom.* **7,** 1099–1105 (1996).
9. Jemal, M.; Xia, Y.-Q. "The Need for Adequate Chromatographic Separation in the Quantitative Determination of Drugs in Biological Samples by High Performance Liquid Chromatography with Tandem Mass Spectrometry," *Rapid Commun. Mass Spectrom.* **13**(2), 97–106 (1999).
10. Miller-Stein, C.; Bonfiglio, R.; Olah, T.V.; King, R.C. "Rapid Method Development of Quantitative LC-MS/MS Assays for Drug Discovery," *Am. Pharm. Rev.* **3**(3), 54–61 (2000).

11. Zheng, J.J.; Lynch, E.D.; Unger, S.E. "Comparison of SPE and Fast LC to Eliminate Mass Spectrometric Matrix Effects from Microsomal Incubation Products," *J. Pharm. Biomed. Anal.* **28**(2), 279–285 (2002).
12. Mei, H.; Hsieh, Y.; Nardo, C.; Xu, X.; Wang, S.; Ng K.; Korfmacher, W.A. "Investigation of Matrix Effects in Bioanalytical High-Performance Liquid Chromatography/Tandem Mass Spectrometric Assays: Application to Drug Discovery," *Rapid Commun. Mass Spectrom.* **17**(1), 97–103 (2004).
13. Korfmacher, W.A.; Cox, K.A.; Bryant, M.S.; Veals, J.; Ng, K.; Watkins, R.; Lin, C.-C. "HPLC-API/MS/MS: A Powerful Tool for Integrating Drug Metabolism into the Drug Discovery Process," *Drug Disc. Today* **2**(12), 532–537 (1997).
14. Hsieh, Y.; Bryant, M.S.; Brisson, J.-M.; Ng, K.; Korfmacher, W.A. "Direct Cocktail Analysis of Drug Discovery Compounds in Pooled Plasma Samples using Liquid Chromatography–Tandem Mass Spectrometry," *J. Chromatogr. B* **767**(2), 353–362 (2002).
15. Watt, A.P.; Morrison, D.; Locker, K.L.; Evans, D.C. "Higher Throughput Bioanalysis by Automation of a Protein Precipitation Assay using a 96-Well Format with Detection by LC-MS/MS," *Anal. Chem.* **72,** 979–984 (2000).
16. Berna, M.; Murphy, A.T.; Wilken, B.; Ackermann, B. "Collection, Storage and Filtration of In Vivo Study Samples using 96-Well Filter Plates to Facilitate Automated Sample Preparation and LC/MS/MS Analysis," *Anal. Chem.* **74,** 1197–1201 (2002).
17. Murphy, A.T.; Berna, M.J.; Holsapple, J.L.; Ackermann, B.L. "Effects of Flow Rate on High-Throughput Quantitative Analysis of Protein-Precipitated Plasma using Liquid Chromatography/Tandem Mass Spectrometry," *Rapid Commun. Mass Spectrom.* **16,** 537–543 (2002).
18. Locker, K.L.; Morrison, D.; Watt, A.P. "Quantitative Determination for L-775,606, a Selective 5-Hydroxytryptamine 1D Agonist, in Rat Plasma using Automated Sample Preparation and Detection by Liquid Chromatography–Tandem Mass Spectrometry," *J. Chromatogr. B* **750**(1), 13–23 (2001).
19. Bakhtiar, R.; Lohne, J.; Ramos, L.; Khemani, L.; Hayes, M.; Tse, F. "High-Throughput Quantification of the Anti-Leukemia Drug STI571 (Gleevec™) and its Main Metabolite (CGP 74588) in Human Plasma Using Liquid Chromatography-Tandem Mass Spectrometry," *J. Chromatogr. B* **768,** 325–340 (2002).
20. Sadagopan, N.; Cohen, L.; Roberts, B.; Collard, W.; Omer, C. "Liquid Chromatography–Tandem Mass Spectrometric Quantitation of Cyclophosphamide and its Hydroxy Metabolite in Plasma and Tissue for Determination of Tissue Distribution," *J. Chromatogr. B* **759,** 277–284 (2001).
21. Shou, W.Z.; Bu, H.-Z.; Addison, T.; Jiang, X.; Naidong, W. "Development and Validation of a Liquid Chromatographic Tandem Mass Spectrometry (LC/MS/MS) Method for the Determination of Ribavarin in Human Plasma and Serum," *J. Pharm. Biomed. Anal.* **29**(1–2), 83–94 (2002).

22. Ramos, L.; Brignol, N.; Bakhtiar, R.; Ray, T.; McMahon, L.M.; Tse, F.L.S. "High-Throughput Approaches to the Quantitative Analysis of Ketoconazole, a Potent Inhibitor of Cytochrome P450 3A4, in Human Plasma," *Rapid Commun. Mass Spectrom.* **14**(23), 2282–2293 (2000).
23. Biddlecombe, R.A.; Pleasance, S. "Automated Protein Precipitation by Filtration in the 96-Well Format," *J. Chromatogr. B* **734,** 257–265 (1999).
24. De Nardi, C.; Braggio, S.; Ferrari, L.; Fontana, S. "Development and Validation of a High-Performance Liquid Chromatography–Tandem Mass Spectrometry Assay for the Determination of Sanfetrinem in Human Plasma," *J. Chromatogr. B* **762,** 193–201 (2001).
25. Walter, R.E.; Cramer, J.A; Tse, F.L.S. "Comparison of Manual Protein Precipitation (PPT) versus a New Small Volume PPT 96-Well Filter Plate to Decrease Sample Preparation Time," *J. Pharm. Biomed. Anal.* **25,** 331–337 (2001).
26. Rouan, M.C.; Buffet, C.; Masson, L.; Marfil, F.; Humbert, H.; Maurer, G. "Practice of Solid-Phase Extraction and Protein Precipitation in the 96-Well Format Combined with High-Performance Liquid Chromatography-Ultraviolet for the Analysis of Drugs in Plasma and Brain," *J. Chromatogr. B* **754,** 45–55 (2001).
27. Rouan, M.C.; Buffet, C.; Marfil, F.; Humbert, H.; Maurer, G. "Plasma Deproteinization by Precipitation and Filtration in the 96-Well Format," *J. Pharm. Biomed. Anal.* **25,** 995–1000 (2001).
28. Mazenko, R.S.; Skarbek, A.; Woolf, E.J.; Simpson, R.C.; Matuszewski, B. "Sample Preparation via Solid Phase Extraction in the 96-Well Form for HPLC/UV Detection-Based Biofluid Assays. Application to the Determination of a Novel Cyclooxygenase II Inhibitor in Human Plasma and Urine," *J. Liq. Chrom.* **24**(17), 2601–2614 (2001).
29. Nicholson, J.K.; Connelly, J.; Lindon, J.C.; Holmes, E. "Metabonomics: A Platform for Studying Drug Toxicity and Gene Function," *Nature Rev.* **1,** 153–161 (2002).
30. Palmer, D.S.; Rosborough, D.; Perkins, H.; Bolton, T.; Rock, G.; Ganz, P.R. "Characterization of Factors Affecting the Stability of Frozen Heparinized Plasma," *Vox Sang* **65**(4), 258–270 (1993).
31. McCauley-Myers, D.L.; Eichhold, T.H.; Bailey, R.E.; Dobrozsi, D.J.; Best, K.J.; Hayes II, J.W.; Hoke II, S.H. "Rapid Bioanalytical Determination of Dextromethorphan in Canine Plasma by Dilute-and-Shoot Preparation Combined with One Minute per Sample LC-MS/MS Analysis to Optimize Formulations for Drug Delivery," *J. Pharm. Biomed. Anal.* **23,** 825–835 (2000).
32. Zhang, N.; Hoffman, K.L.; Li, W.; Rossi, D.T. "Semi-Automated 96-Well Liquid-Liquid Extraction for Quantitation of Drugs in Biological Fluids," *J. Pharm. Biomed. Anal.* **22,** 131–138 (2000).
33. Ke, J.; Yancey, M.; Zhang, S.; Lowes, S.; Henion, J. "Quantitative Liquid Chromatographic-Tandem Mass Spectrometric Determination of Reserpine in

FVB/N Mouse Plasma using a "Chelating" Agent (Disodium EDTA) for Releasing Protein-Bound Analytes during 96-Well Liquid/Liquid Extraction," *J. Chromatogr. B* **742,** 369–380 (2000).

34. Ramos, L.; Bakhtiar, R.; Tse, F.L.S. "Liquid-Liquid Extraction using 96-Well Plate Format in Conjunction with Liquid Chromatography/Tandem Mass Spectrometry for Quantitative Determination of Methylphenidate (Ritalin) in Human Plasma," *Rapid Commun. Mass Spectrom.* **14**(9), 740–745 (2000).
35. Jemal, M.; Teitz, D.; Ouyang, Z.; Khan, S. "Comparison of Plasma Sample Purification by Manual Liquid-Liquid Extraction, Automated 96-Well Liquid-Liquid Extraction and Automated 96-Well Solid Phase Extraction for Analysis by High-Performance Liquid Chromatography with Tandem Mass Spectrometry," *J. Chromatogr. B* **732,** 501–508 (1999).
36. Hoke II, S.H.; Tomlinson II, J.A.; Bolden, R.D.; Morand, K.L.; Pinkston, J.D.; Wehmeyer, K.R. "Increasing Bioanalytical Throughput using pcSFC-MS/MS: 10 Minutes per 96-Well Plate," *Anal. Chem.* **73,** 3083–3088 (2001).
37. Zweigenbaum, J.; Heinig, K.; Steinborner, S.; Wachs, T.; Henion, J. "High-Throughput Bioanalytical LC/MS/MS Determination of Benzodiazepines in Human Urine: 1000 Samples per 12 Hours," *Anal. Chem.* **71,** 2294–2300 (1999).
38. Zweigenbaum, J.; Henion, J. "Bioanalytical High-Throughput Selected Reaction Monitoring-LC/MS Determination of Selected Estrogen Receptor Modulators in Human Plasma: 2000 Samples/Day," *Anal. Chem.* **72,** 2446–2454 (2000).
39. Onorato, J.M.; Henion, J.D.; Lefebvre, P.M.; Kiplinger, J.P. "Selected Reaction Monitoring LC-MS Determination of Idoxifene and Its Pyrrolidinone Metabolite in Human Plasma using Robotic High-Throughput, Sequential Sample Injection," *Anal. Chem.* **73,** 119–125 (2001).
40. Shen, Z.; Wang, S.; Bakhtiar, R. "Enantiomeric Separation and Quantification of Fluoxetine (Prozac) in Human Plasma by Liquid Chromatography/Tandem Mass Spectrometry Using Liquid-Liquid Extraction in 96-Well Plate Format," *Rapid Commun. Mass Spectrom.* **16,** 332–338 (2002).
41. Brignol, N.; McMahon, L.M.; Luo, S.; Tse, F.L.S. "High-Throughput Semi-Automated 96-Well Liquid/Liquid Extraction and Liquid Chromatography/-Mass Spectrometric Analysis of Everolimus and Cyclosporin A in Whole Blood," *Rapid Commun. Mass Spectrom.* **15,** 898–907 (2001).
42. Bolden, R.D.; Hoke II, S.H.; Eichhold, T.H.; McCauley-Myers, D.L.; Wehmeyer, K.R. "Semi-Automated Liquid-Liquid Back-Extraction in a 96-Well Format to Decrease Sample Preparation Time for the Determination of Dextromethorphan and Dextrorphan in Human Plasma," *J. Chromatogr. B* **772**(1), 1–10 (2002).
43. Steinborner, S.; Henion, J. "Liquid-Liquid Extraction in the 96-Well Plate Format with SRM LC/MS Quantitative Determination of Methotrexate and Its Major Metabolite in Human Plasma," *Anal. Chem.* **71,** 2340–2345 (1999).

44. Susanto, F.; Humfeld, S.; Reinauer, H. "A Simple High-Performance Liquid Chromatographic Determination of Mexiletine in Human Plasma at Therapeutic Levels," *Chromatographia* **21,** 41–43 (1986).

45. Plomp, T.A.; Engels, M.; Robles de Medina, E.O.; Maes, R.A. "Simultaneous Determination of Amiodarone and its Major Metabolite Desethylamiodarone in Plasma, Urine and Tissues by High-Performance Liquid Chromatography," *J. Chromatogr.* **273,** 379–392 (1983).

46. Wesley, J.F.; Lasky, F.D. "Simultaneous Analysis of Antiarrhythmic Drugs and Metabolites by High Performance Liquid Chromatography: Interference Studies and Comparisons with other Methods," *Clin. Biochem.* **15**(6), 284–290 (1982).

47. Ruud-Christensen, M. "High-Performance Liquid Chromatographic Determination of (R)- and (S)-Proxyphylline in Human Plasma," *J. Chromatogr.* **491,** 355–366 (1989).

48. Van de Wiele, M.; De Wasch, K.; Vercammen, J.; Courtheyn, D.; De Brabander, H.; Impens, S. "Determination of 16B-Hydroxystanozolol in Urine and Feces by Liquid Chromatography-Multiple Mass Spectrometry," *J. Chromatogr. A* **904,** 203–209 (2000).

49. Bendriss, E.; Markoglou, N.; Wainer, I.W. "High-Performance Liquid Chromatography Assay for Simultaneous Determination of Dextromethorphan and Its Main Metabolites in Urine and in Microsomal Preparations," *J. Chromatogr. B* **754,** 209–215 (2001).

50. Zhao, J.J.; Xie, I.H.; Yang, A.Y.; Roadcap, B.A.; Rogers, J.D. "Quantitation of Simvastatin and its β-Hydroxy Acid in Human Plasma by Liquid-Liquid Cartridge Extraction and Liquid Chromatography/Tandem Mass Spectrometry," *J. Mass Spectrom.* **35,** 1133–1143 (2000).

51. Peng, S.X.; Henson, C.; Strojnowski, M.J.; Golebiowski, A.; Klopfenstein, S.R. "Automated High-Throughput Liquid-Liquid Extraction for Initial Purification of Combinatorial Libraries," *Anal. Chem.* **72,** 261–266 (2000).

52. Peng, S.X.; Branch, T.M.; King, S.L. "Fully Automated 96-Well Liquid-Liquid Extraction for Analysis of Biological Samples by Liquid Chromatography with Tandem Mass Spectrometry," *Anal. Chem.* **73,** 708–714 (2001).

53. Wang, A.Q.; Fisher, A.L.; Hsieh, J.; Cairns, A.M.; Rogers, J.D.; Musson, D.G. "Determination of a β3-Agonist in Human Plasma by LC/MS/MS with Semi-Automated 48-Well Diatomaceous Earth Plate," *J. Pharm. Biomed. Anal.* **26,** 357–365 (2001).

54. Wells, D.A. *High Throughput Bioanalytical Sample Preparation: Methods and Automation Strategies*, Elsevier Science, Amsterdam (2003).

55. Hennion, M.-C. "Solid-Phase Extraction: Method Development, Sorbents and Coupling with Liquid Chromatography," *J. Chromatogr. A* **856,** 3–54 (1999).

56. Hennion, M.-C.; Cau-Dit-Coumes, C.; Pichon, V. "Trace Analysis of Polar Organic Pollutants in Aqueous Samples Tools for the Rapid Prediction and Optimisation of the Solid-Phase Extraction Parameters," *J. Chromatogr. A* **823**(1–2), 147–161 (1998).

57. Bouvier, E.S.P.; Iraneta, P.C.; Neue, U.D.; McDonald, P.D.; Phillips, D.J.; Capparella, M.; Cheng, Y.-F. "Polymeric Reversed-Phase SPE Sorbents-Characterization of a Hydrophilic-Lipophilic Balanced SPE Sorbent," *LC-GC* **16**(Supplement), S53–S57 (1998).
58. Bouvier, E.S.P.; Martin, D.M.; Iraneta, P.C.; Capparella, M.; Cheng, Y.-F.; Phillips, D.J. "A Novel Polymeric Reversed-Phase Sorbent for SPE," *LC-GC* **15**(2), 152–158 (1997).
59. Gilar, M.; Bouvier, E.S.P.; Compton, B.J. "Advances in Sample Preparation in Electromigration, Chromatographic and Mass Spectrometric Separation Methods," *J. Chromatogr. A* **909,** 111–135 (2001).
60. Huck, C.W.; Bonn, G.K. "Recent Developments in Polymer-Based Sorbents for Solid-Phase Extraction," *J. Chromatogr. A* **885,** 51–72 (2000).
61. Oertel, R.; Richter, K.; Fauler, J.; Kirch, W. "Increasing Sample Throughput in Pharmacological Studies by using Dual-Column Liquid Chromatography with Tandem Mass Spectrometry," *J. Chromatogr. A* **948,** 187–192 (2002).
62. Millership, J.S.; Hare, L.G.; Farry, M.; Collier, P.S.; McElnay, J.C.; Shields, M.D.; Carson, D.J. "The Use of Hydrophilic Lipophilic Balanced (HLB) Copolymer SPE Cartridges for the Extraction of Diclofenac from Small Volume Pediatric Plasma Samples," *J. Pharm. Biomed. Anal.* **25,** 871–879 (2001).
63. Wu, J.-T.; Zeng, H.; Deng, Y.; Unger, S.E. "High-Speed Liquid Chromatography/Tandem Mass Spectrometry using a Monolithic Column for High-Throughput Bioanalysis," *Rapid Commun. Mass Spectrom.* **15,** 1113–1119 (2001).
64. Zimmer, D.; Pickard, V.; Czembor, W.; Müller, C. "Comparison of Turbulent Flow Chromatography with Automated Solid-Phase Extraction in 96-Well Plates and Liquid-Liquid Extraction used as Plasma Separation Techniques for Liquid Chromatography–Tandem Mass Spectrometry," *J. Chromatogr. A* **854,** 23–35 (1999).
65. Wu, J.-T.; Zeng, H.; Qian, M.; Brogdon, B.L.; Unger, S.E. "Direct Plasma Sample Injection in Multiple-Component LC-MS-MS Assays for High-Throughput Pharmacokinetic Screening," *Anal. Chem.* **72,** 61–67 (2000).
66. Ayrton, J.; Dear, G.J.; Leavens, W.J.; Mallet, D.N.; Plumb, R.S. "Use of Generic Fast Gradient Liquid Chromatography–Tandem Mass Spectroscopy in Quantitative Bioanalysis," *J. Chromatogr. B* **709,** 243–254 (1998).
67. King, R.C.; Miller-Stein, C.; Magiera, D.J.; Brann, J. "Description and Validation of a Staggered Parallel High Performance Liquid Chromatography System for Good Laboratory Practice Level Quantitative Analysis by Liquid Chromatography/Tandem Mass Spectrometry," *Rapid Commun. Mass Spectrom.* **16,** 43–52 (2002).
68. Deng, Y.; Wu, J.-T.; Lloyd, T.L.; Chi, C.L.; Olah, T.V.; Unger, S.E. "High-Speed Gradient Parallel Liquid Chromatography Tandem Mass Spectrometry with Fully Automated Sample Preparation for Bioanalysis: 30 Seconds per Sample from Plasma," *Rapid Commun. Mass Spectrom.* **16,** 1116–1123 (2002).

69. Joyce, K.B.; Jones, A.E.; Scott, R.J.; Biddlecombe, R.A.; Pleasance, S. "Determination of the Enantiomers of Salbutamol and its 4-O-Sulphate Metabolites in Biological Matrices by Chiral Liquid Chromatography Tandem Mass Spectrometry," *Rapid Commun. Mass Spectrom.* **12**(23), 1899–1910 (1998).
70. Yin, H.; Racha, J.; Li, S.-Y.; Olejnik, N.; Satoh, H.; Moore, D. "Automated High Throughput Human CYP Isoform Activity Assay using SPE-LC/MS Method: Application in CYP Inhibition Evaluation," *Xenobiotica* **30**(2), 141–154 (2000).
71. Cheng, Y.-F.; Neue, U.D.; Bean, L. "Straightforward Solid-Phase Extraction Method for the Determination of Verapamil and its Metabolite in Plasma in a 96-Well Extraction Plate," *J. Chromatogr. A* **828**(1–2), 273–281 (1998).
72. Rose, M.J.; Merschman, S.A.; Eisenhandler, R.; Woolf, E.J.; Yeh, K.C.; Lin, L.; Fang, W.; Hsieh, J.; Braun, M.P.; Gatto, G.J.; Matuszewski, B.K. "High-Throughput Simultaneous Determination of the HIV Protease Inhibitors Indinavir and L-756423 in Human Plasma Using Semi-Automated 96-Well Solid Phase Extraction and LC-MS/MS," *J. Pharm. Biomed. Anal.* **24**(2), 291–305 (2000).
73. Simpson, R.C.; Skarbek, A.; Matuszewski, B.K. "Quantitative Determination of a Selective Alpha-1a Receptor Antagonist in Human Plasma by HPLC with MS-MS," *J. Chromatogr. B* **775**(2), 133–142 (2002).
74. Eerkes, A.; Addison, T.; Naidong, W. "Simultaneous Assay of Sildenafil and Desmethylsildenafil in Human Plasma using Liquid Chromatography-Tandem Mass Spectrometry on Silica Column with Aqueous-Organic Mobile Phase," *J. Chromatogr. B* **768**(2), 277–284 (2002).
75. Naidong, W.; Bu, H.; Chen, Y.-L.; Shou, W.Z.; Jiang, X.; Halls, T.D.J. "Simultaneous Development of Six LC/MS/MS Methods for the Determination of Multiple Analytes in Human Plasma," *J. Pharm. Biomed. Anal.* **28**(6), 1115–1126 (2002).
76. Shou, W.Z.; Jiang, X.; Beato, B.D.; Naidong, W. "A Highly Automated 96-Well Solid Phase Extraction and Liquid Chromatography/Tandem Mass Spectrometry Method for the Determination of Fentanyl in Human Plasma," *Rapid Commun. Mass Spectrom.* **15**(7), 466–476 (2001).
77. Chen, Y.-L.; Hanson, G.D.; Jiang, X.; Naidong, W. "Simultaneous Determination of Hydrocodone and Hydromorphone in Human Plasma by Liquid Chromatography with Tandem Mass Spectrometric Detection," *J. Chromatogr. B* **769,** 55–64 (2002).
78. Lensmeyer, G. "Solid-Phase Extraction Disks: Second-Generation Technology for Drug Extractions," in Wong, S.; Sunshine, I. eds., *Handbook of Analytical Therapeutic Drug Monitoring and Toxicology*, CRC Press, Boca Raton, FL 137–148 (1997).
79. Wells, D.A.; Lensmeyer, G.L.; Wiebe, D.A. "Particle-Loaded Membranes as an Alternative to Traditional Packed-Column Sorbents for Drug Extraction: In-Depth Comparative Study," *J. Chrom. Sci.* **33,** 386–392 (1995).

80. Plumb, R.S.; Gray, R.D.M.; Jones, C.M. "Use of Reduced Sorbent Bed and Disk Membrane Solid-Phase Extraction for the Analysis of Pharmaceutical Compounds in Biological Fluids, with Application in the 96-Well Format," *J. Chromatogr. B* **694,** 123–133 (1997).
81. Mallet, C.R.; Lu, Z.; Fisk, R.; Mazzeo, J.R.; Neue, U.D. "Performance of an Ultra-Low Elution-Volume 96-Well Plate: Drug Discovery and Development Applications," *Rapid Commun. Mass Spectrom.* **17,** 163–170 (2003).
82. Wells, D.A. "96-Well Plate Products for Solid-Phase Extraction," *LC/GC North America*, **17**(7), 600–610 (1999).
83. Wells, D.A. "Accessory Products for Solid-Phase Extraction Using 96-Well Plates," *LC/GC North America*, **17**(9), 808–822 (1999).
84. Janiszewski, J.S.; Swyden, M.C.; Fouda, H.G. "High-Throughput Method Development Approaches for Bioanalytical Mass Spectrometry," *J. Chrom. Sci.* **38,** 255–258 (2000).
85. Jemal, M.; Huang, M.; Mao, Y.; Whigan, D.; Schuster, A. "Liquid Chromatography/Electrospray Tandem Mass Spectrometry Method for the Quantitation of Fosinoprilat in Human Serum using Automated 96-Well Solid-Phase Extraction for Sample Preparation," *Rapid Commun. Mass Spectrom.* **14**(12), 1023–1028 (2000).
86. Rule, G.; Henion, J. "High-Throughput Sample Preparation and Analysis Using 96-Well Membrane Solid-Phase Extraction and Liquid Chromatography–Tandem Mass Spectrometry for the Determination of Steroids in Human Urine," *J. Am. Soc. Mass Spectrom.* **10,** 1322–1327 (1999).
87. Janiszewski, J.; Schneider, R.P.; Hoffmaster, K.; Swyden, M.; Wells, D.; Fouda, H. "Automated Sample Preparation Using Membrane Microtiter Extraction for Bioanalytical Mass Spectrometry," *Rapid Commun. Mass Spectrom.* **11**(9), 1033–1037 (1997).
88. Simpson, H.; Berthemy, A.; Buhrman, D.; Burton, R.; Newton, J.; Kealy, M.; Wells, D.; Wu, D. "High Throughput Liquid Chromatography/Mass Spectrometry Bioanalysis Using 96-Well Disk Solid Phase Extraction Plate for the Sample Preparation," *Rapid Commun. Mass Spectrom.* **12**(2), 75–82 (1998).
89. Bakhtiar, R.; Khemani, L.; Hayes, M.; Bedman, T.; Tse, F. "Quantification of the Anti-Leukemia Drug STI571 (Gleevec) and its Metabolite (CGP 74588) in Monkey Plasma Using a Semi-Automated Solid Phase Extraction Procedure and Liquid Chromatography-Tandem Mass Spectrometry," *J. Pharm. Biomed. Anal.* **28,** 1183–1194 (2002).
90. Schütze, D.; Boss, B.; Schmid, J. "Liquid Chromatographic–Tandem Mass Spectrometric Method for the Analysis of a Neurokinin-1 Antagonist and Its Metabolite Using Automated Solid-Phase Sample Preparation and Automated Data Handling and Reporting," *J. Chromatogr. B* **748,** 55–64 (2000).
91. Fraier, D.; Frigerio, E.; Brianceschi, G.; Casati, M.; Benecchi, A.; James, C. "Determination of MAG-Camptothecin, a New Polymer-Bound Camptothecin

Derivative, and Free Camptothecin in Dog Plasma by HPLC with Fluorimetric Detection," *J. Pharm. Biomed. Anal.* **22**(3), 505–514 (2000).

92. Rule, G.; Chapple, M.; Henion, J. "A 384-Well Solid-Phase Extraction for LC/MS/MS Determination of Methotrexate and Its 7-Hydroxy Metabolite in Human Urine and Plasma," *Anal. Chem.* **73,** 439–443 (2001).
93. Biddlecombe, R.A.; Benevides, C.; Pleasance, S. "A Clinical Trial on a Plate? The Potential of 384-Well Format Solid Phase Extraction for High-Throughput Bioanalysis Using Liquid Chromatography/Tandem Mass Spectrometry," *Rapid Commun. Mass Spectrom.* **15**(1), 33–40 (2001).
94. Wells, D.A.; Lloyd, T.L. "Automation of Sample Preparation for Pharmaceutical and Clinical Analysis," in Pawliszyn, J., ed., *Sampling and Sample Preparation for Field and Laboratory: Fundamentals and New Directions in Sample Preparation*, Elsevier Science, Amsterdam, Chapter 24, 837–868 (2002).
95. Rossi, D.T. "Sample Preparation and Handling for LC/MS in Drug Discovery," in Rossi, D.T.; Sinz, M.W., eds., *Mass Spectrometry in Drug Discovery*, Dekker, New York (2002).
96. Smith, G.A.; Lloyd, T.L. "Automated Solid-Phase Extraction and Sample Preparation: Finding the Right Solution for your Laboratory," *LC-GC North America* **16**(12), 21–27 (1998).
97. Rossi, D.T.; Zhang, N. "Automating Solid-Phase Extraction: Current Aspects and Future Prospects," *J. Chromatogr. A* **885**(1–2), 97–113 (2000).
98. Hempenius, J.; Steenvoorden, R.J.J.M.; Lagerwerf, F.M.; Wieling, J.; Jonkman, J.H.G. "High-Throughput Solid-Phase Extraction Technology and Turbo Ionspray LC-MS-MS Applied to the Determination of Haloperidol in Human Plasma," *J. Pharm. Biomed. Anal.* **20,** 889–898 (1999).
99. Ooms, J.A.B.; Van Gils, G.J.M.; Duinkerken, A.R.; Halmingh, O. "Development and Validation of Protocols for Solid-Phase Extraction Coupled to LC and LC-MS," *Amer. Lab* **32,** 52–57 (2000).
100. McLoughlin, D.A.; Olah, T.V.; Gilbert, J.D. "A Direct Technique for the Simultaneous Determination of 10 Drug Candidates in Plasma by Liquid Chromatography-Atmospheric Pressure Chemical Ionization Mass Spectrometry Interfaced to a Prospekt Solid-Phase Extraction System," *J. Pharm. Biomed. Anal.* **15,** 1893–1901 (1997).
101. Beaudry, F.; Le Blanc, J.C.Y.; Coutu, M.; Brown, N.K. "In Vivo Pharmacokinetic Screening in Cassette Dosing Experiments: The Use of On-Line Prospekt Liquid Chromatography-Atmospheric Pressure Chemical Ionization Tandem Mass Spectrometry Technology in Drug Discovery," *Rapid Commun. Mass Spectrom.* **12**(17), 1216–1222 (1998).
102. Marchese, A.; McHugh, C.; Kehler, J.; Bi, H. "Determination of Pranlukast and Its Metabolites in Human Plasma by LC/MS/MS with Prospekt On-Line Solid-Phase Extraction," *J. Mass Spectrom.* **33,** 1071–1079 (1998).
103. Hsieh, J.Y.-K.; Lin, L.; Matuszewski, B.K. "High-Throughput Liquid Chromatographic Determination of Rofecoxib in Human Plasma Using a Fully

Automated On-Line Solid-Phase Extraction System," *J. Liquid Chrom. Rel. Technol.* **24**(6), 799–812 (2001).

104. Sola, J.; Prunonosa, J.; Colom, H.; Peraire, C.; Obach, R. "Determination of Ketorolac in Human Plasma by High Performance Liquid Chromatography after Automated On-Line Solid-Phase Extraction," *J. Liquid Chrom. Rel. Technol.* **19**(1), 89–99 (1996).
105. Kurita, A.; Kaneda, N. "High-Performance Liquid Chromatographic Method for the Simultaneous Determination of the Camptothecin Derivative Irinotecan Hydrochloride, CPT-11, and Its Metabolites SN-38 and SN-38 Glucuronide in Rat Plasma," *J. Chromatogr. B* **724**(2), 335–344 (1999).
106. Koster, E.; Ooms, B. "Recent Developments in On-Line SPE for HPLC and LC-MS in Bioanalysis," *LC-GC Europe* **14**(12), 55–57 (2001).
107. Jemal, M.; Ouyang, Z.; Xia, Y.-Q.; Powell, M.L. "A Versatile System of High-Flow High Performance Liquid Chromatography with Tandem Mass Spectrometry for Rapid Direct-Injection Analysis of Plasma Samples for Quantitation of a β-Lactam Drug Candidate and its Open Ring Biotransformation Product," *Rapid Commun. Mass Spectrom.* **13**(14), 1462–1471 (1999).
108. Svennberg, H.; Lagerstrom, P.-O. "Evaluation of an On-Line Solid-Phase Extraction Method for Determination of Almokalant, an Antiarrhythmic Drug, by Liquid Chromatography," *J. Chromatogr. B* **689**(2), 371–377 (1997).
109. Crescenzi, C.; Bayoudh, S.; Cormack, P.A.G.; Klein, T.; Ensing, K. "Determination of Clenbuterol in Bovine Liver by Combining Matrix Solid-Phase Dispersion and Molecularly Imprinted Solid-Phase Extraction Followed by Liquid Chromatography/Electrospray Ion Trap Multiple-Stage Mass Spectrometry," *Analyst* **125**(9), 1515–1517 (2000).
110. Yritia, M.; Parra, P.; Iglesias, E.; Barbanoj, J.M. "Quantitation of Nifedipine in Human Plasma by On-Line Solid-Phase Extraction and High-Performance Liquid Chromatography," *J. Chromatogr. A* **870,** 115–119 (2000).
111. Hedenmo, M.; Eriksson, B.-M. "Liquid Chromatographic Determination of the Macrolide Antibiotics Roxithromycin and Clarithromycin in Plasma by Automated Solid-Phase Extraction and Electrochemical Detection," *J. Chromatogr. A* **692**(1–2), 161–166 (1995).
112. Herman, J.L. "Generic Method for On-Line Extraction of Drug Substances in the Presence of Biological Matrices using Turbulent Flow Chromatography," *Rapid Commun. Mass Spectrom.* **16,** 421–426 (2002).
113. Grant, R.P.; Cameron, C.; Mackenzie-McMurter, S. "Generic Serial and Parallel On-Line Direct-Injection Using Turbulent Flow Liquid Chromatography/Tandem Mass Spectrometry," *Rapid Commun. Mass Spectrom.* **16,** 1785–1792 (2002).
114. Jemal, M.; Xia, Y.-Q.; Whigan, D.B. "The Use of High-Flow High-Performance Liquid Chromatography Coupled with Positive and Negative Ion Electrospray Tandem Mass Spectrometry for Quantitative Bioanalysis via Direct Injection of the Plasma Serum Samples," *Rapid Commun. Mass Spectrom.* **12**(19), 1389–1399 (1998).

115. Lim, H.K.; Chan, K.W.; Sisenwine, S.; Scatina, J.A. "Simultaneous Screen for Microsomal Stability and Metabolite Profile by Direct Injection Turbulent-Laminar Flow LC-LC and Automated Tandem Mass Spectrometry," *Anal. Chem.* **73**(9), 2140–2146 (2001).
116. Schafer, C.; Lubda, D. "Alkyl Diol Silica: Restricted Access Pre-Column Packings for Fast Liquid Chromatography-Integrated Sample Preparation of Biological Fluids," *J. Chromatogr. A* **909,** 73–78 (2001).
117. Boos, K.-S.; Rudolphi, A. "The Use of Restricted Access Media in HPLC. Part I—Classification and Review," *LC-GC* **15**(7), 602–611 (1997).
118. Rudolphi, A.; Boos, K.-S. "The Use of Restricted Access Media in HPLC. Part II—Applications," *LC-GC* **15**(9), 814–823 (1997).
119. Gundersen, T.E.; Blomhoff, R. "Qualitative and Quantitative Liquid Chromatographic Determination of Natural Retinoids in Biological Samples," *J. Chromatogr. A* **935,** 13–43 (2001).
120. Majors, R.E. "New Chromatography Columns and Accessories at the 2002 Pittsburgh Conference, Part II," *LC-GC* **20**(4), 332–344 (2002).
121. Yu, C.; Cohen, L.H. "Tissue Sample Preparation—Not the Same Old Grind," *LC/GC Europe* **17**(2), 2–6 (2004).

18

AUTOMATION

JOHN D. LAYCOCK AND THOMAS HARTMANN

Abstract Automation is an essential component of industrialized processes, which need to be productive, efficient, and of high quality. The use of robotic liquid handling robots for high throughput sample preparation in 96-well plates has become commonplace in drug discovery laboratories utilizing mass spectrometry. However, potential bottleneck remain in the areas of sampling, formatting, and aliquoting prior to sample preparation. Also, data handling bottlenecks can exist at the data reporting and warehousing. The strategies for an integrated workflow from study request all the way to final data summary reporting are examined. The following areas of the process flow of mass spectrometry (MS) based analysis are discussed: sample handling & tracking, sample preparation, data acquisition, MS front-end automation, data analysis, reporting and visualization, data archiving and retrieving. The development and use of custom software for interfacing between commercial database systems, robotics, and automated LC/MS/MS systems is discussed.

Keywords automation, robotics; robotic liquid handling, scripting, Visual Basic, LIMS, XML, LC/MS/MS.

INTRODUCTION

Automation is an essential component of industrialized processes. These processes need to be productive, efficient, and of high quality. Automation can increase the throughput of a laboratory by parallel processing, process integration, extended hours of operation, and reduced cycle times. It is commonly expected that automation will free resources from repetitive tasks, create

Integrated Strategies for Drug Discovery Using Mass Spectrometry, Edited by Mike S. Lee

walkaway time, and provide greater consistency than manual processes. With the increased pressure to improve productivity and reduce cost in the drug-screening process, automation plays a critical role in the analytical laboratory.

There are a remarkable number of emerging technologies that exist across today's drug-discovery landscape that have the potential to improve the productivity in the laboratory. When viewed from a historical perspective [1], it is remarkable to follow the rapid pace at which technology development has occurred. Within a short time, mass spectrometry (MS) has become the tool of choice for many pharmaceutical drug-discovery analyses. It is well beyond the scope of this chapter to discuss the advancements of MS technology comprehensively. Instead, we present a systematic view of the automation of mass spectrometry–based analytical processes. We focus mostly in the area of small-molecule screening and lead optimization, and cite examples of practical approaches to improve laboratory productivity through the implementation of mass spectrometry–related automation. The topic of proteomics and automation are not discussed in any detail; however, the topic has been recently reviewed [2,3]. As we explain later, the approach and process would be quite similar across most MS laboratories that employ automation in the drug-discovery process.

When both process and data automation are considered, the range of scope and number of applicable technologies are numerous. Automation tools that have a positive impact can be as simple as a programmable handheld multichannel pipette or an Excel macro to a corporate compound repository with millions of new chemical entities or multitier enterprise laboratory information management systems (LIMS) or other database applications. Of particular interest of this chapter is to review some of the commercially available and emerging technologies, both hardware and software, that are practical and simple enough to be implemented by the laboratory scientist with an interest in engineering and software development. The authors cite simple, but effective, examples where work-flow efficiency was improved dramatically with custom programming and in-depth knowledge of laboratory practices and the associated work flow.

Automation goes hand-in-hand with technology development and subsequent improvements in efficiency. The scientist involved in automation projects in the drug-discovery mass spectrometry laboratory must integrate a diverse collection of analytical instruments, robotic workstations, computer workstations, commercial or in-house software packages, fileservers, and multiple databases. These systems deal with various data types and software applications that should be integrated seamlessly to function efficiently [4]. Vendors and information technology departments may tend to support broad strategic objectives, regulatory compliance, and mainstream applications. Mass spectrometry automation projects may have specific need, and it can often be difficult to justify budget and resource priorities. Integrating and bridging the patchwork of systems and workstations requires some competency in software, process, and automation engineering. In addition, an understanding of emerging information technology solutions is important to adapt a progressive approach to data integration.

The first part of this chapter focuses on automation technologies as they relate to the logically separated steps in the analytical process of mass spectrometry–based analysis. Within each section we discuss some of the emerging and niche technologies that will likely have impact on automated mass spectrometry. The second part focuses on automation strategies and tools that include robotic liquid handling, LIMS, programming automation, and the role of information technologies in automation. Due to its popularity among automation scientists, the use of Visual Basic is discussed.

THE PROCESS FLOW OF MASS SPECTROMETRY BASED ANALYSIS

The typical steps of the analysis process that might be employed in the genomics, proteomics, high throughput, in vitro absorption, distribution, metabolism, and excretion (ADME), or in vivo pharmacokinetics (PK) screening mass spectrometry laboratory is shown in Figure 18.1. The simplified process view (top) of sample analysis can be segmented into a series of cones that represent logical portions in the process of analysis (bottom). Within any of these defined logical segments of the process flow, various automation tools and technologies are available to specific needs. These tools span the various levels in the diagram, which have been defined at the hardware, software, and systems levels. While discussion of automation could be limited to automation hardware that is directly coupled to the mass spectrometer analysis steps, an integrated approach must consider the entire process flow below, since maximum output is limited by process bottlenecks. Automation should address primarily repetitive and time-consuming steps in the whole process. Particular attention should be paid to data and sample handling between distinct process steps to avoid excess data entry or sample identification and reformatting. We have taken the approach to discuss the problem on the basis of the extended analytical process from start to finish. This approach is primarily due to the power of mass spectrometry–based analysis and the tendency to create a bottleneck in other portions of the process.

Sample Handling and Tracking

Standardization of sample collection to robot-friendly labware is a critical component in the successful automation of tube handling, and automated pipetting procedures on robotic workstations. The robotic software typically

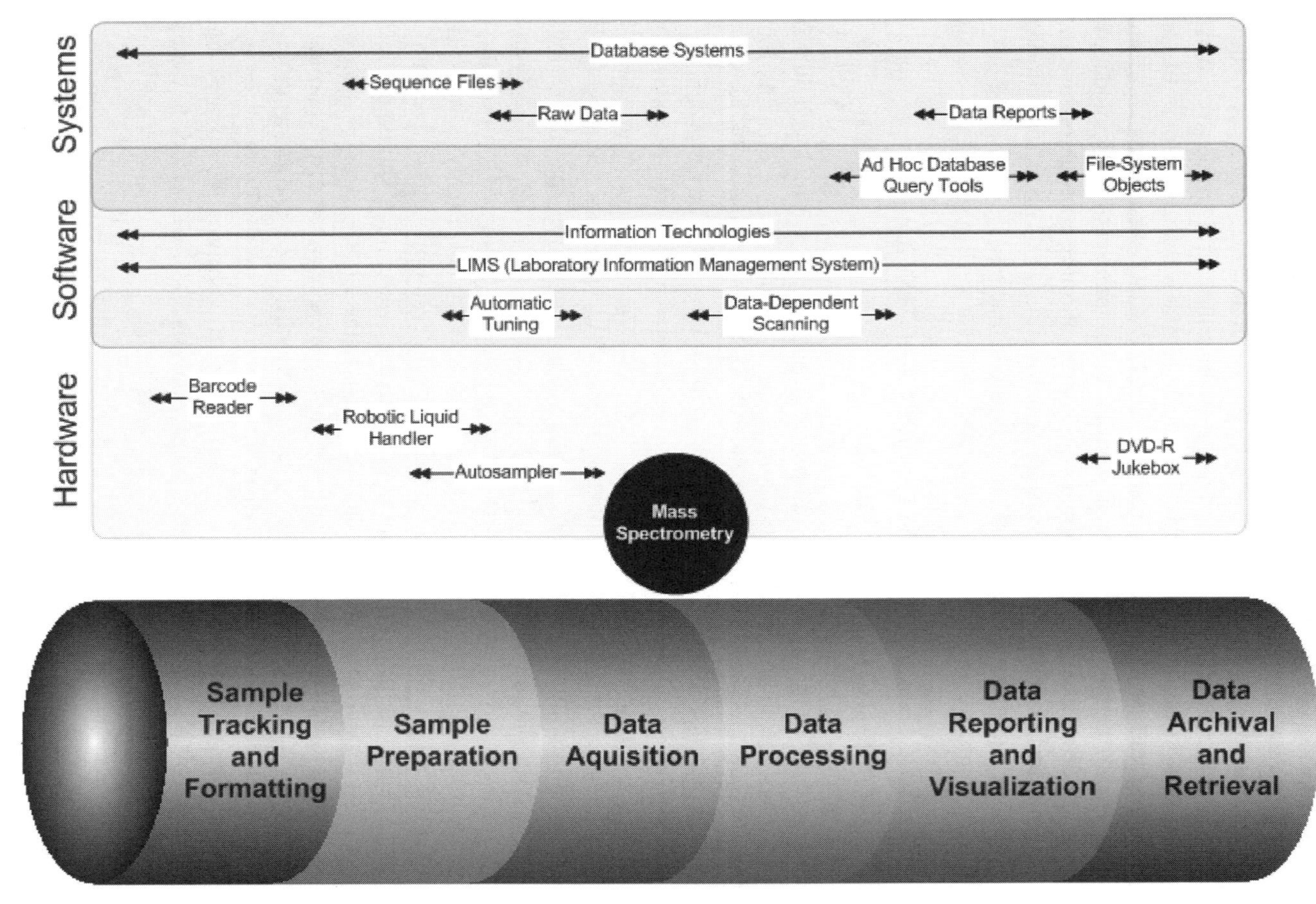

FIGURE 18.1 The mass spectrometry–based analytical process flow depicting systems related to process and data automation.

requires that the end user "teach" the robot to handle and access tubes and plates. For this reason, standardization of the upstream sample-collection procedures is required. This has been reported for the streamlined sample collection for bioanalysis where protocol-based study samples are collected [5,6]. Modular 96-well formats have become widely adapted in many laboratories that require a discrete, manual sampling event due to their compatibility with robotics and convenient sealing procedures (non-screw-cap strip or mat seals). These modular 96-well formats remove the significant bottleneck that exists with manual sample reformatting and uncapping of individual tubes. The approach to plate these samples on the final assay plate can be either static or dynamic, and depends on the requirements of the laboratory. If an in vitro experiment is performed, then it is feasible to integrate the entire sample-collection and tracking work flow into successive robotic procedures where sample locations are assigned by the user. The objective of the robotic procedure is to present sample plates that can seamlessly be analyzed according to a mass spectrometer analysis sequence.

One particularly challenging automation problem in bioanalysis is the pipetting of plasma samples received from drug-discovery pharmacokinetic screening studies. Blood samples from animal studies must first be collected in microtubes containing an appropriate anticoagulant. The formation of fibrin clots in plasma specimens presents a formidable challenge to achieving accurate, fully automated robotic liquid handling. Samples that contain fibrin clots require significant user intervention, and in many cases, manual pipetting of the samples are necessary where the robotic pipetting failed. There have been reports of the significance of anticoagulant selection. Sadagopan et al. reported that ethylenediaminetetraacetic acid (EDTA) improves the plasma collection [7]. Other recent reports have brought attention to the critical steps during sample collection that minimize the formation of fibrin clots [8]. Sample-collection conditions (such as types of anticoagulants, number of tube inversions, whole-blood storage time on wet-ice, flash freezing versus wet-ice storage with slow freezing) were evaluated. The most critical factors for clot-free sample collection were found to be a sufficient number of blood collection tube inversions and minimal storage time in whole blood before centrifugation.

Using this optimized-collection approach along with the use of disposable wide-bore tips and liquid-handling parameters, accurate, unattended liquid handling of plasma has been achieved in our laboratory [8]. Another approach demonstrated the use of 96-well polypropylene filter plates for collection, storage, and filtration prior to robotic liquid handling [9]. Recently, the use of an Apricot™ 96-well pipettor to transfer plasma from Microtainer™ collection tubes to 96-well plates was reported [10]. Other new technologies involve the use of robotic systems such as the Culex® (Bioanalytical Systems Inc, West Lafayette, Indiana) for automatic unattended collection of biofluid samples from small-animal studies using blood collection in conjunction with microdialysis [11–13].

A widely used two-dimensional (2D) symbology for small containers is Data Matrix ECC 200. This bar code has built-in error correction, which allows the correct decoding even of partially defective bar code. An imaging device is necessary to capture the bar code, which then can be decoded, independent of orientation, by integrated circuitry (fast, expensive) or by software. The definition of this bar code is available from www.aimglobal.org. Several commercial packages are available from tube manufacturers and image-analysis software companies. An example of a low-cost custom-software solution for reading 2D barcoded tubes in racks with standard flatbed scanners has been reported [14]. An example of these 2D bar codes scanned from a flatbed scanner can be seen in Figure 18.2. Reliable and cost-efficient radio-frequency identification devices (RFID) are used in highly industrialized processes and have potential application in pharmaceutical laboratories [15].

Linear bar codes on plates and tubes are commonly used in laboratories for sample identification and tracking purposes. Depending on the throughput, automated sample tracking may be helpful or necessary. Automated clinical systems identify linear bar codes on tubes by presenting bar codes individually to a reader. This method often requires the removal of individual tubes from their racks or the rotation of the tube, due to the size of the tube or the location of bar code. In pharmaceutical compound management and high throughput screening laboratories, tubes that can be arrayed in 96-well format that adhere to the Society for Biomolecular Screening (SBS) standard, with a unique 2D bar code at the bottom of each tube, are used to simplify tube-handling operations. This procedure allows the usage of standard laboratory robotic equipment to handle tubes and plates with identical tools. Furthermore, the bar codes can be decoded "on the fly" while tubes are located on a workstation, and thereby eliminates the need for individual tube handling. Together, these properties improve throughput and reliability of tube-based sample handling. In the past, tubes were individually sealed with a variety of lids. Automated cappers and decappers have been bulky and expensive in the past, but more modular devices, which can be integrated into workstations, have come to the market in recent months. Alternatively, piercable lids are available, and require the least sophisticated automation technology. However, piercable lids pose cross-contamination risks, which may not be compatible with regulatory compliance requirements. Finally, screw caps are available that can be handled by automated tools. However, the technology to do this reliably is more involved than with the other options. In addition, to remove the screw caps, tubes need to be handled one at a time, which is feasible but time-consuming.

Cherry-picking is a common practice in high throughput screening, which allows the analysis of selected samples from a larger collection. Sample identification and tracking is an absolute requirement for this type of operation.

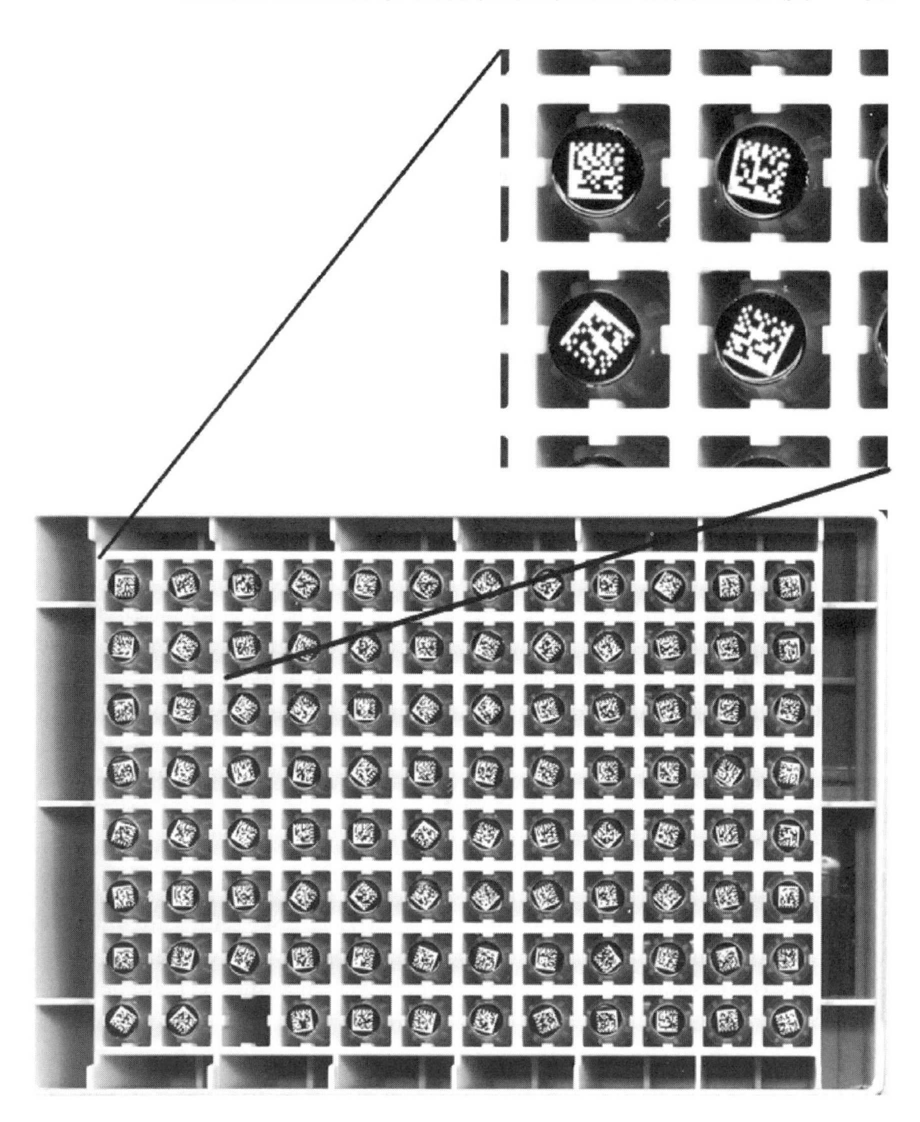

FIGURE 18.2 An image of the modular 96-well tube format with the DataMatrix™ 2D barcode identification is shown. The tubes are the Matrix ScrewTop ScreenMates® that come equipped with an interlocking wrench systems for ease of tube access and decapping.

At a minimum an electronic manifest within a set of samples must identify each sample uniquely and define the plate and well location. Individually labeled sample containers are essential for this purpose. 2D barcoded tubes are, as previously discussed, ideally suited to provide this functionality. In addition, to retrieve and store samples rapidly, modular devices from companies, such as TekCel (Massachusetts Hopkinton, AAA) or The Technology Partnership (Melbourn, UK), have become available, which may fit the medium storage requirements of pharmaceutical MS laboratories. In conjunction with a LIMS, these devices provide the sorted output of samples as requested, which eliminates sample mixup due to operator error. Alternatively, the bar codes could be used in real time for dynamic automated sample pipetting.

Sample Preparation

The use of automated sample-preparation techniques have become commonplace in the laboratory. It has been approximately 10 years since the introduction of solid-phase extraction (SPE) sorbents into flowthrough 96-well plates, which is now routinely performed on robotic workstations. The SPE approach has also been reported in higher density 384-well formats [16], but the approach has not since been widely adapted. Other sample-preparation schemes that use liquid–liquid extraction (LLE) [17] and protein-precipitation extraction (PPE) and protein-precipitation filtration [18–20] (PPF) have been reported. Combined use of automated sample preparation approaches, fast chromatography, and multiplexing to allow for the analysis of up to 2000 bioanalytical samples in 24 hours, has been reported [21]. For more detailed reading on high throughput sample-preparation technologies, we refer the reader to other chapters in this volume [Chapter 11: "Advances in High Throughput Quantitative Drug-Discovery Bioanalysis"; Chapter 12: "New Strategies for the Implementation and Support of Bioanalysis in a Drug Metabolism Environment"; Chapter 17: "Sample Preparation for Drug-Discovery Bioanalysis"] and review articles on the topic [22–26] .

Method development to provide robust analyte-specific procedures from complex biomatrices has received significant attention. Modular on-line and off-line approaches have been reported that focus on a systematic and integrated approach to select optimal extraction and separation chemistries [27]. These approaches have automated data analysis, whereby the feedback loop from experimentation to decision can be significantly reduced [18]. Wells has comprehensively discussed the use of automated strategies for the rapid determination of optimal extraction chemistries to maximize analyte recovery and selectivity [28]. The book includes up-to-date information on available robotic platforms for automated sample preparation and comprehensive information on labware.

Zip-Tips® (Billerica, Massachusetts) are automation-compatible pipette tips that contain a chromatography sorbent held in the tip with a membrane. These tips allow for microscale solid-phase extraction to purify and preconcentrate samples. Some examples of this approach are sample cleanup step with isoelectric point (pI)–based fractionation of peptides prior to matrix-assisted laser-desorption/short ionization (MALDI) spotting [29] or chip-based ESI analysis of colon adenocarcinoma (Caco-2) screening samples [30]. The preliminary results of induction-based fluidics in which nanoliter volumes of liquid was spotted from electrically charged ZipTips onto MALDI plates has been reported [31]. Recently, the direct elution of protein digests extracts from the Zip-Tips into an on-line chip-based ESI-MS has been reported [32].

Formats where chromatography is performed off-line and fractions are collected via chip-based electrospray ionization mass spectrometry (ESI-MS) or spotted LC/MALDI-MS approaches have received significant attention. A multichannel, noncontact deposition approach has been reported [33]. This approach has been reported for MALDI spotting techniques with a reported zero chromatographic dispersion [34]. Optimized spray deposition techniques and ionization conditions in MALDI time-of-flight mass spectrometry (MALDI-TOF-MS) can provide attamole sensitivity [35]. Recent reports of improved quantitation in small-molecule MALDI analysis due to improved laser speed and rastering techniques from desorption/ionization on direct ionization off silicon (DIOS) [36]. This type of sample analysis may provide a high-speed alternative to LC-MS analysis.

Data Acquisition

There has been significant effort and progress by the mass spectrometer vendors in the last decade to create software that is user-friendly and automates many of the routine functions. One particular example is the use of autotuning and calibration scripts that are used to tune and calibrate against a series of reference peaks to be found in a tuning standard.

In drug discovery, conditions for many compounds must be quickly determined for the creation of optimized multiple-reaction monitoring (MRM) tables for quantitative analysis of small-molecule drug candidates in ADME and PK screening [37]. An automated quantitative optimization package is an absolute requirement for quantitative bioanalysis involved in drug-candidate screening. It also has been reported that predictive models can be used to help generate MRM for compounds that have been synthesized onto the same scaffold.

There have been significant advances to harvest the power of the mass spectrometry–based analysis technique. The ability to perform real-time scanning experiments with real-time automated dynamic feedback has become an essential component of qualitative analysis, such as biotransformation or proteomics applications. Most commercial mass spectrometry software packages provide the ability to perform data-dependent scanning procedures. Metabolic profiling and proteomics generate large quantities of data that must be quickly and efficiently interpreted and filtered such that the data can be quickly reported in a way that will support rapid decisions.

One area that has had a significant number of recent reports is extending the data-dependent capabilities of the mass spectrometer with custom software and algorithms. There have been several reports in the area of automated high throughput determination of metabolites [38–40]. A custom software solution that automatically evaluates and reinjects samples with poor microsomal stability for metabolic profiling was recently reported [41]. A novel algorithm is performed to determine potential metabolites with summarized results automatically tabulated in an Excel worksheet.

There are many applications of real-time data-dependent scanning functions of tandem mass spectrometry (MS/MS) instruments that include metabolite identification and proteomics. The tandem mass spectrometer has the ability to generate massive quantities of data in these data-dependent modes, and has been typically limited by the ability of software and informatics that can render the data intelligible. Two examples of these increasingly complex applications are De Novo sequencing algorithms for peptides [42] and oligonucleotides [43]. Development of algorithms, data mining, and multivariate statistical approaches such as principal-component analysis (PCA) to deconvolute complex multidimensional sets in proteomics have been reported [44–46]. Similar multivariate approaches have been used for the determination of unknown metabolites [47,48]. A metabonomics approach with LC-TOF-MS has been demonstrated to provide data complementary to existing nuclear magnetic resonance (NMR) approaches [49].

Mass Spectrometry Front-End Automation

Since the advent of hyphenated MS techniques, gas chromatography (GC)-MS in the 1980s and LC-MS in the 1990s autosamplers have become a necessity at the front of MS-based instrument systems. Indeed the autosampler is a critical component of any modern LC-MS–based analysis system. Autosamplers have evolved to meet the increased demand requirements of automated well-based MS analysis. The primary figures of merit for autosampling devices are robustness, speed, lack of memory effect, swept volume, plate capacity, integration with MS software, and flexibility:

- *Robustness*: A primary goal of any front-end technology is to perform unattended and error-free and with relatively low maintenance.
- *Speed*: Current paradigms that use fast LC with microbore columns that contain traditional packing or monolithic stationary phases are currently capable of achieving adequate chromatographic separation in the time frame of ~10 seconds.
- *Lack of memory effect (carryover)*: A rule-of-thumb for carryover is that 0.2% is acceptable. This assumption is based on the FDA guidance for bioanalytical validation where the maximum response of a blank sample relative to the lower limit of quantitation is 30%. It is typical to run a quantitation curve that spans three orders of magnitude.

- *Swept volume*: Microbore chromatography applications require that the swept or dead volume be minimal relative to the chromatographic volume.
- *Integration*: The autosampler must have integrated control with the mass spectrometer software and include scheduling functionality for multiplexed systems. This feature is also important for regulated studies where sample well information metadata should be stored in the raw data.
- *Flexibility*: The autosampler system should be configurable, and have extended functionality and programmability for applications such as data-dependent cherry-picking.
- *Capacity*: The autosampler should have sufficient plate capacity to match the required throughput.

Current autosamplers have evolved to include integrated sample preparation and multiplexing functionality. Autosampling technology has been integrated with column switching (CS) technologies with multiplexed, parallel sample-preparation capabilities [50–53].

The development of a chip-based electrospray device from microfabricated nozzles has been reported [54], and was later commercialized [55] as the Nanomate™ (Advion BioSciences, Ithaca, New York). There have been several reports of the applications of ESI chips for front-end analysis. The application of this technique has been reported in several different areas, which include quantitative bioanalysis [56,57], metabolite identification [58,59], and proteomics [60]. The Advion Nanomate100 chip-based ESI-MS system has integrated the entire autosampler platform sans LC into an area the size of the ion source.

Data Analysis, Reporting, and Visualization

To provide meaningful drug-discovery information, the raw data from a mass spectrometer must typically go through a series of processing steps. Data processing of mass spectrometer data can have very different characteristics and be highly dependent on the application. For example, vendor software packages that perform quantitative analysis of specific MRM transistions based on peak integration and calibration regression are well established. In the absence of a LIMS system, a manual process typically would involve being exported to a text-based file. The text file could be opened with Microsoft Excel and is manually reformatted, sorted, and filtered. This type of activity would be a good candidate for automation with Visual Basic for Applications (see the following paragraphs).

To interface postacquisition data into databases requires the ability to query the database and provide exactly the data that are needed into a specified report format. Often, even with commercial LIMS applications, the provided reports do not provide enough information or place it in the correct format. In this case, there are a number of approaches that can be used to create custom queries. Many LIMS vendors are providing integrated custom reporting tools based on commercially available query engines such as Crystal Reports.

Structured Query Language (SQL) is an industry-standard language for interaction with databases. A number of powerful commands and functions are available; however, an analyst may only need to read data for reports, and therefore only needs to use the SELECT command. Ad hoc database query tools, such as BRIO Intelligence Designer (Sunnyvale, California), can allow end users with no in-depth knowledge of SQL to rapidly build and execute database queries through a graphical interface. This tool is also an excellent way for those new to database queries to learn SQL, or extract lengthy SQL commands for insertion into other applications. In addition, it is possible to create query reports that span multiple database systems. This approach can be particularly useful in drug-discovery studies where automated reporting is required for streamlining the data release to clients.

Typically, for BRIO™ application development a database administrator can provide a catalog file that contains the necessary connection information for the database of interest. Permission must also typically be granted for access to the database outside of the normal data-access application. BRIO files can be deployed as Web applications provided the client machine has the BRIO plug-in for Microsoft Internet Explorer. The graphing options within BRIO are somewhat limited for complex scientific charting and visualization. However, ad hoc queries have the flexibility and power to create custom data pivot tables that can be automatically exported

Spotfire (Somerville, Massachusetts) is a powerful data-visualization tool that can accept data from a number of different data sources. Spotfire has gained in popularity in high throughput screening areas and is capable of integrating with chemical-structure databases. Examples of Spotfire for visualization of experimental ADME, in silico, and chemical data have been reported [61].

Data Archiving and Retrieving

The issues around electronic records have received high visibility recently, particularly due to the significant attention to FDA's 21 CFR Part 11 [62]. The significance of 21 CFR Part 11 is normally regarded to cover only studies to be issued with regulatory filings. However, the software tools and approaches that are available to secure electronic records and enable paperless laboratories are applicable throughout drug discovery. Similar documentation requirements, concerns, and concepts must also be considered for legal defensibility of patents for new chemical entities. One area that has received significant interest due to the possibility of streamlined documentation is the electronic notebook [63]. There are concerted efforts in this direction within the Collaborative

Electronic Notebook Systems Association (www.censa.org), an industry organization that improves and promotes tools for electronic record keeping.

There are a large number of software products available to assist with the archival and retrieval of laboratory MS data. Regardless of regulatory requirements, it makes business sense that all data be catalogued and archived so as to be readily retrievable to reconstruct and document experimental data and results. These archival tools are designed to scan specified locations in the file system for Windows- or UNIX-based operating systems. Files are automatically archived based on preset conditions that are compared to the attributes of target files such as date created, date modified, or the archive bit (indicates the file is ready for archiving). The archive system is designed to secure and catalog data files with tracking information recorded to a database. The archive "engine" can be configured to remove data and write the data to durable media such as CD-R or DVD-R in jukeboxes. Products currently available to perform automated data archival include NuGenesis® Scientific Data Management System (Waters Corporation, Milford, Massachusetts), Business Process Management (Scientific Software, Pleasanton, California), and eRecordManager (Thermo LabSystems, Beverly, Massachusetts).

The systems just discussed also provide visual data-capture functions that have received significant attention recently. These systems capture relevant scientific content (usually visual) that can be catalogued, indexed, and later retrieved independent of the instrument specific application. When properly configured and implemented, these data-capture systems can be implemented to provide automated capture and retrieval of reports where queries can easily retrieve output from particular experiments. A virtual print driver captures enhanced metafiles and stores these files an Oracle database. The use of this technology in an open-access LC-MS laboratory has been reported [64]. A general overview of the technology and implementation strategies has been published [65]. These data management tools differ from LIMS in that they are less rigid in terms of work flow and may be better suited to the earlier stages of drug discovery.

Duckworth has reported the advantages and disadvantages of various alternatives to the standard raw-data file archival (i.e., graphical, textual, XML) that will provide robust long-term alternatives [66]. With current data-capture technologies and transfer standards there is a trade-off between transportability and information content.

AUTOMATION STRATEGIES

Robotic Liquid Handling

A robotic liquid handler is the centerpiece of automation in drug-discovery laboratories and is an important component of the sample-preparation process.

To automate sample preparation, a liquid-handling robot (i.e., liquid-handler) as commonly used in high throughput screening laboratories [67] is essential. This device transfers aliquots of sample or solvent and should eliminate the need for manual pipetting. In practice, workstations with a multichannel pipetting arm are sufficient to achieve the throughput required for a typical MS lab. If sample collection and distribution is not standardized to the SBS microplate format (www.sbs.org), then it is necessary to use a liquid handler that can access samples individually. Commercial devices with 1 to 16 independent probes are available. Independent probes are required for applications such as the preparation of serial dilutions or cherry-picking. Robotic liquid-handling arms with fully independent tip-positioning systems that have independent y- and z-axis movement are available for applications that require flexible pipetting. Liquid handlers can also be purchased equipped with accessories such as wash stations, vacuum boxes, heat exchangers, or plate readers that are fitted on the deck of the platform application-specific tasks. New features such as real-time optical position calibration can be useful, especially when disposable tips are used with high-density plate formats.

To minimize the risks of cross-contamination many laboratories will use disposable tips instead of fixed, washable tips. The use of disposable tips is and has been an area of constant discussion within the automation community. Fixed tip operations are inherently more reliable and cost effective, and in many cases, suitable washing procedures can be identified that can reduce the levels of carryover below detection limits. Many liquid handlers provide a mechanism of liquid detection that can be used to limit the exposure of tips to liquid, and thereby help prevent cross-contamination. However, depending on the nature of the compounds to be analyzed, compliance and liability issues, the use of disposable tips may be required. A number of different approaches to liquid handling, such as positive displacement, system liquid, and air pressure, have been incorporated into liquid handlers currently on the market. With the various pipetting approaches there are various ways to perform liquid sensing to detect empty sample wells or pipetting errors. The most common approach to liquid sensing is capacitive detection between the pipette tip and the metal surface of the deck. This approach requires the use of carbonized conducting tips. Another approach is the use of pressure sensing, in which an expected pressure increase can be predicted and measured. If the actual pressure during pipetting deviates significantly, then an error is indicated.

It is necessary to adjust the hardware and the liquid-handling parameters (syringe size, pipetting speed, break-off speed, air gaps, tip touching, clot detection, calibration factors, etc.) of the pipetting robot for accurate and precise liquid transfer. Many robots can achieve a coefficient of variation (CV) below 5% in the 1- to 100-μL range when properly programmed and calibrated.

It is especially important to perform calibration experiments with plasma or other high-viscosity samples. Automated procedures that include on-line gravimetric and density determinations have been reported [68,69]. Not all workstations provide the same level of tuning of liquid-class parameters. Some vendors offer multivariate analysis packages that help users set up the appropriate liquid-class experiments and then use statistical approaches such as design of experiment (DOE) to guide the determination of optimal liquid-handling parameters [70].

Modern systems should provide a visual interface for rapid method development that can be mastered by a properly trained staff. A more dynamic integration of the robot into a work flow usually requires some form of scripting or high-level programming skills. The programmability and user interface of a robot is a critical factor to obtain user acceptance. The reader can find more details on the robotic platforms currently on the market in recently published review articles [71,72].

Laboratory Management Information Systems

Data analysis and report generation are prime examples of where productivity and efficiency of a laboratory can be greatly improved. This efficiency gain is most easily achieved with integrated or connected data systems. The transfer of data between instruments or informatics systems should be integrated and require minimal human intervention. In addition to the time spent, manual data transfer also increases the risk for errors, which ultimately may lead to inaccurate results.

Most analytical laboratories use a LIMS to manage work flow and data analysis. In essence, a LIMS is a specialized database application that facilitates project tracking, data analysis, and result archival. Over the last 40 years, LIMSs have evolved into custom applications that support specialized needs and requirements of analytical laboratories. The ideal LIMS would be a system that is configurable and expandable in response to the changing needs of the laboratory. In reality, most LIMS require custom programming, which usually involves extended requirement and implementation phases [73]. Therefore, it is important to choose the best-suited LIMS from the start. Even after significant market consolidation in the last few years, there are still a large number of LIMS vendors to choose from, as well as consulting firms that offer services to survey the LIMS market.

Many LIMSs support the control of a wide variety of analytical instruments, such as HPLC and spectrophotometers. Few offer the integration of MS instrument control. The import of spectra is commonly supported, but is usually not vendor neutral. Data can be securely stored in the LIMS database, so that it is not necessary to transfer raw data into corporate databases. Recently, many vendors provide support of 21 CFR Part 11 and good laboratory practices (GLP) compliance for their systems. Most laboratories will perform data analysis within the LIMS system, create reports, and store information in a document management system. LIMS are just part of a number of information systems used in a modern lab. Data exchange between programs has become greatly simplified through a number of commonly available transfer protocol specifications and markup languages (OLEDB, XML).

One area where LIMSs have not provided out-of-the-box support is in the automation of sample preparation. This may be due to the fact, that analytical labs do not have sample-preparation workstations or that there is no standard for these workstations. Nevertheless, the authors think that sample preparation can be automated, providing similar benefits as in high throughput screening, for example. The use of portable devices such as pocket has been reported in the laboratory [74]. Commercially available applications such as LimsLink (Labtronics, Inc.) can upload laboratory instrument data from handheld devices to any LIMS.

Programming Automation

The modern automated workstation provides software with intuitive and simple graphical interfaces that are easy to learn. Most automated workstation software packages have simple drag-and-drop or wizard-based programming functions. Also, in some cases when robotic scripts have been developed and put into production, users are presented with a very simple interface. The software conceals the underlying complexity of the robotic programming from the casual user. If additional script complexity is required, more complex programming constructs, such as variable declaration, loops, Boolean logic, and user prompts should be supported. These programming features allow for the development of dynamic and flexible robotic scripts. Some vendors provide sophisticated simulation modes with 3D visualization of script programs. This visualization feature can be valuable during script development when access to production robots is limited.

Full flexibility can be achieved if a vendor/supplier worklisting construct is provided. Typically a worklist is delimited text files or a database source that contains robotic command functions and all associated information (i.e., arguments) to perform one robotic step that will drive individual robotics. To some degree, each line of the text file approximates a list of commands that are sent to the workstation. An additional and flexible way to program automated workstations is by use of vendor-provided programmable components (i.e., DCOM, ActiveX, Named Pipes) that expose interfaces to control the software from an independent application. Often, maximum flexibility and complexity can be achieved by using a combination of the vendor-supplied worklists and programmable interfaces.

Scripting Robotics

Static scripting is the easiest to achieve and can typically performed within the vendor-supplied scripting language. However, static scripting is the most inflexible and requires that other portions of the work flow be adapted to fit the script requirements. Our experience with this approach is that additional up-front efforts to create and verify sequence files to the script must be made. If a LIMS system is used, then the analysis sequence file may make use of a template for this purpose. The number of samples can be restricted and dilutions or pooling may not be available. One of the areas to consider is the source of samples for analysis (see earlier). One approach for dynamic scripting, for example, is to develop a custom Visual Basic program that combines aspects of stand-alone application, OLE-automation of Gemini™ software components, and prewritten scripts. The program creates a text file (referred to as the worklisting file) in a delimited format that describes parameters for each robotic step. The software has easy, programmable built-in functions that interpret and execute the commands in the file. The robot requires each of the critical pieces of information to perform a particular command. For example, for an aspiration step the robot must receive the deck-location information, which includes carrier position, rack type and location, tube number, and the pipetting parameters. In a custom program developed in our laboratories, two text files are needed to perform the fully automated cherry-picking routine. The first file is a manifest file that describes the plate and tube location for each uniquely identified sample. The second file is a sequence file that is directly exported from the LIMS and contains sample identification, volume, dilution, tube location, and calibration curve information. The files are parsed and the application automatically creates the worklist file that is semicolon delimited in the following format:

```
A,D,or W;Rack Label;;Rack Name;Well #;;volume;
liquid class
Where A=Aspirate, D=Dispense, and W=wash

For Example:
A;S1;;Matrix w/ Costar tubes, LS;2;;50;Plasma
D;D1;;Costar Modular 96 well P;19;;50;Plasma
W;
A;S1;;Matrix w/ Costar tubes, LS;10;;50;Plasma
D;D1;;Costar Modular 96 well P;18;;50;Plasma
W;
```

In addition to the liquid-handling instructions, we have incorporated functionalities such as metrics tracking and error notification by e-mail or paging systems. A similar approach on the Packard™ has been reported [75].

THE ROLE OF INFORMATION TECHNOLOGIES IN AUTOMATION

The workstation and hardware tools previously described typically have very specific uses in the analytical process. The application of information technologies to the development of integrated automation is ubiquitous to the analytical process. There have been an increasing number of automation engineers and scientists who have discovered the joys and productivity enhancements that can be realized though custom-application development. This increase is in large part due to the emergence of high-level and object-based programming environments that provide for rapid development of applications. Visual Basic (VB) has become a well-established programming language for automation due to its accessibility and ability to quickly create applications. A survey performed at the 2001 International Symposium on Laboratory Automation and Robotics (ISLAR) indicated the following breakdown for the programming environments used with automation: Visual Basic 41%, Visual Basic for Applications (VBA) 34%, Java 14%, and C++ 10%. These results are due to the fact that by and large, most laboratory workstations are programmed to operate on Microsoft Windows workstations. Data can be conveniently reported and summarized with Microsoft Office Applications, since these applications are usually readily available and support data exchanged. VB and VBA are discussed in some detail below.

The component object model (COM) has been the bread and butter of Microsoft's modular programming approaches. Essentially, each component is an encapsulated unit of code that is compiled into a binary format (i.e., .dll, .ocx, .exe). The component exposes its essential properties and methods in a simplified fashion. There is no need to understand underlying complexities during implementation of these objects. A Visual Basic (or any object-based) application essentially operates as a "glue" language and combines the functionalities of the individual components at the top level. One of the primary challenges to develop these types of applications is to have in-depth knowledge of the application or its component object model. The term object linking and embedding (OLE) automation has been around for some time and was renamed in its third generation to ActiveX® by Microsoft. This technology can be defined as the ability of an application to expose interfaces to other programming languages. This can be more specifically defined as automation clients and servers, which we will define with a brief example below. Once this is understood, it is straightforward to exchange data programmatically from one application to another.

Many mass spectrometry and automation vendors provide software development kits (SDKs) or "cookbooks" that provide the required documentation to use the vendor-specific application programming interfaces (API). The level of documentation and support provided by the vendor should be considered a high-priority criterion for selection of software if customization and component reuse are intended. If the vendor has taken the time to provide good documentation of the components and the programmer has a good working knowledge of the application and desired output, then the process to develop an application can be straightforward and rapid. Also, vendors can provide training or consulting services for the API they provide.

Visual Basic

First we discuss VBA, a powerful development environment used directly from within the Microsoft Office suite of programs. Typically, Word or Excel is most frequently used, since many LIMS and instrument software directly export reports in these formats. One of the primary advantages with VBA is the rich, intuitive object model that is provided by each of the Microsoft Office applications. The implementation of VBA as a cost-effective and first-line approach to data automation in drug discovery has been reported [76]. Bisset's book entitled *Practical Pharmaceutical Automation* contains many examples of Excel VBA in the laboratory setting [67].

Microsoft Office allows the option to record "macros" for performing repetitive tasks. Although macros are somewhat limited for the development of true VBA applications, macro recording can be used effectively to learn VBA or create templates of code that can be generalized with the addition of variables. If VBA for Microsoft Office is enabled, you only need to press ALT-F11 to open the VBA Integrated Development Environment (IDE). A search of the on-line help for Application Object will provide an interactive object hierarchy that describes (with detailed code examples) each object or collection. There are literally thousands of objects within the Microsoft Office application object models. Another very useful tool within the VBA environment is the Object Browser, which can be invoked with the F2 key. The Object Browser can be particularly useful to learn the objects of external components that have been referenced from within VB or VBA. This method is an excellent way to get started with new object types and follow along with the documentation. An example is shown in Figure 18.3 from the Excel 2000 VBA IDE. In this example, the Excel is the automation client and Sciex Analyst™ method component

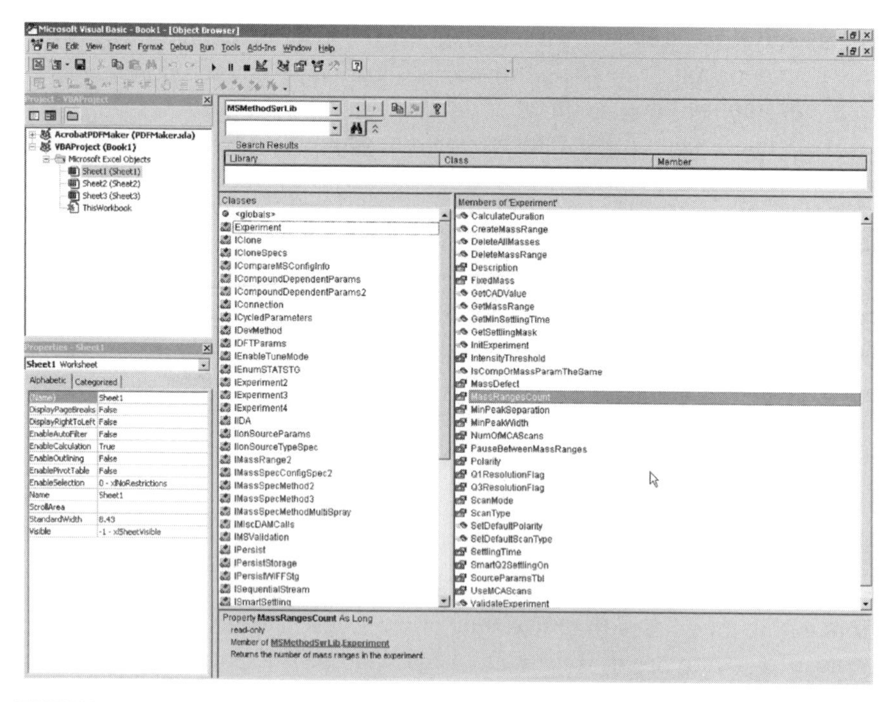

FIGURE 18.3 The Object Browser of the Excel Visual Basic for Applications Integrated Development Environment is set up to explore the Mass Spectrometer Method Server Library (MSMethodSvr.dll) provided with Sciex Analyst version 1.3.

is the automation server. Russo and Echols have authored the essential guide to VB in the laboratory; it describes the significance of object-based programming and guides the reader through the essential concepts to understand how to program with VB [77].

VB version 6 has been the gold standard since its release in 1998, and there have been many reports in the mass spectrometry literature of its use. Several recent reports have demonstrated the utility of custom-application development with VB to streamline many aspects of the work flow with improved automation and integration. Xu et al. have reported the development of custom applications that support high throughput mass-directed parallel purification [78,79]. Choi et al. have has reported the development of custom VB application and databases in combinatorial library screening applications [80,81], and microsomal stability screening [82]. Whalen et al. have developed a suite of applications to aid in high throughput bioanalytical screening [83–85]. Bean et al. have reported enterprise Web applications for simplifying open-access mass spectrometers [86,87]. The latter was developed with Microsoft .NET, which is discussed below in the following subsection.

Microsoft .NET

Microsoft .NET framework is an entirely new platform introduced in early 2002. A detailed discussion of the .NET Framework is well beyond the scope of this chapter and at this stage is indeed a daunting prospect. We attempt here to supply the reader with some key features that make .NET worth considering. Visual Basic 6 (VB 6) is a platform unto itself. However, in .NET VB is just one of a number languages such as C++ and C# provided for programming the .NET framework. All .NET languages fully support object-oriented programming concepts such as class inheritance and polymorphism. VB.NET offers a substantial increase in the number of built-in user-interface components. Many core Windows functions are included in the .NET framework and are easily accessible from the IDE. Many of these functions would require advanced knowledge of Win32 API programming in VB 6 in order to access them. VB.NET allows programmers with experience developing Windows applications to quickly migrate toward Web development with ASP.NET. ASP.NET is a Web programming technology that substantially improves previous Microsoft Active Server Pages programming. Most notably ASP.NET provides real-time debugging and diagnostics that VB programmers are familiar with. Real-time debugging of Web applications is a vast improvement over the debugging of client and server side scripts that was required for ASP. Developing ASP.NET in Visual Studio requires that Microsoft's Internet Information Server Software be running. Some other exciting technologies supported in .NET are XML Web Services and the .NET compact framework for developing applications for portable devices. ActiveX Data Objects .NET (ADO.NET) provides a rich and powerful object model to access, sort, and manipulate data. Despite all of this additional power and functionality, there are some omissions. The most notable omission for automation scientists is that the widely used serial port control (MSCOM) was not provided in the .NET framework. Interoperability with preexisting components is supported since .NET is backwards compatible with COM and Win32 API. However, interoperability within .NET can be more complex than standard COM programming [88]. The net outcome of the preceding is less simplicity and accessibility for casual programmers and a longer learning curve. However, the integrated development environment provides many more integrated components and controls than what are available in VBasic 6.

The emergence of the Microsoft .NET platform has a significant potential impact on the use of VB. Based on the large population of experienced laboratory programmers implementing VB, it is likely that eventually a migration to VB.NET will occur. It has yet to be seen how this might impact automation applications, though adoption appears to be slow to occur. Interested readers can find a wealth of current .NET information on the Microsoft Web site (www.microsoft.com/net/).

Practical Extraction and Report Language (Perl)

Practical Extraction and Report Language (Perl) is another programming language that deserves mention, since it has received significant attention in the bioinformatics and proteomics communities. Perl was developed in the mid-1980s and was once the most common language used for Web development. The strength of Perl lies in its ability to process text and to find patterns. As such it has been utilized in the bioinformatics community in the form of the BioPerl project, an open-source international collaborative resource with object-based bioinformatics libraries [89,90]. There have been reports of the use of Perl in peptide-fragment ion screening and other proteomic mass spectrometry applications [91–94]. ActiveState (Vancouver, BC, Canda) has created ActivePerl, currently in version 5.8, for the development of Perl applications for Microsoft Windows. In addition, ActiveState has recently released Visual Perl and similarly, Visual Python, which allows for integrated programming of these popular open-source languages within the Microsoft Visual Studio .NET IDE.

Extensible Markup Language

Markup languages are not new. Markup languages are text-based data formats derived from Standard Generalized Markup Language (SGML), which originated in the 1960s as a way to describe complex electronic documents. Hypertext Markup Language (HTML), a much simpler language, was invented to represent human-readable Internet content. However, the limitations of HTML led to the development of two intermediary languages, dynamic and extensible versions (DHTML and XHTML, respectively). Extensible Markup Language (XML) was created in 1996 as a subset of SGML. XML relies on self-described tags to define elements, attributes that sit in hierarchical nodes. XML has established itself due to its simplicity, portability, and platform independence. Parsers for XML have been written in almost every major programming language, including VB.NET, which is tightly integrated with XML [95]. Web services use XML as the transport protocol for transfer of data across systems or platforms.

The extensible nature of XML has led to the creation of numerous custom markup languages that are developed and refined by worldwide consortiums. These consortiums create standards whereby the data descriptions in a particular field of interest can be harmonized. Scientific disciplines that have benefited from early adoption of XML are mathematics with MathML and chemistry with Chemical Markup Language (CML). The CML consortium has released XML Schema Definitions (XSD), which allow for standardized exchange of data [96,97]. For more details on the use of XML in the area of bioinformatics and data integration as well as comparison to other technologies, we suggest the excellent review article by Achard et al. [98]. Russo and Rubin have published an introduction to the use of XML in the laboratory that includes a detailed description on the installation and use of the Microsoft XML parser [99].

XML has been implemented for various specific markup languages related to analytical instrument and laboratory data. One example is the Generalized Analytical Markup Language (GAML). GAML is used and promoted by eRecordManager (Thermo LabSystems, Beverly, Massachusetts). Duckworth has proposed XML as a standard for long-term storage of analytical data and outlines the limitations of some of the current American Society for Testing and Materials (ASTM) standards such as JCAMP and AnDi [100]. SpectroML is another recently proposed standard [101]. GAML and SpectroML served as the stating points for the Analytical Information Markup Language [102] (AnIML). Development of AnIML is ongoing and has been sanctioned by the ASTM under Subcommittee E13.15. AnIML is based on a layered approach and allows for the inclusion of core and technique layers [103]. AnIML has been designed to handle multidimensional data, sample information, and process information. For more details on-line information is available at http://animl.sourceforge.net/.

The Software Development Life Cycle

It is useful to understand the steps of the software development life cycle (SDLC) during a project. In fact, this approach is commonly used for implementation and validation even when commercial off-the-shelf software is involved [104]. It is also important to consider the economic aspects of the automation project [105]. A 10-step guide to success that emphasizes clearly defined project control in outsourced automation/information technology projects has been published [106].

CONCLUSIONS

Mass spectrometry is currently integrated into the analytical laboratory by using existing generic laboratory automation technologies or by adapting and developing customized hardware and software for MS-related uses. The integration tools have matured over the years to enable automated MS analysis at the current scale.

Recent shifts in paradigms of the drug development process, which include the concepts of compound profiling, the advent of novel assays, such as specific transporter assays, and the progress in assay miniaturization, will lead to an increased demand for MS analysis. This trend has been recognized in the industry. The first commercially available fully integrated MS systems have appearred on the market. We expect in the future, that the automation of MS analyses will progress and mature, similar to the advancements we have seen in the automation of high-throughput screening efforts in the past.

ACKNOWLEDGMENTS

The authors wish to thank Greg Woo of Amgen, Inc. Information Systems for his review and comments on this manuscript. We would also like to thank the members of the Mass Spectrometry Group in the Department of Pharmacokinetics and Drug Metabolism at Amgen in their support during the preparation of this manuscript.

REFERENCES

1. Crouch, S.R.; Atkinson, T.V. "The Amazing Evolution of Computerized Instruments," *Anal. Chem.* **72**(17), 597A–603A (2000).
2. Quadroni, M.; James, P. "Proteomics and Automation," *Electrophoresis* **20** (4–5), 664–677 (1999).
3. Gygi, S.P.; Aebersold, R. "Mass Spectrometry and Proteomics," *Curr. Opin. Chem. Biol.* **4**(5), 489–494 (2000).
4. Fish, M. "Effective Access and Retrieval of Laboratory Data to Enable Knowledge Management," *Am. Lab.* **35**(18), 25–30 (2003).
5. Rossi, D.T. "Integrating Automation and LC/MS for Drug Discovery Bioanalysis," *J. Auto. Meth. Manage. Chem.* **24**(1), 1–7 (2002).
6. Zhang, N.Y.; Rogers, K.; Gajda, K.; Kagel, J.R.; Rossi, D.T. "Integrated Sample Collection and Handling for Drug Discovery Bioanalysis," *J. Pharm. Biomed. Anal.* **23**(2–3), 551–560 (2000).
7. Sadagopan, N.P.; Li, W.; Cook, J.A.; Galvan, B.; Weller, D.L.; Fountain, S.T.; Cohen, L.H. "Investigation of EDTA Anticoagulant in Plasma to Improve the Throughput of Liquid Chromatography/Tandem Mass Spectrometric Assays," *Rapid Commun. Mass Spectrom.* **17**(10), 1065–1070 (2003).
8. Laycock, J.D.; Jaramilla, J.; Zhang, L.; Smithson, A.; Mallard, S.; Miller, K.J. "Optimized Plasma Sample Collection Procedures to Minimize Clot Formation Prior to Robotic Liquid Handling," "paper presented at" LabAutomation2003, February 2–5, 2003, Palm Springs, CA.

9. Berna, M.; Murphy, A.T.; Wilken, B.; Ackermann, B.L. "Collection, Storage, and Filtration of In Vivo Study Samples Using 96-Well Filter Plates To Facilitate Automated Sample Preparation and LC/MS/MS Analysis," *Anal. Chem.* **74**(5), 1197–1201 (2002).
10. Villa, J.S.; Cass, R.T.; Carr, D.E.; Adams, S.M.; Shaw, J.P.; Schmidt Jr., D.E. "Increasing the Efficiency of Pharmacokinetic Sample Procurement, Preparation and Analysis by Liquid Chromatography/Tandem Mass Spectrometry," *Rapid Commun. Mass Spectrom.* **18**(10), 1066–1072 (2004).
11. Gunaratna, P.C.; Kissinger, P.T.; Kissinger, C.B.; Gitzen, J.F. "An Automated Blood Sampler for Simultaneous Sampling of Systemic Blood and Brain Microdialysates for Drug Absorption, Distribution, Metabolism, and Elimination Studies," *J. Pharmacol. Toxicol. Methods* **49**(1), 57–64 (2004).
12. Garrison, K.E.; Pasas, S.A.; Cooper, J.D.; Davies, M.I. "A Review of Membrane Sampling From Biological Tissues with Applications in Pharmacokinetics, Metabolism and Pharmacodynamics," *Eur. J. Pharm. Sci.* **17**(1–2), 1–12
13. Xie, F.M.; Bruntlett, C.S.; Zhu, Y.X.; Kissinger, C.B.; Kissinger, P.T. "Good Preclinical Bioanalytical Chemistry Requires Proper Sampling From Laboratory Animals: Automation of Blood and Microdialysis Sampling Improves the Productivity of LC/MSMS," *Anal. Sci.* **19**(4), 479–485 (2003).
14. Neil, W.; Michalczyk, S.; Russo, M. "Construction of a Low Cost Custom 2-D Bar Code Reader for Tube Racks," *J. Assoc. Lab. Autom.* **6,** 96–102 (2002).
15. Want, R. "RFID—A Key to Automating Everything," *Sci. Am.* **290**(1), 56–65 (2004).
16. Rule, G.; Chapple, M.; Henion, J. "A 384-Well Solid-Phase Extraction for LC/MS/MS Determination of Methotrexate and Its 7-Hydroxy Metabolite in Human Urine and Plasma," *Anal. Chem.* **73**(3), 439–443 (2001).
17. Peng, S.X.; Branch, T.M.; King, S.L. "Fully Automated 96-Well Liquid–Liquid Extraction for Analysis of Biological Samples by Liquid Chromatography with Tandem Mass Spectrometry," *Anal. Chem.* **73**(3), 708–714 (2001).
18. Zhang, L.; Laycock, J.D.; Hayos, J.; Flynn, J.; Yesionek, G.; Miller, K.J. "Automated Strategies for Protein Precipitation Filtration and Solid Phase Extraction (SPE) Optimization on the TECAN Robotic Sample Processor—Applications in Quantitative LC-MS/MS Bioanalysis," "paper presented at" LabAutomation 2004, February 1–5, 2004, San Jose, CA.
19. Watt, A.P.; Morrison, D.; Locker, K.L.; Evans, D.C. "Higher Throughput Bioanalysis by Automation of a Protein Precipitation Assay Using a 96-Well Format with Detection by LC-MS/MS," *Anal. Chem.* **72**(5), 979–984 (2000).
20. Rouan, M.C.; Buffet, C.; Marfil, F.; Humbert, H.; Maurer, G. "Plasma Deproteinization by Precipitation and Filtration in the 96-Well Format," *J. Pharm. Biomed. Anal.* **25**(5–6), 995–1000 (2001).
21. Zweigenbaum, J.; Henion, J. "Bioanalytical High Throughput Selected Reaction Monitoring-LC/MS Determination of Selected Estrogen Receptor

Modulators in Human Plasma: 2000 Samples/Day," *Anal. Chem.* **72**(11), 2446–2454 (2000).

22. Kataoka, H. "New Trends in Sample Preparation for Clinical and Pharmaceutical Analysis," *Trends Anal. Chem.* **22**(4), 232–244 (2003).
23. O'Connor, D. "Automated Sample Preparation and LC-MS for High-Throughput ADME Quantification," *Curr. Opin. Drug Disc. Devel.* **5**(1), 52–58 (2002).
24. Majors, R.E. "New Designs and Formats in Solid-Phase Extraction Simple Preparation," *LC GC North America* **19**(7), 678+ (2001).
25. Majors, R.E. "Trends in Sample Preparation," *LC GC North America* **20**(12), 1098+ (2002).
26. Henion, J.; Brewer, E.; Rule, G. "Sample Preparation for LC/MS/MS: Analyzing Biological and Environmental Samples," *Anal. Chem.* **70**(19), 650A–656A (1998).
27. Kallal, T.; Grant, R.P.; Sanchez, A. "Accelerating Bioanalytical LC-MS/MS Method Development Using an Automated Modular Strategy," in *Proceedings of the 52nd ASMS Conference on Mass Spectrometry and Allied Topics*, Nashville, Tennessee, May 23–27, 2004.
28. Wells, D.A. *High Throughput Bioanalytical Sample Preparation Methods and Automation Strategies*, Elsevier Science, Amsterdam (2003).
29. Baczek, T. "Fractionation of Peptides in Proteomics With the Use of PI-Based Approach and ZipTip Pipette Tips," *J. Pharm. Biomed. Anal.* **34**(5), 851–860 (2004).
30. Van Pelt, C.K.; Zhang, S.; Fung, E.; Chu, I.H.; Liu, T.T.; Li, C.; Korfmacher, W.A.; Henion, J. "A Fully Automated Nanoelectrospray Tandem Mass Spectrometric Method for Analysis of Caco-2 Samples," *Rapid Commun. Mass Spectrom.* **17**(14), 1573–1578 (2003).
31. Sauter Jr., A.D.; Sauter, III, A.D. " 'Electric' Zip Tips™ , Preliminary Results," *J. Assoc. Lab. Autom.* **7**(2), 52–55 (2002).
32. Huang, X.; Prince, A.; Corso, T.N.; Van Pelt, C.; Howe, K.; Zhang, S. "Direct Elution of Preconcentrated Protein Digests from Packed Pipette Tips to Chip-Based NanoESI-MS/MS for Enhanced Protein Identification Sensitivity," in *Proceedings of the 52nd ASMS Conference on Mass Spectrometry and Allied Topics*, Nashville, Tennessee, May 23–27, 2004.
33. Ericson, C.; Phung, T.; Horn, D.M.; Peters, E.C.; Fitchett, J.R.; Ficaro, S.B.; Salomon, A.R.; Brill, L.M.; Brock, A. "An Automated Noncontact Deposition Interface for Liquid Chromatography Matrix-Assisted Laser Desorption/Ionization Mass Spectrometry," *Anal. Chem.* **75**(10), 2309–2315 (2003).
34. Patterson, D.; van Soest, R.; van Gils, M.; Schwartz, H.; Swart, R.; Dragan, I.; Chervet, J.P. "Interfacing Capillary/Nano LC With MALDI/MS for High-Throughput Proteomics," *J. Assoc. Lab. Autom.* **8**(2), 34–35 (2003).

35. Wei, H.; Nolkrantz, K.; Powell, D.H.; Woods, J.H.; Ko, M.-C.; Kennedy, R.T. "Electrospray Sample Deposition for Matrix-Assisted Laser Desorption/Ionization (MALDI) and Atmospheric Pressure MALDI Mass Spectrometry With Attomole Detection Limits," *Rapid Commun. Mass Spectrom.* **18,** 1193–1200 (2004).
36. Wall, D.B.; Finch, J.W.; Cohen, S.A. "Quantification of Codeine by Desorption/Ionization on Silicon Time-of-Flight Mass Spectrometry and Comparisons with Liquid Chromatography/Mass Spectrometry," *Rapid Commun. Mass Spectrom.* **18,** 1403–1406 (2004).
37. Hiller, D.L.; Zuzel, T.J.; Williams, J.A.; Cole, R.O. "Rapid Scanning Technique for the Determination of Optimal Tandem Mass Spectrometric Conditions for Quantitative Analysis," *Rapid Commun. Mass Spectrom.* **11**(6), 593–597 (1997).
38. Rourick, R.A.; Jenkins, K.M.; Walsh, J.; Xu, R.; Cai, Z.; Kassel, D.B. "Integration of Custom LC/MS Automated Data Processing Strategies for the Rapid Assessment of Metabolic Stability and Metabolic Identification in Drug Discovery," in *Proceedings of the 50th ASMS Conference on Mass Spectrometry and Allied Topics*, Orlando, Florida, June 2–6, 2002.
39. Whitney, J.L.; Hail, M.E.; Detlefsen, D.J.; Nugent, K.D. "An Approach for Automatic Metabolic Stability and Detailed Metabolite Profiling on a Single Low-Level In-Vitro Assay Sample," in *Proceedings of the 52nd ASMS Conference on Mass Spectrometry and Allied Topics*, Nashville, Tennessee, May 23–27, 2004.
40. Jones, D.S.; Johnson, L.; Bæk, C.; Egestad, B.; Bayliss, M.; Lashin, V.; Jones, R.C. "A Generic Co-Ordinated Chemical, Instrumental and Software Strategy for the Identification of Drug Metabolites in Rat Urine," in *Proceedings of the 52nd ASMS Conference on Mass Spectrometry and Allied Topics*, Nashville, Tennessee, May 23–27, 2004.
41. Detlefsen, D.J.; Whitney, J.L.; Hail, M.E.; Josephs, J.L.; Sanders, M.; Nugent, K.D. "A Total Analysis Solution for Metabolic Stability and Detailed Metabolite Profiling," in *Proceedings of the 51st ASMS Conference on Mass Spectrometry and Allied Topics*, Quebec, Canada, June 8–12, 2003.
42. Zhang, Z.Q.; McElvain, J.S. "De Novo Peptide Sequencing by Two-Dimensional Fragment Correlation Mass Spectrometry," *Anal. Chem.* **72**(11), 2337–2350 (2000).
43. Oberacher, H.; Mayr, B.M.; Huber, C.G. "Automated De Novo Sequencing of Nucleic Acids by Liquid Chromatography–Tandem Mass Spectrometry," *J. Am. Soc. Mass Spectrom.* **15**(1), 32–42 (2004).
44. Lee, K.R.; Lin, X.; Park, D.C.; Eslava, S. "Megavariate Data Analysis of Mass Spectrometric Proteomics Data Using Latent Variable Projection Method," *Proteomics* **3**(9), 1680–1686 (2004).
45. Kleno, T.G.; Leonardsen, L.R.; Kjeldal, H.O.; Laursen, S.M.; Jensen, O.N.; Baunsgaard, D. "Mechanisms of Hydrazine Toxicity in Rat Liver Investigated by Proteomics and Multivariate Data Analysis," *Proteomics* **4**(3), 868–880 (2004).

46. MacCoss, M.J.; Wu, C.C.; Liu, H.B.; Sadygov, R.; Yates, J.R. "A Correlation Algorithm for the Automated Quantitative Analysis of Shotgun Proteomics Data," *Anal. Chem.* **75**(24), 6912–6921 (2003).
47. Idborg, H.; Edlund, P.O.; Jacobsson, S.P. "Multivariate Approaches for Efficient Detection of Potential Metabolites from Liquid Chromatography/Mass Spectrometry Data," *Rapid Commun. Mass Spectrom.* **18**(9), 944–954 (2004).
48. Plumb, R.S.; Stumpf, C.L.; Granger, J.H.; Castro-Perez, J.; Haselden, J.N.; Dear, G.J. "Use of Liquid Chromatography/Time-of-Flight Mass Spectrometry and Multivariate Statistical Analysis Shows Promise for the Detection of Drug Metabolites in Biological Fluids," *Rapid Commun. Mass Spectrom.* **17**(23), 2632–2638 (2003).
49. Plumb, R.S.; Granger, J.; Stumpf, C.; Wilson, I.D.; Evans, J.A.; Lenz, E.M. "Metabonomic Analysis of Mouse Urine by Liquid–Chromatography-Time of Flight Mass Spectrometry (LC-TOFMS): Detection of Strain, Diurnal and Gender Differences," *Analyst* **128**(7), 819–823 (2003).
50. Herman, J.L. "Generic Method for on-Line Extraction of Drug Substances in the Presence of Biological Matrices Using Turbulent Flow Chromatography," *Rapid Commun. Mass Spectrom.* **16**(5), 421–426 (2002).
51. Zeng, H.; Wu, J.-T.; Unger, S.E. "The Investigation and the Use of High Flow Column-Switching LC/MS/MS As a High-Throughput Approach for Direct Plasma Sample Analysis of Single and Multiple Components in Pharmacokinetic Studies," *J. Pharm. Biomed. Anal.* **27**(6), 967–982 (2002).
52. King, R.C.; Miller-Stein, C.; Magiera, D.J.; Brann, J.A. [E-mail: richard_king@merck.com] "Description and Validation of a Staggered Parallel High Performance Liquid Chromatography System for Good Laboratory Practice Level Quantitative Analysis by Liquid Chromatography/Tandem Mass Spectrometry," *Rapid Commun. Mass Spectrom.* **16**(1), 43–52 (2002).
53. Grant, R.P.; Cameron, C.; Mackenzie-McMurter, S. "Generic Serial and Parallel On-Line Direct-Injection Using Turbulent Flow Liquid Chromatography/Tandem Mass Spectrometry," *Rapid Commun. Mass Spectrom.* **16**(18), 1785–1792 (2002).
54. Schultz, G.A.; Corso, T.N.; Prosser, S.J.; Zhang, S. "A Fully Integrated Monolithic Microchip Electrospray Device for Mass Spectrometry," *Anal. Chem.* **72**(17), 4058–4063 (2000).
55. Van Pelt, C.K.; Zhang, S.; Kapron, J.; Huang, X.; Henion, J. "Chip-Based Automated Nanoelectrospray Mass Spectrometry," *Am. Lab.* **35**(12), 14+ (2003).
56. Dethy, J.M.; Ackermann, B.L.; Delatour, C.; Henion, J.D.; Schultz, G.A. "Demonstration of Direct Bioanalysis of Drugs in Plasma Using Nanoelectrospray Infusion From a Silicon Chip Coupled With Tandem Mass Spectrometry," *Anal. Chem.* **75**(4), 805–811 (2003).
57. Kapron, J.T.; Pace, E.; Van Pelt, C.K.; Henion, J. "Quantitation of Midazolam in Human Plasma by Automated Chip-Based Infusion Nanoelectrospray

Tandem Mass Spectrometry," *Rapid Commun. Mass Spectrom.* **17**(18), 2019–2026 (2003).

58. Borts, D.J.; Cook, S.T.; Bowers, G.D.; O'Mara, M.J.; Quinn, K.E. "Use of Automated Nanoelectrospray for Drug Metabolite Structure Elucidation," in *Proceedings of the 52nd ASMS Conference on Mass Spectrometry and Allied Topics*, Nashville, Tennessee, May 23–27, 2004.
59. Staack, R.F.; Varesio, E.; Hopfgartner, G. "LC-MS Fraction Collection Combined with Chip-MS Infusion for High Throughput Metabolites Identification," in *Proceedings of the 52nd ASMS Conference on Mass Spectrometry and Allied Topics*, Nashville, Tennessee, May 23–27, 2004.
60. Van Pelt, C.K.; Zhang, S.; Henion, J. "Characterization of a Fully Automated Nanoelectrospray System With Mass Spectrometric Detection for Proteomic Analyses," *J. Biomol. Tech.* **13**(2), 72–84 (2002).
61. Stoner, C.L.; Gifford, E.; Stankovic, C.; Lepsy, C.S.; Brodfuehrer, J.; Prasad, J.V.N.V.; Surendran, N. "Implementation of an ADME Enabling Selection and Visualization Tool for Drug Discovery," *J. Pharm. Sci.* **93**(5), 1131–1141 (2003).
62. McDowall, R.D. "The Impact of 21 CFR 11 (Electronic Records and Electronic Signatures Final Rule) on Bioanalysis," *Chromatographia* **55**(Suppl S), S85–S90 (2002).
63. Bruce, S. "A Look at the State of Electronic Lab Notebook Technology," *Sci. Comp. Instrum.* January (2003). Retreived online from www.scimag.com.
64. McLellan, T.; Liao, Q.; Trudel, S.; Phillips, D.; Kelly, M. "Automated Data Capture and Retrieval: Linking Open Access Analysis to a Scientific Desktop Application," in *Proceedings of the 49th ASMS Conference on Mass Spectrometry and Allied Topics*, Chicago, Illinois, May 27–31, 2001.
65. Helfrich, J.P. "A Scientific Data Management Platform for the Biopharmaceutical Industry," *Am. Biotechnol. Lab.* **21**(7), 26–27 (2003).
66. Duckworth, J. "An XML Data Model for Analytical Instruments," Collaborative Electronic Notebook Systems Association, "paper presented at" End User Meeting, July 10, 2001. Available from www.censa.org.
67. Bissett, B. *Practical Pharmaceutical Laboratory Automation*, CRC Press, Boca Raton, FL (2003).
68. Xie, I.H.; Wang, M.H.; Carpenter, R.; Wu, H.Y. "Automated Calibration of TECAN Genesis Liquid Handling Workstation Utilizing an Online Balance and Density Meter," *Assay Drug Dev. Technol.* **2**(1), 71–84 (2004).
69. Bruner, J.; Birkemo, L.; Jordan, K.; Smith, G.; Ormand, J. "Collecting Sample Weight Data On Various Liquid Handling Robots," *J. Assoc. Lab. Autom.* **6**(5), 64–66 (2001).
70. Vessey, A.; Porter, G. "Optimization of Liquid-Handling Precision with Neptune Software on a Tecan Genesis," *J. Assoc. Lab. Autom.* **4,** 81–84 (2003).
71. Harris, C.M. "Finding the Right Robot for MALDI," *Anal. Chem.* **73**(15), 447A–451A (2001).

72. Felton, M.J. "Liquid Handling: Dispensing Reliability," *Anal. Chem.* **75**(17), 397A–399A (2003).
73. Mansfield, P.; Boother, J. "Configurable: The Most Widely Used but Misunderstood Word in LIMS," *Am. Lab.* **36**(9), 18–20 (2004).
74. Pavlis, R.; Bolton, S. "A New Era in Instrument Automation," *Am. Lab.* **33**(6), 24–27 (2001).
75. Kern, R.E.; Fitzgerald, G. "Visual Basic Applications for Simplifying the Use of the Packard MPII for HPLC/MS/MS Analysis of PK Samples," in *Proceedings of the 50th ASMS Conference on Mass Spectrometry and Allied Topics*, Orlando, Florida, June 2–6, 2002.
76. Milgram, K.E.; Jenkins, K.; Yates, D.; Harr, J. "Using Microsoft Excel and Visual Basic for Applications (VBA) As a Cost-Effective Solution for Processing, Validating and Reviewing LC/MS/MS Results in a High-Throughput Screening Environment," in *Proceedings of the 51st ASMS Conference on Mass Spectrometry and Allied Topics*, Quebec, Canada, June 8–12, 2003.
77. Russo, M.F.; Echols, M.M. *Automating Science and Engineering Laboratories with Visual Basic*, Wiley-Interscience, New York (1999).
78. Xu, R.; Cai, Z.; Brailsford, A.; Smith, B.W.; Kassel, D.B. "Development of a Visual Basic Application to Automate Data Acquisition, Processing, and Sample Tracking on a Multiplexed Quadrupole LC/MS," in *Proceedings of the 49th ASMS Conference on Mass Spectrometry and Allied Topics*, May 27–31, 2001, Chicago, Il.
79. Xu, R.; Wang, T.; Isbell, J.; Cai, Z.; Sykes, C.; Brailsford, A.; Kassel, D.B. "High-Throughput Mass-Directed Parallel Purification Incorporating a Multiplexed Single Quadrupole Mass Spectrometer," *Anal. Chem.* **74**(13), 3055–3062 (2003).
80. Choi, B.K.; Hercules, D.M. "Application of Visual Basic for Automation of LC-MS Analysis," *J. Assoc. Lab. Autom.* **5**(6), 102–105 (2000).
81. Choi, B.K.; Hercules, D.M.; Sepetov, N.; Issakova, O.; Gusev, A.I. "Intelligent Automation of LC-MS Analysis for the Characterization of Compound Libraries," *LC GC North America* **20**(2), 152–162 (2002).
82. Choi, B.K.; Tong, S.; Yates, N.A. "Automated LC-MS Analysis and Online Reporting for High-Throughput Microsome Stability Screening," in *Proceedings of the 50th ASMS Conference on Mass Spectrometry and Allied Topics*, Orlando, FL, June 2–6, 2002.
83. Whalen, K.M.; Cole, M.J.; Janiszewski, J.S.; Olech, R.; Johnson, D.; Bracknell, K.; Shirley, J. "Software Tools for the Acquisition, Storage and Distribution of MS/MS Parameters in a High-Throughput Screening Environment," in *Proceedings of the 50th ASMS Conference on Mass Spectrometry and Allied Topics*, Orlando, Florida, June 2–6, 2002.
84. Whalen, K.M.; Janiszewski, J.S.; Cole, M.J. "Evalution II: Software Automation for Industrialized High-Throughput Bioanalytical Mass Spectrometry in Support of ADME-HTS," in *Proceedings of the 52nd ASMS Conference*

on Mass Spectrometry and Allied Topics, Nashville, Tennessee, May 23–27, 2004.

85. Janiszewski, J.S.; Rogers, K.J.; Whalen, K.M.; Cole, M.J.; Liston, T.E.; Duchoslav, E.; Fouda, H.G. "A High Capacity LC/MS System for the Bioanalysis of Samples Generated From Plate-Based Metabolic Screening," *Anal. Chem.* **73**(7), 1495–1501 (2001).
86. Bean, M.F.; Jin, Q.K.; Carr, S.A.; Hemling, M.E. "Instrument-Independent LCMS Data Processing With Web-Based Review and Revision of Results," in *Proceedings of the 49th ASMS Conference on Mass Spectrometry and Allied Topics*, Chicago, Illinois, May 27–31, 2001.
87. Bean, M.F.; Jin, Q.K.; Smeltz, D.L.; Quinn, C.J.; Hemling, M.E. "Open-Access MUX LCMS and Enterprise-Level Analysis Queue Management," in *Proceedings of the 50th ASMS Conference on Mass Spectrometry and Allied Topics*, Orlando, Florida, June 2–6, 2002.
88. Appleman, D. *Moving to VB.NET: Strategies, Concepts, and Code*, Springer-Verlag, New York (2003).
89. Stajich, J.E.; Block, D.; Boulez, K.; Brenner, S.E.; Chervitz, S.A.; Dagdigian, C.; Fuellen, G.; Gilbert, J.G.R.; Korf, I.; Lapp, H.; Lehvaslaiho, H.; Matsalla, C.; Mungall, C.J.; Osborne, B.I.; Pocock, M.R.; Schattner, P.; Senger, M.; Stein, L.D.; Stupka, E.; Wilkinson, M.D.; Birney, E. "The Bioperl Toolkit: Perl Modules for the Life Sciences," *Genome Res.* **12**(10), 1611–1618 (2003).
90. Perkel, J.M. "Bioperl Unveils Version 1–4," *Scientist* **18**(4), 32+ (2004).
91. Moore, R.E.; Young, M.K.; Lee, T.D. "Method for Screening Peptide Fragment Ion Mass Spectra Prior to Database Searching," *J. Am. Soc. Mass Spectrom.* **5**, 422–426 (2000).
92. Dondetti, V.; Dezube, R.; Yang, X.; Maynard, D.M.; Markey, S.P.; Geer, L.; Epstein, J.; Kowalak, J.A. "DBParser: A Perl Program for Proteome Data Analysis," in *Proceedings of the 51st ASMS Conference on Mass Spectrometry and Allied Topics*, Quebec, Canada, June 8–12, 2003.
93. Sweeney, M.J.; Turck, C.W.; Pallat, H.; Robinson, T.; Motchnick, P. "MassSpec.Pm—A Perl Module for Mass Spectrometric Applications," in *Proceedings of the 50th ASMS Conference on Mass Spectrometry and Allied Topics*, Orlando, Florida, June 2–6, 2002.
94. Zhang, Y.; Cocklin, R.R.; Bidasee, K.R.; Wang, M. "Rapid Determination of Advanced Glycation End Products of Proteins Using MALDI-TOF-MS and PERL Script Peptide Searching Algorithm," *J. Biomol. Tech.* **14**(3), 224–230 (2003).
95. Stephens, R.; Hochgurtel, B. *Visual Basic .NET and XML*, Wiley, New York (2002).
96. Murray-Rust, P.; Rzepa, H.S. "Chemical Markup, XML, and the Worldwide Web. 1. Basic Principles," *J. Chem. Inf. Comput. Sci.* **39**(6), 928–942 (1999).
97. Murray-Rust, P.; Rzepa, H.S. "Markup Languages—How to Structure Chemistry-Related Documents," *Chem. Int.* **24**(4), 9–13 (2002).

98. Achard, F.; Vaysseix, G.; Barillot, E. "XML, Bioinformatics and Data Integration," *Bioinformatics* **17**(2), 115–125 (2001).
99. Russo, M.F.; Rubin, A.E. "An Introduction to Using XML for the Management of Laboratory Data," *J. Assoc. Lab. Autom.* **6**(6), 89–94 (2001).
100. Duckworth, J.; Smith, K.; Kuehl, D. "XML As a Data Format Standard for the Long-Term Storage of Analytical Data," *Am. Lab.* **34**(5), 16–49 (2002).
101. Rühl, M.A.; Schäfer, R.; Kramer, G.W. "Spectro ML—A Markup Language for Molecular Spectrometry Data," *J. Assoc. Lab. Autom.* **6**(6), 76–82 (2001).
102. Schaefer, B.A.; Kramer, G. "Creating the Analytical Information Markup Language (AnIML)," "paper presented at" LabAutomation 2004, February 1–5, 2004, San Jose, CA.
103. Julian, R. "Proposal for a 'Multi-Layer' XML-Schema Approach to Data Interchange Standards," "paper presented at" Pittcon, Orlando, FL, March 11, 2003.
104. Browne, D.; Thompson, T.; Mole, D.; McDowall, R.D. "Some Experiences with a Mass Spectrometry Data System Validation," *Chromatographia* **55**(Suppl S), S75–S78 (2002).
105. Gurevitch, D. "Economic Justification of Laboratory Automation," *J. Assoc. Lab. Autom.* **9**(1), 33–43 (2004).
106. Hilt, L.; Berlin, J.; Weeks, T. "Project Control for Laboratory Automation Outsourced to Consultants: A 10-Step Process to Optimize the Effectiveness of Custom Information Technology Development," *J. Assoc. Lab. Autom.* **8**(1), 31–37 (2004).

INDEX

Integrated Strategies for Drug Discovery Using Mass Spectrometry, Edited by Mike S. Lee